TRAFFIC AND GRANULAR FLOW '99

Springer
Berlin
Heidelberg
New York
Barcelona
Hong Kong
London
Milan
Paris
Singapore
Tokyo

D. Helbing H. J. Herrmann
M. Schreckenberg D. E. Wolf
Editors

Traffic and Granular Flow '99

Social, Traffic, and Granular Dynamics

With 213 Figures, 20 in Colour

Springer

Editors

Dirk Helbing
Institut für Wirtschaft und Verkehr
Technische Universität Dresden
01062 Dresden, Germany
e-mail: helbing@trafficforum.de

Hans J. Herrmann
ICA 1
Universität Stuttgart
Pfaffenwaldring 57/III
70550 Stuttgart, Germany
e-mail: hans@ica1.uni-stuttgart.de

Michael Schreckenberg
Physik von Transport und Verkehr
Gerhard-Mercator-Universität Duisburg
Lotharstraße 1
47048 Duisburg, Germany
e-mail: schreckenberg@uni-duisburg.de

Dietrich E. Wolf
Theoretische Physik
Gerhard-Mercator-Universität Duisburg
Lotharstraße 1
47048 Duisburg, Germany
e-mail: wolf@rs1.comphys.uni-duisburg.de

Mathematics Subject Classification (1991):
92Hxx, 92Bxx, 90Axx, 93Exx, 82-XX, 82D99

Cataloging-in-Publication Data applied for

Die Deutsche Bibliothek - CIP-Einheitsaufnahme

Traffic and granular flow '99 : social, traffic, and granular dynamics / D. Helbing ... ed.. - Berlin ; Heidelberg ; New York ; Barcelona ; Hong Kong ; London ; Milan ; Paris ; Singapore ; Tokyo : Springer, 2000
ISBN 3-540-67091-2

ISBN 3-540-67091-2 Springer-Verlag Berlin Heidelberg New York

Springer-Verlag is a company in the BertelsmannSpringer publishing group.

Printed in Germany

Cover design: *design & production* GmbH, Heidelberg
Typeset by the authors using a Springer LATEX macro package
SPIN 10750893 46/3143LK - 5 4 3 2 1 0 Printed on acid-free paper

Welcome Address

Ladies and Gentlemen,

I wish you a very good morning.

On behalf of the University of Stuttgart and in particular on behalf of our rector Prof. Pritschow who is—due to a mission abroad with our minister—unfortunately not present today, I have the pleasure to convey a very warm welcome to all of you to your conference on "Social, Traffic and Granular Dynamics".

This conference is—after two previous events in Jülich 1995 and Duisburg 1997—the third conference on the same or rather similar topic to be held this time here in Stuttgart. One of the reasons for the choice of this location may be that Germany—as a highly populated and in terms of traffic pretty crowded community—plays with its car industry one of the leading roles in our worldwide mobile society. And South Germany, in particular the Stuttgart area is a centre of such activities. This is, on the one hand, due to the industrial concentration because, for example, DaimlerChrysler as an important car manufacturer and Bosch as a major supplier are located, here, and some other world-leading automobile producers are not far away. But, on the other hand, also this university is since many decades highly devoted to the topic of mobility by providing a close cooperation between science and application through various chairs, some of which are also sponsored by the car industry. And this university is at present the only one in Germany at which, 2 years ago, a full 5 years' university degree course on vehicle and motor engineering was installed. And we noticed with great satisfaction that, for this new course, already more than 50 freshmen signed up in the first year. Now, in the second year, the number of interested freshmen already doubled!

But there is also a second reason why this important conference has been brought to Stuttgart: You may have noticed from the programme that, apart from fundamental and technological topics, this conference will especially concentrate on the social and economical aspects. This area was and is the domain of our colleague Prof. Wolfgang Weidlich, an area in which he has been working highly successfully for many years. In this respect, it is also worthwhile mentioning that he took part in a long years' special research project on "natural constructions" financed by our federal government, and in other projects, in which traffic research was approached in an interdisciplinary, multi-partner cooperation involving not only university institutions, but also leading industries like DaimlerChrysler and Siemens. We, the university, are very proud that the organisers succeeded in establishing this conference here in Stuttgart as an important recognition of the successful work of Prof. Weidlich — and this almost at the end of his professional career, close to the beginning of his retirement.

My dear colleague Prof. Weidlich, although there will be time tonight during the welcome dinner for a special laudation in your honour, I would like to take the opportunity on behalf of this university to thank you here and now for your important work and to congratulate you to your great achievements. This conference with most valuable contributions from so many external experts, in particular from abroad, can be regarded as an overwhelming recognition of what you did for science and for our society. Many thanks, again, and my best wishes to your future—and this always in best health!

Ladies and Gentlemen, noteworthy about your conference is also that, according to the programme, about 50 % of the oral contributions will be presented by guest speakers from institutions from outside of Germany. This internationality shows not only the worldwide expertise in the main topic of this conference, but it is also fully in line with the intension of this university, namely to improve the international cooperation in education and research, in science and technology within Europe and beyond. We at the University of Stuttgart have devoted ourselves to increasingly open our house to foreign students and scientists by incorporating, for example, the English language in teaching courses, by creating the possibility of PhD examinations in non-German language, mainly English, and by offering Master courses already in various future-orientated disciplines—and this together with guest lecturers from external partner institutions and universities abroad. Therefore we particularly welcome conferences, symposia, and workshops such as your meeting here. I wish you enlightening papers and talks, a mutually beneficial exchange in knowledge and experience, in short, a successful conference!

Ladies and Gentlemen, have some most interesting and joyful days.

September 27, 1999

Klaus Hein
(Prorector of the
University of Stuttgart)

Preface

Loosely speaking, the motivation to organize an international workshop on the subject "Social, Traffic, and Granular Dynamics" could be summarized by the question: "Are there common phenomena and laws in the dynamic behavior of stock markets, traffic, and granular materials?". Although some people have problems to see such interdisciplinary connections between natural, engineering, and social sciences, from the point of view of statistical physics and the theory of complex systems there are good reasons for them.

Non-linear interactions in many-particle systems are known to produce a lot of interesting phenomena, including chaotic motion. In recent years, a particular attention has been received by *driven* many-particle systems, which continuously take up energy from outside the system (like in falling or vibrated granular media) or from internal resources (like for so-called "motorized" or "self-propelled" particles). This energy is usually absorbed ("dissipated") by some kind of friction mechanism, so that the systems are characterized by a competition between gaining and losing energy. The latter is associated with a spatio-temporal redistribution of energy due to the non-linear interactions among the particles, resulting in a fascinating variety of spatio-temporal patterns (of motion). These depend on the respective interactions, energy supply, boundary conditions, etc.

In sand falling through a vertical pipe, for example, one can observe the formation of density waves similar to stop-and-go waves in traffic flow. Sheared or vibrated granular media may show a segregation of different grain sizes into layers. A vibration of granular materials can also display convection patterns like in heated fluids. One has even observed collective oscillating excitations ("oscillons") when vibrating metal spheres. Some of these phenomena are of great technological relevance, as for the construction of silos, funnels, or conveyor-belts.

A prominent example for *motorized* particles are vehicles and pedestrians, which are the subject of the rapidly growing field of "Traffic Physics". Related questions are conditions and mechanisms of the formation of usual or "phantom" traffic jams and of stop-and-go traffic. Of particular interest are the transitions between the different kinds of traffic states. Detailed empirical and simulation studies have recently revealed various unexpected properties of traffic flow, including self organized characteristic constants like the outflow from traffic jams. Moreover, one has discovered various kinds of congestion, conditions for their appearance and for transitions among them. Surprisingly, most congested states are not caused by traffic volumes exceeding the potential street capacity, but by disturbances of homogeneous traffic flow. This is, because the *dynamical* capacity of streets is given by the characteristic outflow from traffic jams, which is considerably lower than the *potential* street capacity for uniform flow. The reason is that the time gaps of vehicles leaving a traffic jam are greater in accelerating than in steadily flowing traffic, which implies a reduced traffic flow downstream of congested traffic.

From the above, it is obvious that traffic flow can be optimized and made more efficient by any measure which manages to suppress the undesired disturbances of traffic flow. This includes intelligent speed and on-ramp controls as well as driver assistant systems based on communicating vehicles, implying an enormous technological potential for telematics. Researchers are presently also developing methods for simulation-based traffic forecasts.

In crowds of oppositely moving pedestrians, similar to granular layering phenomena, one typically observes the formation of lanes of uniform walking direction, even if the pedestrian interactions are symmetrical with respect to the left-hand and the right-hand side. In contrast, "panicking" pedestrians are likely to get stuck in a blocked situation, based on a phenomenon called "freezing by heating". At doors, one observes a bunch-wise passing of the bottleneck and oscillatory changes of the passing direction, which is comparable to the granular dynamics in the "ticking hour glass". The corresponding simulation programs for pedestrian streams can be used for constructing optimized pedestrian facilities and buildings that will not be deadly traps in emergency situations.

Other examples for self-propelled particles are animals and bacteria. So-called "active walker" models have been developed for the collective behavior of social insects like ants and for the dynamics of flocks of birds, which has implications for air traffic control. Another fascinating field is the rich pattern formation behavior of bacterial colonies, based on chemotaxis, which can be reproduced by "communicating walker" as well as reaction-diffusion models. In the near future, this kind of research may help to develop improved treatments of various diseases.

On an abstract level, social and economic systems can be viewed as driven multi-particle systems as well. However, because of the more sophisticated capabilities of interacting individuals, it is more adequate to call them multi-agent systems. Scientist try to understand social phenomena like the formation and behavior of groups, organizations, and companies in terms of elementary, nonlinear interactions of individuals, similar to the formation and dynamics of atoms, molecules and solid bodies on the basis of interactions of elementary particles. A topic of particular recent interest is the behavior of currency and stock markets, because of the good data situation (and, of course, hopes to make a lot of money). In a certain way, the up and down of stock markets resembles the oscillatory passing of bottlenecks by oppositely pushing pedestrians, who may be compared to optimistic and pessimistic traders (driving the stock prices into opposite directions). Among other things, researchers are interested in possibilities to stabilize markets. Other topics covered by the overlapping fields of "Econophysics" and "Evolutionary Economics" are business cycle theory, models of innovation, social dilemmas, opinion formation and leadership, communication structures and their impact on the efficiency of companies, Internet congestion and how to avoid it, market-based control, optimization, etc.

The workshop on "Traffic and Granular Flow" in Stuttgart ("TGF '99") was the third one with this title in a series of international conferences on driven particle systems. While the first conference 1995 in Jülich tried to work out the similarities and differences of traffic and granular flows, the second conference 1997 in Duisburg wanted to shed light on the connections with collective motion in biological systems. On occasion of the emeritation of Wolfgang Weidlich, one of the pioneers in the field of "Quantitative Sociodynamics", the workshop in the year 1999 was held in Stuttgart and intended to stress the connections with socio-economic systems.

The conference "Social, Traffic, and Granular Dynamics" aimed at working out the common structures, mechanisms, phenomena, and methods in the different fields and disciplines studying many-particle or multi-agent systems. Some of the topics relevant to several or all of these fields are listed below:

- Driven Motion
- Non-Linear Waves
- Binary Interactions
- Boltzmann Equations
- Cellular Automata
- Phase Transitions
- Metastable States
- Correlated Motion
- Complex Dynamics
- Stochastic Models
- Scaling Laws
- Decision Behavior
- Game Theory
- Optimization

We believe, however, that this conference was not only interesting from a theoretical point of view. As outlined above, studying "Social, Traffic, and Granular Dynamics" has many applications and can actually make relevant contributions to understanding and solving the problems of today's societies. If we manage to communicate this to the public, we are sure people are willing to invest more money in this kind of research. Also, physics may gain some fresh taste that is likely to attract the interest of students. Thus, apart from disseminating important quantitative results on social, traffic, and granular dynamics, focusing on the interrelations among these fields, the aim of the conference was to stimulate interdisciplinary research, to simulate international research cooperation, and to build up a community committed to this research area.

At this place, we would like to thank the sponsors that have made this workshop possible. Generous financial support has been provided by the German Research Foundation (DFG), by Siemens, DaimlerChrysler, the Ministry of Science, Research and Arts Baden-Württemberg, the University of Stuttgart, its Physics Faculty, the Landesgirokasse, the Department of Physics at the Gerhard-Mercator University Duisburg, the ICA 1 and the II. Institute of Theoretical Physics at the University of Stuttgart, and Wolfgang Weidlich

Last but not least, we are very grateful to the conference secretaries Wiltrud Meyer-Haake and Sabine von Viebahn, as well as to the technical assistants Illés Farkas, Ansgar Hennecke, Lutz Neubert, Benno Tilch, and Martin Treiber for their selfless and tireless support in organizing and running the conference, Birgit Dahm-Courths, Hubert Klüpfel, and Joachim Wahle for preparing this conference proceedings and finally the advertising agency **z.B.** for the cover picture.

Duisburg, March 2000

Dirk Helbing
Hans J. Herrmann
Michael Schreckenberg
Dietrich E. Wolf

Table of Contents

Biology, Internet, Transport Theory

Traffic Data and Applications

Modelling of Traffic Flow

Granular Dynamics

List of Participants

1. **María A. Aguirre**, Grupo de Medios Porosos, Fac. de Ingeniería, Universidad de Buenos Aires, Paseo Colon 850, 1063 Capital Federal, Argentina, maaguir@tron.fi.uba.ar
2. **Sonal Ahuja**, Delhi Metro Rail Corporation Ltd., B-1, MIG Flats, Ashok Vihar -IV, 110 052 Delhi, India, paradi101@hotmail.com
3. **Robert Barlovic**, Physik von Transport und Verkehr, Universität Duisburg, Lotharstr. 1, 47048 Duisburg, Germany, barlovic@uni-duisburg.de
4. **Eshel Ben-Jacob**, School of Physics and Astronomy, Tel-Aviv University, Levanon Street, 69978 Tel Aviv, Israel, eshel@albert.tau.ac.il
5. **Peter Berg**, Department of Mathematics, University of Bristol, University Walk, BS8 1TW Bristol, Great Britain, Peter.Berg@bristol.ac.uk
6. **Rainer Berkemer**, ITV Denkendorf - Management Research, Körschtalstr. 26, 73770 Denkendorf, Germany, rainer.berkemer@itvd.uni-stuttgart.de
7. **Bibudhananda Biswal**, Sri Venkateswara College, University of Delhi, Benito Juarez Raod, 110 021 New Delhi, India, biswal@ica1.uni-stuttgart.de
8. **Chris Blokhuis**, Television VPRO, Holland
9. **Elmar Brockfeld**, Universität Osnabrück, Hermannstr. 19, 49080 Osnabrück, Germany, ebrockfe@uos.de
10. **Luciana Bruno**, Grupo de Medios Porosos, Paseo Colon 850, 1063 Buenos Aires, Argentina, lbruno@tron.fi.uba.ar
11. **Christian Caron**, Springer-Verlag, Tiergartenstr. 17, 69121 Heidelberg, Germany, caron@springer.de
12. **Susanne Cheybani**, Department of Theoretical Physics, Technical University of Budapest, 1111 Budapest, Hungary. cheybani@dirac.phy.bme.hu
13. **Roland Chrobok**, Physik von Transport und Verkehr, Universität Duisburg, Lotharstr. 1, 47048 Duisburg, Germany, chrobok@traffic.uni-duisburg.de
14. **Eric Clement**, Laboratoire des Milieux Désordonnés et Hétérogènes, Case 86, Université Paris 6, 4 Place Jussieu, 75252 Cedex 05 Paris, France, erc@ccr.jussieu.fr
15. **Werner Ebeling**, Institut für Theoretische Physik, Humboldt-Universität Berlin, Invalidenstr. 110, 10115 Berlin, Germany, werner@summa.physik.hu-berlin.de
16. **Nils Eissfeldt**, ZPR/ZAIK, Universität zu Köln, Weyertal 80, 50931 Köln, Germany, eissfeldt@zpr.uni-koeln.de
17. **Udo Erdmann**, Humboldt-Universität Berlin, Invalidenstraße 110, 10115 Berlin, Germany, udo.erdmann@physik.hu-berlin.de

18. **Illes Farkas**, Department of Biological Physics, Eötvös University, Pázmány Péter Sétány 1A, 1117 Budapest, Hungary, fij@elte.hu
19. **Mohammad-Ebrahim Fouladvand**, Institute for Studies in Physics and Mathematics, Farmaniah, P.O. Box 19395-5531 Tehran, Iran, foolad@theory.ipm.ac.ir
20. **Gerardo Físcher**, Instituto de Matemática Aplicada de San Luis, Ejército de los Andes 950, RA-5700 San Luis, S.L., Argentina, midgard@unsl.edu.ar
21. **Serge Galam**, Laboratoire des Milieux Désordonnés et Hétérogènes (LMDH), Case 86, Université Paris 6, T13, 4 Place Jussieu, 75252 Cedex 05 Paris, France, galam@ccr.jussieu.fr
22. **Isaac Goldhirsch**, Department of Fluid Mechanics and Heat Transfer, Tel-Aviv University, 69978 Tel-Aviv, Israel, isaac@newton.eng.tau.ac.il
23. **Ido Golding**, Department of Physics, Tel-Aviv University, Ramat Aviv, 69978 Ramat Aviv, Israel, golding@gina.tau.ac.il
24. **Siegfried Grossmann**, Fachbereich Physik, Universität Marburg, Renthof 6, 35032 Marburg, Germany, grossmann_s@physik.uni-marburg.de
25. **Marco Günther**, Institut für Techno- und Wirtschaftsmathematik, Erwin-Schrödinger-Straße, 67653 Kaiserslautern, Germany, guenther@itwm.uni-kl.de
26. **Günter Haag**, Steinbeis-Transferzentrum Angewandte Systemanalyse, Schönbergstr. 22, 70599 Stuttgart, Germany, haag@sofo.uni-stuttgart.de
27. **K.P. Hadeler**, Universität Tübingen, Auf der Morgenstelle 10, 72076 Tübingen, Germany, hadeler@uni-tuebingen.de
28. **Hermann Haken**, Institut für Theoretische Physik 1, Universität Stuttgart, Pfaffenwaldring 57, 70569 Stuttgart, Germany, haken@theo.physik.uni-stuttgart.de
29. **Kathrin Happe**, Institut 410 B, Universität Hohenheim, 70593 Stuttgart, Germany, khappe@uni-hohenheim.de
30. **Claudia Heinen**, Institut für Mechanik, Kaiserstr. 12, 76131 Karlsruhe, Germany, ub63@rz.uni-karlsruhe.de
31. **Dirk Helbing**, II. Institut für Theoretische Physik, Universität Stuttgart, Pfaffenwaldring 57/III, 70550 Stuttgart, Germany, helbing@theo2.physik.uni-stuttgart.de
32. **Ansgar Hennecke**, II. Institut für Theoretische Physik, Universität Stuttgart, Pfaffenwaldring 57, 70550 Stuttgart, Germany, ansgar@theo2.physik.de
33. **Hans J. Herrmann**, Institut für Computer-Anwendungen 1, Universität Stuttgart, Pfaffenwaldring 27, 70569 Stuttgart, Germany, hans@ica1.uni-stuttgart.de
34. **Georg Hertkorn**, Deutsches Zentrum für Luft- und Raumfahrt, Linder Höhe, 51147 Köln, Germany, Georg.Hertkorn@dlr.de
35. **Rudolf Hilfer**, ICA 1, Stuttgart, Germany, R.Hilfer@ica1.uni-stuttgart.de

36. **Janusz Holyst**, Institute of Physics, Warsaw University of Technology, Koszykowa 75, 00-662 Warsaw, Poland, jholyst@if.pw.edu.pl
37. **Bernardo Huberman**, Xerox PARC, 3333 Coyote Hill Road, CA 94304 Palo Alto, USA, huberman@parc.xerox.com
38. **Jerzy Hubert**, Niewodniczanski Institute of Physics, Radzikowskiego 152, 31-342 Krakow, Poland, hubert@alf.ifj.edu.pl
39. **Torsten Huisinga**, Physik von Transport und Verkehr, Universität Duisburg, Lotharstr. 1, 47048 Duisburg, Germany, huisinga@uni-duisburg.de
40. **Peter Hänggi**, Universität Augsburg, Universitätsstr. 1, 86135 Augsburg, Germany, hanggi@physik.uni-augsburg.de
41. **Yuji Igarashi**, Faculty of Education, Niigata University, Ikarashi 2-8050, 950-2181 Niigata, Japan, igarashi@ed.niigata-u.ac.jp
42. **Shio Inagaki**, Ibaraki University, Hakamazuka 3-12-8-108, 310-0055 Mito, Japan, shio@mail.ne.jp
43. **Imre M. Janosi**, Department of Physics of Complex Systems, Eötvös University, Pazmany Peter Setany 1, 1117 Budapest, Hungary, janosi@krumpli.elte.hu
44. **James T. Jenkins**, Department of Theoretical and Applied Mechanics, Cornell University, 211 Kimball Hall, 14853 Ithaca, New York, USA, jtj2@cornell.edu
45. **Oliver Kaumann**, Physik von Transport und Verkehr, Universität Duisburg, Lotharstr. 1, 47048 Duisburg, Germany, kaumann@uni-duisburg.de
46. **Jevgenijs Kaupuzs**, Institute of Mathematics and Computer Science, University of Latvia, 29 Rainja Boulevard, 1459 Riga, Latvia, kaupuzs@com.latnet.lv
47. **Boris Kerner**, DaimlerChrysler AG, HPC: E224, 70546 Stuttgart, Germany, boris.kerner@daimlerchrysler.com
48. **Janos Kertesz**, Department of Theoretical Physics, Technical University of Budapest, Budafoki ut 8, 1111 Budapest, Hungary, kertesz@phy.bme.hu
49. **Eberhard Keyl**, Soziologie, Universität Tübingen, Friedhofweg 4, 73061 Rosswaelden, Germany, keyl@gmx.de
50. **Achim Kittel**, Fachbereich Physik, Universität Oldenburg, Carl-von-Ossietzky-Strasse 9-11, 26129 Oldenburg, Germany, kittel@uni-oldenburg.de
51. **Kai-Oliver Klauck**, Institut für Theoretische Physik, Universität zu Köln, Zülpicher Str. 77, 50937 Köln, Germany, kok@thp.uni-koeln.de
52. **Hubert Klüpfel**, Physik von Transport und Verkehr, Universität Duisburg, Lotharstr. 1, 47048 Duisburg, Germany, kluepfel@traffic.uni-duisburg.de
53. **Wolfgang Knospe**, Physik von Transport und Verkehr, Universität Duisburg, Lotharstr. 1, 47048 Duisburg, Germany, knospe@uni-duisburg.de

54. **Steffen Kraemer**, Universität Stuttgart, Pfaffenwaldring 57, 70550 Stuttgart, Germany, s.kraemer@physik.uni-stuttgart.de
55. **Jan Krawczyk**, H. Niewodniczanski Institute of Nuclear Physics, Radzikowskiego 152, 31-342 Krakow, Poland, krawczyk@ifj.edu.pl
56. **Stefan Kriso**, Bosch GmbH, Auf Hart 74, 71706 Markgröningen, Germany, Stefan.Kriso@de.bosch.com
57. **Christof Krülle**, Experimental-Physik IV, Universität Bayreuth, Universitätsstr., 95440 Bayreuth, Germany, christof.kruelle@uni-bayreuth.de
58. **Christina Kuttler**, Universiät Tübingen, Auf der Morgenstelle 10, 72076 Tübingen, Germany, christina.kuttler@uni-tuebingen.de
59. **Reinhart D. Kühne**, Institut für Straßen- und Verkehrswesen, Seidenstr. 36, 70174 Stuttgart, Germany, kuehne@isvs.uni-stuttgart.de
60. **Marc Laetzel**, Institut für Computer-Anwendungen (ICA I), Pfaffenwaldring 27, 70569 Stuttgart, Germany, m.laetzel@ica1.uni-stuttgart.de
61. **Ha Youn Lee**, Department of Physics and Center for Theoretical Physics, Seoul National University, Seoul National University, 151-742 Seoul, Korea, agnes@phya.snu.ac.kr
62. **Heiko Lehmann**, GMD-First, Rudower Chaussee 5, Geb. 13.10, 12489 Berlin, Germany, heiko@first.gmd.de
63. **Stefan Luding**, Institut für Computer-Anwendungen 1, Universität Stuttgart, Pfaffenwaldring 27, 70569 Stuttgart, Germany, lui@ica1.uni-stuttgart.de
64. **Reinhard Mahnke**, Universität Rostock, Fachbereich Physik, Universitätsplatz 1, 18051 Rostock, Germany, mahnke@darss.mpg.uni-rostock.de
65. **Hernan Makse**, Schlumberger-Doll Research, Old Quarry Road, 06877 Ridgefield, USA, makse@ridgefield.sdr.slb.com
66. **Ezio Marchi**, Instituto de Matemática Aplicada de San Luis, Ejército de los Andes 950, RA-5700 San Luis, S.L., Argentina, emarchi@unsl.edu.ar, midgard@unsl.edu.ar
67. **Thorsten Materne**, FB Mathematik, TU Darmstadt, Schlossgartenstr. 7, 64289 Darmstadt, Germany, materne@math.fu-berlin.de
68. **Hans-Georg Matuttis**, Institut für Computer-Anwendungen 1, Universität Stuttgart, Pfaffenwaldring 27, 70569 Stuttgart, Germany, hg@ica1.uni-stuttgart.de
69. **Eckart Mayer**, Max-Planck-Institut für Dynamik komplexer technischer Systeme, Magdeburg, Leipziger Strasse 44, 39120 Magdeburg, Germany, mayer@mpi-magdeburg.mpg.de
70. **Jürgen Meier**, Institut für Regelungstechnik und Systemdynamik, Universität Stuttgart, Pfaffenwaldring 9, 70550 Stuttgart, Germany, meier@isr.uni-stuttgart.de
71. **Marcus Metzler**, ZPR/ZAIK, Universität zu Köln, Weyertal 80, 50931 Köln, Germany, mocm@zpr.uni-koeln.de

72. **Jean-Pierre Minier**, Electricité de France, Recherche et Développement, 6 Quai Watier, 78400 Chatou, France, Jean-Pierre.Minier@der.edf.fr
73. **Peter Molnar**, Center for Theoretical Studies of Physical Systems, Clark Atlanta University, James P. Brawley Drive, Georgia 30314 Atlanta, USA, pmolnar@cau.edu
74. **Jose Daniel Munoz**, Universität Stuttgart, Allmandring 20C26, 70569 Stuttgart, Germany
75. **Lutz Neubert**, Physik von Transport und Verkehr, Universität Duisburg, Lotharstr. 1, 47048 Duisburg, Germany, neubert@uni-duisburg.de
76. **Tadeusz Platkowski**, Department of Mathematics, Informatics and Mechanics, University of Warsaw, Banacha 2, 02-097 Warsaw, Poland, tplatk@mimuw.edu.pl
77. **Florian Plenge**, Max-Planck-Institut für Kernphysik, Friedrich-Ebert-Anlage 53, 69117 Heidelberg, Germany, plenge@mpi-hd.mpg.de
78. **Andreas Pottmeier**, Physik von Transport und Verkehr, Universität Duisburg, Lotharstr. 1, 47048 Duisburg, Germany, pottmeier@uni-duisburg.de
79. **Gerald Ristow**, Institut für Theoretische Physik, Universität des Saarlandes, Renthof 6, 66041 Saarbrücken, Germany, ristow@lusi.uni-sb.de
80. **Stephan Rosswog**, Deutsches Zentrum für Luft- und Raumfahrt (DLR), Linder Höhe, 51147 Köln, Germany, stephan.rosswog@dlr.de
81. **Thomas Rytz**, Könizstr. 215, 3097 Liebefeld/Bern, Switzerland, rytz@iamexwi.unibe.ch
82. **Leonid Safonov**, Bar-Ilan University, 52900 Ramat-Gan, Israel, leonid@moria.ph.biu.ac.il
83. **Andreas Schadschneider**, Institut für Theoretische Physik , Universität zu Köln, Zülpicher Str. 77, 50937 Köln, Germany, as@thp.uni-koeln.de
84. **Karin Schagen**, Television VPRO, The Netherlands, karsch@vpro.nl,lank@wxs.nl
85. **Alexander Schinner**, Universität Magdeburg, Universitätsplatz 2, 39106 Magdeburg, Germany, alexander.schinner@physik.uni-magdeburg.de
86. **Matthias Schmidt**, GMD-FIRST, Kekulestr. 7, 12489 Berlin, Germany, schmidt@first.gmd.de
87. **Claudius Schnörr**, DDG, Niederkasseler Lohweg 20, 40547 Düsseldorf, Germany, Claudius.Schnoerr@ddg.de
88. **Michael Schreckenberg**, Physik von Transport und Verkehr, Universität Duisburg, Lotharstr. 1, 47048 Duisburg, Germany, schreckenberg@uni-duisburg.de
89. **Thomas Schreiber**, Fachbereich Physik, Universität Wuppertal, Gaussstr. 20, 42097 Wuppertal, Germany, schreibe@theorie.physik.uni-wuppertal.de
90. **Peter Schuster**, Institut für Theoretische Chemie, Universität Wien, Waehringer Str. 17, 10900 Wien, Austria, pks@tbi.univie.ac.at

91. **Frank Schweitzer**, GMD - Institut für Autonome Intelligente Systeme, Schloss Birlinghoven, 53754 Sankt Augustin, Germany, schweitzer@gmd.de, frank@physik.hu-berlin.de
92. **Bernd Schürmann**, Siemens AG, ZFET SN 4, 81730 München, Germany, bernd.schuermann@mchp.siemens.de
93. **Gunter Schütz**, IFF, Forschungszentrum Jülich, 52425 Jülich, Germany, g.schuetz@fz-juelich.de
94. **Vital Sever**, Ul. bratov Uèakar 26, 1000 Ljubljana, Slovenija, vital.sever@siol.net
95. **Rudolf Sollacher**, Siemens AG, Otto-Hahn-Ring 6, 81739 München, Germany, Rudolf.Sollacher@mchp.siemens.de
96. **Michael Sonis**, Bar-Ilan University, Department of Geography, 52900 Ramat-Gan, Israel, sonism@mail.biu.ac.il
97. **H. Eugene Stanley**, Center for Polymer Studies and Department of Physics, Boston University, 590 Commonwealth Avenue, 02215 Boston, MA, USA, hes@bu.edu
98. **André Stebens**, Zukunft NRW Logistik, Universität Duisburg, Geibelstr. 41, 47048 Duisburg, Germany, stebens@math.uni-duisburg.de
99. **G. Keith Still**, Orchid Fractal Engineering Limited, "Rosemary Cottage", Plumgarths, Crook Road, LA8 8LX Near Kendal, Cumbria, UK, keithstill@cs.com
100. **Yuki Sugiyama**, Division of Mathematical Science, City College of Mie, Ishinden Nakano, 514-0112 Tsu, Mie, Japan, genbey@eken.phys.nagoya-u.ac.jp
101. **Benno Tilch**, II. Institut für Theoretische Physik, Universität Stuttgart, Pfaffenwaldring 57, 70550 Stuttgart, Germany, benno@theo2.physik.uni-stuttgart.de
102. **Elad Tomer**, Bar-Ilan University, 52900 Ramat-Gan, Israel, tomer@alon.cc.biu.ac.il
103. **Martin Treiber**, II. Institut für Theoretische Physik, Universität Stuttgart, Pfaffenwaldring 57, 70550 Stuttgart, Germany, treiber@theo2.physik.uni-stuttgart.de
104. **Tamas Vicsek**, Department of Biological Physics, Eötvös University, Pazmany P. Stny 1A, 1117 Budapest, Hungary, h845vic@ella.hu
105. **Ivo Voragen**, Television VPRO, The Netherlands, karsch@vpro.nl
106. **Peter Vortisch**, PTV AG, Stumpfstr. 1, 76131 Karlsruhe, Germany, Peter.Vortisch@ptv.de
107. **Peter Wagner**, Deutsches Zentrum für Luft- und Raumfahrt (DLR), Linder Höhe, 51147 Köln, Germany, peter.wagner@dlr.de
108. **Joachim Wahle**, Physik von Transport und Verkehr, Universität Duisburg, Lotharstr. 1, 47048 Duisburg, Germany, wahle@traffic.uni-duisburg.de
109. **Thomas Waldeer**, Fachhochschule Braunschweig/Wolfenbüttel, Karl-Scharfenberg-Str. 55, 38229 Salzgitter, Germany, waldeer@fh-wolfenbuettel.de

110. **Wolfgang Weidlich**, II. Institut für Theoretische Physik, Universität Stuttgart, Pfaffenwaldring 57, 70550 Stuttgart, Germany, office@theo2.physik.uni-stuttgart.de
111. **Petra Weis**, Bauhaus Universität Weimar, Professur Verkehrsplanung u. Verkehrstechnik, Marienstr. 13C, 99423 Weimar, Germany, petra.weis@uni-weimar.de
112. **Katarzyna Winkowska-Nowak**, ISS, University of Warsaw, u. Stawki 5/7, 00-183 Warsaw, Poland, kasia@theta1.ifpan.edu.pl
113. **Dietrich Wolf**, Universität Duisburg, Lotharstr. 1, 47048 Duisburg, Germany, d.wolf@uni-duisburg.de
114. **Kim Youngho**, Fachgebiet Verkehrstechnik und Verkehrsplanung, TU Müchen, Arcisstr. 21, 80333 Müchen, Germany, ykim@fgv.tum.de
115. **Marguerite Zarrillo**, University of Massachusetts Dartmouth, Physics Department 285 Old Westport Road, 02747-2300 North Dartmouth, Massachusetts, USA, MZarrillo@UMassD.edu

Social Dynamics

Recurrence in Physical and Social Systems

W. Weidlich

Universität Stuttgart, II. Institut für Theoretische Physik, Pfaffenwaldring 57/III,
70550 Stuttgart, Germany

Abstract. The problem of the relation between recurrence and irreversibility is an old and universal one: it has been discussed by philosophers, physicists, historians, and social scientists. After briefly mentioning philosophical formulations of the problem, the controversy, being deeply inherent in the notions of Statistical Physics, is discussed in terms of Poincare's recurrence theorem versus irreversible equations such as the Boltzmann equation. The conclusion is that the neglection of certain correlations, i.e., an approximation, leads from recurrence to irreversibility. Thereupon the inverse problem is considered, in which manner recurrent, in particular periodic or quasi-periodic sub-processes can appear to be embedded in a globally irreversible process. Some approaches how to trace and recognize embedded recurrent (and even periodic) sub-processes are discussed. Finally, selected examples of model-based (quasi-) periodic processes in social systems are presented. They belong to the sectors demography (migration) and sociology (group dynamics).

1 The General Problem of Recurrence vs. Irreversibility

The problem, whether the evolution of complex systems takes place by traversing irreversibly a sequence of never repeating states or, alternatively, by returning after some period of time to states which had already been reached before, is an old one and a very universal one. The problem relates to physical and to social systems as well and it has been formulated in more philosophical or more mathematical terms.

1.1 Philosophical Formulations

Let us begin with a few parsimonious remarks about how ancient and modern philosophers speak about irreversibility and recurrence:

Heraclitus, living about 460 b.c., one of the ancient Greek philosophers of nature, stressed the evolution and steady transformation of all things: "Everything flows and nothing stays ($\pi\alpha\nu\tau\alpha\ \rho\epsilon i$)" and "nobody climbs down a second time into the same river". This is a clear statement of the irreversibility of all events!

And let us now hear, how more than two-thousand years later, the German philosopher Friedrich Nietzsche (1844-1900) develops in his work "Also sprach Zarathustra" his alternative view:

"See this gateway", spoke Zarathustra, "which has two faces! Two lanes meet here, and nobody yet went to their end. This long lane backwards lasts an eternity; and that long lane forward means another eternity; and here at this gateway they join. The name of the gateway is written aloft: 'The instant'. And see: From this instant there runs backwards a long eternal lane; behind us there lies an eternity!

Mustn't, whatever can go of all things, have already walked along this lane? Mustn't, whatever can occur of all things, have already pre-appeared, and have been pre-done?

And this slow spider creeping in the moonlight, and the moonshine itself, and you and myself in the gateway, whispering about eternal things, mustn't we already have been here? And this very instant, mustn't it have pre-existed?

And aren't all things so closely intertwined that this instant engenders all coming things, including once more itself?

Because, whatsoever can go of all things, mustn't it return and proceed on its way along the other lane, in that long gruesome lane forward? Mustn't we thus recur in all eternity?" So far F. Nietzsche's allegation of "eternal recurrence"!

1.2 Recurrence versus Irreversibility in Physics

Let us become sober now: Although allegations like that of Heraclitus or Nietzsche are poetic and deep, they cannot be considered as really conclusive, the more as, exactly spoken, the statements of irreversibility and recurrence contradict each other! Therefore mathematicians and physicists, working with quantitative methods, are doomed to solve the riddle at least for those systems, whose dynamics can be formulated in exact mathematical terms.

We will now briefly review the controversy concerning "recurrence versus irreversibility" in classical mechanics at the end of the last century when Statistical Mechanics was beginning to emerge.

The Recurrence Theorem of Henri Poincaré (1854-1912) H. Poincaré proved exactly that recurrence holds, i.e., that every state must – sooner or later – return in systems of classical mechanics possessing a phase space of finite volume $\phi(\Gamma) < \infty$. (An example of such a system is a gas enclosed in a box and having a finite total inner energy.)

Poincaré's proof of the recurrence theorem is ingenious and simultaneously simple enough to be reproduced here. He considers systems described by canonical coordinates $q_k(t), p_k(t); k = 1, 2, \dots, f$, which obey Hamilton's equations

$$\frac{dq_k}{dt} = \frac{\partial H(\mathbf{q}, \mathbf{p})}{\partial p_k}; \qquad \frac{dp_k}{dt} = -\frac{\partial H(\mathbf{q}, \mathbf{p})}{\partial q_k} \tag{1}$$

Formally, the motion takes place in the $2f$-dimensional phase space Γ with points $P(q_1 \dots q_f; p_1 \dots p_f) \in \Gamma$. Then, the Liouville-Theorem states that the motion in Γ according to Hamilton's equations (1) is that of an incompressible fluid. This includes that the volume $\phi(g)$ of a sub-domain $g \subset \Gamma$ "swimming" with the moving systems remains constant though its shape may change dramatically.

Following Poincaré, let us now consider a domain $g(0) \subset \Gamma$ in phase space Γ of volume $\phi(g(0))$ comprising an ensemble of systems in states at time 0. At a time $t > 0$ these systems whose states have evolved according to (1) are now contained in domain $g(t)$ which is called *a future phase* of $g(0)$ and which has the *same volume* as $g(0)$, i.e., $\phi(g(t)) = \phi(g(0))$, because of the Liouville-Theorem.

Furthermore, let us introduce $G(0) \subset \Gamma$, the domain comprising *all future phases* of $g(0)$, i.e., all domains $g(t)$ with $0 \leq t < \infty$, and also the domain $G(t_1) \subset \Gamma$, where $t_1 > 0$, comprising *all future phases* of $g(t_1)$, i.e., all domains $g(t)$ with $t_1 \leq t < \infty$.

Evidently by definition $G(0)$ must include $G(t_1)$, i.e., $G(0) \supseteq G(t_1)$ (*). On the other hand, $G(t_1)$ is a future phase of $G(0)$, so that the Liouville theorem must also hold for $G(0)$ and $G(t_1)$; this means $\phi(G(t_1)) = \phi(G(0))$ (**). [Because of the presumption $\phi(\Gamma) < \infty$, the volumes $\phi(G(0))$ and $\phi(G(t_1))$ must be finite, since $G(t_1) \subseteq G(0) \subset \Gamma$].

From (*) and (**) now follows that $G(t_1)$ and $G(0)$ can only differ by a domain of volume zero. Therefore $G(t_1)$, which comprises all future phases of $g(t_1)$ must also comprise $g(0) \subset G(0)$ – except at most for a domain of volume zero. Since $g(t_1)$ is by definition a future phase of $g(0)$, it follows that the union of all future *phases of* $g(0)$ *must comprise* $g(0)$ *itself!* This means that the systems contained in $g(0)$ in their initial states at $t = 0$ *must* – sooner or later – *return* to states within $g(0)$, q.e.d.

Remark: This recurrence theorem of Poincaré states that recurrence *must take place*, but it does *not* state *in which time intervals* the return to states in the vicinity of the original state takes place! These intervals – the Poincaré recurrence times – turn out to be extremely long in concrete cases. Figure 1 illustrates (in 2-dimensional instead of $2f$-dimensional space) what happens during the evolution of $g(0)$.

The Boltzmann Equation Ludwig Boltzmann (1844-1906) treated the evolution of a gas with a somewhat different method, but also starting from the motion of individual gas atoms. He was interested in the evolution with time of the distribution of the atoms of a gas with respect to their positions $\mathbf{x}$ and velocities $\mathbf{v}$. The distribution function $f(\mathbf{x}, \mathbf{v}; t)$ has the meaning to find $f(\mathbf{x}, \mathbf{v}; t) d^3x d^3v$ atoms in the space interval d^3x and velocity interval d^3v at time t. The equation of motion of the distribution derived by Boltzmann has the general form:

$$\frac{\partial f(\mathbf{x}, \mathbf{v}; t)}{\partial t} = \left(\frac{\partial f}{\partial t}\right)_{\text{conv.}} + \left(\frac{\partial f}{\partial t}\right)_{\text{coll.}} \tag{2}$$

where $\left(\frac{\partial f}{\partial t}\right)_{\text{conv.}}$ are the changes per time of $f(\mathbf{x}, \mathbf{v}; t)$ because of the free motion (*convection*) of all particles and $\left(\frac{\partial f}{\partial t}\right)_{\text{coll.}}$ are the changes per time of $f(\mathbf{x}, \mathbf{v}; t)$ because of the *collisions* of particles. Boltzmann found the following expressions for $\left(\frac{\partial f}{\partial t}\right)_{\text{conv.}}$ and $\left(\frac{\partial f}{\partial t}\right)_{\text{coll.}}$:

$$\left(\frac{\partial f(\mathbf{x}, \mathbf{v}; t)}{\partial t}\right)_{\text{conv.}} = -\mathbf{v} \cdot \boldsymbol{\nabla}_{(x)} f(\mathbf{x}, \mathbf{v}; t) - \frac{\mathbf{F}}{m} \cdot \boldsymbol{\nabla}_{(v)} f(\mathbf{x}, \mathbf{v}; t) \tag{3}$$

where $\mathbf{F} = -\boldsymbol{\nabla}_{(x)} V(\mathbf{x})$ is the force (derived from a potential $V(\mathbf{x})$) exerted on the particles of mass m and

$$\left(\frac{\partial f(\mathbf{x},\mathbf{v};t)}{\partial t}\right)_{\text{coll.}} = \int d^3v_2 d^3v_1' d^3v_2' \delta^3(\mathbf{v}_1' + \mathbf{v}_2' - \mathbf{v} - \mathbf{v}_2) \cdot \\ \cdot \delta(v_1'^2 + v_2'^2 - \mathbf{v}^2 - \mathbf{v}_2^2) s(\mathbf{v},\mathbf{v}_2;v_1',v_2') \cdot \\ \cdot \big[f(\mathbf{x},\mathbf{v}_1';t) f(\mathbf{x},\mathbf{v}_2';t) - f(\mathbf{x},\mathbf{v};t) f(\mathbf{x},\mathbf{v}_2;t)\big]. \tag{4}$$

By inserting (3) and (4) into (2) one obtains the explicit form of the Boltzmann equation.

Although this nonlinear partial differential equation for $f(\mathbf{x},\mathbf{v};t)$ looks complicated, it possesses a stationary, i.e., time-independent equilibrium solution which reads:

$$f_{st}(\mathbf{x},\mathbf{v}) = c \cdot \exp\Big[-\frac{\big(V(x) + \frac{m}{2}\mathbf{v}^2\big)}{kT}\Big] \tag{5}$$

where T is a parameter to be interpreted as the absolute temperature, and k is the Boltzmann constant.

Even more remarkable is the fact, which could be proved by Boltzmann in his famous H-theorem that *all* time-dependent solutions of (2) which may start from arbitrary initial conditions, must necessarily approach the equilibrium distribution (5).

Comparing the results of H. Poincaré and L. Boltzmann we are however led to the conclusion that they clearly *contradict each other*: Boltzmann's equation involves the *irreversible motion* of any initial state of a system of atoms into states corresponding to, and being compatible with the equilibrium distribution (5), whereas Poincaré's theorem states that the same system of atoms obeying the same dynamics must sooner or later return to its initial state!

Where is the mistake leading to this contradiction?

A close analysis of the derivation of the Boltzmann equation shows that an extremely plausible but somewhat hidden assumption (the so-called Stosszahlansatz) enters the derivation of the form of the collision term (4).

The assumption means that the colliding atoms are *uncorrelated* before their collision. Apart from extremely rare exceptions this assumption is practically always fulfilled, but it is just this highly plausible approximation which leads to an equation describing an *irreversible process* and *neglecting the very rare events of recurrence* of initial states.

The result of the analysis of the Poincaré-Boltzmann controversy can now be summarized as follows:

a) There exists no contradiction.
b) The transition from *recurrence* to *irreversibility* comes about by the only approximately valid assumption of the *uncorrelatedness of the particles before their collision.* This assumption determines the form of the collision term (4).

The Inverse Problem We now investigate the inverse problem, whether a *globally irreversible process* can still contain *recurrent* (i.e., quasi-periodic) *subprocesses.* This is indeed possible if we allow of different levels of description. We now discuss different ways of how recurrence (i.e., *periodicity* or at least *quasi-periodicity*) can be *embedded* into a globally irreversible process.

a) **The way from irreversibility to recurrence by omission of dimensions**
   This is most easily demonstrated by projecting a helical (irreversible) motion onto a periodic motion in a lower-dimensional space (Fig. 2). The procedure of projecting onto periodic processes in subspaces is of course not restricted to a trajectory in a *three-dimensional* space only. Indeed, periodic or quasi-

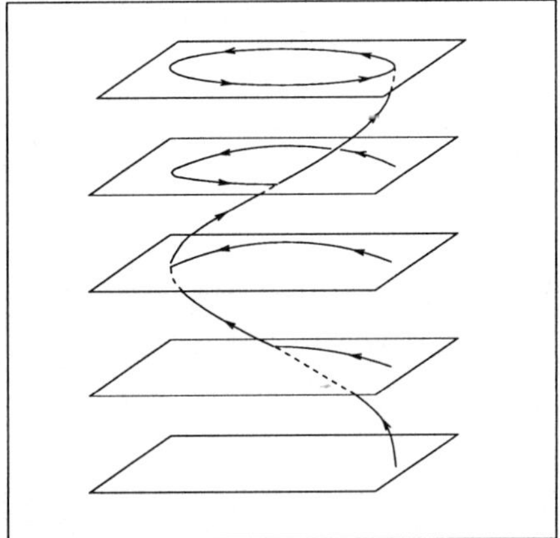

Fig. 1. The omission of one dimension leads from an irreversible helical motion to a periodic motion in a subspace.

   periodic subprocesses being part of globally irreversible processes are very common. Think of the trivial case of the rotating wheels of a traveling car or of the less trivial case of Volterra-Lotka-cycles in the evolution of predator populations interacting with prey populations.
b) **The way from irreversibility to recurrence by transition to a more coarse-grained description**
   Whether one detects a recurrence of states of a system or not, also depends on the *window of perception.* Processes which are irreversible on a microscopically fine-grained scale of description may exhibit recurrence on a more macroscopically coarse-grained descriptive scale.
   This is demonstrated in Fig. 3. The small (large) cells represent a microscopic (macroscopic) description of system states. The trajectory has returned two times to the macro-cell of a coarse-grained description, but not to the initial micro-cell of a fine-grained description. As an application let us mention a

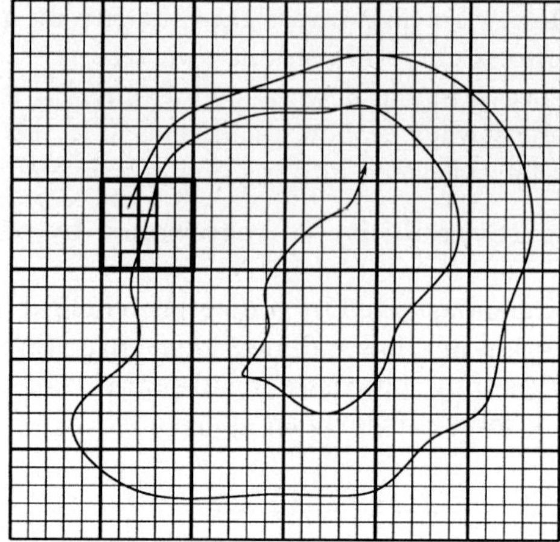

Fig. 2. The trajectory does not return to the initial micro-cell, but it does return to the initial macro-cell.

political example: Recently, Milosovic was compared to Hitler by a group of people, whereas another group did not accept this comparison. It may be that both groups are right. However, the one group used a relatively coarse-grained set of indicators and found recurrence of typical events, whereas the other group used a more fine-grained set of criteria and found more differences than coincidences that means no recurrence of the same type of events.

The cases a) and b) show that a reduction of the complexity of description (i.e., either omission of dimensions or use of a coarse-grained set of variables) can lead to the detection of recurrence phenomena, which otherwise remains concealed behind too much complexity.

It is remarkable that on the one side one came from recurrence to irreversibility via a small but important approximation (i.e., the neglection of correlations) but on the other side one can come from irreversible processes to embedded recurrent ones by again making an approximation (reduction of complexity).

c) **Individual Quasi-Periodic Stochastic Trajectories Belonging to a Probabilistically and Irreversibly Evolving Ensemble**

At last we compare the *ensemble-description* via a *probability distribution* with the description of individual systems by *stochastic trajectories.* The remarkable fact is, that the *probability distribution* $P(j;t)$, which obeys a master equation, of the type

$$\frac{dP(j;t)}{dt} = \sum_i w(j \mid i)P(i;t) - \sum_i w(i \mid j)P(j;t) \tag{6}$$

irreversibly approaches a stationary state $P_{st}(j)$, whereas the stochastic trajectories, whose ensemble is described by $P(j;t)$, execute a stochastic hopping process which *must* – sooner or later – *recur* to the initial state! That means: in the probabilistic frame of description *irreversibility* is connected with the *ensemble description*, which does not prevent the *individual trajectories* from exhibiting *recurrence*! Figure 4 illustrates the fact that stochastic trajectories must – probabilistically sooner or later – return to the initial state: If a chain of non-vanishing transition rates exists from each cell i_0 to each cell i_n (and vice versa from each cell i_n to each cell i_0), then a stochastic trajectory starting from i_0 will – sooner or later – reach i_n and will – sooner or later – return to i_0. This means there exists a probabilistic periodicity. Nevertheless the recurrent motion of each individual stochastic trajectory takes place "under the cover" of the probability distribution which comprises the whole ensemble of trajectories and which approaches irreversibly a stationary state!

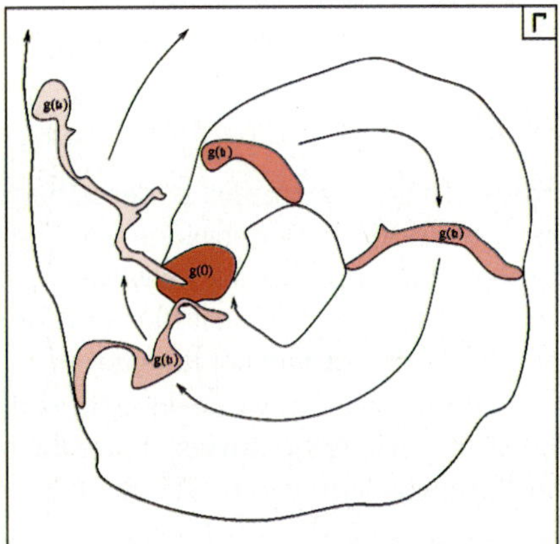

Fig. 3. The evolution in phase space Γ of an ensemble of systems contained in domain $g(0)$ at time $t = 0$.

2 Recurrence and (Quasi-) Periodicity in Social Systems

Reduction of Complexity of Social Systems In Sociodynamics, a general method of designing mathematical models for capturing the dynamics of collective phenomena in the human society [1,2], we reduce the complexity of the system under consideration by the following sequential steps:

Firstly, one introduces appropriate (material and personal) macro-variables characterizing the social sector to be modeled.

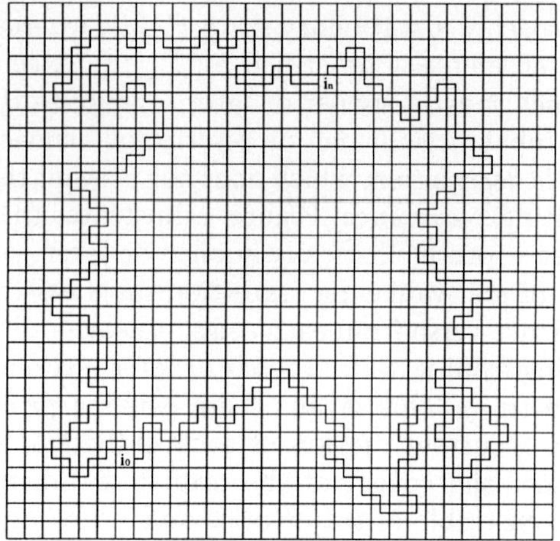

Fig. 4. If non-vanishing transition rates between neighboring cells exist, the hopping process from i_0 to i_n and back from i_n to i_0 indicated in the Figure has a finite probability. It is one of the possibilities to leave i_0 and to return to i_0.

Secondly, the dynamics of these variables is described in probabilistic terms connecting the microlevel of individual decisions and actions with the macro-level of the macro-variables. This amounts to setting up a master equation for the probability distribution of the macro-variables and of stochastic trajectories describing the stochastic evolution of the social system.

Thirdly, one makes use of quasi-mean value equations describing the (smoothed out) mean path of the stochastic trajectories. The quasi-mean value equations are coupled nonlinear differential equations. Their solutions also include limit cycles and strange attractors, thus demonstrating the appearance of (quasi-) periodic evolution in appropriately chosen subsets of macro-variables of the society. The kind of periodicity, treated in Sociodynamics belongs to recurrence phenomena of the type considered in section 1.3c.

Selected Examples It is not possible in this short lecture to give account of the detailed mathematical structure of the models designed for several classes of social phenomena. We will however present some results, i.e., computer solutions of the model equations after selecting appropriate scenarios which correspond to typical sets of trend-parameters. We focus on those scenarios which exhibit periodicity or quasi-periodicity. The variables will be shortly explained and a social interpretation of the periodic phenomena will be given.

Model 1) Migration of Interacting Populations. The Special Case of Two Populations Migrating Between Two Regions

The variables describing the migratory dynamics of two populations $\mathcal{P}^\mu, \mathcal{P}^\nu$ between two regions $1, 2$ are $\{n_1^\mu, n_2^\mu; n_1^\nu, n_2^\nu\}$. Assuming equal constant total populations of $\mathcal{P}^\mu$ and $\mathcal{P}^\nu$ $(n_1^\mu + n_2^\mu = 2N\,; n_1^\nu + n_2^\nu = 2N)$ the relevant scaled variables are

$$x = \frac{n_1^\mu - n_2^\mu}{2N}; \qquad -1 \le x \le +1$$
$$y = \frac{n_1^\nu - n_2^\nu}{2N}; \qquad -1 \le y \le +1 \tag{7}$$

describing whether the majority of $\mathcal{P}^\mu$, respectively of $\mathcal{P}^\nu$ lives in region 1 or 2. The model leads to the following quasi-mean value equations

$$\frac{dx}{d\tau} = \sinh(\tilde{\kappa}x + \tilde{\kappa}^{\mu\nu}y) - x\cosh(\tilde{\kappa}x + \tilde{\kappa}^{\mu\nu}y)$$
$$\frac{dy}{d\tau} = \sinh(\tilde{\kappa}^{\nu\mu}x + \tilde{\kappa}y) - y\cosh(\tilde{\kappa}^{\nu\mu}x + \tilde{\kappa}y), \tag{8}$$

where $\tilde{\kappa}, \tilde{\kappa}^{\mu\nu}, \tilde{\kappa}^{\nu\mu}$ are sociologically interpretable "trendparameters".
We consider only one scenario, which is however a dramatic one: In this case, both groups have a *strong internal* agglomeration trend $\tilde{\kappa} = 1,2$ (i.e., people of $\mathcal{P}^\mu$ wish to live together with people of their own group; the same holds for $\mathcal{P}^\nu$). However, there also exists a *strong asymmetric interaction* between $\mathcal{P}^{(\mu)}$ and $\mathcal{P}^{(\nu)}$: $\tilde{\kappa}^{\mu\nu} = -1.0$ ($\mathcal{P}^\mu$ does not want to live together with $\mathcal{P}^\nu$) and $\tilde{\kappa}^{\nu\mu} = +1.0$ ($\mathcal{P}^\nu$ wants to live together with $\mathcal{P}^\mu$).
Interpretation of the solution
The solution approaches a limit cycle describing a never ending evasion-invasion dynamics with the following phases:

Quadrant 1: The majority of $\mathcal{P}^\mu$ as well as $\mathcal{P}^\nu$ lives in district 1, but $\mathcal{P}^\mu$ wishes to evade into district 2 because of its segregation trend.
Quadrant 2: The majority of $\mathcal{P}^\mu$ has agglomerated in district 2, leaving behind $\mathcal{P}^\nu$. However, now $\mathcal{P}^\nu$ wishes to invade district 2 because it wishes to live together with $\mathcal{P}^\mu$.
Quadrant 3: Now $\mathcal{P}^\mu$ and $\mathcal{P}^\nu$ live together in district 2 but $\mathcal{P}^\mu$ wishes to evade back into district 1 because of its segregation trend.
Quadrant 4: $\mathcal{P}^\mu$ has now agglomerated in district 1, leaving behind $\mathcal{P}^\nu$. However, $\mathcal{P}^\nu$ now wishes to come to district 1, too, in order to join $\mathcal{P}^\mu$.

This migratory process describes in oversimplified manner the rather unpleasant process observed in several multi-cultural metro-poles all over the world: the sequential erosion of districts by asymmetric evasion-invasion interaction of populations of different cultural, ethnic or economic background.

Model 2) Group Dynamics: Special Case: Political Parties with Reciprocal Undermining Activity

The model describes the rise and fall of interacting social groups in terms of the number of their members and further variables such as the internal degree of solidarity. For two interacting groups 1, 2 we only consider here the numbers $N_1(t)$ and $N_2(t)$ of their members which may vary with time. Decisive trend-parameters are internal faith-confirming coefficients (w_{11} and w_{22}) and on the

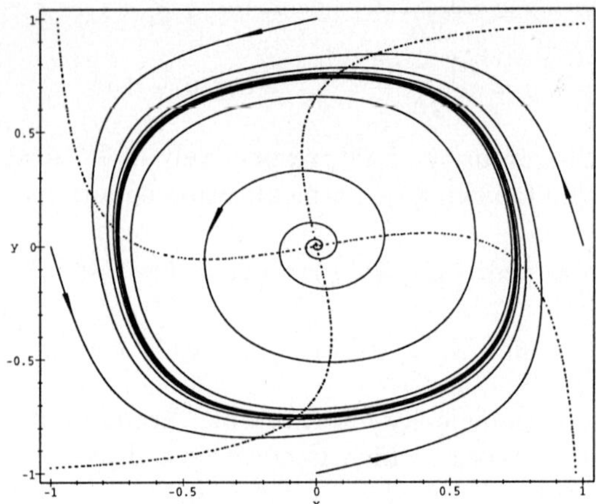

Fig. 5. a) Very strong internal agglomeration trend $\tilde{\kappa} = 1.2$ and strong asymmetric interaction $\tilde{\kappa}^{\mu\nu} = -1.0$ and $\tilde{\kappa}^{\nu\mu} = +1.0$. The origin $(0,0)$ is an unstable focus. All flux-lines approach a limit cycle.

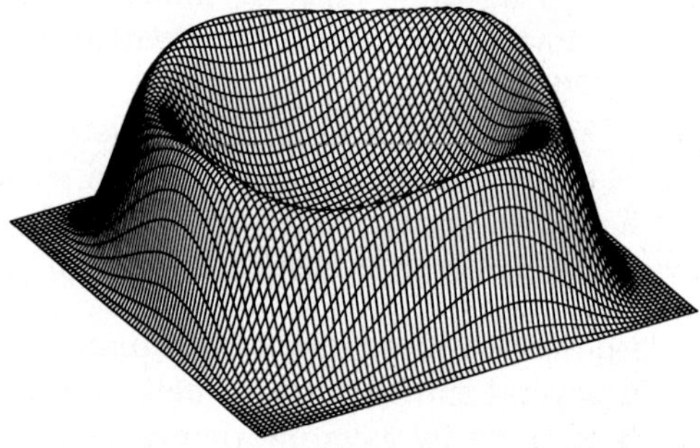

Fig. 5. b) Parameters as in Fig. 5a, $2N = 80$. The quadrumodal stationary probability has four maxima corresponding to meta-stable situations and ridges between the maxima along the limit cycle.

other hand mutual faith-undermining coefficients (w_{12} and w_{21}). For large mutual undermining activity (which can, e.g., exist between fiercely antagonistic political parties) the membership dynamics undergoes a phase-transition from stability or regular oscillations to a chaotic dynamics exhibited in Figs. 6a/b.

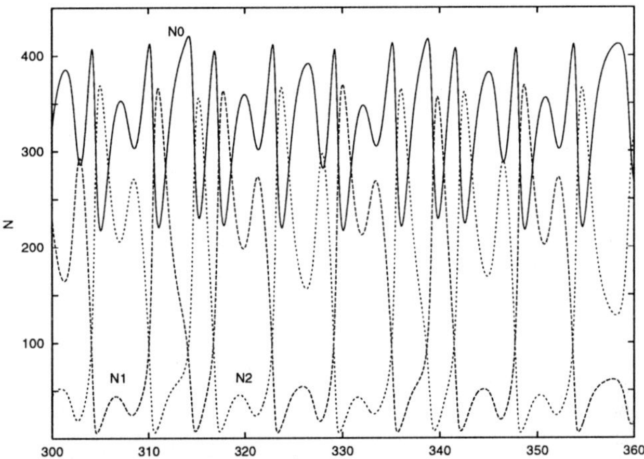

Fig. 6. a) The *non-periodic, irregular* and *chaotic evolution* with time of the population numbers $N_1(t), N_2(t), N_0(t)$ of G_1, G_2 and the crowd, respectively, is exhibited.

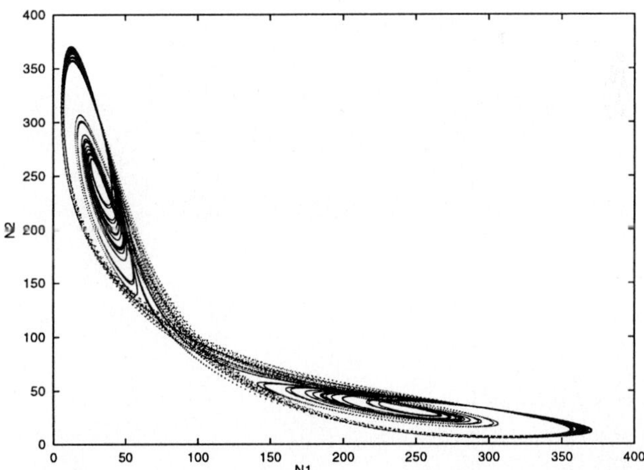

Fig. 6. b) An even better insight into the chaotic character of this evolution yields the phase-portrait of the trajectory in the N_1/N_2 plane.

References

1. W. Weidlich, *Physics and Social Science – The Approach of Synergetics,* Phys. Rep. **204**, 1-163 (1991).
2. W. Weidlich, *Sociodynamics – A Systematic Approach to Mathematical Modelling in the Social Sciences,* to appear, (Gordon & Breach, 2000).

Econophysics: What Can Physicists Contribute to Economics?

H.E. Stanley[1], L.A. Nunes Amaral[1], P. Gopikrishnan[1], V. Plerou[1,2], and B. Rosenow[1,3]

[1] Center for Polymer Studies and Department of Physics, Boston University, Boston, MA 02215, USA
[2] Department of Physics, Boston College, Chestnut Hill, MA 02617, USA
[3] Institut für Theoretische Physik, Universität zu Köln, 50937 Köln, Germany

Abstract. In recent years, a considerable number of physicists have started applying physics concepts and methods to understand economic phenomena. The term "Econophysics" is sometimes used to describe this work. Economic fluctuations can have many repercussions, and understanding fluctuations is a topic that many physicists have contributed to in recent years. Further, economic systems are examples of complex interacting systems for which a huge amount of data exist and it is possible that the experience gained by physicists in studying fluctuations in physical systems might yield new results in economics. Much recent work in econophysics is focused on understanding the peculiar statistical properties of price fluctuations in financial time series. In this talk, we discuss three recent results. The first result concerns the probability distribution of stock price fluctuations. This distribution decreases with increasing fluctuations with a power-law tail well outside the Lévy stable regime and describes fluctuations that differ by as much as 8 orders of magnitude. Further, this non stable distribution preserves its functional form for fluctuations on time scales that differ by 3 orders of magnitude, from 1 min up to approximately 10 days. The second result concerns the accurate quantification of volatility correlations in financial time series. While price fluctuations themselves have rapidly decaying correlations, the volatility estimated by using either the absolute value or the square of the price fluctuations has correlations that decay as a power-law and persist for several months. The third result bears on the application of random matrix theory to understand the correlations among price fluctuations of any two different stocks. We compare the statistics of the cross-correlation matrix constructed from price fluctuations of the leading 1000 stocks and a matrix with independent random elements, i.e., a random matrix. Contrary to first expectations, we find little or no deviation from the universal predictions of random matrix theory for all but a few of the largest eigenvalues of the cross-correlation matrix.

1 Introduction

The analysis of financial data using concepts and methods developed for physical systems has a long tradition [1–4] and has recently attracted the interest of physicists [5–9]. Possible reasons for this interest include the scientific challenge of understanding the dynamics of a strongly fluctuating complex system with a large number of interacting elements. Moreover, economic fluctuations could have many repercussions and understanding fluctuations in physical systems is a topic where many physicists have contributed. In addition, it is possible that the experience gained by studying fluctuations in physical systems might yield new results in economics.

One can ask how physicists can contribute to the search for solutions to the puzzles posed by modern economics that economists themselves have not yet solved? One approach – in the spirit of experimental physics – is to begin empirically, with real data that one can analyze in some detail, but without prior models. In economic systems such as financial markets, one has available a great deal of real data. Moreover, if one has at one's disposal the tools of statistical physics and the computing power to carry out any number of approaches, this abundance of data is to great advantage. Thus, for many physicists, studying the economy means studying a wealth of data on a strongly fluctuating complex system. Indeed, physicists in increasing numbers are finding problems posed by economics sufficiently challenging to engage their attention [10–26].

Recent studies attempt to uncover and explain the peculiar statistical properties of financial time series such as stock prices, stock market indices or currency exchange rates. The dynamics of financial markets is difficult to understand not only because of the complexity of its internal elements but also due to the many intractable external factors acting on it, which may differ from market to market. Remarkably, the statistical properties of certain observables appear to be similar for quite different markets [27], consistent with the possibility that there may exist "universal" mechanisms.

The most challenging difficulty in the study of financial markets is that the nature of the interactions between the different elements comprising the system is unknown, as is the way in which external factors affect it. Therefore, as a starting point, one may resort to empirical studies to help uncover the regularities or "empirical laws" that may govern financial markets [28]. The interactions between the different elements comprising financial markets generate many observables such as the transaction price, the share volume traded, the trading frequency, and the values of market indices. Recent empirical studies are based on the analysis of price fluctuations. This talk reviews recent results on (a) the distribution of stock price fluctuations and its scaling properties, (b) time-correlations in financial time series, and (c) correlations among the price fluctuations of different stocks. Space limitations restrict us to focusing mainly on our group's work; a more balanced account can be found in two recent books [5,6], other articles in these proceedings, and two other recent international conferences[7,9]. Recent work in this field also focuses on applications such as risk control, derivative pricing, and portfolio selection [29], which shall not be discussed in this talk. The interested reader should consult, for example, [5,6,30].

2 What is the Question?

The key question for physicists entering this field – or virtually any other field physicists might enter – is "how do I quantify things?" There are many things to quantify in economics.

An appropriate place to begin is with simple fluctuations. Consider a prototype fluctuating quantity: the stock market and stock-price averages. If we make a graph on log-linear paper of the value of the stock index as a function of time, we see fluctuations – sometimes even dramatic fluctuations such as the negative 25 percent fluctuation on Black Monday in 1987. When we first examine these fluctuations, those of us who have worked in statistical mechanics naturally think of a simple one-dimensional random walk – with the displacement of the random walk on the y-axis and the time on the x-axis. When we do such a plot on a graph, we immediately see that the curve produced does not agree with the empirical data for stock-price fluctuations.

If this approach is unhelpful, what do we do next? How are we to understand these fluctuations? For that matter, why do we want to?

3 Why Do We Care?

There are many practical reasons for wanting to understand stock-price fluctuations. The first reason is painfully obvious: although not everyone owns stocks, everyone is powerfully affected by stock markets and financial systems. If a country goes bankrupt and the food supply fails, the poorest citizens – who probably are not stockholders – still suffer. Beginning to understand financial fluctuations means beginning to understand risk, and if we can begin to quantify risk, then perhaps we can develop ways to manage risk.

The second reason for wanting to understand stock-price fluctuations is intellectual. The economy is a complex system, but is unlike other complex systems we study, and offers unique intellectual opportunities. In most research on complex systems, we start with a general statement, invent some sort of theoretical model to test the general statement, compare the data produced by the model with the general statement, and only then, perhaps, compare the theoretical data with whatever real-world data might be available. When the subject of research is the economy, however, the situation is totally different. The economy is data-dominated. There are huge quantities of data on the economy. Virtually *every* economic transaction today is recorded. Many of these data are available free on the Internet, and others can be obtained at a minimal cost. This "complex system" has already been extensively analyzed, of course, and most of this analysis has been done by economists. What can we physicists add to what has already been done? How might our approach differ from that of economists?

The third reason for wanting to understand stock-price fluctuations is simply that people are interested in the topic. Within the physics community, econophysics is a fast-growing subfield. The business community has become interested, and recently there was an article on econophysics in the *Wall Street Journal.*

4 What Do We Do?

We start with the initial set of data: a simple time series giving the value of some stock average as a function of time. How can we improve on the biased random walk model to describe this set of data? We add a fundamental variable: the change (g) in the value of this index over some time window (Δt). Obviously the value of g depends on the size of Δt, and on where in time I look at g. If what we see in this initial data set does not conform to the behavior of a simple random walk, perhaps we can find some correlations. We have seen in other problems, e.g., DNA base-pair sequences, that a time-series that resembles the one before us is an indication of the presence of some long-range correlation.

To start off, we study the correlation function of g and, at the same time, the correlation function of the absolute value of g. Why? Because in financial market data, absolute values tend to cluster.

Our results show that when we examine the correlations in the price changes themselves, g, using a log-linear scale (with the logarithmic y-axis showing the price changes and the linear x-axis showing the time), an approximate straight line is produced (see [31] and citations therein). This line extends out to about 20 minutes on the linear time scale, at which point it hits the noise level. A straight line on log-linear paper indicates an exponential decrease, which is consistent with these data in this autocorrelation function – with a time constant of about 4 minutes. In order to make any money with this observed correlation, one would have to react on a time scale significantly shorter than four minutes.

When we examine the correlations in the ***absolute value*** of g, we discover not exponential behavior but power-law behavior. On the logarithmic y-axis we now have the absolute values of g. This time the data are only approximately straight over a little more than one decade. Although they are quite noisy, they still approximate a power-law slope of 0.3.

To study this in more detail, we can analyze the power spectrum of the correlation function. The slope of the power-law changes at approximately one day – one slope with a value of 0.3 for time scales shorter than approximately one day and another slope with a value of 0.9 for time scales longer than approximately one day. The crossover at approximately one day is a genuine property of the data. If we shuffle the data before we do the analysis, we end up with white noise. If we analyze the integral of this correlation function by adding up the fluctuations, we end up with a quantity that has a power law and a crossover at approximately one day.

To summarize thus far: because the experimental data display correlations with a very short range ($\approx$ 2 minutes), we cannot understand the time-series and the fluctuations – and we cannot quantify them – simply by using second-order correlation functions to explain the data.

What has been done traditionally in this field? In 1963, Benoit Mandelbrot published a paper in which he analyzed the fluctuations in the market price of cotton [4]. Available to him were 1000 data points in three different data sets, which he used to analyze the histogram of the price changes of this single commodity. He plotted the cumulative distribution function and then turned his attention to the behavior of the tails of the distribution. He was familiar with Pareto's work on power-law tails, so he decided to use log-log plots of his data sets. For the three data sets, he ended up with six curves—three for the positive tails and three for the negative. Since the six all approximate a straight line, we know the tails display a power-law behavior. Mandelbrot described the tails using an exponent, α, of approximately 1.7. This exponent falls within the range of that predicted by a Lévy distribution, so Mandelbrot concluded that fluctuations in the price of cotton could be described by a Lévy distribution. He also concluded that the exponent was the same for time windows ranging from one day to one month.

Although Mandelbrot's paper was seminal in the field, because he had only daily and monthly data – and not shorter time-windows of Δt – the data fluctuations span only two orders of magnitude. Since we now have much more data available to us, how do we extend the work Mandelbrot started?

This has been done by Rosario Mantegna, who has followed the same prescription as Mandelbrot and has analyzed not fluctuations in cotton prices, but fluctuations in the S&P 500 stock average, the weighted average of the 500 largest firms in the US. He studied them not in daily time intervals, but in 1-minute time intervals (roughly 300 times smaller) [18]. His data span 6 years. Thus he was dealing with not 1000 records, but on the order of magnitude of 1 million records – three orders of magnitude more data.

He has found that, out to approximately 5 or 6 standard deviations, the S&P 500 data conform nicely to Mandelbrot's Lévy distribution (but with a parameter $\alpha \approx 1.4$ instead of $\alpha \approx 1.7$). Beyond approximately 5 or 6 standard deviations, the data deviate below the prediction of the Lévy distribution. This is probably a good thing; if the Lévy distribution held out to, say, 20 standard deviations, you would have many more Black Mondays. The fact that the data are truncated is also good for an intellectual reason: Lévy distributions have an infinite variance, and this truncation restores the finite variance. So the data are consistent with a truncated Lévy distribution ("flight") [32].

How robust is this truncated Lévy flight behavior? Skeltorp used the same approach to study the behavior of the OVX index of the Norwegian stock market with equivalent time-interval ranges. The results are the same – the same Lévy distribution seems to hold for 4 to 5 standard deviations, and then the data begin to fall below the Lévy distribution. So this behavior does indeed seem robust.

Mantegna's work has been extended by Gopikrishnan, who, instead of working with stock averages, worked with individual stocks ([33] and citations therein). Instead of the 1000 data points of Mandelbrot, or the 1 million data points, of Mantegna, Gopikrishnan used 100 million data points. He took the 1000 largest stocks, which, if simply averaged, look benign—but individually they exhibit much larger fluctuations. When working with individual stocks, Gopikrishnan found the most useful minimum time window to be 5 minutes. Because the plots from these short time windows of individual stocks are extremely noisy, the cumulative distribution function (the integral of the pdf) is used. If you plot the cumulative distribution function on a sheet of log-log paper, you see a noisy but approximate straight line with a slope of ≈ -3. The slope for each stock will be slightly different. If we make a histogram of all of the slopes of the stocks, we get a bell-shaped curve centered around the value of 3.

This is interesting because, in the previous work by Mandelbrot in which he describes cotton-price fluctuations as a Lévy distribution, the bell-shaped curve is centered around some value between 0 and 2 – which is one of the characteristics of a Lévy distribution. The fact that Gopikrishnan's bell-shaped curve is centered around 3 means that the data are *not* described by a Lévy distribution in the tails. Thus this surprising discovery disagrees with the classic work of Mandelbrot.

To check this discovery, Gopikrishnan and his collaborators spent a year and a half checking their work. They found that the cumulative distribution function on log-log paper is a function of the scale of returns of 100 standard deviations follows a power law, i.e., has a slope of ≈ -3, which – irrespective of error bars – excludes the value 2. The data are described by this power law distribution out to 10^{-8}.

One physical implication of this is that "shocks" in the economy are not isolated points that have to be added to the data. They are actually part of the data set. If we return to the stock average provided by the S&P 500, we see the same cumulative distribution function with the Lévy distribution in the center and a crossover – not to an exponentially truncated tail, but to a power-law truncated tail.

In conclusion, we can say that the price fluctuations of individual stocks are consistent with this relatively simple power-law distribution and are at odds with Mandelbrot's Lévy distribution. They are consistent with the power-law distribution over fully 100 standard deviations. This distribution preserves its functional form in time windows ranging from one minute up to almost 4 orders of magnitude. We can also say that the amplitudes of price changes, i.e., the absolute values of price changes, display long-range correlations – even when the price changes themselves do not.

5 Distribution of Price Fluctuations: Details

The recent availability of "high frequency" data allows one to study economic time series on a wide range of time scales varying from seconds up to a few months. For example, our recent work [18,31,34] involves the analysis of the S&P 500 index, an index of the New York Stock Exchange that consists of 500 companies representative of the US economy. It is a market-value (stock price times number of shares outstanding) weighted index, with each stock's weight in the index proportionate to its market value [18]. The S&P 500 index is one of the most widely used benchmarks of U.S. equity performance. We analyzed high frequency data for the 13-year period 1984-96 with a recording frequency of one minute or shorter and the daily records for the 35-year period 1962–1996.

The S&P 500 index $Z(t)$ from 1962-96 has an overall upward drift – interrupted by drastic events such as the market crash of October 19, 1987. One analyzes the difference in logarithm of the index, often called the return $G(t) \equiv \log_e Z(t+\Delta t) - \log_e Z(t)$, where Δt is the time scale investigated. One only counts the number of minutes during the opening hours of the stock market. One finds that when one analyzes returns on short time scales, large events are much more likely to occur, in contrast to a sequence of Gaussian distributed random numbers of the same variance. As one analyzes returns on larger time scales, this difference is apparently much less pronounced. In order to understand this process, one starts by analyzing the probability distribution of returns on a given time scale Δt, which in our study, varies from 1 min up to a few months.

The nature of the distribution of price fluctuations in financial time series is a long standing open problem in finance which dates back to the turn of the century. In 1900, Bachelier proposed the first model for the stochastic process of returns – an uncorrelated random walk with independent, identically Gaussian distributed (*i.i.d*) random variables [1]. This model is natural if one considers the return over a time scale Δt to be the result of many independent "shocks", which then lead by the central limit theorem to a Gaussian distribution of returns [1]. However, empirical studies [4,18,19] show that the distribution of returns has pronounced tails in striking contrast to that of a Gaussian. Despite this empirical fact, the Gaussian assumption for the distribution of returns is widely used in theoretical finance because of the simplifications it provides in analytical calculation; indeed, it is one of the assumptions used in the classic Black-Scholes option pricing formula [35].

In his pioneering analysis of cotton prices, Mandelbrot observed that in addition to being non-Gaussian, the process of returns shows another interesting property: "time scaling" – that is, the distributions of returns for various choices of Δt, ranging from 1 day up to 1 month have similar functional forms[4]. Motivated by (i) pronounced tails, and (ii) a stable functional form for different time scales, Mandelbrot [4] proposed that the distribution of returns is consistent with a Lévy stable distribution [2,3].

Conclusive results on the distribution of returns are difficult to obtain, and require a large amount of data to study the rare events that give rise to the tails. More recently, the availability of high frequency data on financial market indices, and the advent of improved computing capabilities, has facilitated the probing of the asymptotic behavior of the distribution. For example, Mantegna and Stanley [18] analyzed approximately 1 million records of the S&P 500 index. They report that the central part of the distribution of S&P 500 returns appears to be well fit by a Lévy distribution, but the asymptotic behavior of the distribution of returns shows faster decay than predicted by a Lévy distribution. To quantify this, Hence, [18] proposed a truncated Lévy distribution—a Lévy distribution in the central part followed by an approximately exponential truncation—as a model for the distribution of returns. The exponential truncation ensures the existence of a finite second moment, and hence the truncated Lévy distribution is not a stable distribution [32,36]. The truncated Lévy process with *i.i.d.* random variables has slow convergence to Gaussian behavior due to the Lévy distribution in the center, which could explain the observed time scaling for a considerable range of time scales [18].

Recent studies [33,37] on considerably larger time series using larger databases show quite different asymptotic behavior for the distribution of returns. Our recent work [33] analyzed three different data bases covering securities from the three major US stock markets. In total, we analyzed approximately 40 million records of stock prices sampled at 5 min intervals for the 1000 leading US stocks for the 2-year period 1994-95 and 35 million daily records for 16,000 US stocks for the 35-year period 1962-96. We study the probability distribution of returns for individual stocks over a time interval Δt, where Δt varies approximately over a factor of 10^4 – from 1 min up to more than 1 month. We also conduct a parallel study of the S&P 500 index.

Our key finding is that the *cumulative* distribution of returns for both individual companies and the S&P 500 index can be well described by a power law asymptotic behavior, characterized by an exponent $\alpha \approx 3$, well outside the stable Lévy regime $0 < \alpha < 2$. Further, it is found that the distribution, although not a stable distribution, retains its functional form for time scales up to approximately 16 days for individual stocks and approximately 4 days for the S&P 500 index. For larger time scales our results are consistent with break-down of scaling behavior, i.e., convergence to Gaussian [33]. Similar results have also been found for currency exchange data [37].

6 Correlations in Financial Time Series

In addition to the probability distribution, an aspect of equal importance for the characterization of any stochastic process is the quantification of correlations. Studies of the autocorrelation function of the returns show exponential decay with characteristic decay times of only 4 min [38] consistent with the efficient market hypothesis [39]. This is paradoxical, for in the previous section we have seen that the distribution of returns, in spite of being a non-stable distribution, preserves its shape for a wide range of Δt. Hence, there has to be some sort of correlations or dependencies that prevent the central limit theorem to take over sooner and preserve the scaling behavior.

Indeed, lack of linear correlation does not imply independent returns, since there may exist higher-order correlations. Recently, Liu and his collaborators found that the amplitude of the returns, the absolute value or the square – closely related to what is referred to in economics as the *volatility* [40] – shows long-range correlations [21,23,31,34,41,42] with persistence [43] up to several months. They analyzed the correlations in the absolute value of the returns [31,34] of the S&P 500 index using traditional correlation function estimates, power spectrum and the recently-developed detrended fluctuation analysis (DFA). All the three methods show the existence of power-law correlations with a cross-over at approximately 1.5 days. For the S&P 500 index, DFA estimates for the exponents characterizing the power law correlations are $\alpha_1 = 0.66$ for short time scales smaller than ≈ 1.5 days and $\alpha_2 = 0.93$ for longer time scales up to a year. For individual companies, the same methods yield $\alpha_1 = 0.60$ and $\alpha_2 = 0.74$, respectively. The power spectrum gives consistent estimates of the two power-law exponents.

The long memory in the amplitude of returns suggests that it is useful to define a subsidiary process, referred to as the volatility. Volatility of a certain stock measures how much it is likely to fluctuate. It can also be related to the amount of information arriving at any time. The volatility can be estimated for example by the local average of the absolute values or the squares of the returns. In their recent work on the statistical properties of volatility Liu *et al.* [31,42] show that the volatility correlations show asymptotic $1/f$ behavior [34,31,42]. Using the same data bases as above, Liu and his collaborators also study the *cumulative* distribution of volatility [34,42] and find that it is consistent with a power-law asymptotic behavior, characterized by an exponent $\mu \approx 3$, just the same as that for the distribution of returns. For individual companies also, one finds a similar power law asymptotic behavior [31]. In addition, it is also found that the volatility distribution scales for a range of time intervals just as the distribution of returns.

7 Correlations among Different Units

Recently, the problem of understanding the correlations among the returns of different stocks has been addressed by applying methods of random matrix theory to the cross-correlation matrix [44,45]. Aside from scientific interest, the study of correlations between the returns of different stocks is also of practical relevance in quantifying the risk of a given portfolio [29]. Consider, for example, the equal-time correlation of stock returns for a given pair of companies. Since the market conditions may not be stationary, and the historical records are finite, it is not clear if a measured correlation of returns of two stocks is just due to "noise" or genuinely arises from the interactions among the two companies. Moreover, unlike most physical systems, there is no "algorithm" to calculate the "interaction strength" between two companies (as there is for, say, two spins in a magnet). The problem is that although every pair of companies should interact either directly or indirectly, the precise nature of interaction is unknown.

In some ways, the problem of interpreting the correlations between individual stock-returns is reminiscent of the difficulties experienced by physicists in the fifties, in interpreting the spectra of complex nuclei. Large amounts of spectroscopic data on the energy levels were becoming available but were too complex to be explained by model calculations because the exact nature of the interactions were unknown. Random matrix theory (RMT) was developed in this context, to deal with the statistics of energy levels of complex quantum systems [46–48]. With the minimal assumption of a random Hamiltonian, given by a real symmetric matrix with independent random elements, a series of remarkable predictions were made and successfully tested on the spectra of complex nuclei [46]. RMT predictions represent an average over all possible interactions

[47]. Deviations from the *universal* predictions of RMT identify system-specific, non-random properties of the system under consideration, providing clues about the underlying interactions [48]

Recently, Plerou and her collaborators analyzed the cross-correlation matrix $\mathsf{C} \equiv C_{ij} \equiv \langle G_i G_j \rangle - \langle G_i \rangle \langle G_j \rangle / \sigma_i \sigma_j$ of the returns at 30-minute intervals of the largest 1000 US stocks for the 2-year period 1994–95. They analyze the statistical properties of C by applying techniques of random matrix theory (RMT) [44,45]. First, they test the eigenvalue statistics of the cross-correlation matrix for universal properties of real symmetric random matrices such as the Wigner surmise for the eigenvalue spacing distribution and eigenvalue correlations. Remarkably, they find that eigenvalue statistics of the correlation matrix agree well with the universal predictions of random matrix theory for real symmetric random matrices, in contrast to our naive expectations for a strongly interacting system.

Deviations from RMT predictions represent genuine correlations. In order to investigate deviations, we compute the distribution of the eigenvalues of the C and compare with the prediction [44] for uncorrelated time series [49]. We find that the statistics of all but a few of the largest eigenvalues in the spectrum of C agree with the predictions of random matrix theory, but there are deviations for a few of the largest eigenvalues [44,45]. The deviations of the largest few eigenvalues from the random matrix result are also found when one analyzes the distribution of eigenvector components. Specifically, the largest eigenvalue which deviates significantly (25 times larger than random matrix bound) has almost all components participating equally and thus represents the correlations that pervade through the entire market. This result is in agreement with the results of Laloux and collaborators [44] for the eigenvalue distribution of C on a daily time scale.

8 How Economic Organizations Grow and Shrink

Most current research on the economy starts off by dividing the economy into sectors: food, automotive, computer, entertainment, and so on. Then the interactions between firms within a sector are analyzed, e.g., how does the behavior of General Motors affect the behavior of Ford? The assumption is that firms that compete directly affect each other's behavior much more strongly than firms that do not.

Physicists look at this model and immediately see a similarity between it and something we were playing with 30 years ago during the early days of critical phenomena research, when we divided a set of spins into cluster subsystems. We treated exactly the interactions of a spin within the small cluster of spins surrounding it, and approximated the interactions between spins in that cluster and spins in more distant clusters. We eventually abandoned these "effective field theories of magnetism" because they failed to give information that was in accord with accurate experiments.

What is the analog with research on the economy? Direct interactions within sectors (clusters) are obvious; if General Motors has quality-control problems, their customers will start buying Fords. Indirect interactions between sectors (clusters) may not be immediately obvious, but they are there; Ford has to hire more workers to meet the rising demand for their cars and the McDonald's outlet across the street from the assembly plant has to expand to accommodate the much larger lunchtime crowd.

It is a kind of spin-glass with both ferromagnetic and anti-ferromagnetic interactions, both short-range and long-range. Unlike the spin-glass, however, we have no *a priori* way to plausibly choose which interactions to study.

There are data available on the approximately 4000 publicly-traded firms listed in the stock markets – by law they are required to make a great deal of information public. One category of data that is often of interest is whether a firm is growing or shrinking. The sales of a company this year divided by the sales last year is one possible measure of the growth rate.

In making a histogram of growth rates, we could put all 4000 companies together and make one single histogram; as was done by Gibrat 70 years ago. Instead, however, we divide the 4000 companies into 20 "bins" according to each company's size. When we do that, we get different histograms for different sizes. Smaller companies can grow – or shrink – more rapidly than larger companies. It is highly unlikely that Ford would grow or shrink by a factor of 10 in a single year. On the other hand, a small company can – and often will – grow or shrink by a factor of 10 in a single year. So each of the 10 histograms has a different characteristic standard deviation or "width." The width is found to be a decreasing function of the size of the firm – the larger the firm, the smaller the width.

If we make a plot with the width on the y-axis and the firm size – measured in amount of sales – on the x-axis, we find we have an approximate straight line over 8 or 9 decades. If we make another plot, this time using number of employees as the measure of firm size, we find we have another approximate straight line, this time over roughly 5 decades. Remarkably, the slopes of these two straight lines are identical. After checking this empirical law of economics by applying it to a number of financial markets in countries other than those in the US, we find that it is quite a robust law [50].

This law has also been applied to not just the financial status of firms and markets, but also to the overall economies of entire countries. Data obtained through the Harvard Institute for International Development on the economies of 152 countries have been analyzed using the same procedures, and virtually the same results as those obtained for business firms were produced [51]. The law has also been applied to data related to changes in the size of university research budgets [52], as well as to data recording the changing populations of various species of birds [53], and similar results were found in each case.

To approach this problem, M.H.R. Stanley and M.A. Salinger first located and secured a database – called COMPUSTAT – that lists the annual sales of every firm in the United States. With this information, M.H.R. Stanley and coworkers calculated histograms of how firm sizes change from one year to the next [54]. They find that the distribution of growth rates of firm sales has the same functional form regardless of industry or market capitalization. Moreover, the width of these distributions decrease with increasing sales as a power-law with an exponent approximately 1/6. Recently, similar statistical properties were found for the GDP of countries [51] and for university research fundings [52]. Hence, it is not impossible to imagine that there are some very general principles of complex organizations at work here, because similar empirical laws appear to hold for data on a range of systems that at first sight might not seem to be so closely related. Buldyrev *et al.* models this firm structure as an approximate Cayley tree, in which each subunit of a firm reacts to its directives from above with a certain probability distribution [55]. More recently, Amaral *et al.* [56] have proposed a microscopic model that reproduces both the exponent and the distribution function. Takayasu and Okuyama [50] extended the empirical results to a wide range of countries.

9 Open Questions

Econophysics is a field wherein questions are as difficult to pose as to answer. The empirical results shown above clearly beckon explanation. For example, in first two sections, we have looked mainly at two empirical results: (i) the distribution of fluctuations, which shows a power law behavior well outside the stable Lévy regime, and yet preserves its shape – scales – for a range of time scales and (ii) the long range correlations in the amplitude of price fluctuations. How are the two related?

Previous explanations of scaling relied on Lévy stable [4] and exponentially-truncated Lévy processes [6,18]. However, the empirical data that we analyze are not consistent with either of these two processes. In order to confirm that the scaling is *not* due to a stable distribution, one can randomize the time series of 1 min returns, thereby creating a new time series which contains *statistically-independent* returns. By adding up n consecutive returns of the shuffled series, one can construct the n min returns. Both the distribution and its moments show a rapid convergence to Gaussian behavior with increasing n, showing that the time dependencies, specifically volatility correlations are intimately connected to the observed scaling behavior [33].

Using the statistical properties summarized above, can we attempt to deduce a statistical description of the process which gives rise to this output? For example, the standard ARCH model [28,57] reproduces the power-law distribution of returns; however it assumes finite memory on past events and hence is not consistent with long-range correlations in volatility. On the other hand, the distribution of volatility and that of returns which have similar asymptotic behavior, however support the central ARCH hypothesis that $g(t) = \epsilon\, v(t)$, where ϵ is an *i.i.d.* Gaussian random variable independent of the volatility $v(t)$, and $g(t)$ denotes the returns. A consistent statistical description may involve extending the traditional ARCH model to include long-range volatility correlations [58].

A more fundamental question would be to understand the above results starting from a microscopic setting. Researchers have also studied microscopic models that might give rise to the empirically observed statistical properties of returns [5,10]. For example, Lux and Marchesi [10] recently simulated a microscopic model of financial markets with two types of traders, what they refer to as 'fundamentalist' and 'noise' traders. Their results reproduce the power-law tail for the distribution of returns and also the long range correlations in volatility.

In the last section, we found evidence for different modes of correlations between different companies. For example, the largest eigenvalue of the cross-correlation matrix showed correlations that pervade the entire market. Could it be that the above observed scaling properties are related to how correlations propagate from one unit to the other such as occur in critical phenomena? Researchers have studied economic data from the physics perspective of a complex system with each unit depending on the other. Specifically, the possibility that all the companies in a given economy might interact, more or less, like a spin glass. In a spin glass, each spin interacts with every other spin – but not with the same coupling and not even with the same sign. For example, if the stock price of a given business firm A decrease by, e.g., 10% , this will have an impact in the economy. Some of these will be favorable – firm B, which competes with A, may experience an increase in market share. Others will be negative – service industries that provide personal services for firm A employees may experience a drop-off in sales as employee salaries will surely decline. There must be positive and negative correlations for almost any economic change. Can we view the economy as a complicated spin glass?

10 Conclusion

In this presentation we examine two new empirical laws of economics. The first empirical law concerns the probability distribution function of price changes of stocks. When we examine that distribution function 100 standard deviations into the tails, we discover an apparent power law with an exponent of approximately -4 (100 standard deviations – two decades – corresponds to eight decades). This is data about events that occur with a very small probability. The second empirical law of economics – a very robust law that is applicable not only to business firms but also to the economies of entire countries and even to university research budgets – concerns how the width of a histogram of changes of the size of these entities (firms, countries, research budgets) is also a power law (of $\approx 1/6$ or 0.2) over approximately eight decades of the abscissa of the graph.

Acknowledgement. We conclude by thanking all our collaborators and colleagues from whom we learned a great deal. These include the researchers and faculty visitors to our research group with whom we have enjoyed the pleasure of scientific collaboration. Those whose research provided the basis of this short report include (in addition to the authors): S. V. Buldyrev, D. Canning, P. Cizeau, X. Gabaix, S. Havlin, P. Ch. Ivanov, R. N. Mantegna, C.-K. Peng, M.

A. Salinger, and M. H. R. Stanley. In addition to the above, we also thank M. Barthélemy, J.-P. Bouchaud, D. Sornette, D. Stauffer, S. Solomon, and J. Voit for very helpful discussions and comments.

References

1. L. Bachelier, Ann. Sci. École Norm. Sup. **3**, 21 (1900).
2. V. Pareto, *Cours d'Économie Politique*, Lausanne and Paris, 1897.
3. P. Lévy, *Théorie de l'Addition des Variables Aléatoires*, Gauthier-Villars, Paris, 1937.
4. B.B. Mandelbrot, J. Business **36**, 294 (1963).
5. J.P. Bouchaud and M. Potters, *Theorie des Risques Financiéres*, (Alea-Saclay, Eyrolles, 1998).
6. R.N. Mantegna and H.E. Stanley, *An Introduction to Econophysics: Correlations and Complexity in Finance*, (Cambridge University Press, Cambridge, 1999.)
7. I. Kondor and J. Kértesz, (Eds.), *Econophysics: An Emerging Science*, (Kluwer, Dordrecht, 1999).
8. R.N. Mantegna, (Ed.), *Proc. of the Int. Workshop on Econophysics and Statistical Finance*, Physica A [special issue] **269**, 1 (1999).
9. K.B. Lauritsen, (Ed.), *Application of Physics in Financial Analysis*, Int. J. Theor. Appl. Finance [special issue], (1999).
10. T. Lux and M. Marchesi, Nature **297**, 498 (1999); T. Lux, J. Econ. Behav. Organizat. **33**, 143 (1998) ; J. Econ. Dyn. Control **22**, 1 (1997); Appl. Econ. Lett. **3**, 701 (1996).
11. M. Levy, H. Levy and S. Solomon, Econ. Lett. **45**, 103 (1994); M. Levy and S. Solomon, Int. J. Mod. Phys. C **7**, 65 (1996).
12. J.-P. Bouchaud and R. Cont, Eur. Phys. J. B **6**, 543 (1998); R. Cont and J.-P. Bouchaud, cond-mat/9712318.
13. M. Potters, R. Cont, and J.-P. Bouchaud, Europhys. Lett. **41**, 239 (1998).
14. J.-P. Bouchaud and D. Sornette, J. Phys. I France **4**, 863 (1994); D. Sornette, A. Johansen, and J.-P. Bouchaud, J. Phys. I France **6**, 167 (1996); D. Sornette and A. Johansen, Physica A **261** , 581 (1998); D. Sornette, Physica A **256**, 251 (1998); A. Johansen and D. Sornette, Risk **1**, 91 (1999), cond-mat/9901035; cond-mat/9811292.
15. D. Stauffer, Ann. Phys.-Berlin **7**, 529 (1998); D. Stauffer and T.J.P. Penna, Physica A **256**, 284 (1998); D. Chowdhury and D. Stauffer, Eur. Phys. J. B **8**, 477 (1999); I. Chang and D. Stauffer, Physica A **264**, 1 (1999); D. Stauffer and T.J.P. Penna, Physica A **256**, 284 (1998); D. Stauffer, P.M.C. de Oliveria, and A.T.Bernardes, Int. J. Theor. Appl. Finance **2**, 83 (1999).
16. M. Marsili and Y.-C. Zhang, Phys. Rev. Lett. **80**, 2741 (1998); G. Caldarelli, M. Marsili, and Y.-C. Zhang, Europhys. Lett. **40**, 479 (1997); S. Galluccio, G. Calderelli, M. Marsili, and Y.-C. Zhang, Physica A **245**, 423 (1997); M. Marsili, S. Maslov, and Y.-C. Zhang, Physica A **253**, 403 (1998); S. Maslov and Y.-C. Zhang, Physica A **262**, 232 (1999); D. Challet and Y.C. Zhang, Physica A **256**, 514 (1998).
17. P. Bak, M. Paczuski, and M. Shubik, Physica A **246**, 430 (1997).
18. R.N. Mantegna and H.E. Stanley, Nature **376**, 46 (1995).
19. S. Ghashghaie, W. Breymann, J. Peinke, P. Talkner, and Y. Dodge, Nature **381**, 767 (1996); see also R.N. Mantegna and H.E. Stanley, Nature **383**, 587 (1996); Physica A **239**, 255 (1997).

20. A. Arneodo, J.-F. Muzy, and D. Sornette, Eur. Phys. J. B **2**, 277 (1998).
21. M. Pasquini and M. Serva, cond-mat/9810232; cond-mat/9903334.
22. M.M. Dacorogna, U.A. Muller, R.J. Nagler, R.B. Olsen, and O.V. Pictet, J. Int. Money and Finance **12**, 413 (1993); see also the Olsen website at http://www.olsen.ch for related papers.
23. R. Cont, *Statistical Finance: Empirical study and theoretical modeling of price variations in financial markets*, PhD Thesis, (Université de Paris XI, 1998); cond-mat/9705075; R. Cont, M. Potters, and J.-P. Bouchaud, in: *Scale Invariance and Beyond*, B. Dubrulle, F. Graner, and D. Sornette, (Eds.), (Springer, Berlin, 1997).
24. N. Vandewalle and M. Ausloos, Int. J. Mod. Phys. C **9**, 711 (1998); Eur. Phys. J. B **4**, 257 (1998); N. Vandewalle, P. Boveroux, A. Minguet, and M. Ausloos, Physica A **255**, 201 (1998).
25. H. Takayasu, A.H. Sato, and M. Takayasu, Phys. Rev. Lett. **79**, 966 (1997); H. Takayasu, H. Miura, T. Hirabayashi, and K. Hamada, Physica A **184**, 127 (1992).
26. K.N. Ilinski and A.S. Stepanenko, cond-mat/9806138; cond-mat/9902046; K.N. Ilinski, cond-mat/9903142.
27. R.N. Mantegna, Physica A **179**, 232 (1991); J. Skjeltrop, *University of Oslo Thesis*, Oslo, Norway, 1998.
28. A. Pagan, J. Empirical Finance **3**, 15 (1996).
29. G.W. Kim and H.M. Markowitz, J. Portfolio Management **16**, 45 (1989); E.J. Elton and M.J. Gruber, *Modern Portfolio Theory and Investment Analysis*, (J. Wiley, New York, 1995).
30. J.-P. Bouchaud, cond-mat/9806101; D.F. Wang, cond-mat/9807066; cond-mat/9809045; J.-P. Bouchaud, G. Iori, and D. Sornette, Risk **93**, 61 (1996); M. Rosa-Clot and S. Taddei, cond-mat/9901277; cond-mat/9901279; B.E. Baaquie, cond-mat/9708178; R. Cont, cond-mat/9902018.
31. Y. Liu, P. Gopikrishnan, P. Cizeau, M. Meyer, C.-K. Peng, and H.E. Stanley, Phys. Rev. E **60**, 1390 (1999); cond-mat/9903369.
32. R.N. Mantegna and H.E. Stanley, Phys. Rev. Lett. **73**, 2946 (1994).
33. P. Gopikrishnan, M. Meyer, L.A.N. Amaral, and H.E. Stanley, Eur. Phys. J. B**3**, 139 (1998); P. Gopikrishnan, V. Plerou, L.A.N. Amaral, M. Meyer, and H.E. Stanley, Phys. Rev. E **60**, 5305 (1999); V. Plerou, P. Gopikrishnan, L.A.N. Amaral, M. Meyer, and H.E. Stanley, Phys. Rev. E**60**, 6519 (1999).
34. Y. Liu, P. Cizeau, M. Meyer, C.-K. Peng, and H.E. Stanley, Physica A **245**, 437 (1997).
35. F. Black and M. Scholes, J. Pol. Economy **81**, 637 (1973).
36. I. Koponen, Phys. Rev. E **52**, 1197 (1995).
37. M.M. Dacorogna, U.A. Muller, O.V. Pictet, and C.G. de Vries (Olsen group, www.olsen.ch, 199x).
38. Z. Ding, C.W.J. Granger, and R.F. Engle, J. Emp. Finance **1**, 83 (1983); For foreign exchange markets the decay time is even smaller[22].
39. E.-F. Fama, J. Finance **25**, 383 (1970).
40. T. Bollerslev, R.Y. Chou, and K.F. Kroner, J. Econometrics **52**, 5 (1992); G.W. Schwert, J. Finance **44**, 1115 (1989); A.R. Gallant, P.E. Rossi, and G. Tauchen, Rev. of Financial Studies **5**, 199 (1992); B. Le Baron, J. Business **65**, 199 (1992); K. Chan, K.C. Chan, and G.A. Karolyi, Rev. of Financial Studies **4**, 657 (1991).
41. C.W.J. Granger and Z. Ding, J. Econometrics **73**, 61 (1996).
42. P. Cizeau, Y. Liu, M. Meyer, C.-K. Peng, and H.E. Stanley, Physica A **245**, 441 (1997).
43. J. Beran, *Statistics for Long-Memory Processes,* (Chapman & Hall, NY, 1994).

44. S. Galluccio, J.-P. Bouchaud, and M. Potters, Physica A **259**, 449 (1998); L. Laloux, P. Cizeau, J.-P. Bouchaud, and M. Potters, cond-mat/9810255; Risk **3**, 69 (1999); Phys. Rev. Lett. **83**, 1467 (1999).
45. V. Plerou, P. Gopikrishnan, B. Rosenow, L.A.N. Amaral, and H.E. Stanley, cond-mat/9903369); Phys. Rev. Lett. **83**, 1471 (1999).
46. E.P. Wigner, Ann. Math. **53**, 36 (1951); E.P. Wigner, in: *Conference on Neutron Physics by Time-of-flight,* (Gatlinburg, Tennessee, 1956).
47. F.J. Dyson, J. Math. Phys. **3**, 140 (1962); F.J. Dyson and M.L. Mehta, J. Math. Phys. **4**, 701 (1963); M.L. Mehta and F.J. Dyson, J. Math. Phys. **4**, 713 (1963).
48. T. Guhr, A. Müller–Groeling, and H.A. Weidenmüller, Phys. Rep. **299**, 190 (1998); M. L. Mehta, *Random Matrices*, (Academic Press, Boston, 1991).
49. A.M. Sengupta and P.P. Mitra, cond-mat/9709283.
50. H. Takayasu and K.Okuyama, Fractals **6**, 67 (1998).
51. Y. Lee, L.A.N. Amaral, D. Canning, M. Meyer, and H.E. Stanley, Phys. Rev. Lett. **81**, 3275 (1998).
52. V. Plerou, L.A.N. Amaral, P. Gopikrishnan, M. Meyer, and H.E. Stanley, Nature **400**, 433 (1999).
53. T. Keitt and H.E. Stanley, Nature **393**, 257 (1998).
54. M.H.R. Stanley, L.A.N. Amaral, S.V. Buldyrev, S. Havlin, H. Leschhorn, P. Maass, M.A. Salinger, and H.E. Stanley, Nature **379**, 804 (1996); L.A.N. Amaral, S.V. Buldyrev, S. Havlin, H. Leschhorn, P. Maass, M.A. Salinger, H.E. Stanley, and M.H.R. Stanley, J. Phys. I France **7**, 621 (1997).
55. S.V. Buldyrev, L.A.N. Amaral, S. Havlin, H. Leschhorn, P. Maass, M.A. Salinger, H.E. Stanley, and M.H.R. Stanley, J. Phys. I France **7**, 635 (1997).
56. L.A.N. Amaral, S.V. Buldyrev, S. Havlin, M.A. Salinger, and H.E. Stanley, Phys. Rev. Lett. **80**, 1385 (1998).
57. R. F. Engle, Econometrica **50**, 987 (1982).
58. R. Ballie, J. Econometrics **17**, 223 (1991).

Catastrophe Effects and Optimal Extensions of Transportation Flows in the Developing Urban System: A Review

M. Sonis

Department of Geography, Bar-Ilan University, 52900 Ramat-Gan, Israel

Abstract. This paper deals with description of all possible minimal cost extensions of transportation flows in the developing urban system.
The enumeration of such extensions is based on competitive exclusion behavioral rules for suppliers and demanders connected by means of minimal cost solution of the classical Linear Programming Transportation Problem: in such a way the complete set of all topological structures for all possible optimal extensions of transportation network is constructed in the form of all maximal tries without cycles with exactly $m+n-1$ basic sells (where m is a number of suppliers and n is a number of demanders). It is important to underline that there exist only finite number of such trees.
The catastrophe effect analysis in this enumeration set is based on the polyhedral form of general sensitivity analysis for classical minimal cost transportation problem: for each preset topological structure of the minimal cost flow there is a polyhedral cone K in the space of supply-demand and the polyhedral wedge W in the space of transportation costs, such that the choice of arbitrary supply-demand within the cone K and the choice of arbitrary set of transportation costs within the wedge W will give the existing minimal cost flow with a preset topological structure.
The finite numbers of Cartesian products $K \times W$ represents the domains of structural stability of minimal cost flows with preset topological structures. Eventually the vector method of potentials is elaborated for the construction of the domains of structural stability $K \times W$.
As a practical numeric example the expanding set of bounded Christaller-Loesh and Beckman-McPherson Central Place System are considered in detail.

1 Introduction

This paper reviews a series of studies concerning the catastrophe effects for the solutions of the classical Transportation Linear Programming [19–25] and concentrates especially on the description of all possible minimal cost extensions of transportation flows in the developing urban systems and in hierarchical Central place systems [11,21,22,25]. The description of such kind of extensions is very important practically and theoretically, because we are living within such expending urban systems. As a practical numeric example the expanding set of bounded Christaller-Loesh and Beckman-McPherson Central Place System are considered in detail.

This series of papers was published during last decades as occasional papers usually in obscure places, but in my opinion they are presenting the same line of reasoning. We will start from the Catastrophe effects or discontinuous behavior of optimal solutions of the classical Linear Programming optimization problem. The special form of Transportation matrix with the help of ideas of the method of potentials immediately gives the possibility for constructing the conditions of optimality of the transportation flows with preset admissible topographic structure. These results are parallel to the catastrophe effects investigated by T. Puu in the case of continuous Beckmann transportation model [15–18].

2 Catastrophe Effects in Linear Programming

The catastrophe effect analysis is based on the polyhedral form of general sensitivity analysis for classical Linear programming Problem: let LP and D be primal and dual Linear Programming Problems:

$$\begin{array}{ll} \mathrm{LP}: & AX = b,\ X > 0,\ CX \to \min, \\ \mathrm{D}: & YA \leq c,\ Yb \to \max. \end{array} \tag{1}$$

Let B be an invertible submatrix of A with the inverse C, such that $Cb \geq 0$ and $dCa \leq c$, where d is obtained from c by inserting zeroes in the place of $A \setminus B$. Then the optimal solutions of LP and D are non-zero vectors and non zero components of solution (basic components) have a form:

$$X = Cb; \quad Y = dC. \tag{2}$$

The proof of this statement is straightforward ([24], p. 86). This statement gives the complete description of the domains of structural stability of optimal solutions for primal and dual Linear Programming problem. The domains of structural stability are the domains of permissible change of production costs c and the permissible fluctuations of the resources b under which optimal solutions correspond to the same basis B, i.e., to the same optimal assortment of production. If vectors b and c are changed, the sets

$$K = b : Cb \geq 0; \quad W = c : dCA \leq c \tag{3}$$

are the polyhedral cone in the space of resources b and the polyhedral wedge in the space of the costs c. Since the matrix A contains a finite number of invertible submatrices B the space of resources and the space of costs are decomposable into a finite number of domains of the form of Cartesian products $K_1 \times W_1, K_2 \times W_2, \cdots, K_p \times W_p$, where each product $K \times W$ corresponds to the same preserved basic components of optimal solutions. The catastrophe effects can be described in the following manner: if we are considering the continuous motion of some point $b \times c$ within the Cartesian product of the space of resources $\times$ space of costs then if this point moves continuously within some Cartesian product $K \times W$ the basis (i.e., optimal assortment of production) will be the same. If this point intersects the boundary between two different Cartesian products then the previous optimal solution will be degenerated on the boundary and then will be replaced by the optimal solution with the different assortment of production and different combination of costs. Thus, the continuous change in resources and costs will give the discontinuous catastrophic change in optimal solutions.

3 Catastrophe Effects in Transportation Problem

Consider the classical discrete minimal cost transportation problem which includes n suppliers with a_j units of supply for the jth *supplier* ($j = 1, 2, \cdots, n$) and m demanders with b_i units of demand for the ith demander ($i = 1, 2, \cdots, m$), and let c_{ij} be the cost of transportation of one unit of production from jth supplier to ith demander. It is necessary to find the transportation flow between suppliers and demanders (under the condition of balance between total supply and total demand) which minimizes the total cost of transportation. As it is known from the theory of transportation problems ([9], Chap. 15, 16), the topological structure of optimal minimal cost flow between suppliers and demanders is the maximal tree, that is the maximal connected subgraph of the transportation network without cycles, which includes $m - n + 1$ arcs exactly. We will call such subgraphs the *basic* subgraphs.

The transfer of catastrophe effects theory from the general Linear programming problem to the case of Transportation problem gives the following results: for each preset basic subgraph there is the domain of structural stability of optimal minimal cost solution. This domain describes all optimal solutions with the same preset basic subgraph. The domain of structural stability corresponds to the Cartesian product $K \times W$ of the polyhedral convex positive cone K in the space of supply-demand and the polyhedral wedge in the space of transportation costs. The cone K gives a description of all possible perturbations of supply-demand under which the transportation flow with a preset basic subgraph structure exists, and the wedge W gives a description of the possible changes in transportation costs which preserve the optimality of flow with a preset given structure.

The faces of $K \times W$ are the closed hyperplanes in the supply-demand space or in space of transportation costs. On the face of the cone K the optimal transportation flow divides into two or more independent flows which are the optimal solutions of transportation problems of smaller dimension. If one moves out of the cone K then the condition of existence of admissible flow with preset basic subgraph fails to hold and the structure of basic subgraph must be changed by substituting the some arc of the subgraph for another, giving the different basic subgraph. If one moves out of wedge W staying within the cone K then there is the admissible transportation flow with preset given basic subgraph, but the condition of optimality fails to hold and the other basic subgraph should be found, which satisfies the condition of optimality.

As a result it is possible to cover the Cartesian product of positive part of supply-demand space $\times$ space of costs by the finite number of domains - Cartesian products $K_1 \times W_1, K_2 \times W_2, \ldots, K_p \times W_p$, such that each of them preserves the same basic subgraph, i.e., represent the same topographical structure of optimal flow. It is important to stress that the set of cones $K_1, K_2, \ldots, K_p$ has empty intersections, while this is not true for the set of wedges $W_1, W_2, \cdots, W_p$.

4 Vector Generalization of Method of Potentials

Next the block-scheme for the construction of the Cartesian product $K \times W$ will be presented which is the vector form generalization of the well-known method of potentials (MODI-method) of Kantorovich and Gavurin [13] and Dantzug [8] ([20]):
Step 1. To start with, construct the usual computation table for the solution of standard classical transportation problem- the Hichkok Problem ([10], [9], Ch. 14-3). This computation table includes m rows (demand) and n columns (supply). Mark the cells (i_k, j_k), $k = 1, 2, \cdots, m+n-1$, which correspond to the preset basic subgraph. The cost vector $d = (d_{i_1 j_1}, d_{i_2 j_2}, \cdots)$ corresponds to the preset given structure of basic subgraph which id desired to preserve.
Step 2. In marked cells insert the unit vector-columns e_k, $k = 1, 2, \cdots, m+n-1$ of the identity matrix of order $m+n-1$.
Step 3. Construct the vector potentials

$$u_1(=0), u_2, \cdots, u_n, v_1, v_2, \ldots, v_n, \tag{4}$$

so that $v_{i_k} - u_{j_k} = e_k$, where e_k is the unit vector-column corresponding to the marked cell (i_k, j_k).
Step 4. In non-marked cells (i, j) of the computing table insert vectors $v_i - u_j$.
Step 5. Construct the square matrix with the following columns:

$$C = [v_m, v_m - u_2, \ldots, v_m - u_n, v_1 - v_m, v_2 - v_m, \cdots, v_{m-1} - v_m]. \tag{5}$$

The cone K which gives the condition for the existence of a transportation flow with the arcs corresponding preset basic subgraph, i.e., to the marked cells (i_k, j_k), $k = 1, 2, \ldots, m+n-1$, is given by inequality $Cb \geq 0$ (cf., (2), (3)), where b is the vector-column of supply-demand and the volumes of flow on the marked arcs are the components of vector Cb.
Step 6. Next construct the matrix W which includes the vectors from non-marked cells of the computation table. The wedge W which describes the conditions of optimality for the flow corresponding to the preset basic subgraph is given by the inequality $dW \leq c$, where d is the vector with coordinates corresponding to non marked cells.

5 Enumeration of All Possible Basic Subgraphs with the Help of Competitive Exclusion Behavioral Rules for Suppliers and Demanders

The enumeration of basic subgraphs, giving the spatial representation of the minimal cost flows based on competitive exclusion behavioral rules for suppliers and demanders connected by means of minimal cost solution of the classical Linear Programming Transportation Problem: in such away the complete set of all topological structures for all possible optimal extensions of transportation network is constructed in the form of all maximal tries without cycles with exactly $m + n - 1$ basic sells on the computing table. For the formulation of the behavioral rules let us consider an arbitrary subset of the set of all basic sells. Let us call suppliers and demanders connected by this subset the "old" suppliers and demanders, and other suppliers and demanders the "new" suppliers and demanders. The following three behavioral rules present the competitive exclusion effect:

1. each "new" demander can be served by only one "old" supplier;
2. each "new" supplier can serve only one "old" demander;
3. if a "new" demander is served by "old" and "new" suppliers, then this "new" supplier cannot serve any other "old" demander.

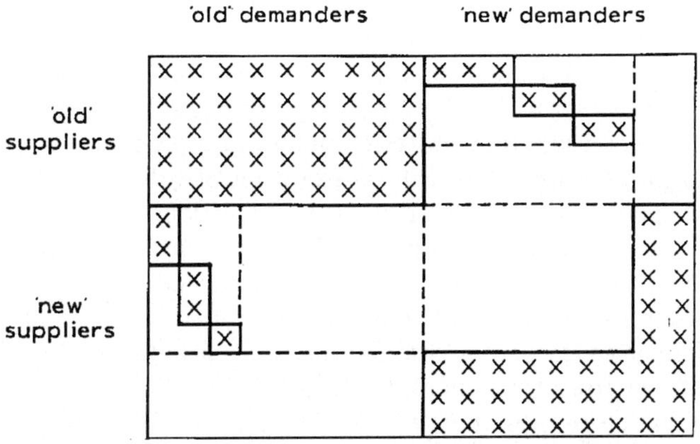

Fig. 1. Geometric scheme of competitive behavior of suppliers and demanders on the computation table (shaded cells may correspond to basic subgraphs of non-zero components of optimal flows).

Thus, in classical minimal cost transportation problem the *global collective minimization of costs* implies the totally antagonistic *competitive exclusion individual behavior* of suppliers and demanders.

These behavioral rules allow constructing the geometric and numeric algorithm of enumeration of all basic subgraphs presenting geographical the transportation network carrying the optimal transportation flows under various requirement on supply-demand and transportation costs. It is possible to reconstruct the computation table with the help of these rules in the geometrical schematic form presented in Fig. 1. This form essentially simplifies the basic subgraph enumeration procedure.

6 Comparison with the Catastrophe Effects in the Beckmann Continuous Transportation Problem

First attempt to generalize the Linear Programming Transportation problem to the case of the continuous flows on the plane was undertaken in the beginning of 40ties by Kantorovich, in the form of mechanical model of mass dislocation [12]. His model has no direct economic meaning and did not attract much consideration. A really meaningful and economically rich model of the optimization of continuous flows in the two-dimensional plane was developed in the beginning of 50ties by Beckmann [1]. In the following paper [2] the basis to what is now called Spatial Economics was established. The success of Beckmann is based on innovative incorporation of the three approaches coming from economics, hydrodynamics and planarity of flows and the sophisticated use of the variational methods of two-dimensional system of differential equations in location theory. Building on this two seminal papers Beckmann in collaboration with T. Puu developed further the general two-dimensional theory of spatial economics [4–6]. The series of papers by T. Puu [15–18] connected with the Beckmann model of continuous flows in two-dimensional space the consideration of optimality were replaces by the consideration of structural stability, showing under what conditions spatial flow and location patterns undergo fundamental structural change under small perturbations. The Puu description of solutions of the continuous transportation problem includes the optimal solutions which are topologically equivalent to the laminar flows on the plane with a finite number of isolated singular points of two types: stable and unstable nodes and ordinary saddles. In the Beckmann model of continuous flows the optimal flow lines are solutions to the two-dimensional system of differential equations, so the concept of the structural stability from the generic theory of differential equations can be used.

Moreover, since the directions of the optimal flow are the gradients to some spatial potential function, the general theory for classification of structural changes in catastrophe theory is applicable.

The problems being discussed here in Chapters 3 and 5 are analogous to Puu's problems for continuous flows:
(a) what kind of perturbations are admissible;
(b) what type of structure of flow is likely to be preserved under these perturbations:
(c) under what type of restrictions on supply-demand are their flows with a preset-structure;
(d) under what type of cost perturbations wills these flows are optimal minimal cost flows: and
(e) how do the transition from one type of structure to another occur?

The answers on these questions in the case of discrete transportation problems are:
(a) While in the continuous case only small perturbation are admissible in the discrete case the arbitrary changes in transportation costs and in supply and demand are considered in the form of finite number of Cartesian products of

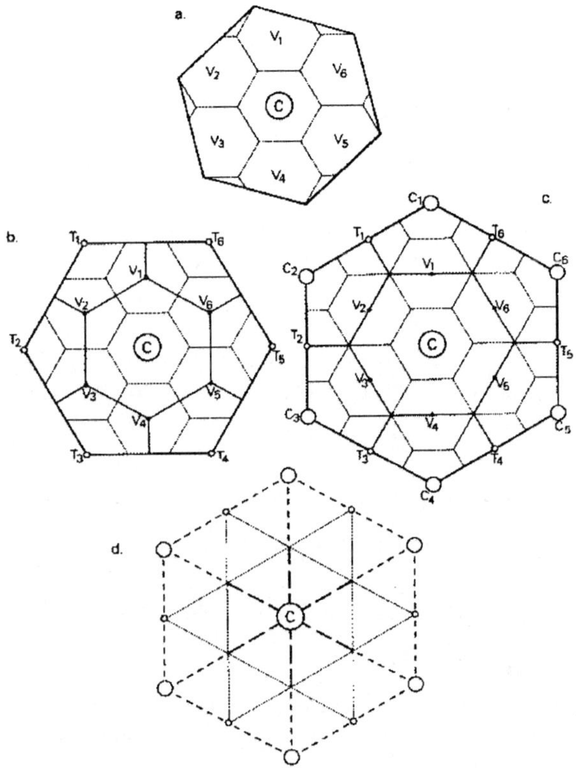

Fig. 2. Expanding chain of Beckmann-McPherson Central place Systems: (a) Two-tier system with a nesting factor $k_1 = 7$; (b) Three-tier system with nesting factors $k_1 = k_2 = 3$; (c) three-tier system with nesting factors $k_1 = 3, k_2 = 4$; (d) transportation networks and their extensions.

infinite polyhedral cones and wedges $K \times W$which represent the domains of structural stability of minimal cost flows with preset topological structures of basic subgraphs.
(b) Non zero-components of optimal flow correspond to the basic subgraph of transportation network, which is thc maximal tree without cycles, which includes $m-n+1$ arcs exactly. The enumeration of all basis subgraphs is the combinatorial problem resolved with the help of competitive exclusion behavioral rules for suppliers and demanders;
(c) Restrictions on supply-demand are presented in the form of positive polyhedral cones K;
(d) Cost perturbations lies within the polyhedral wedge W;
(e) The transition from one Cartesian product to another one means also the transition from one basic subgraph to another with the help of replacement of one arc of the subgraph by another arc. The vectorial method of potentials is elaborated for the construction of the domains of structural stability $K \times W$.

These answers can be use for the consideration of each practical transportation problem. Moreover, this methodology can be used for the consideration of a developing urban system with extension of transportation network.

As a practical numerical example, the optimal extensions of transportation flow within the expanding set of the bounded central place systems will be reviewed.

7 Structure of the Classical Central Place System and its Beckmann-McPherson Generalization

The spatial description of the original Chrisraller Central Place model is based on three generic geometric properties of central places associated this Central Place system:
1. The first property is that all hinterland areas of the central places at the same hierarchical level form a hexagonal covering of the plane with the centers on the homogeneous triangular lattice;
2. The second property is that the size of the hinterland areas increases from the smallest (on the lower tier of Central Place hierarchy) to the largest (on the highest tier of hierarchy) by a constant nesting factor k. This nesting factor expresses one of the Christaller's three principles, namely, marketing ($k = 3$), transportation ($k = 4$) and administrative ($k = 7$) principles;
3. The third property is that the center of a hinterland area of a given size is also the center of an hinterland of each smaller size [7,22].

The Beckmann-McPherson [3] Central Place model differs from the Christaller framework by applying variable nesting factors and by using the Loeshian principle of all possible coverings of the plane by hexagons of variable integer sizes. Their centers are the vertices of the initial Christaller triangular lattice [14].

In this paper we will consider a stylized example of the Beckmann-McPherson Central place system which can be considered as a developing set of Beckmann-McPherson Central place systems including a single largest Central place and

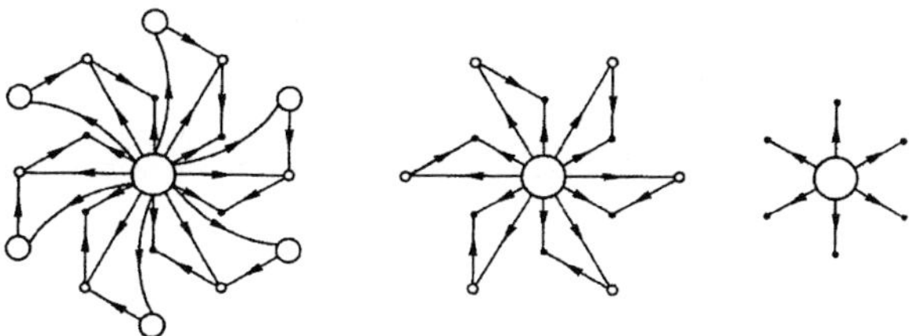

Fig. 3. An example of extensions of minimal costs flows within the expanding chain of Beckmann-McPherson hierarchical Central Place system.

the finite number of hierarchical levels including villages, towns and intermediate cities. (The reader should not associate the properties of real villages, towns and cities to the central places in this stylized example.) Next important property of the transportation flows is assumed to hold: the case of "top-down" transportation flows where each central is sending flows to each dependent central place within its own hinterland area and exchanging production with the closest central places of the same hierarchical tier.

The definition of the developing chain of Central place a systems is as follows: the expanding chain presents an ordered sequence of bounded Beckmann-McPherson hierarchical systems such that each previous central place system includes all central places for all previous systems. Figure 2 presents an example of the expanding chain of three bounded Beckmann-McPherson hierarchical systems.

The first system (a) includes the central city C and the nearby villages $V_1, V_2, \cdots, V_6$ arranged in two hierarchical tiers with the nesting factor $k_1 = 7$; the "top-down" flows from the central city to villages are assumed. The second system (b) includes in addition towns $T_1, T_2, \cdots, T_6$ in such a way, that a three-tier hierarchy with nesting factors $k_1 = k_2 = 3$ is generated. A central city can supply the towns and villages, and each town can supply the closest cities, towns and villages. The third Beckmann-McPherson system (c) contains six cities $C_1, C_2, \cdots, C_6$ that can supply the closest cities, towns and villages. The central places are organized in the three-tier hierarchy with the nesting factors $k_1 = 3, k_2 = 4$.

Next we will present the result of the enumeration of cases when the minimum cost transportation flow in central place systems in the expanding chain are fully incorporated on the minimal cost flow for the next central place system.

The minimal cost transportation flow for the first (a) Beckmann-McPherson system includes the supply of villages by central city. The second (b) Beckmann-McPherson system allows 8 possibilities for enlarging the previous minimal cost flow ([25], p.10), for each of them there are 11 possibilities to expand to the minimal cost flow in the third (c) Beckman-McPherson system. One of these possibilities is presented in Fig. 3.

References

1. M.J. Beckmann, *A continuous transportation model*, Econometrica **20**, 643-660 (1952).
2. M.J. Beckmann, *The Partial Equilibrium in Continuous Space Market*, Weltwirtschaftl. Archiv **71**, 73-89 (1953).
3. M.J. Beckman and J.C McPherson, *City size distribution in the Central Place hierarchy: an alternative approach*, J. Regional Sci. **10**, 243-248 (1970).
4. M.J. Beckman and T. Puu, *The continuous transportation model,* (Res. Rep., Int. Institute for Appl. Sys. Analysis, Laxenburg, Austria, 1980).
5. M.J. Beckman and T. Puu, *Spatial Economics: Potential, Density and Flow*, (North Holland Publ. Co., Amsterdam, 1985).
6. M.J. Beckman and T. Puu, *Spatial Structures*, (Springer Verlag, Berlin, 1900).
7. W. Christaller, *Die Zentralen Orte in Süddeutschland*, (Fischer, Jena, 1933); English translation: *Central places in Southern Germany*, (Prentice Hall, Englewood cliffs, NY, 1966).
8. G.B. Dantzig, *Application of the simplex method to a transportation problem,* in: *Cowles Commission Monograph 13: Activity Analysis of Production and Allocation,* T.C. Koopmans, (Ed.), pp. 359-373 (John Waley, NY, 1951).
9. G.B. Dantzig, *Linear Programming and Extensions*, (Princeton University Press, Princeton, NY, 1963).
10. F.L. Hichkok, *The Distribution of products from several sources to numerous localities,* J. Math. Physics **20**, 224-230 (1941).
11. J. Huff, D.A. Griffith, M. Sonis, L. Leifer, and D. Straussfogel, *Dynamic Central Place Theory: An Appraisal and Future Prospects,* in: *Transformations through Space and Time,* D.A. Griffith and R.A. Haining, (Eds.), pp. 121-151 (Martinus Nijhoff, Amsterdam, 1986).
12. L.V. Kantorovitch, *On dislocation of masses,* Russian Doklady of the USSR Academy of Sciences **37**, 227-229 (1942).
13. L.V. Kantorovitch and M.K. Gavurin, *Application of mathematical methods to the analysis of commodity flows,* in: *Problems in the rise of effectiveness of Transport Activity*, pp. 110-138, in Russian, (Academy of Sciences, USSR, Moscow, 1949).
14. A. Loesh, *Die Raeumliche Ordnung der Wirtschaft*, (Fischer, Jena, 1940); English translation: *The Economics of Location*, (Yale University Press, New Haven, Connecticut, 1954).
15. T. Puu, *Regional Modeling and Structural Stability,* Environment and Planning A **11**, 1431-1438 (1979).
16. T. Puu, *Structural Stability and Change in geographical space*, Environment and Planning A **13**, 979-989 (1980).
17. T. Puu, *Catastrophic structural change in the continuous regional model,* Regional Sci. and Urban Economics **11**, 317-333 (1981).
18. T. Puu, *Structural Stability as a modeling instrument in spatial economics,* J. Computational and Appl. Math. **22**, 369-379 (1988).
19. M. Sonis, *Sensitivity analysis and Multiparametric Programming,* (Tel Aviv University, Department of Pure Mathematics, unpublished manuscript, 1971).
20. M. Sonis, *Domains of Structural Stability for minimal cost Discrete Flows with reference to Hierarchical Central-Place Models,* Environment and Planning A **14**, 455-469 (1982).
21. M. Sonis, *Transportation Flows within Central Place Systems,* pp. 639-660 (Proc. of the Conference of the Italian Operation Research Association, Pescara, Italy, Vol. II, 1984).

22. M. Sonis, *Hierarchical structure of Central Place System - the barycentric calculus and decomposition principle*, Sistemi Urbani **1**, 3-28 (1985).
23. M. Sonis, *A contribution to the Central Place Theory: superimposed Hierarchies, Structural Stability, Structural changes and Catastrophes in Central place Hierarchical Dynamics*, in: *Space-Structure-Economy: A Tribute To August Loesch,* R. Funck and A. Kuklinsky, (Eds.), Karlsruhe Papers in Economic Policy Research **3**, 159-176 (1986).
24. M. Sonis, *Transportation Flows within Central-Place Systems,* in: *Transformations through Space and Time,* D.A. Griffith and R.A. Haining, (Eds.), pp. 81-103 (Martinus Nijhoff, Amsterdam, 1986).
25. M. Sonis, *Optimal extensions of the transportation flows and competition between suppliers and demanders*, Sistemi Urbani **1**, 3-15 (1993).

Evolutionary Models of Innovation Dynamics

W. Ebeling[1] and A. Scharnhorst[2]

[1] Institut für Theoretische Physik, Humboldt-Universitt Berlin, Invalidenstr. 110, 10115 Berlin, Germany
[2] Wissenschaftszentrum Berlin für Sozialforschung, Reichpietschufer 50, 10785 Berlin, Germany

Abstract. We investigate the dynamics of social processes and, in particular, processes of innovation in socio-economical evolution, technological change, and scientific research in discrete and continuous state spaces. The discrete description is based on the occupation number formalism and transition probabilities (Master equation formalism). Our special attention is devoted to the creation and survival of the NEW (i.e., new behavior, new technologies, new ideas etc.). The second (continuous) model is based on the idea that evolution is hill-climbing in an adaptive landscape over a continuous characteristics space. The behavior of an individual, a technological product, or a scientific problem is described by a large number of characteristics covering behavioral aspects, technology-inherent and economic parameters, or thematic dimensions. Further, we define a real-valued multi-modal fitness function/functional and a population density over the characteristics space. The evolutionary dynamics including competition and innovations is modeled by reaction-diffusion equations of Fisher-Eigen or Lotka-Volterra type.

1 Introduction

In the last two decades considerable progress has been achieved in understanding socio-cultural and socio-economical processes as evolutionary phenomena [2,4,5,25,27–29]. In this context several dynamical models have been developed [11]. Here our main interest is devoted to modeling processes of innovation, i.e., the historical process of the first appearance of a new behavior, a new idea, a new technology etc. In the simplest approach the problem of technological or scientific change has been approached on the basis of substitution models [20]. The stochastic aspects of this process were investigated only recently [7]. The stochastic character of social, scientific or technological innovations is basically connected with the appearance of one or a few representatives of the new: the elementary step of the evolution of technology/science is the replacement of old by new elements (technologies, ideas, inventions etc.). The new elements are subject to the competition process with the already existing old elements [13].

In an earlier work by the present authors a model of the replacement dynamics was developed which included the stochastic effects due to the initially small number of the new elements [6,7]. Here, after summarizing the replacement dynamics the modeling of innovations will be approached in a completely different way by studying the dynamics in a fitness landscape. We use the conceptual setting of the dynamics in fitness landscapes to implement a new approach to innovation processes [16]. Since the mathematical model of the evolution of social/technological/scientific innovations which will be developed is formally equivalent to a model developed for biological or ecological populations [11,19,13] we may concentrate here on the qualitative aspects and refer with respect to mathematical details to the above mentioned papers. In this way, we will present two types of models for evolutionary processes which are in some sense complementary: (i) discrete dynamics in occupation number spaces and (ii) continuous dynamics in fitness landscapes.

Between both types of models close mathematical relations exists which are comparable with the relations between the Heisenberg mechanics of matrices and the Schrödinger wave mechanics in quantum theory.

2 Discrete Dynamics in Occupation Number Space

2.1 Substitution Dynamics

The basic step of the evolution of a social system (society/economy/science) is the replacement of old by new elements (behavior, technologies, inventions, ideas, theories etc.) in the competition process. We describe here a model for the replacement dynamics which includes the substitution of an initially small number of new elements into the game.

Let us assume that N_0 is the total number of elements of a given (old) type at a given time. The old elements are the masters in the game. First, we have $N_0 = N = N_{total}$. Initially, the number of new elements is zero: $N_1 = 0$. Assuming that all possible processes are substitutions we have $N_0 + N_1 = N = const$ at all times. This restriction defines a competitive condition. The NEW can only emerge and grow in the system by replacement of elements of the old type.

In social systems, the process of replacement can be realized in different ways. In technological evolution, users or adopters, existent only in a limited number, replace old technical products by new ones. The users are the elements of the economic system, the used products define the different types present in the system. In science, scientists create new topics or themes. The change of problems under investigation is called *f*ield-mobility. The scientists are the elements and the scientific fields actually occupied are the types. The competition process proceeds between the types (populations).

Initially we have $N_0 = N$, all elements belong to the master species. Now we introduce the transition probability that in the following time slice the transitions $N_1 \to N_1 + 1$ and $N_0 \to N_0 - 1$ occur simultaneously. This transition probability describes the probability of a certain individual decision. For simplicity, we will assume that it may be represented as a Taylor series in N_0, N_1 (the latter is assumed to be small)

$$W(N_0 - 1, N_1 + 1 \mid N_0, N_1) = a_1 N_1 N_0 + b_1 \frac{N_1}{N} N_1 N_0 + \dots \tag{1}$$

$$W(N_0 + 1, N_1 - 1 \mid N_0, N_1) = a_0 N_1 N_0 + b_0 \frac{N_0}{N} N_1 N_0 + \dots \tag{2}$$

The first term corresponds just to a linear growth. The second contribution is of more complicated nature, it means that the growth rate depends on the fraction NEW/OLD. In this way, we are able to describe imitation and other non-linear effects. If we introduce the mean fraction: $x_1 = \langle N_1 \rangle / N$ then (up to higher order corrections) the following growth law results

$$\frac{d}{dt}x_1 = (a_1 - a_0)\, x_1(1 - x_1) + b_1 x_1^2(1 - x_1) - b_0 x_1(1 - x_1)^2 \tag{3}$$

Let us restrict ourselves first to the linear case $b_0 = b_1 = 0$. Then, the outcome of the deterministic game is rather simple. If the NEW has a selection advantage $(a_1 > a_0)$, it will win the game and dominate the hole population, if it performs worse, it will disappear again. The outcome of the stochastic game is much more complicated as is shown in an earlier work [6]. If the NEW has "better" characteristics (growth rate) only a probability to win the game can be formulated. For instance, if the NEW is twice as good as the existent type, the probability that a substitution will occur is still only 50%. On the other hand, in the stochastic game also slightly worse mutations will have a chance of survival, at least for several generations.

The game including quadratic terms is still more complicated. The NEW has to overcome a threshold before it is able to win. This threshold also depends on the total number of elements participating in the game (size of the system). One can show that in the stochastic game the NEW is able to tunnel through the threshold. In small systems, the NEW has a better chance to survive. Therefore, the principle of a dynamic niche can be formulated [7].

So far we studied just one substitution. In a previous work we considered Poisson sequences of substitutions [14]. We assumed that substitutions of the OLD by the NEW occur at discrete times which are Poisson distributed. At each of the Poisson distributed birth times a newcomer representing a new quality is born. The birth of the new quality is due to "self-reproduction" of one element of the population at time. Of course, the notation of "self-reproduction" should not be understood in its biological meaning, it stands in a socio-economic context for teaching, forming schools, imitating, copying and related processes [4,5,13]. In our model "self-reproduction" means that the quality Q is "handed over" to the newcomer, who develops it further. In this way, the dynamics consists of discrete events, the substitutions and the continuous development of the "handed over" quality to a new level of saturation. The "inside intervals dynamics" is characterized as a pure growth process described by equations given above. Further, we developed a theory including aging effects [14].

2.2 Evolution Dynamics in the Occupation Number Space

So far we considered the competition between the NEW and one OLD species. In general, the situation is much more complicated, we have got many OLD species and the NEW has to compete with all of them. In order to handle such processes we introduce a general evolution dynamics based on the concept of a stochastic population. A population is, by definition, a set of numbers

$N_1, N_2, N_3, \ldots, N_i, \ldots$. This set spans the occupation number space which is a discrete lattice in the positive cone $N_i \geq 0$. The dynamics of social processes in occupation number spaces was developed by many authors and, in particular, we mention the contribution of the Weidlich school [28].

The basic point of introducing a dynamics in the occupation number space is the transition probabilities. Here we use a rather simple *ansatz* based on the assumption that the transition probabilities are polynomials in the numbers N_i. Weidlich and his group use in fact a more complicated non-linear (exponential) ansatz. Starting from the simplifying view that evolution is basically replacement we restrict the dynamics to a plane (the simplex) on the positive cone $N_1 + N_2 + N_3 + \ldots = N = const$. For the elementary step of a replacement of i by j we assume the probability

$$\begin{aligned} W\left(N_1, ..., N_i - 1, ...N_j + 1, ... \mid N_1, ..., N_i, ...N_j, ...\right) \\ = \quad A_{ij} N_i N_j + B_{ij} N_i N_j^2 + ... \end{aligned} \tag{4}$$

With the restriction of the dynamics to the simplex, all possible processes turn out to be exchange processes in which two occupation numbers simultaneously change. In a further generalization we can skip this condition. Then, competition can be introduced into the system by non-linear exchange terms of the type $A_{ij} N_i N_j$.

The Master equation formalism allows to link assumptions about the rules of dynamic behavior on the micro-level (expressed by the transition rates) with trends concerning ensemble or time averages on the macro-level. In dependence on the nature of the system under consideration we can define elementary processes like (self-)reproduction, decline, exchange, imitation, (self-)inhibition, (self-)supporting, and innovation. The elementary processes affect either the change of one occupation number ($N_i \rightarrow N_i \pm 1$) or of two occupation numbers simultaneously ($N_i, N_j \rightarrow N_i + 1, N_j - 1$). The actions as well as the populations are described in a discrete way. The uncertainty about individual actions finds its expression in the probability assumptions about the elementary processes. But, the coefficients in the transition rates ensure that the statistics of actions (observed over large ensembles or time ranges) will follow certain common rules or trends. Further, the stochastic discrete game allows to describe the singularity of the emergence of the NEW, what in every case is a zero-to-one transition. Then, an innovation appears as the occupation of a yet not-occupied type and is still an unforeseeable, uncertain event, because the time of invention as well as the invented type can be modeled as stochastic fluctuations.

2.3 Interdisciplinary Applications

Let us first summarize a few of the main ideas developed so far.

(i) It was assumed that one basic step of the evolution of technology/science is the replacement of old by new elements (technologies, ideas, inventions etc.) by mutations or substitutions.

(ii) The second basic element is the competition process which gives support to the better quality and eliminates the worse quality.
(iii) The third important feature is the influence of stochastic factors which makes the outcome of the competition uncertain, what may lead to rather complicated situations.

The discrete dynamics of occupation number spaces have been applied in different areas of social systems. Urban growth processes, competition of firms, travel dynamics, substitution of technologies, formation of political opinions as well as scientific inventions have been modeled in this framework [24].

Let us summarize only few results in this area. In the case of the science system, the scientists with their education and written communication (articles) are the elements of a system of growing and declining scientific fields. The latter form the types or populations. The emergence of a new problem, field, or paradigm can only be understood as outcome of a game of competition between existent and potential scientific fields. With help of the evolutionary model framework set up before processes of education, of retirement, and thematic changes in the working career of scientists are bundled together as elementary processes of a game whose outcome produces growth curves of scientific fields measured by numbers of publications or authors. Well known effects in science history as fashion in science, prematurity, and suddenly occurring breakthrough can be explained by means of the interplay of individual decisions and non-linear effects [5].

In the case of technological evolution survival conditions for new technologies can be formulated. The substitution model proposed allows to calculate survival probabilities from analytic expressions. Thereby, the relationship between growth properties, the starting size of the NEW for market competition, and the size of the market (competing adopters or users) can be analyzed. Non-linear growth properties (like hyperbolic growth potential) create barriers for the entry of a new technology and explain the existence of log-in situations in technological change. Such situations can be overcome by protecting the introduction of a new technology. If the new technology is first introduced in small markets the barrier of entry can be tunneled through. By a step–wise increase of the number of participating competitors (dynamic niche) the new technology can overtake the market also in the case of hyper-selection.

In general one can state, that such models are not suited for forecasting of innovations and inventions, but they produce statements of survival conditions of the NEW and they allow to design scenarios of a successful implementation of the NEW.

3 Continuous Dynamics in Fitness Landscapes

3.1 State Space and Fitness Landscapes

Now, we will switch to a completely different model which is based on the idea that evolution is hill-climbing in an adaptive landscape over a continuous characteristics space. The idea of the evolutionary landscape goes back to theoretical biology and was originally created by Wright [30]. The concept was developed further by Conrad [8], and mathematically developed later by several authors [13,18,19]. In recent times, the concept of fitness landscapes found also very fruitful applications in the description of socio-economical processes [15].

We will assume that the properties and the dynamics of social behavior, economical-technological development, or scientific search is described by a large number of attributes, features, or characteristics covering behavioral characteristics, technology-inherent aspects, economic parameters, and thematic dimensions. Further, we assume that these characteristics are metrizable and can be expressed by quantitative variables. These quantities are real numbers. All together they span a characteristics space which is a real Euclidean vector space, in analogy to the phenotype space in biology. Further, we define a real-valued multi-modal fitness function/functional and a population density over the characteristics space. In this way, we have two real valued functions or functionals which are defined on the characteristics space (see Fig. 1). The fitness landscape has a more or less complicated structure with many hills and valleys. The hills correspond to the positive, favorable combinations of the parameters, the valleys to the unfavorable combinations. The population density attempts to follow the structure of the fitness landscape, i.e., it will develop maxima at places where the fitness is large and minima where the fitness has a deep valley. However, this is a process which needs time and, therefore, at a given time not all but just a few of the fitness hills will be populated. In the course of evolution more favorable hills may be occupied, however, due to the high dimensionality of the space a total occupation will never be reached, the evolution never reaches a final target. Progress is always relative and limited. Innovation corresponds to the occupation of hills which were not populated so far. Now let us give a mathematical description of the ideas given above. We will assume that the characteristics consist of d variables $\overrightarrow{q} = (q_1, q_2, ..., q_d)$ which are real numbers. The set of numbers q_i forms an abstract vector space Q. The individual properties q_i are the coordinates in the characteristics space Q which has the dimension d. As a rule, we have many characteristics, i.e., $d >> 1$ is a large number. The set of coordinates $q_1, q_2, ..., q_d$ (the components of the vector $\overrightarrow{q}$) characterizes the given entity which is subject to evolution. Any point in Q is a potential state of evolution. The change of properties corresponds to a trajectory in the space Q. Assuming that the set of state points is dense we may introduce the density function $x(\overrightarrow{q}, t)$. The population density x is a real non-negative function on Q. As a rule $x(\overrightarrow{q})$ has a complex structure, it has many peaks and valleys, and in many parts of Q it is simply zero, what means that these combinations of the q_i are (not yet or no more) realized at given time. The density function $x(\overrightarrow{q}, t)$ takes over the role of the occupation numbers N_i in the discrete setting. Populations are groups of elements with similar properties, we may think about a broad peak of the density function. Evolution means change of $x(\overrightarrow{q})$ in time. The laws of this change may be very complicated. For simplicity, we have to restrict ourselves here to rather simple mathematics. We will assume that the evolutionary dynamics including competition and mutations/innovations is de-

Elements of a geometrically oriented evolution theory

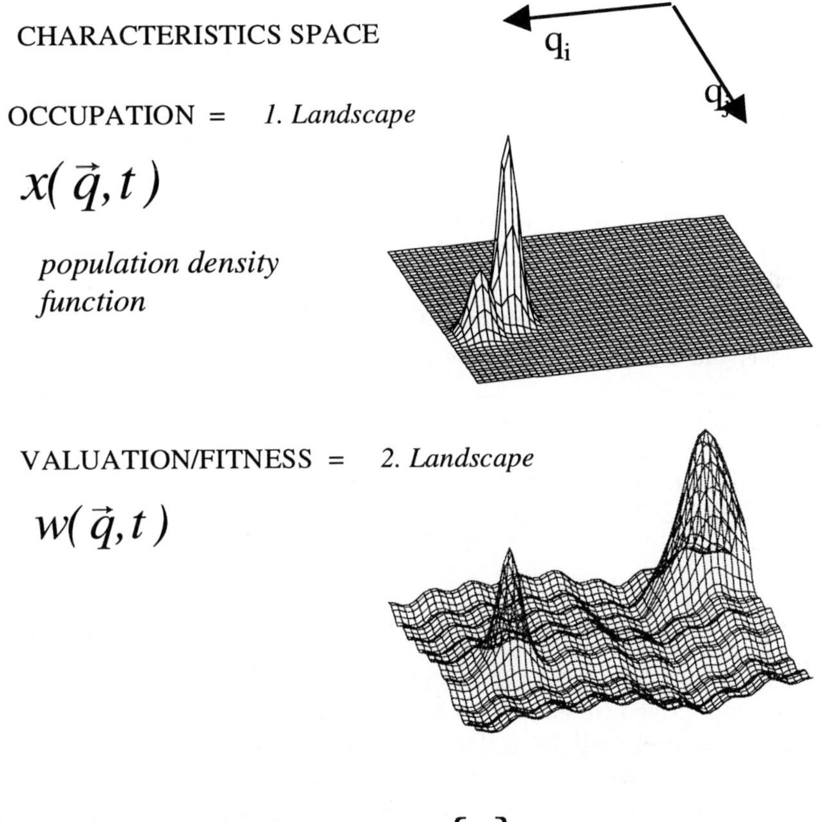

$$\partial_t x(\vec{q},t) = x(\vec{q},t)\, w(\vec{q};\{x\}) + Mx(\vec{q},t)$$

selection *mutation*

Fig. 1. Elements of G_O_E_THE.

termined by some state function which we call fitness. Fitness is in our model a real-valued function denoted by $w(\vec{q},t)$. This function associates each point in the space Q with some real value w which determines the growth rates. The fitness function forms a (second) landscape.

3.2 Evolutionary Dynamics in the Phenotype Space

In order to come to a concrete dynamics we recall the general concept that evolution is hill-climbing in an adaptive landscape over a continuous characteristics space. In other words, any local element in the space attempts to move to domains of the space where the fitness is higher. An evolutionary dynamics which formulates this idea in mathematical terms may be based on reaction- diffusion equations of Fisher-Eigen or Lotka-Volterra type.

The general *ansatz* for the dynamics reads [12,18,19]:

$$\partial_t x(\vec{q},t) = x(\vec{q},t)\, w(\vec{q};\{x\}) + Mx(\vec{q},t) \qquad (5)$$

The basic assumption which leads to the Fisher-Eigen model is that the local growth rate is proportional to the difference between a local fitness value v and the social average $\langle v \rangle$.

$$w = v(\vec{q}) - \langle v \rangle\,, \qquad (6)$$

where $\langle v \rangle$ is the average of v. In the framework of this model the form of the fitness landscape is fixed, only the reference level $\langle v \rangle$ is changing in time. For the mutation term we may assume a diffusion approximation corresponding to $Mx(\vec{q},t) = D\Delta x(\vec{q},t)$.

If the fitness itself depends on the population density we may use the Lotka-Volterra type dynamics. Here the value of w is determined as a functional of the density x:

$$w(\vec{q};\{x(\vec{q},t)\}) = a(\vec{q}) + \int b(\vec{q},\vec{q}')\, x(\vec{q}',t)\, d\vec{q}'. \qquad (7)$$

In this case the fitness consists of two contributions, the first one, $a(\vec{q})$, evaluates the reproductive fitness and the second one describes the interaction of the species by the kernel $b(\vec{q},\vec{q}')$. The fitness becomes time-dependent. For this case Conrad introduces the term "adaptive landscape" [8].

If the population density is concentrated in certain regions of Q ("islands") then these "islands" can be related to the original classified populations. The "selective value" $v(\vec{q})$ is linked to the net reproduction or growth rate. The choice of a diffusion-like mutation operator corresponds to the assumption, that the mutation rates are symmetric, homogeneous and of short range. For social processes it seems to be of particular interest to consider inhomogeneous and "directed" mutations which can be modeled by means of more elaborate mutation operators. Here, for the sake of mathematical simplicity we restrict ourselves to the diffusion approach.

In contrast to discrete models, the vanishing, merging, division, and emergence of technologies are expressed by changes in the shape of the function $x(\vec{q})$

without having to consider changes in the taxonomy of the model. This results in a greater mathematical complexity of the model. As mentioned above, the population density follows the shape of the fitness landscape. If we assume $w(\vec{q})$ to be a random function, then the shape of $x(\vec{q},t)$ is sensitive to statistical properties of this function given by the probability density functional $P[w(\vec{q})]$.

In (5) the term given by (6) describes the selection process. This becomes evident if we consider the temporal evolution of populations without mutations. With increasing time the population is concentrated in islands which correspond to particularly high values of the random function $v(\vec{q})$. These islands of high density are surrounded by regions of low density. This means, that the selection process leads to a concentration of the distribution around the maxima. The diffusion process, on the other hand, leads to a widening of the distribution and an extension of its tails. The mathematical solution of (5) in the presence of mutations is more complicated. Using the analogy with the Schrödinger equation for an electron in a random field some approximate expression for the time-dependent solution $x(\vec{q},t)$ can be derived [12,13,19]. It can be shown that the existence of populations corresponds to the problem of the existence of localized states (according to the localization problem in random potentials). A localized state can be understood as a distribution of individuals around a dominant type. It is well known that for high-dimensional spaces ($d \geq 4$) and δ-correlated potentials $v(\vec{q})$, there are no localized states at all. Therefore, the existence of a correlation length greater than zero seems to be a necessary condition for the emergence of distinguishable parts (or populations) in the population density function. This is in accordance with the smoothness postulate of an adaptive landscape formulated by Conrad [8].

An evolutionary system can only develop a strategy for search processes in a landscape with correlations. Furthermore, this entails that we have to distinguish between the lack of information about the environment (which can be modeled by means of a random fitness function with certain statistical properties) and complete irregularity (stochasticity) where in principle no extrapolation from the local knowledge is possible. The most important result of the mathematical analysis [13] is that there exists a characteristic finite jump – the so-called "evolutionary quantum". Evolution in correlated valuation landscapes proceeds jump-like. In the case of technological evolution it is a well-determined fact that longer periods of smooth evolution of technologies are sometimes interrupted by jumps. In the model framework "incremental" and "radical" innovations are both part of the system dynamics driven by mutations. From the perspective of the system the "big jumps" occur as the radical innovations changing the composition of the system. The drift stands for continuous change (incremental innovations). It was shown by Zhang, Engel and one of the present authors [31,17,16] that there exists a definite scaling between the distance $|\delta q|$ of the jump in the characteristics space and the characteristic time τ for the jump.

3.3 Interdisciplinary Applications

The landscape picture and continuous dynamical evolution models seems to be particularly interesting for the description of search processes in social systems. In the following, the potential of such an approach will be demonstrated in the case of technological evolution and the development of the international scientific community.

The concept of technological trajectories and path-dependency of technological evolution already comprises a space concept [21]. With the concept of a characteristics space of output indicators characterizing technical and service characteristics of technological products, it is possible to make technological trajectories visible [22]. Let us consider as an example the development of the aircraft. Saviotti and Bowman analyzed Jane's Encyclopedia of Aviation and found for the period of time from 1915 to 1983 several characteristic groups and temporal trajectories [23]. The phenotypic characteristics space is built by two indicators: the maximum take-off weight of an aircraft and the speed. The different models (or types) of aircraft are the elements. Each constructed aircraft is located at a certain point in this space. When similar aircraft models have been developed by different firms (similar in accordance with the two parameters considered) these points are occupied by more than one model. From these data a population density function can be constructed. From 1915 to 1929 6 different models occupy 2 locations near the origin of the coordinate system. In the following 15 years, about 60 models appeared at 19 different locations, but all in the range of $\{100 \leq \text{speed} \leq 600; 100 \leq \text{maximal take-off weight} \leq 64000\}$, i.e., still in lower left corner. At the end of the time period considered (Fig. 2) we see a differentiation process of types of aircraft models over time. The group of aircraft models concentrated at the beginning in the lower left corner moves along two different trajectories.

In general, what we see from empirical data are the occupation landscapes. The model framework introduced above produces explanations for the extension of a population (variety of product models in a certain limited area), as well as for processes of drift and specialization. We assume that the population density function (first landscape) follows an usually unknown but existent valuation function (second landscape). From the shape of the observable first landscape and its changes shape we can draw conclusions with respect to the shape of the second one. We can try to find indicators which might reflect the valuation landscape as for instance the price of a product. Depending from the choice of variable for the population density function the focus of the described selection processes changes. For instance, if we consider the number of sold products the competition process takes place at the market. In the case of aircraft development where the constructed product models are counted independently of their market shares, the competition process rather reflect engineering properties. Let us consider now the system of the international scientific community which can be visualized by means of bibliometric indicators like publications, citations etc. comprised in data bases like the Science Citation Index (SCI). As agents on the international scene national science systems appear. If we consider a space of

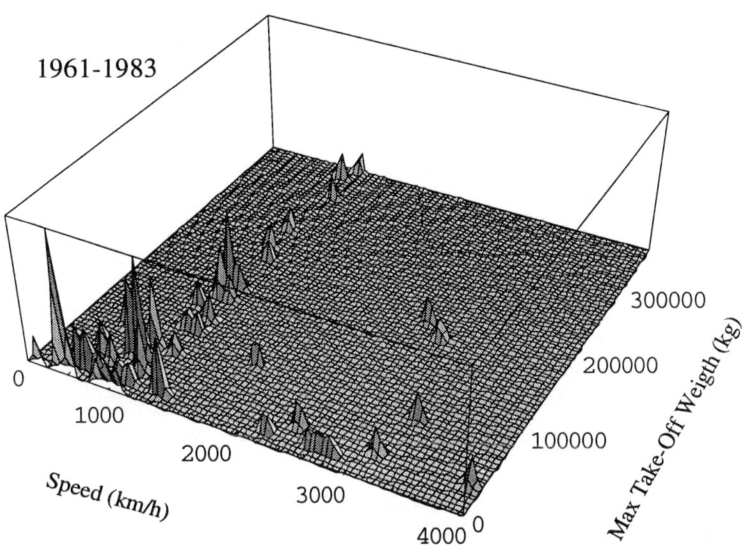

Fig. 2. Aircraft models in a characteristics space (1961-1983).

publication profiles (e.g., the share of publications in life sciences L[%] and the share of publications in physics P[%] of the total number of publications of a country)[3] each country is located at a special point. An occupation landscape can be constructed. Accordingly to the model framework countries search for optimal research strategies and compete for recognition. In the space of publication profiles, groups of countries and isolated countries are observable. They change their position and composition in time. The emergence and strengthening of centers is visible as well as the drift of peripheral countries (see Fig. 3). In the last two decades rather an increase in science stratification instead of a convergence is observable.

4 Summary and Conclusions

Evolution is described as a sequence of self-organization processes in which innovations play a central role. Since the emergence of the new paradigm of self-organization and evolution in the natural sciences, a continuous knowledge transfer of methods to social science applications has taken place. In this paper two classes of models: (i) the discrete dynamics on occupation number spaces and (ii) the continuous dynamics on fitness landscapes have been reviewed. Different interdisciplinary applications are discussed.

Concerning social evolution, the adaptive landscape concept seems to be important for the understanding of an "innovative environment" [10] and the

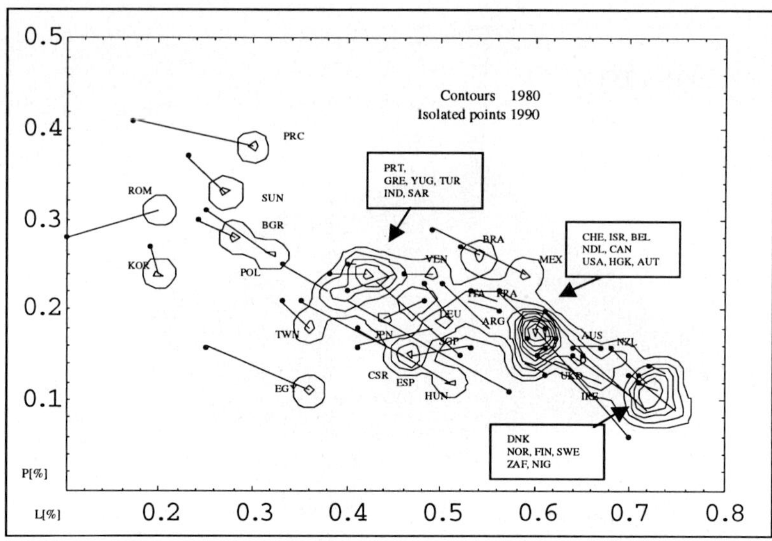

Fig. 3. Countries in the space of publication profiles - changes in their research strategy between 1980 and 1990.

relation between "blind" and "directed" search strategies or the nature of change [1]. "Irreversibility is not only the result of imperfect information and sequence of decisions but is due to the fact that the world genuinely changes and it changes as a consequence of the very actions of the agents." [10].

A further objective of the present paper is to link the concept of an adaptive landscape to complex search processes. Therefore, the structure of the fitness landscape and, in particular, its different statistical properties must be considered in greater detail. The fitness function is assumed to be a correlated random potential. Thus, links to mathematical techniques used in statistical physics can be established. In the continuous approach the existence of populations or groups is not an initial assumption (like in discrete descriptions) but the result of search processes in a random landscape. On a long time scale, hill-climbing proceeds by small but discrete steps. Transitions from an initial region with a certain given fitness to regions whose fitness is better just by a "quantum" are preferred. Surprisingly, this discontinuous character of evolution is the result of a continuous approach. This indicates that the step-wise character of social evolution as observed is not a direct consequence of the discreteness of mutations but rather a general feature of the selection-mutation processes.

Thus, the modeling framework developed here seems to be useful for the description of social systems which behave in a complicated unknown environment.

References

1. P.M. Allen and M. Lesser, *Evolutionary human systems: learning, ignorance and subjectivity*, in: *Evolutionary Theories of Economic and Technological Change*, P.P. Saviotti and J.S. Metcalfe, (Eds.), pp. 160–171 (Harwood Academic Publishers, Chur, 1991).
2. P.M. Allen, *Coherence, chaos and evolution in the social context*, Futures **26,** 583–597 (1994).
3. M. Bonitz, E. Bruckner, and A. Scharnhorst, *The science strategy index*, Scientometrics **26,** 37-50 (1993)
4. E. Bruckner, W. Ebeling, and A. Scharnhorst, *Stochastic dynamics of instabilities in evolutionary systems*, Sys. Dyn. Rev. **5,** 176–191 (1989).
5. E. Bruckner, W. Ebeling, and A. Scharnhorst, *The application of evolution models in scientometrics,* Scientometrics **6,** 21–41 (1990).
6. E. Bruckner, W. Ebeling, M.A. Jiménez-Montaño, and A. Scharnhorst, *Hyperselection and innovation described by a stochastic model of technological evolution*, in: *Evolutionary economics and chaos theory*, L. Leydesdorff and P. van den Besselaar, (Eds.), pp. 79–90 (Pinter, London 1994).
7. E. Bruckner, W. Ebeling, M.A. Jiménez-Montaño, and A. Scharnhorst, *Nonlinear stochastic effects of substitution – an evolutionary approach*, J. Evolutionary Econ. **6,** 1–30 (1996).
8. M. Conrad and W. Ebeling, *M.V. Volkenstein, evolutionary thinking and the structure of fitness landscapes,* BioSystems **27,** 125–128 (1992).
9. G. Dosi, *Technological paradigms and technological trajectories,* Res. Policy **11,** 147–162 (1982).
10. G. Dosi and J.S. Metcalfe, *On some notions of irreversibility in economics*, in: *Evolutionary Theories of Economic and Technological Change,* P.P. Saviotti and J.S.Metcalfe, (Eds.), pp. 133–159 (Harwood Academic Publishers, Chur, 1991).
11. W. Ebeling and R. Feistel, *Physik der Selbstorganisation und Evolution,* (Akademie-Verlag, Berlin, 1982).
12. W. Ebeling, A. Engel, B. Esser, and R. Feistel, *Diffusion and reaction in random media and models of evolution processes,* J. Stat. Phys. **37,** 369–384 (1984).
13. W. Ebeling, A. Engel, and R. Feistel, *Physik der Evolutionsprozesse,* (Akademie-Verlag, Berlin, 1990).
14. W. Ebeling, M.A. Jiménez-Montaño, and Karmeshu, *Dynamics of innovations in technology and science based on individual development*, in: *Self-organization of complex structures*, F. Schweitzer, (Ed.), pp. 407 414 (Gordon, and Breach, Amsterdam 1997).
15. W. Ebeling, Karmeshu, and A. Scharnhorst, *Economic and technological search processes in a complex adaptive landscape*, in: *Econophysics,* J.Kertesz and I. Kondor, (Eds.), (Kluwer, Dordrecht, in press)
16. W. Ebeling, A. Scharnhorst, M.A. Jiménez-Montaño, and Karmeshu, *Evolutions- und Innovationsdynamik als Suchprozess in komplexen adaptiven Landschaften,* in: *Komplexe Systeme und nichtlineare Dynamik in Natur und Gesellschaft,* K. Mainzer, (Ed.), pp. 446–473 (Springer, Berlin et al., 1999).
17. A. Engel and W.Ebeling, *Comment on "Diffusion in a random potential: hopping as a dynamical consequence of localization"*, Phys. Rev. Lett. **59,** 1979 (1987).
18. R. Feistel and W. Ebeling, *Models of Darwin processes and evolution principles,* BioSystems **15,** 291–299 (1982).

19. R. Feistel and W. Ebeling, *Evolution of Complex Systems,* (Kluwer, Dordrecht, 1989).
20. J.C. Fisher and R.H. Pry, *A simple substitution model of technological change,* Techno. Forecasting and Soc. Change **3,** 75–88 (1971).
21. R.R. Nelson and S.G. Winter, *In search of useful theory of innovation,* Res. Policy **5,** 36–76 (1977).
22. P.P. Saviotti and J.S. Metcalfe, *A theoretical approach to the construction of technological output indicators,* Res. Policy **13,** 141–151 (1984).
23. P.P. Saviotti and A. Bowman, *Indicators of output of technology,* in: *Science and technology policy in the 1980s and beyond,* M. Gibbons, P. Gummett, and B. Udgaonkar, (Eds.), pp. 117–147 (Longman, London, 1984).
24. F. Schweitzer, (Ed.), *Self-organization of complex structures,* (Gordon and Breach, Amsterdam, 1997)
25. F. Schweitzer and G. Silverberg, (Eds.), *Selbstorganisation und Ökonomie,* (Duncker-Humblot, Berlin, 1998).
26. A. Scharnhorst, *Citations – networks, science landscapes and evolutionary strategies,* Scientometrics **43,** 95–106 (1998).
27. W. Weidlich and G. Haag, *Concepts and models in quantitative sociology,* (Springer, Berlin et al., 1983).
28. W. Weidlich, *Physics and social science – the approach of synergetics,* Phys. Rep. **204,** 1–163 (1991).
29. U. Witt, (Ed.), *Evolutionary Economics,* (Elgar, Aldershot, 1993).
30. S. Wright, *The roles of mutation, inbreeding, crossbreeding and selection in evolution,* Proc. of the 6th Int. Congress on Genetics **1,** 356–366 (1932).
31. Y.C. Zhang, *Diffusion in a random potential: hopping as a dynamical consequence of localization,* Phys. Rev. Lett. **56,** 2113-2116 (1986).

Universality of Group Decision Making

S. Galam

Laboratoire des Milieux Désordonnés et Hétérogènes, Tour 13, Case 86, 4 place Jussieu, 75252 Paris Cedex 05, France

Abstract. Group decision making is assumed to obey some universal features which are independent of both the social nature of the group making the decision and the nature of the decision itself. On this basis a simple magnetic like model is built. Pair interactions are introduced to measure the degree of exchange among individuals while discussing. An external uniform field is included to account for a possible pressure from outside. Individual biases with respect to the issue at stake are also included using local random fields. A unique postulate of minimum conflict is assumed. The model is then solved with emphasize on its psycho-sociological implications. Counter-intuitive results are obtained. At this stage no new physical technicality is involved. Instead, the full psycho-sociological implications of the model are drawn. Few cases are then detailed to enlight them.

1 What It Is About?

Every day up to millions of people get together to make up a decision about some issue of interest for the group or the collectivity it represents. The spectrum of group nature is almost infinite. It includes professional groups having to decide on some technical issue, as well as all kind of friendly groups. The nature of the decision itself is even more diversified. It can be a high rank political issue related to some military retaliation, like recently about bombing Yugoslavia. It can be also a public jury to decide about someone muder culpability. But it can equally be causal with some school board to decide on the cafeteria setting.

A priori to decide about bombing a country or painting a dining room seems and is of a totally different nature. Especially, the decision cost and the associated human consequences. Nevertheless, it may be that the respective groups undergoing the decision making do obey an identical process of group decision making. At least that is the hypothesis behind our approach. In this work, we are assuming the existence of some universal mechanisms which produce group decision making.

Of course, we are dealing with the character of the decision rather than with the nature of the decision itself. We aim to determine for a given issue the psycho-sociological conditions under which a group either polarize or get to a compromise [1].

Modeling a complex social situation using Statistical Physics have started long ago [2]. For instance to study a strike process [3]. They are getting more numerous in recent years. Among others, we can cite voting in political organizations [4], group power dynamics [5], social impact [6], outbreak of cooperation

[7], stock market [8], and more recently traffic flows [9] and sexual reproduction [10].

To keep the presentation simple, we used a model in which a group of N persons has to make up a decision with respect to two options "yes" and "no". The model is articulated around a *Postulate* of minimum conflict. Competing interactions are also present.

Formally, we are using a random field Ising ferromagnetic model in an external magnetic field at zero temperature. However here the system is finite in size. Moreover, "random fields" may have a non-zero configurational average. Results may also depend on the field configuration.

Our model does not aim to novelty in Statistical Physics. It aimed instead to draw the psycho-sociological implications of a human-like version of a physical model.

Section 2 considers the simplest situation with only pair interactions. A measure of the group conflict is determined. The concept of a *symmetry breaking choice* is introduced. Section 3 deals with quantizing the anticipation effect at work in the group process of making up a collective choice. Surrounding pressure as well as individual biases are included in Sect. 4. Given a class of individual conflicts we study mechanisms by which either a compromise or a polarization of the group is produced. Examples illustrate the model in Sect. 5. Last Section comments on the possibility to extend the model to non-rational behavior, i.e., in physical terms, to non-zero temperatures.

2 Pair Interactions

We start from the simplest situation with a collection of N persons. The decision is between two answers yes or no. Each individual choice is represented by a variable c_i, where $i = 1, 2, ..., N$ and $c_i = \pm 1$ where $c_i = 1$ is associated to answer yes and $c_i = -1$ to answer no.

The aggregated collective choice of the N person group is just the sum of each individual choice,

$$C = \frac{1}{N} \sum_{i=1}^{N} c_i \ . \tag{1}$$

From (1) it is seen that aggregation enlarges drastically the spectrum of possible choice configuration. It actually increases from 2 at the individual level up to 2^N at the group level. Nevertheless, to materialize this spectrum a structure is necessary to collect individual answers, to sum them up, and to display the net result.

Moreover, to go beyond the initial two-fold answer requires some complex internal transformation in order to associate a meaning to each one of the 2^N answers. However, the use of some rules, for instance a majority rule, can bring the collective choice back to the individual one with only two answers.

At this stage yes and no are equi-probable. Therefore, the collective choice of N isolated individuals, is indeed zero on average, with fluctuations of order

$1/\sqrt{N}$. The result $C = 0$ creates a new qualitative choice which did not exist at the individual level.

$C = 0$ can thus be understood as the perfect compromise choice with a fully symmetrical group decision.

We now introduce interactions. For a pair i and j of persons, there exist four different choice configurations which are,

a $c_i = c_j = +1$,
b $c_i = c_j = -1$,
c $c_i = -c_j = +1$,
d $c_i = -c_j = -1$.

In configurations (a, b) i and j are making the same choice. In configurations (c, d) they are at conflict making opposite choices. However, agreement or conflict materializes only if i and j are both aware of the other's choice, in other words, only once they are interacting. Let us denote E the exchange amplitude of this interaction. The product $-Ec_ic_j$ then measures the degree of conflict of a given configuration. It is either $+E$ (configurations c, d) or $-E$ (configurations a, b). An agreement being a negative conflict. Both cases do not differentiate which choice is actually made, in accordance with the a priori symmetrical nature of the individual choice.

Restricting interaction to pairs the overall group conflict is then measured by the function,

$$G_E \equiv -E \sum_{<i,j>} c_i c_j \ , \tag{2}$$

where we have assumed that the exchange amplitude E is constant for all interacting (i, j) pairs. We call G_E the group exchange conflict function and $< i, j >$ represents all interacting pairs.

The exchange conflict function G_E measures the conflict amplitude in a group for each one of the 2^N decisional configurations. It discriminates among various possible choices, but does not indicate which one is chosen by the group. For the group decision dynamics to operate, it is necessary to invoke a criterion to select which among the possible states is actually selected by the group. Along the minimum energy principle, we introduce a *Postulate* to determine the group dynamics direction. It reads,

"Each individual selects the choice which minimizes its own conflict".

Justification of this *Postulate* is beyond the scope of the present work. It will be motivated a posteriori by the results obtained from the model. Minimum conflict means maximum agreement.

From a random distribution of choices, the decision making dynamics to reduce individual conflicts leads to an extreme polarization of the group. Its collective choice is then $C = \pm 1$. The sign, i.e., the polarization direction, is random.

In real life situations, above polarization process is a rather complex phenomenon. Monte Carlo simulations on zero-temperature dynamics on the Ising model showed indeed non trivial behavior at all dimensions [11]. We conclude,

Symmetrical groups polarize themselves towards an extreme choice. This direction choice being indeed arbitrary.

This polarization effect which results from group member interactions is identical to the *Spontaneous Symmetry Breaking* phenomenon well known in the physics of collective phenomena [12]. Individual local interactions make the group to behave as one *super-person* [13]. That *super-person* chooses between two possible choices with equi-probability likewise the isolated individual.

Here, perfect compromise has disappeared. Polarization effect in social systems was clearly evident in data reported in [14]. However, until now, most theoretical explanations have been unconvincing in connecting choices at respectively the individual level [15] and the group level [16]. Our finding is that polarization effect arises quite naturally from interactions.

3 Anticipating Effect

We now formalize the internal group dynamics which proceeds from initial individual choices towards the final collective choice. The exchange term must be modified to account for the emergent group decision. We first rewrite G_E as,

$$G_E = -\frac{E}{2} \sum_{i=1}^{N} \left\{ \sum_{j=1}^{n} c_j \right\} c_i \,, \tag{3}$$

where n is the number of persons one individual interacts with. To keep the presentation simple this number is taken as a constant, the same for everyone In case everyone interacts with everyone $n = N$.

Now we modify (3) to account for the process of group formation. People do anticipate the emergence of a collective choice. Each individual i will thus try to project through its partner's choices c_j (the people i discusses with), its expectation of the overall final group decision.

Individual i then extrapolates the j's choice c_j to the expected collective choice the group will eventually make without its own choice. Within this process, individual i perceives the j's choice as given by the transformation,

$$c_j \rightarrow \frac{1}{N-1}(NC - c_i) \,, \tag{4}$$

where C is the collective choice. Trough the anticipating process (3) becomes,

$$G_E^g = -\frac{E}{2} \sum_{i=1}^{N} \left\{ \sum_{j=1}^{n} \frac{1}{N-1}(NC - c_i) \right\} c_i \,, \tag{5}$$

and,

$$G_E^g = -\frac{nE}{2(N-1)} \left\{ NC \sum_{i=1}^{N} c_i - \sum_{i=1}^{N} c_i^2 \right\} \,, \tag{6}$$

where superscript g denotes the acting anticipation process. Using the property $c_i^2 = 1$ we get,

$$G_E^g = \gamma - \gamma C \sum_{i=1}^{N} c_i \,, \tag{7}$$

where $\gamma \equiv \frac{nNE}{2(N-1)}$ is a constant independent of the group choice. As such first term of (7) is irrelevant to the collective choice. It is worth noting that C is not yet the final decision. Rather it is the expected final collective choice. We can rewrite (7) in the form,

$$G_E^g = -S_g \sum_{i=1}^{N} c_i + \gamma \,, \tag{8}$$

where

$$S_g \equiv \gamma C \tag{9}$$

acts as a group field which couples to each individual choice. The field notion is a natural way to account for some pressure towards a definite choice. Within our convention of minimum conflict the product $S_g c_i$ measures that influence. A positive field S_g favors a positive choice $+1$, while -1 is associated to a negative field. The conflict amplitude is given by S_g.

We have indeed a self-consistent expression since on one hand, individual E wants to go along the virtual field S_g, and on the other hand it contributes directly to this virtual field through its dependence on the collective choice C. Rewriting (8) as

$$G_E^g = -\gamma C^2 + \gamma \tag{10}$$

shows that minimizing G_E^g results in minimizing C^2 which is obtained by $C^2 = 1$. It is an extreme polarization with either $C = +1$ or $C = -1$.

At this stage of the model our "group formation process" is formally identical to the so called mean field transformation in Statistical Physics. While in physics, it is an approximation, here it is not and embodies the social mechanism of anticipation.

4 Individual Bias

We now introduce the possibility of an external pressure applied to the group. Individual biases are also considered.

4.1 External Bias

The existence of an external pressure differentiates between the two possible choices, pushing indeed towards one of them. We call this quantity the social field S. Each person's conflict with S is then represented by the product $-Sc_i$. Agreement is associated with $Sc_i > 0$, i.e., the choice is made along the field with S and c_i having the same sign. At contrast $Sc_i < 0$ represents a conflict between the individual and the social pressure. The total external pressure group conflict measure is,

$$G_S \equiv - \sum_{i=1}^{N} S c_i \, . \tag{11}$$

Applying the *Postulate* to the sum $G_E^g + G_S$ still results in an extreme polarization but now its direction is no longer random. The group choice is $C = +1$ for $S > 0$, and $C = -1$ with $S < 0$. Under external pressure, even extremely weak, the group and the individual behave identically. They both follow the pressure induced by the external pressure. The *Symmetry Breaking* choice is no longer random.

Here, the *super-person* represented by the whole group is identical to the individual person. They are both aligned along the field. This result is at contrast with the symmetrical state, where the individual loses its freedom of choice in favor of the group choice freedom.

4.2 The Representational State

Bias accounts for individual representations [17] which are well established in social sciences. It results from cultural values, beliefs and personal experiences. Following social literature they are called "individual representations" [17]. A representation varies in both, direction and amplitude, from one person to another. It is a characteristic of each person.

The representational effect can be materialized using an internal social field S_i attached to individual i. Similar to a social field S it acts on one person only. The product $-S_i c_i$. It is negative for a choice made along the representation (internal agreement with personal values), and positive otherwise (internal conflict with personal values). The group representation conflict measure is given by,

$$G_R \equiv - \sum_{i=1}^{N} S_i c_i \, . \tag{12}$$

Individual representation distribution is therefore required to determined the final group collective choice. Its effect is enhanced in the isolated-person case with both zero exchange amplitude and social external field. There, from the *Postulate*, final decision is found to result from every individual following its own representation. It gives,

$$C = \frac{1}{N} \sum_{i=1}^{N} \frac{S_i}{|S_i|} \, , \tag{13}$$

where the $|\ldots|$ denotes absolute value.

This equation illustrates the qualitative change driven by the existence of representations. Actual C value can now vary over the whole spectrum of values $-1, -\frac{(N-1)}{N}, \ldots, 0, \ldots, \frac{(N-1)}{N}, +1$. Compromise $C = 0$ can again be an outcome.

Individual representations are thus instrumental for making the whole model relevant to real situations in which collective choices are far more richer than $C = \pm 1$.

In others words, prior to group formation, individuals have their own representations which determine their a priori answers to the initial question. All these representations result in either yes or no. Then, in the process of group formation, people start to interact through the yes and no distribution in the group.

However, to reach a collective choice, due to the existence of opposite representations, people must construct new answers in addition to the initial yes and no. Answers are thus enriched during group formation, due to driving representations. On the other hand, within the neutral state groups do not produce new answers.

Once the final decision is reached, each group member identifies with the collective choice triggering its new individual choice to $d_i = C$ which may differ from the initial c_i. Group formation has qualitatively modified individual representations.

Note our qualitative departure from usual Statistical Physics. Here, we are not considering an average individual position, but a well defined and fixed individual position. This position results from the group forming. In most cases d_i is different from ± 1. We are thus passing from a class of Ising variables $c_i = \pm 1$ to o continuous variable $-1 \leq d \geq +1$.

4.3 The Frustrated Individual

Adding together all above effects results in an extended group internal conflict function $G = G_E^g + G_S + G_R$ which is,

$$G = -E \sum_{i,j} c_i c_j - S \sum_{i=1}^{N} c_i - \sum_{i=1}^{N} S_i c_i \,. \tag{14}$$

The extended form of (14) makes minimizing G a more difficult task since competing effects are now present. A given individual wants to minimize its overall own conflict which combines,

- Interacting group members: the individual wants to come up with the same final decision as preferred by interaction partners.
- External social field: the individual wants to comply to the external pressure from immediate surroundings.
- Internal social field: the individual wants to comply to the internal pressure from its personal representations.

These three elements are not necessarily satisfied simultaneously. From the *Postulate*, the individual wants to minimize its overall personal conflict. Such a scheme could result in simultaneous agreement with some of above items, and conflict with others. It will appear clearer rewriting (14) as,

$$G = \gamma - \sum_{i=1}^{N} S_i^r c_i \,, \tag{15}$$

where,

$$S_i^r = S_g + S + S_i \,, \tag{16}$$

is the resulting field applied to individual i in the group formation process. Minimum individual conflicts are achieved when each individual follows his resulting field sign. If $S^r_{g,i} > 0$, then $c_i = 1$ and $c_i = -1$ for $S^r_{g,i} < 0$. The case $S^r_{g,i} = 0$ results in an undetermination of i choice as in the isolated case.

Satisfying S_i^r sign does imply satisfying simultaneously S, S_i, and S_g signs. This competing effect is the signature of the psychological complexity involved in the decision making process. Each person first follows its resulting field S_i^r which in turn produces a collective choice C. This collective choice is then integrated back at each individual level with $c_i \rightarrow d = C$.

5 Illustration of the Model

We now illustrate the model in two different specific cases.

5.1 Two Balanced Opposite Biases Case

We consider a N person group evenly divided in two opposite biases and no external social field, i.e., $S = 0$. Half the persons have a positive representation $S_i = +S_0$, and the other half have a negative representation with the same amplitude $S_j = -S_0$. Overall the group has thus no net representation.

Interactions are of amplitude E and each person discusses with n other persons. In small, face-to-face groups, everyone usually interacts with everyone else, so then $n = N$. Corresponding internal conflict function is,

$$G = \gamma - \frac{\gamma}{N} C^2 - \frac{N}{2}(S_0 c_i^+ - S_0 c_j^-) \,, \tag{17}$$

where c_i^+ and c_j^- are attached to persons with respectively positive and negative representation. The constant $\gamma \equiv \frac{nIN}{2(N-1)}$ has been introduced earlier in the group formation section.

The collective choice may be written as $C = \frac{N}{2}(c_i^+ + c_j^-)$. Actual choice is the one which maximizes G. In this case it is easily singled out, since there exist only two different kinds of persons symbolized by c_i^+ and c_j^-. Four choice configurations are possible,

a $c_i^+ = +1$; $c_j^- = +1$; $C = N$,
b $c_i^+ = -1$; $c_j^- = -1$; $C = -N$,
c $c_i^+ = +1$; $c_j^- = -1$; $C = 0$,

d $c_i^+ = -1$; $c_j^- = +1$; $C = 0$.

The first two (a and b) are agreement and others (c and d) are conflict. Associated internal conflict functions are,

a $G(a) = -\gamma + N\gamma$,
b $G(a) = G(b)$,
c $G(c) = -\gamma + NS_0$,
d $G(d) = -\gamma - NS_0$.

Clearly $G(d) < G(c)$, reducing the choice to either (a and b) or (c). In case $\gamma > S_0$, we have $G(a,b) > G(c)$, indicating that the interaction strength proportional to nI is stronger than S_0. The group then polarizes with $C = \pm N$. The direction of the extreme choice occurs at random.

Half of the members are fully satisfied with both their representation and their partners while the other half is in conflict with its own representation. This result means in particular that the "losing" subgroup has to build a new representation which embodies some level of internal conflict. The "winning" part does not modify its initial representation. In this case, no new answer was built. We have $c_i \to d = \pm 1$.

On the other hand, strong representation, i.e., $\gamma < S_0$ favors compromise, with the collective choice $C = 0$. Each member E starts from a personal representation to decide eventually through weak interactions on a medium compromise, with the creation of a new answer $d = 0$. Again, this compromise choice did not exist prior to the group formation. It is the result of cooperation between the group level and the individual level.

Within a balanced representation group, exchange favors a compromise. Weak exchange result in an extreme polarization along a random direction.

5.2 Two Unbalanced Opposite Biases Case

We now go back to the previous example, but consider a stronger positive representation. This is done by writing the negative representation fields as $S_j = -\alpha S_0$, with $0 < \alpha < 1$. Respective numbers of positive and negative representations are equal.

Only the internal conflict function values are changed to become respectively,

a $G(a) = -\gamma + N\gamma + \frac{N}{2}(1-\alpha)S_0$,
b $G(b) = -\gamma + N\gamma - \frac{N}{2}(1-\alpha)S_0$,
c $G(c) = -\gamma + \frac{N}{2}(1+\alpha)S_0$,
d $G(d) = -\gamma - \frac{N}{2}(1+\alpha)S_0$.

Since $0 < \alpha < 1$, $G(a) < G(b)$ and $G(d) < G(c)$, always. However, in the case $\gamma > S_0$ the polarization direction is determined with $C = +N$.

Before we had $\alpha = 1$ which made the direction arbitrary, but now it is the strongest initial representation which wins. The discussion process within the forming group has made the weaker-biased people align themselves with the stronger ones. Here we have $c_i \to d = 1$.

In order for a compromise outcome to be favored, a decrease in exchanges among group members is required. For $\gamma < S_0$, the final choice is $C = 0$ which gives $c_i \to d = 0$.

Within an unbalanced representation group, exchange favors the initially strongest representation. Only a limitation of exchange may produce a compromise.

6 Conclusion

A simple Ising-like model has been presented to describe group decision making. It is indeed a modified version of the random field Ising ferromagnetic model in an external magnetic field at zero temperature. However, our system is finite in size and fields may have a non-zero configurational average. In principle, results may also depend on the field configuration. Moreover, we crossover in the group decision making process from a class of Ising variables to one continuous variable.

The hypothesis behind our approach is that group decision making obeys universal laws which are independent of the nature of the issue at stake. Our main results with respect to the qualitative properties of group decision making are:

- Exchanges among individuals does not aim to select an issue, but rather to align people along the same issue. The issue itself is random with respect to exchanges.
- Exchanges among individuals does not favor compromise about an issue. On the opposite it produces polarization, i.e., extreme options.
- Reducing exchanges favors compromise.
- External social pressure is extremely efficient on selecting an option.
- Individual bias is a necessary ingredient to both weaken extreme option and oppose an external social pressure.

These theoretical results must be put in parallel to various data obtained from a large number of experimental studies which show groups polarize along an extreme position reflecting the dominant pole of attitudes and not around an average position as a priori expected [16,17].

Our emphasize is on building a conceptual methodology rather than a final complete theory. In a forthcoming paper we will introduce non-rational behavior which is a real life basic feature. It will be analogous to temperature. However, within our model we will define a "local temperature" in a finite system.

References

1. S. Galam and S. Moscovici, *Towards a theory of collective phenomena. I: Consensus and attitude change in groups*, Euro. J. Soc. Psychology **21**, 49-74 (1991).

2. N. A. Chigier and E. A. Stern (Eds.), *Collective Phenomena and the Applications of Physics to Other Fields of Science,* (Brain Research Publications, New York, 1974).
3. S. Galam, Y. Gefen, and Y. Shapir, *Sociophysics: a new approach of sociological collective behavior*, J. Math. Sociology **9**, 1 (1982).
4. S. Galam, *Majority rule, hierarchical structure and democratic totalitarianism*, J. Math. Psychology **30**, 426 (1986); *Social paradoxes of majority rule voting and renormalization group*, J. Stat. Phys. **61**, 943-951 (1990).
5. S. Galam and S. Moscovici, *Towards a theory of collective phenomena. II: Conformity and Power*, Euro. J. Soc. Psychology **24**, 481 (1994); *III: Conflicts and Forms of Power*, Euro. J. Soc. Psychology **25**, 217 (1995).
6. M. Lewenstein, A. Nowak, and B. Latané, *Statistical mechanics of social impact*, Phys. Rev. A **45**, 1 (1992); G.A. Kohring, *Ising models of social impact: the role of cumulative advantage*, J. Phys. I. France **6**, 301 (1996).
7. N.S. Glance and B.A. Huberman, *The outbreak of cooperation*, J. Math. Sociology **17**, 281 (1993).
8. M. Levy, H. Levy, and S.Solomon, *Microscopic simulation of the stock market*, J. Phys. I. France **5**, 1087 (1995).
9. D. Helbing, A. Hennecke, and M. Treiber, *Phase diagram of traffic states in the presence of inhomogeneities*, Phys. Rev. Lett. **82**, 4360 (1999).
10. D. Stauffer, P.M.C. de Oliveira, S. Moss de Oliveira, and R.M. Zorzenon dos Santos, *Monte Carlo Simulations of Sexual Reproduction*, Physica **A 231**, 504 (1996).
11. B. Derrida, A.J. Bray, and C. Godréche, *Non-trivial exponents in the zero temperature dynamics of the 1D Ising and Potts models*, J. Phys. A **27**, L357 (1994); D. Stauffer, *Ising spinodal decomposition at $T = 0$ in one to five dimensions*, J. Phys. A **27**, 5029 (1994).
12. Sh-k Ma, *Modern Theory of Critical Phenomena*, The Benjamin Inc.: Reading MA (1976).
13. J.C. Turner, *Rediscovering the Social Group*, Basil Blackwell: Oxford (1987).
14. S. Moscovici and M. Zavalloni, *The group as a polarizer of attitudes*, Journal of Personality and Social Psychology **12**, 125-135 (1969).
15. E. Burnstein, and A. Vinokur, *Testing two classes of theories about group induced shifts in individual choices*, Journal of Experimental Social Psychology **9**, 123-137 (1973).
16. J. Davis, *Group decision and social interactions: A theory of social decision scheme*, Psychological Review **80**, 97-125 (1973).
17. S. Moscovici, *Social influence and social change*, (Academic Press, London, 1976).

Formation of Opinions under the Influence of Competing Agents – a Mean Field Approach

K. Kacperski[1] and J.A. Hołyst[2]

[1] Max Planck Institute for Physics of Complex Systems, Nöthnitzer Str. 38, 01187 Dresden, Germany
[2] Faculty of Physics, Warsaw University of Technology, Koszykowa 75, 00–662 Warsaw, Poland

Abstract. We study a model of opinion formation based on the theory of social impact and the concept of cellular automata. The case is considered when two strong agents influence the group: a strong leader and an external social impact acting uniformly on every individual. There are two basic stationary states of the system: cluster of the leader's adherents and unification of opinions. In the deterministic limit the variation of parameters like the leader's strength or external impact can change the size of the cluster or, when they reach some critical values, make the system jump to another phase. In the presence of noise (social temperature) the rapid changes can be regarded as the first order phase transitions. When both agents are in a kind of balance, a second order transition and critical behaviour can be observed. Analytical results obtained within a mean field approximation are well reproduced in computer simulations.

1 Introduction

Interdisciplinary research has been drawing much attention in the last decades. Models and methods developed in theoretical physics proved to be fruitful in studying complex systems [1,2], composed of relatively simple mutually interacting elements, coming from domains as diverge as neural networks [3], disease spreading [4], population dynamics [5], etc. But the range of the investigations goes also beyond the natural sciences and includes problems from sociology or economy, e.g., pedestrian motion and traffic [6], migrations [7,8], financial crashes [9].

Another important subject of this kind is the process of opinion formation in social groups or decision making, also at the level of whole countries. One way of its quantitative description consists in a macroscopic approach based on the master equation or the Boltzmann-like equations for global variables [7,10,11]. Alternatively, by making some sociologically motivated assumptions on the mechanisms of interactions between individuals "microscopic" models are constructed and investigated numerically and analytically by means of the methods known from statistical physics [12,13]. One concludes that the variety of the emerging collective phenomena has much in common with the complex social processes.

One of the examples is the class of models based on the concept of cellular automata [14] and the theory of *social impact* formulated by Latané [15], and conformed in a number of sociological studies [16,17]. Different variants of the model were explored numerically [16,18], and many of the observations were than explained in the framework of a mean field approach [19] and recently the Landau theory [20]. An extention of the model introducing the time variance of the social strengths of individuals according to some learning rule has also been studied [21]. The essential outcome of both the theory and simulations is the onset of clusters of minority that can survive within the majority holding the opposite opinion and be persistent throughout long periods of dynamics. It has been indicated that strong individuals play an important role in the formation and stability of the clusters. Motivated by this, in [22–24] we considered a particular case of the model, namely when a strong individual (a leader) is present in the social group. Multi-stability and hysteresis phenomena, as well as different kinds of rapid changes (phase transitions) in the distribution of opinions were encountered.

Here the effects of competition between the strong leader and an *external influence* or *preference* acting homogeneously within the group are studied in terms of a mean field approach. We show that such antagonism can lead not only to sudden changes of opinion, but also to the critical behaviour. The high "social temperature" reveals the dominance of the stronger agent. After introducing the model (Sect. 2) and recalling some results for the deterministic case (Sect. 3) we investigate in detail the noise induced transitions giving rise to the critical behaviour (Sect. 4.1).

2 Model of a Social Group

Our system consists of N individuals (members of a social group); we assume that each of them can share one of two opposite opinions on a certain subject, denoted as $\sigma_i = \pm 1$, $i = 1, 2, \cdots, N$. Individuals can influence each other, and each of them is characterised by the parameter $s_i > 0$ which describes the strength of his/her influence. Every pair of individuals (i, j) is ascribed a distance d_{ij} in a social space. The changes of opinion are determined by the *social impact* exerted on every individual:

$$I_i = -s_i\beta - \sigma_i h - \sum_{j=1,j\neq i}^{N} \frac{s_j\sigma_i\sigma_j}{g(d_{ij})}, \tag{1}$$

where $g(x)$ is an increasing function of social distance, β is a so-called self-support parameter reflecting the inclination of an individual to maintain his/her current opinion, and h is an additional (external) influence which may be regarded as a global preference towards one of the opinions stimulated by mass-media, government policy, etc.

Opinions of individuals change simultaneously (synchronous dynamics) in discrete time steps according to the rule

$$\sigma_i(t+1) = \begin{cases} \sigma_i(t) \text{ with probability } \dfrac{\exp\left(-\dfrac{I_i}{T}\right)}{\exp\left(-\dfrac{I_i}{T}\right) + \exp\left(\dfrac{I_i}{T}\right)} \\ -\sigma_i(t) \text{ with probability } \dfrac{\exp\left(\dfrac{I_i}{T}\right)}{\exp\left(-\dfrac{I_i}{T}\right) + \exp\left(\dfrac{I_i}{T}\right)} \end{cases} \quad (2)$$

analogous to the Glauber dynamics with $-I_i \sigma_i$ corresponding to the local field. The parameter T may be interpreted as a "social temperature" describing a degree of randomness in the behaviour of individuals, but also their average volatility [13]. The impact I_i is a "deterministic" force inclining the individual i to change his/her opinion when $I_i > 0$ or to keep it otherwise.

Our model social space is a 2D disc of radius R with the individuals located in the nodes of quadratic grid; the distance between nearest neighbours equals 1, and each node is occupied with probability ρ which may be also regarded as a constant surface density of individuals. The geometric distance models the social immediacy. Strength parameters s_i of the individuals are positive random numbers with probability distribution $q(s)$ and the mean value $\overline{s}$. In the centre of the disc there is a *strong individual* (whom we will call the "leader"); his/her strength s_L is much greater than that of all the others ($s_L \gg s_i$).

3 Deterministic Limit

Let us first recall the properties of the system without noise, i.e., at $T = 0$ (see [22,23] for details). The dynamical rule (2) becomes then strictly deterministic: $\sigma_i(t+1) = -\text{sign}(I_i\sigma_i)$, so the condition for the stability of a state is $I_i < 0$ for every i. To proceed with the calculations we used the approximation of continuous distribution of individuals replacing the sum in (1) by an integral. In the stationary state the impact is maximal, almost equal zero, at places where individuals of opposite opinion about, so the condition

$$I(x, y) = 0 \quad (3)$$

is the equation for the borders between them.

Considering the possible stationary states in our model we find the trivial unification (with equal opinion for each individual) or, due to the symmetry, a circular cluster of individuals sharing the opinion of the leader surrounded by a ring of their opponents (the majority). These states remain stationary also for small self-support parameter β; for sufficiently large β any configuration may remain "frozen".

As far as the shape of the cluster is known, the impact $I(x, y)$ at its border can be calculated and then (3) is the equation for the radius of stationary clusters. Typically, it has two real solutions: the smaller corresponding to the stable cluster and the larger one - to the unstable cluster, which is actually the separatrix between basins of attraction of the stable cluster and the state of unification. Owing to the nonzero self–support parameter β in (1) reflecting the inertia of individuals in changing their minds (it may be regarded as an analogy of the dry friction in mechanical systems), both solutions form in fact a kind of bands; the states within the bands are "frozen". In this way also the unstable clusters can be observed at $\beta > 0$ and appropriately chosen initial conditions.

When the strength of the leader s_L or the external impact h supporting the opinion of the leader is increased the size of the stable cluster grows, and at some critical point both real solutions of (3) collide. Then the equation has no longer real solutions and the unification is the only stable state. At the critical point one observes a discontinuous phase transition: *cluster* $\rightarrow$ *unification.* If a strong persuasive external impact acts on a unified group, in inverse transition *unification* $\rightarrow$ *cluster* is possible.

4 Noisy Dynamics

In the presence of noise the borders of the stable clusters become diluted, i.e., individuals of both opinions appear all over the group. Considering the dynamics (2) we can conclude that the influence of noise on a single individual depends on the ratio I_i/T. Because of the strong influence of the leader, the supportive (negative) impact is stronger inside the cluster than outside it. Thus, due to noise induced flips of opinions the adherents of the leader appear more often outside the cluster (among the majority of opponents) than the opponents inside it. Moreover, the area outside is much greater than that of the cluster itself, so we observe the effective growth of the minority group. This causes that the supportive impact outside the cluster becomes still weaker and the majority becomes more sensitive to random changes; it is a kind of positive feedback. At a certain value of temperature the process becomes avalanche–like and the former majority disappears. Thus noise induces a jump from one attractor (cluster) to another (unification). Such a transition is possible at every non–zero temperature but its probability remains very small until the noise level exceeds a certain critical value. Our simulations prove that it is indeed a well defined temperature that separates two phases (i.e., two attractors). Similarly the transition unification $\rightarrow$ cluster in the presence of external impact can be induced by noise.

4.1 Mean Field Approximation: Phase Transitions, Critical Dynamics

We can derive analytically the stationary states of the system an nonzero temperature using a kind of mean field approximation. To do this let us calculate the impact on an individual at a distance x from the leader of opinion $\sigma_L = 1$, and sharing the opposite opinion:

$$I(x) = \frac{s_L}{g(x)} + \rho\,\overline{s} \int\limits_{D_R} Pr(r) \frac{1}{g(|\mathbf{r}-\mathbf{x}|)}\, d^2\mathbf{r} -$$

$$\rho\,\overline{s} \int\limits_{D_R} (1 - Pr(r)) \frac{1}{g(|\mathbf{r}-\mathbf{x}|)}\, d^2\mathbf{r} - \beta\overline{s} + h. \tag{4}$$

The integration is performed over the whole space D_R excluding the individual under consideration. $Pr(r)$ denotes the probability of finding a leader's follower at the distance r from the centre of the group; it is in fact determined by (2) from which it follows that it depends on the actual state of the system in the previous time step. We would like, however, to have a stationary function $I(x)$. This can be achieved relatively easy if we neglect the self support term. Let us put $\beta = 0$ in the subsequent calculations and discuss the influence of nonzero self support later.

Now we make an approximation replacing $Pr(r)$ by its stationary mean value p over D_R and putting it outside the integral. This is equivalent to the simple assumption that $Pr(r)$ is uniform. We can expect it to be valid for large temperatures when the dynamics is almost random, or for small leader's strength s_L (because it is in fact the large value of s_L that contributes most to the non-uniformity of $Pr(r)$). From (4) we get

$$I(x) = \frac{s_L}{g(x)} + (2p-1)\rho\overline{s} J_D(x) + h, \tag{5}$$

where $J_D(x) = \int_{D_R} 1/g(|\mathbf{r}-\mathbf{x}|) d^2\mathbf{r}$ is a function dependent only on the size of the group and type of interactions. Note that the impact on an individual sharing the opinion of the leader would be the same as in (5) but with the opposite sign. Due to this fact we can easily derive the expression for the stationary probability $Pr(r)$ from the dynamical rule (2) which gives the transition probabilities. Then from the definition of the mean value p (the mean part of individuals sharing the opinion of the leader) it follows

$$p = \frac{1}{\pi R^2 \rho} \int_0^R \rho \, \Pr(r)\, 2\pi r\, dr = \frac{1}{R^2} \int_0^R \frac{\exp[I(r,p)/T]}{\cosh[I(r,p)/T]} r dr \equiv f(p), \tag{6}$$

where $I(x,p)$ is given by (5). This is an integral equation for p. Note that in the above derivation we set the leader's opinion fixed, independent of the influence of the group and the noise.

In Fig. 1 one can see the graphical solution of (6) for certain set of parameters and $g(r) = r$. At low temperatures there are three solutions: the smallest one corresponding to the stable cluster around the leader, the second – to the unstable cluster which, in fact, is not observed, and the biggest – to the unification. The size of the stable cluster grows with increasing the temperature up to a critical value T_{tr} when it coincides with the unstable solution. At this temperature a transition from a stable cluster to unification occurs (Fig. 2). For

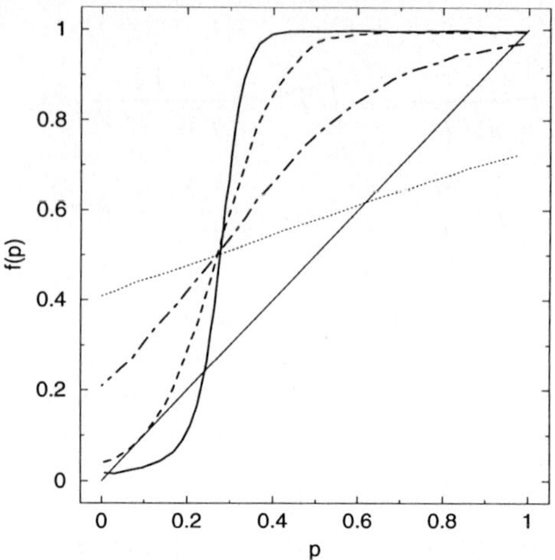

Fig. 1. Numerical solution of (6) for R=20, $\beta = 1$, $\rho = 1$, $\overline{s} = 1$, $s_L = 250$ and $h = 25$. $f(p)$ is the RHS of (6) plotted for different temperatures: solid line - $T = 10$, dashed - $T = 28$, dotted-dashed - $T = 80$, thin dotted - $T = 300$.

$T > T_{tr}$ unification is the only solution, but it is no longer a perfect unification because due to the noise individuals of the opposite opinion appear. When the temperature goes on growing the curve in Fig. 1 becomes more flat and p tends to $1/2$ what means that the dynamics is random and both opinions appear with equal probability.

Figure 2 shows the function $p(T)$ compared with the results of computer simulations. The phase transition mentioned above can be observed. The analytical curve fits the results of simulations quite well, particularly at large temperatures for the reasons mentioned above. Other mean field approximations, e.g., the one used by us in [22] yield better results at low noise levels, specifically for the value of the transition temperature. They consist in taking another function (instead of constant) to replace $Pr(r)$ in (4) (e.g., a stepwise in [22]). In fact, the definition (2) can be used with I given by (5); this would be a "second order" extension of our simple approach. Higher order approaches can be constructed in a similar way; they would give more accurate results, however for the price of more complex equation (6) for p.

Let us thus remain at the"first order" equation (6). Changing the external impact and/or the leader's strength results in a shift of the curves $f(p)$ along the $p-$axis. If we apply a large enough external impact against ($h < 0$) the leader the curves $f(p)$ in Fig. 1 would be shifted to the right so that now, starting from the uniform opinion $p = 1$ and exceeding a critical temperature we observe the transition $unification \to cluster$ which remains the only solution at high noise level.

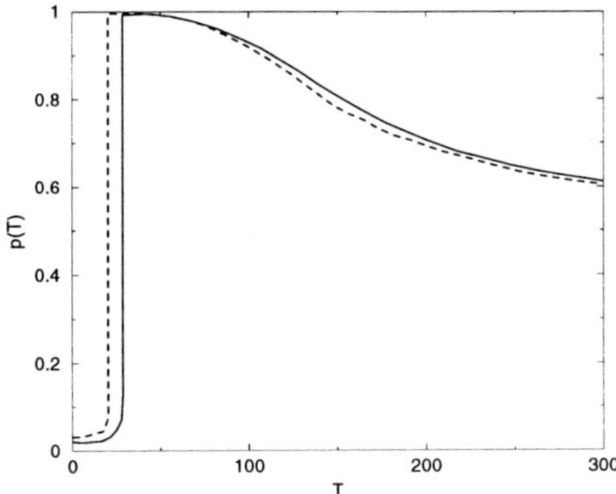

Fig. 2. Fraction p of leader's followers vs. temperature; $s_L = 400$, $h = 0$. Results of our calculations are represented by a solid line and those of computer simulations by a dashed line. Other parameter values as in Fig. 1.

At a certain value of $h = h_b < 0$ the influences of the leader and the external impact are in a way balanced and the curve $f(p)$ becomes roughly symmetric with respect to the bisector p. Let us set the condition for the balance as $f(\frac{1}{2}) = \frac{1}{2}$. With the use of (6) and (5) it can be written as

$$\frac{1}{R^2}\int_0^R \frac{\exp\left[\left(\frac{s_L}{g(x)} + h_b\right)/T\right]}{\cosh\left[\left(\frac{s_L}{g(x)} + h_b\right)/T\right]} x dx = \frac{1}{2}, \tag{7}$$

which gives implicitly h_b as a function of other parameters.

With the increase of temperature at $h = h_b$ (maintained by appropriate changes of s_L) the two stable solutions p_1 and p_2 (corresponding to cluster and unification respectively) converge towards $p = \frac{1}{2}$ (the third, unstable solution). At some critical noise level T_c the solutions coincide and we have only $p = \frac{1}{2}$ which now becomes stable. The condition for this critical temperature can be written as $\left.\frac{df(p)}{dp}\right|_{p=1/2} = 1$. Again using (6) and (5) we get an implicit integral equation for T_c:

$$\frac{2}{T_c R^2}\rho\bar{s}\int_0^R \frac{J(x)}{\cosh^2\left[\left(\frac{s_L}{g(x)} + h_b\right)/T_c\right]} x dx = 1, \tag{8}$$

with h_b given by (7).

In the case of $g(x) = x$ we found that the curve $f(p)$ is almost exactly symmetric with respect to the point $f(\frac{1}{2})$ and thus $p_2 \approx 1 - p_1$. Moreover the balanced external impact h_b depends only weakly on the noise level, and the critical temperature T_c is almost independent of the leader's strength s_L. Figure 3 shows the temperature of the described above transitions *cluster* $\rightarrow$ *unification*

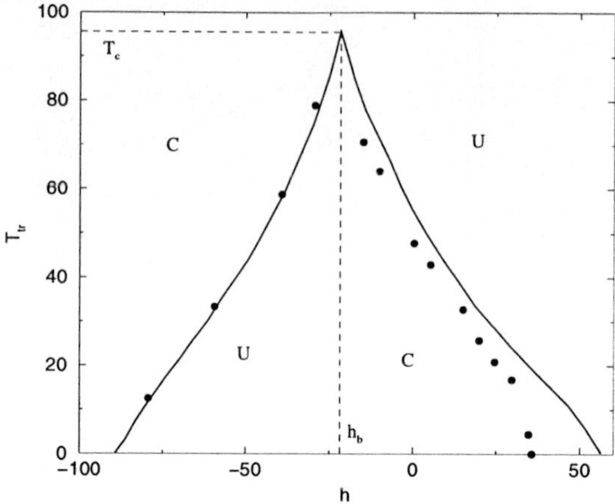

Fig. 3. Transition temperature T_{tr} vs. external impact h at $s_L = 250$ (other parameter values as in Fig. 1). Leader's opinion was fixed (independent of the group). Line corresponds to analytical results (6), points to the computer simulations of the model.

(for $h > h_b$) and *unification* $\rightarrow$ *cluster* (for $h < h_b$) as a function of h. Both curves meet at the critical point $h = h_b$, $T = T_c$. When moving along both curves towards the critical point the magnitude of jump in the majority-minority proportion due to transition decreases and at (h_b, T_c) there is no jump at all.

In analogy to physical systems the transitions occurring while crossing the curves in Fig. 3 from below may be called first-order transitions. When the critical point (h_b, T_c), at which the difference between two different phases disappears, is crossed going from the region below the curves a second-order transition occurs. An example of the dynamics in the neighbourhood of this kind of transition is shown in Fig. 4. As the solutions p_1 and p_2 are approaching $\frac{1}{2}$ increasing fluctuations around them can be observed (Fig. 4a). At T close to T_c, noise induced random jumps between p_1 and p_2 are possible (Fig. 4b). With $T \rightarrow T_c$ the average frequency of jumps increases (Fig. 4c), and at $T = T_c$ we observe the dynamics with $p = \frac{1}{2}$ and large fluctuations (Fig. 4d). When the temperature is further increased the amplitude of fluctuations decreases (Fig. 4e). Large fluctuations in the vicinity of the critical point are a general characteristic feature of critical phenomena in physical systems.

The described above second-order transition through critical point can also be observed in the absence of a leader and the external impact, i.e., when $s_L \approx \overline{s}$ and $h = 0$. Then the two phases corresponding to unifications $U+$ and $U-$ which are stable in low temperatures merge giving rise to the high temperature phase – at average equal numbers of individuals sharing both opinions. In this case the model reduces to the Ising model with long range interactions if we additionally put $s_i = \overline{s}$ for every i.

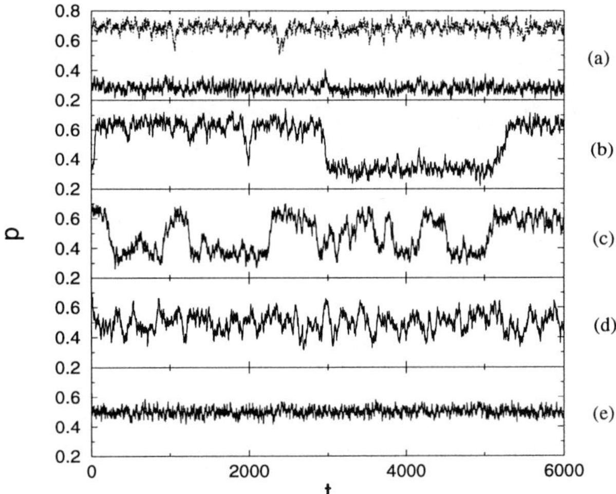

Fig. 4. Time evolution of the number of leader's followers in the vicinity of the critical point, $h = 24.65 \approx h_b$, temperatures: (a) $T = 94$, (b) $T = 97$, (c) $T = 99$, (d) $T = 102 \approx T_c$, (e) $T = 120$; other parameter values as in Fig. 1. The opinion of the leader has been fixed. Fig. (a) shows the results of two runs starting from initial conditions $p(0) > 0.5$ (upper curve) and $p(0) < 0.5$ (lower curve).

It should also be mentioned that the fluctuations in the vicinity of the critical point are relatively slow, e.g., in our simulations significant changes of p happen during a few tens of time steps. This phenomenon known as the critical slowing down is caused by local correlations of opinions. In the temperatures saliently greater than T_c the correlations disappear and we observe fast random fluctuations around $p = \frac{1}{2}$.

The second-order transition occurs also in the opposite direction, i.e., when the temperature is decreased starting from the $p = \frac{1}{2}$ phase at $T > T_c$. At the critical point there is a symmetry breaking; the choice between two symmetric phases depends sensitively on tiny deviations from the balanced external impact h_b and random fluctuations.

One could also describe the global dynamics of the system introducing an effective potential. At low temperatures it would have two minima corresponding to the stable solutions of (6) separated by a maximum corresponding to the unstable solution. For $h = h_b$ both minima would be approximately equal, but otherwise the one corresponding to the dominance of the stronger agent would be global while the other the local one with the corresponding state being *metastable*. However, as we have already mentioned, the probability of escape from the metastable state is very small for $T < T_{tr}$.

In the analytical calculations of this section we neglected the self supportiveness of individuals putting $\beta = 0$. Let us now briefly discuss the effect of a nonzero β. As it was already mentioned, the self support may be regarded as an analogy of dry friction in mechanical systems. It strives to maintain the current state of individuals. Writing the formula (5) for the impact we assumed a stationary state and did not regard how it was achieved. We could do it, as well as to use the formula for the probability in (6) which is independent of the current state only in the case of $\beta = 0$. Due to nonzero self support the mean (stationary) value of the proportion p of the leader's adherents depends on the initial state. One can regard that the curve in Fig. 1 splits into a band and so do the appropriate solutions of (6). It depends on the starting point on which side of the band the system settles. However, at high temperatures noise induced random walk within the band will finally make the system forget its initial conditions and the mean value for $\beta = 0$ will be observed. Accordingly the first order transition occurs when the band looses tangency to the bisector and thus the temperature of transition will be higher than for $\beta = 0$. We can also expect that the critical behaviour will be observed in a broader range of temperatures. Nevertheless, for small β (e.g., $\beta = 1$ in our simulations) the influence of the self support is negligible and the results of our mean field theory remain valid.

5 Summary

Let us summarise the outcomes of our analysis. The influence of two agents on the group: the strong leader and an external impact gives rise to multi-stability in certain ranges of parameters, hysteresis and discontinuous changes in the distribution of opinions in the deterministic case. The situation looks similar in the presence of noise which models the complexity and indeterminism of the process of opinion formation at the level of a particular individual. The noise level ("social temperature") is an additional transition inducing parameter. We have shown that the noise favours the stronger agent; with growing temperature it needs smaller prevalence to convert the majority to its opinion. In the case of a balance (symmetry) between the two agents the appropriate phase transition becomes continuous and we observe characteristic critical behaviour (large fluctuations, critical slowing down).

The phenomena can be understood and described quantitatively in terms of a mean field approach. In general a hierarchy of approximations can be constructed, but our calculations and computer simulations of the model show that already in the first order we get quite good quantitative results.

The model presented here, though presumably not directly applicable to the description of a real social group, may still be useful in explaining the mechanisms underlying often very complicated social processes. Being aware of their complexity we are circumspectful in drawing far reaching sociological conclusions from our results. Let us point out just one general outcome which seems reasonable: the rapid changes of opinions under the influence of strong leaders and some general preferences or prejudices are more probable during the times of big social transformations or turbulence, when people are much confused in their views (high social temperature). Then the opinion supported by the stronger agent prevails and may remain dominant later on in a more quiet period, even after the influence of the agent decreases or even ceases to exist.

Various modifications and extentions of the model are possible; one of the natural ones is the introduction of more than one leaders. A motivating question, also relevant to real life problems, is how to distribute efforts (money, investments, human resources) in order to convince a unified group of people to a given opinion; is it more efficient to concentrate them in one place (one strong leader) or rather to disperse them (a few weaker leaders), if so, in how many parts? These problems will be the subject for future studies.

References

1. H. Haken, *Synergetics. An Introduction,* (Springer, Heidelberg, New York, 1983); *Advanced Synergetics,* (Springer, Heidelberg, New York, 1983).
2. G.A. Cowan, D. Pines, and D. Meltzer, (Eds.), *Complexity. Metaphors, Models, and Reality,* (Addison-Wesley, Santa Fe, 1994).
3. D. Amit, *Modeling Brain Function,* (Cambridge Univ. Press, Cambridge, 1989); E. Domany, J.L. van Hemmen, and K. Schulten, (Eds.), *Models of Neural Networks,* (Springer, Berlin, 1995); A. Browne, (Ed.), *Neural network analysis, architectures and applications,* (Institute of Physics Publishing, Bristol, 1997).
4. A. Johansen, Physica D **78**, 186 (1994); H.C. Tuckwell, L. Toubiana, and J-F. Vibert, Phys. Rev. E. **57**, 2163 (1998).
5. P. Bak and K. Sneppen, Phys. Rev. Lett. **71**, 4083 (1993); A. Pekalski, Physica A **252**, 325 (1998).
6. D. Helbing, Phys. Rev. E **55**, 3735 (1997); Physica A **219**, 375 (1995); D. Helbing and P. Molnar, Phys. Rev. E **51**, 4282 (1995).
7. W. Weidlich and G. Haag, *Concepts and Models of Quantitatively Sociology,* (Springer, Berlin, New York, 1983); W. Weidlich, Phys. Rep. **204**, 1 (1991).
8. J. Fort and V. Méndez, Phys. Rev. Lett. **82**, 867 (1999).
9. D. Sornette and A. Johansen, Physica A **245**, 1 (1997); N. Vandewalle, M. Ausloos, P. Boveroux, and A. Minguet, Eur. Phys. J. B **4**, 139 (1998).
10. W. Weidlich, J. Math. Sociology **18**, 267 (1994).
11. D. Helbing, Physica A **193**, 241 (1993); J. Math. Sociology **19**, 189 (1994); D. Helbing, *Quantitative Sociodynamics,* (Kluwer Academic, Dordrecht, 1995).
12. S. Galam, Physica A **230**, 174 (1996); **238**, 66 (1997).
13. D.B. Bahr and E. Passerini, J. Math. Sociology **23**, 1, 29 (1998).
14. S. Wolfram, *Theory and Application of Cellular Automata,* (World Scientific, Singapore, 1986); H. Gutowitz, (Ed.), *Cellular automata, theory and experiment,* (MIT Press, London, 1991).
15. B. Latané, Am. Psychologist **36**, 343 (1981).
16. R.R. Vallacher and A. Nowak, (Eds.), *Dynamical systems in social psychology,* (San Diego, Academic Press, 1994).
17. E.L. Fink, J. Communication **46**, 4 (1996); B. Latané, J. Communication **46**, 13 (1996).
18. A. Nowak, J. Szamrej, and B. Latané, Psych. Rev. **97**, 362 (1990).
19. M. Lewenstein, A. Nowak, and B. Latané , Phys. Rev. A **45**, 763 (1992).
20. D. Plewczyński, Physica A **261**, 608 (1998).
21. G.A. Kohring, J. Phys. I France **6**, 301 (1996).

22. K. Kacperski and J.A. Hołyst, J. Stat. Phys. **84**, 169 (1996).
23. K. Kacperski and J.A. Hołyst, *Leaders and clusters in a social impact model of opinion formation: the case of external impact*, in: F. Schweitzer, (Ed.), *Self-Organization of Complex Structures: From Individual to Collective Dynamics*, Vol II, (Gordon and Breach, Amsterdam, 1997).
24. K. Kacperski and J.A. Hołyst, Physica A **269**, 511 (1999).

Search for Intelligence by Motion Analysis

I.M. Jánosi

Department of Physics of Complex Systems, Eötvös University, P.O. Box 32, 1518 Budapest, Hungary

Abstract. In this work we analyze an apparently simple question: Is there any sign of intelligence in human motion? Traditional comparative ethology has shown that individual and most collective motions are common in many species of very different levels of intelligence. We show here that an almost unambiguous identification of intellect is feasible by detecting some form of absolutely senseless collective motion of large masses.

For some reason, there is an increasing interest in getting evidence on extraterrestrial intelligence. Since the distance of space travel for humans is restricted to the nearest planets of the solar system in the foreseeable future, alternative methods are developed; e.g., a series of searches has been carried out mainly by radio astronomers seeking directed signals or byproduct radio waves emitted by a civilization of intelligent life forms inhabiting another star system [1]. It is rather astonishing that no evidence of such signals has ever been published in Nature or Physical Review Letters. Nevertheless, the demand for a well established searching methodology is continuously increasing, parallel to the development of rocket technology.

The key issues in any searching method are (i) identification of life and (ii) separating intelligent life from the rest. The solution of the first task seems to be rather easy, excluding the case of plants, like the thinking cactus *Lobivia sapiens* [2]. A comprehensive analysis of relevant background materials [3] and literature gave the result that any extraterrestrial form of life must share a few determining qualities: They have different color (mostly green), different number of limbs (typically from one to a few dozens), they are definitely stinking and evil-minded, intending to kill (and sometimes eat) any life form of different color or other features. Otherwise, the detection of them is simple, because they are *moving* erratically in every direction. Indeed, one of the main aspects of terrestrial animal life is motion, too [4]. The second task, i.e., the identification of intelligence, is far from being so trivial, apart from the surprisingly large number of extraterrestrial beings of rather good command of English. The question naturally arises: Is there any way to separate intellect by means of the analysis of motion which is easy to detect?

In order to get a deeper insight of the problem, let us investigate the reversed situation. Such a typical case is a terrestrial research expedition of Quumbrantapaguians (the inhabitants of the planet Quumbrantapaguia). Since they are using an unknown, but completely different form of communication, speech, writing, art, philosophy, even scientific publications are not observables for them. Similarly, usage of fire for various purposes (which is considered by humans as an important sign of human intelligence [5]) does not mean anything for them, because their metabolism is based on the well-know process of cold fusion [6]. Unfortunately, they do not care with technical civilization either, which by the way does not provide a key to intelligence: As it is known, the frog-like inhabitants with only 281 nerve-cell-like something on the planet B'blowjo produce the most beautiful architecture in the Universe [7]. Nevertheless, the Quumbrantapaguians developed a very efficient life-detect-o-meter which is capable to trace trajectories and measure many aspects of motion for most of the terrestrial animals.

After the evaluation of a huge pile of life-detect-o-meter data, they should conclude first of all that the more complex, the more developed is a given species, the more *seemingly* senseless elements of their motion show up. Indeed, bacteria or algae are moving in a very practical way: Essentially they are approaching good stuff (food, oxygen, light, etc.) or escaping from bad stuff (antibiotics, heat, Eastern Europe, etc.) continuously. A bit more developed species, however, tend to move sometimes less pragmatically by *seemingly* aimless waste of lots of energy. A trivial example is the nuptial dance of many species. In Fig. 1a we show the identified elements of such dance for a ducky. There is no hope to understand the purpose of this motion, especially for an uncivilized Quumbrantapaguian, without the application of modern linguistics [10]. Although the solution is easy: The English term "nuptial dance" refers to a sequence of illogical motion which aims to attract the interest of females. Linguistics, however, is not enough to explain further details, e.g., why the sequence 3-2-3-4-4-3-5-6-7-9 dominates in many cases the dance of mallards (see Fig. 1a). Ethological studies suggest that long sequences of 1-1-1-1-1··· or 4-4-4-4-4··· would be just tiring and boring [11].

There is probably no connection between intellect and nuptial dance. An indirect proof may be based on the observation that many species of relatively low IQ present a complicated dance in order to seduce a partner, while the most intelligent inhabitant of the Earth replaces it simply by wearing Rolex wrist-watch and Ray-Ban sunglasses. It is interesting to note that dance itself, even its most awkward form, the competitive ballroom dancing, has apparently survived evolution in spite of the fact that it showed up already millions of years ago, with the appearance of reptiles (see Fig. 1b).

We can naively speculate that other senseless forms of human motion, like different sports, can help in the identification of intellect. It is easy to see that this idea does not work either. Firstly, if we omit the role of various weird things (balls, clubs, shots, javelins, etc.) in sporting, similarly to the Quumbrantapaguian observers suffering the limitations of their primitive life-detect-o-technics, the rest is absolutely indistinguishable from war dances of tribal cultures in Sumatra [12]. In many cases the situation is even worse, as illustrated in Fig. 2. Careful inspection shows that traced trajectories of individual football players (Fig. 2a) are almost identical with trajectories of swimming Australian whirligig beetles, apart from the different length-scales. (The Australian whirligig beetle is a small bug of infinitesimally low IQ, as proved in Ref. [14], where their whole activity is modeled by solving a sole equation of motion with 6 force terms.)

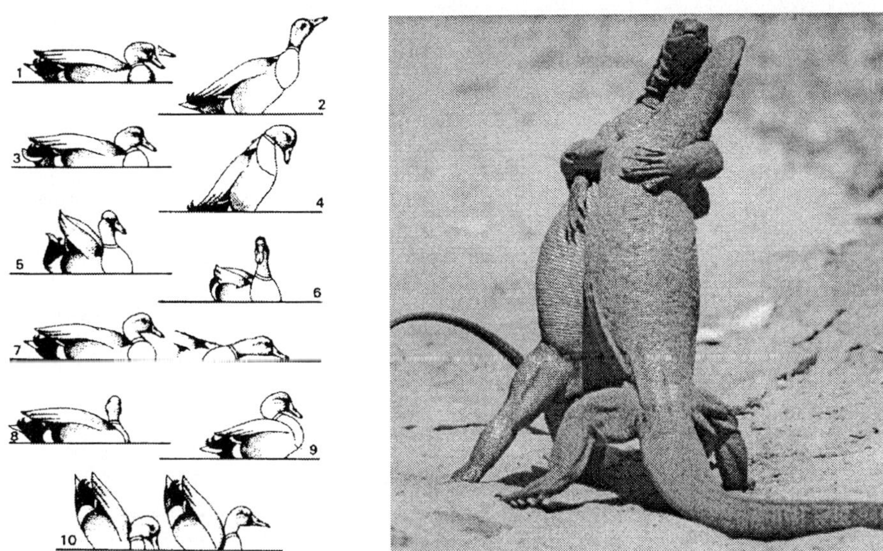

Fig. 1. (a) Elements of the nuptial dance of male mallard (*Anas platyrhynchos*) [8]. **(b)** Two individuals of the lizard *Varanus bengalensis* [9].

One might argue that quantitative trajectory-analysis should find marked differences between humans playing American football and whirligig beetles. One aspect is the possible presence of correlations. Indeed, football players sometimes tend to move in the direction of the falling ball, which results in trajectories headed for a common center. No similar focusing was observed in swarms of whirligig beetles [13]. On the other hand, focusing trajectories are widely observed in groups of Guinea pigs (*Cavia porcellus*) when they try to grab figs occasionally falling down from fig trees. It is strongly probable that the unexplainable strong attraction of some humans to football is rooted in some sort of genetic remembrance of the golden ages when our forefathers happily fought for falling papayas in the middle of a howling mass formed by other sick, old or dead-drunk individuals.

An important feature of the trajectories shown in Fig. 2 and in many other cases is *disorder*. Thus the next idea might be to seek strongly *ordered* trajectories in sporting. Unfortunately, order does not provide the ultimate answer either. If you think, for example, that the primitive but certainly ordered trajectories of swimmers in a swimming pool indicate intellect, visit the cage of wolf (*Canis lupus*) or jackal (*Canis aureus*) in the next zoo. Also, athletes running long distances in a stadium produce nice periodic trajectories, but vultures

(*Sarcoramphus papa*) and other gliding birds exploiting the lift of rising warm air over land circle for long hours, generating the same patterns.

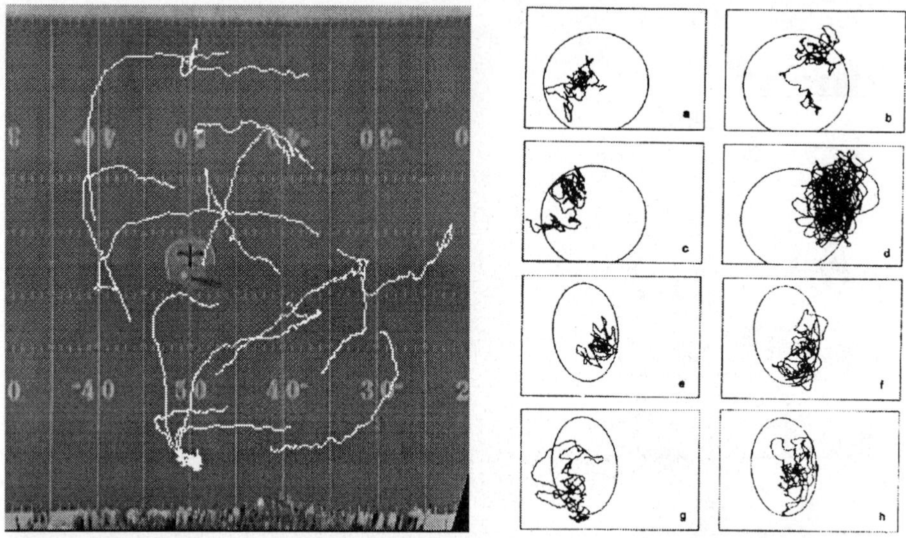

Fig. 2. **(a)** Tagged trajectories of football players. (Made by Aaron Bobick, Stephen Intille, and Anthony Hui, see `http://www.media.mit.edu`.) **(b)** Swimming paths of individual Australian whirligig beetles (*Macrogyrus*) in two different swarms (indicated by ellipses) [13].

Up to this point one might conclude that motion analysis is not a proper tool to identify intellect. Fortunately, the situation is not so bad. We have to recognize only that in the examples above the quality "senseless" is *apparent.* Several humans going in for some, eventually extremely boring sport can argue for sense, which absolutely might remain hidden for other individuals, not to mention the underdeveloped Quumbrantapaguians (by the way, this species is so primitive that even the number of its limbs is not fixed). The task is then to find an *explicitly senseless* form of motion, which is possibly *ordered* and includes *large masses* for an easy detection.

Our comprehensive scientific investigation has revealed for the first time a form of collective human motion meeting the criteria above: *Military parade march.* Let us analyze the decisive characteristics step by step.

- *Large masses.* Out of question, a nice military parade march involves as many individuals as possible. Especially in our century, the number of soldiers in parade marches has exceeded several tens of thousands in many cases.
- *Order.* The typical order of the units is a regular square lattice, as illustrated in Fig. 3. It is interesting to note that other formations like BCC, FCC or rhombohedral arrangements [15] are extremely rare. Probably even the

simple triangular lattice is too complex for most soldiers, because in this case the number of nearest neighbors to be adjusted would be too high (something around 6). It is worth to emphasize that solid lattice structure characterizes *exclusively* military parade march, civilian masses move in a very disordered way, when they tend to make parade marches (c.f. the carnival in Rio).

Fig. 3. Military parade march on China's 50th birthday. (The picture is from Chinese Military Forum, see `http://www.ohyea.com/military/index.html`).

– *The lack of any sense.* This aspect must be absolutely clear for every male who enjoyed the hospitality of military forces for a while. Several hours of exercises make transparent that the sole purpose of such collective motion is to avoid the collapse of the slowly advancing lattice structure, and around 100% of the individual efforts is occupied by the task of keeping the lattice constant as fixed as possible. (This is achieved usually by the synchronous movement of various limbs.) This point might be debated by some generals and dictators claiming that a military parade march demonstrates force, discipline, order, and organization. This argumentation, however, is fully misleading. Military parade march can trigger off threatening emotions only because we have *learned* in our history that people trained to move for rather long in a lattice structure are usually thought other things too, e.g. how to kill with high efficiency persons taking part in military parade marches abroad. The claim that at least the public enjoys such a spectacle is also false. What the public enjoys is far from being the military parade itself but the simple fact that they do not have to work or attend school in those days when such collective motions are visible, and the afternoon program is strongly connected with good foods and alcoholic drinks.

In summary, we could convincingly prove that senseless-motion analysis is a proper tool for the identification of intelligence overall in the Universe. The surprising byproduct of our research is the realization that the supreme commander who organized the first military parade march discovered a very efficient way to demonstrate *our* intellect for other beings equipped with life-detect-o-meters. This remark is a warning for the future, too: The mankind has to invent an equally senseless collective manifestation up to the time of total disarmament, in order to keep sending signs of our intelligent terrestrial life.

References

1. see: `http://www.seti-inst.edu/`
2. D.J. Schwartz, *Magic of Thinking Big,* (Fireside, 1987); T. Hewitt, *Complete Book of Cacti & Succulents,* (DK Publishing, 1997).
3. The X-Files, TM and ©1997, Twentieth Century Fox Film Corporation.
4. A.J. Haggerty and C. Haggerty, *How to Get Your Pet into Show Business,* (IDG Books Worldwide, Foster City, 1994).
5. D.T. Kingsley, *How to Fire an Employee,* (Facts on File Inc., 1984)
6. M. Fleischmann and S. Pons, *Electrochemically induced nuclear fusion of deuterium,* J. Electroanal. Chem. **26**, 301–308, and erratum, **263**, 187 (1989).
7. D. Adams, private communication.
8. A.O. Ramsey, *Behaviour of some hybrids in the mallard group,* Anim. Behav. **9**, 104–113 (1961).
9. B. Scheiba, *Schwimmen, Laufen, Fliegen: die Bewegung der Tiere,* p. 127, (Urania-Verlag, Leipzig, 1990).
10. V.F. Hopper and R.P. Craig, *1001 Pitfalls in English Grammar,* (Barron's Educational Series, 1986).
11. V. Csányi, *Etológia,* p. 180, (Tankönyvkiadó, Budapest, 1994).
12. A. Sibeth, *The Batak : Peoples of the Island of Sumatra (Living With Ancestors),* (Thames & Hudson, 1991).
13. D. Varjú, *The swarms of whirligig beetles: Gross properties and behaviour of individual members,* Recent. Res. Devel. in Biol. Cybernetics **1**, 71–89 (1996).
14. D. Varjú and G. Horváth, *Computer modelling of swimming movements and swarming in whirligig beetles.* Recent. Res. Devel. in Biol. Cybernetics **1**, 57–70 (1996).
15. D.E. Sands, *Introduction to Crystallography,* (Dower Publishing, 1994).

Anticipatory Traffic Forecast Using Multi-Agent Techniques

J. Wahle[1], A.L.C. Bazzan[2], F. Klügl[3], and M. Schreckenberg[1]

[1] Physik von Transport und Verkehr, Gerhard-Mercator-Universität, Duisburg
[2] Instituto de Informática, Universidade do Rio Grande do Sul, Porto Alegre, Brasil
[3] Künstliche Intelligenz, Universität Würzburg, Germany

Abstract. In this contribution, intelligent transportation systems (ITS) and their impact on traffic systems are discussed. Although traffic forecast offers the possibility to rearrange the temporal distribution of traffic patterns, it suffers from a fundamental problem because the reaction of the driver to the forecast is a priori unknown. On the other hand the behaviour of drivers can have a serious impact on the quality of a traffic forecast since it can result in a feedback – an anticipatory forecast is needed. To include such effects we propose a two-layered agent architecture for modelling drivers' behaviour in more detail. The layers distinguish different tasks of road users.

1 Introduction

Advanced Traveller Information Systems (ATIS) are an integral part of Intelligent Transportation Systems (ITS) [1–3]. They provide real-time information about the traffic situation to travellers in order to alleviate traffic congestion and to use the capacity of the existing infrastructure more efficiently. The individual road user can benefit from such systems since anxiety and stress associated with navigating through the network is reduced. There should be a significant overall reduction in travel time, delay and fuel consumption.

Such systems can only be successful if they are able to convince the driver to change his behaviour. Basically, there are four different possibilities: the driver can abandon his trip or try to choose another mean of transportation (modal), an alternative route (spatial) or another departure time (temporal). Since most of the road users have certain habits, there needs to be a personal advantage to change behaviour, like a shorter travel time or a more comfortable trip. Therefore, it is unlikely that somebody abandons his travel because it is connected with some utility (use or pleasure), e.g., enjoying the spare time. One basic condition for a modal change is reliable information about timetables and delays which makes public transportation more attractive (for an experimental investigation about modal choice, see [4]).

Nowadays, the strategy of most ATIS is to change the spatial distribution of traffic patterns, i.e., to provide route guidance. This method is easier than recommending another departure time because in such a case a (short-term) traffic forecast (e.g., [5,6]) or rather anticipatory route guidance is necessary [7].

The outline of this paper is as follows: in the next section we discuss the need of an anticipatory traffic forecast. In the following section we propose and describe a two-layered multi-agent architecture.

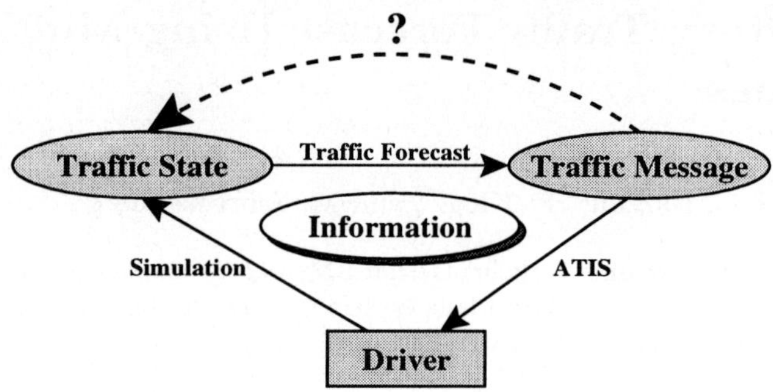

Fig. 1. Impact of a traffic message: according to the actual traffic state, messages are generated using methods of traffic forecast. These messages are transmitted to the driver using Advanced Traveller Information Systems (ATIS). The road user processes the information and changes his plans with regard to the input. The feedback on the actual traffic state needs to be evaluated using a simulation. The most important question is: what is the impact of a traffic message?

2 Anticipatory Traffic Forecast

Intelligent Transportation Systems, especially ATIS, try to affect the behaviour of road users, by providing them with real-time information about the current traffic state. Although they have reached a high technical standard, the reaction of drivers to these systems is fairly unknown. In general, real-time measurements are combined with other data, like historical time series, to generate short-term predictions. These are the basis of recommendations given to the road user by means of communication such as variable message signs or radio broadcasts. Each of these systems is confronted with a fundamental problem: the messages are based on future predictions which themselves are affected by drivers' reactions to the messages they receive. This leads to an undesirable feedback loop, depicted in Fig. 1.

This feedback can be illustrated using a very simple scenario: the day-to-day travel choice of commuters. Every morning the commuters get from home to work. For simplicity, let us assume that there are two possible routes, namely **R** and **A**, connecting these two places. Route **R** is shorter than alternative **A**. From our experience we know that on a usual day there will be an equilibrium. A high number of road users will take **R** because it is shorter and some will take **A** because it is not too crowded.

The commuters have to select between two alternatives – a binary decision. Mostly, their aim is to minimise their travel time but if too many people use the short route, it will be crowded and thus less efficient. Such a scenario is similar to the "El Farol bar problem" [8] or the minority game [9]. In [10] the simple commuter scenario is analysed by means of the minority game.

Now suppose that there is a heavy road work on **R**, the shorter route. A recommendation is given to the drivers to use **A** because of the heavy road work. They have two options: to follow the recommendation or to ignore the it. The basis of their decisions is the raw information, experience of past events, and additional suggestions about the behaviour of other road users. A possible decision-making process may be based on the following consideration: if everyone follows the recommendation, there might be no congestion at all, thus it is better to use **R**.

The basis of the recommendation was the measurement/statistics that usually more drivers take route **R**, thus there might be a congestion. At the moment this message is transmitted to the drivers the basis of the formerly correct assumption about the traffic pattern does not resemble anymore. The feedback loop is closed. Today, this feedback is not very strong since the information about the traffic state is not precise enough. Once a road user has chosen a certain route he will rarely be able to evaluate the other alternatives. But reliable information, which might be available soon, can destabilise a system since it leads to social dilemmas, i.e., situations where there is a contradiction between individual and collective aims [11].

To provide an anticipatory traffic forecast, the reasoning and reaction of the drivers has to be included. Bottom *et al.* [7] propose a framework in which every driver behaves rationally and thus, a fixed point problem has to be solved which is equivalent to find one Nash equilibrium. From experimental game theory it is known that people exhibit bounded rationality and the system does seldomly reach such equilibria [12]. For an anticipatory traffic forecast it is therefore necessary to describe the decision-making of a road user in detail by for instance employing multi-agent techniques (Fig. 2).

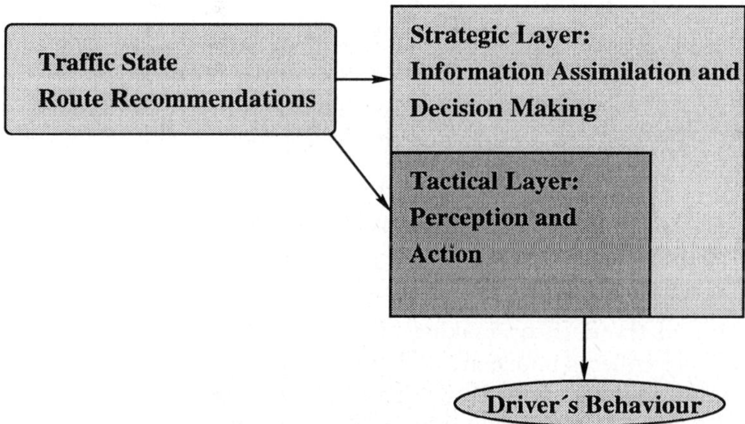

Fig. 2. Schematic sketch of the driver model. Basically, the information processing of a driver can be distinguished by two different time scales. First, a driver needs to react to the traffic situation, i.e., he accelerates, brakes, or changes the lane. Additionally, he collects information and route recommendations which result in his route choice behaviour.

3 Agent-Based Description of a Driver

Multi-agent techniques are a powerful tool to model traffic scenarios since every road user can be naturally identified as an agent [13]. However, there are only a few approaches, mainly related to the field of logistics, as far as pure applications are concerned. In the present work we propose a two-layered agent architecture to model the individual driver. It distinguishes between different tasks of the driver. The basic layer is the tactical layer which describes the task of driving. The more sophisticated problems, like the route choice behaviour are described by the strategic layer. The tasks can be distinguished by their time scale.

3.1 Tactical Layer

The tactical layer (Fig. 2) describes the perception and reaction of the driver-vehicle entity on a short time scale of about one second, the typical reaction time. In principle, every microscopic traffic flow model can be used to describe this layer because, in contrast to macroscopic models, a driver is identified as basic entity and its behaviour, for instance car-following, is modelled.

One example of this class is the Nagel-Schreckenberg model [14]. The cellular automaton can be directly interpreted as a multi-agent system with reactive (sub-cognitive) agents. The driver-vehicle entity (agent) reacts to the perception of its own velocity and the headway gap. The behaviour of an agent is in the simplest case (single lane traffic) specified by the following update rules: it checks the distance to the predecessor within a limited range and calculates the velocity for its movement incorporating a probability- and agent-dependent reduction of the velocity. For such a rather simple behaviour, no cognitive architecture is necessary.

We re-implemented the Nagel-Schreckenberg model using the multi-agent simulation environment SeSAm (Shell for Simulated Multi-Agent Systems), described in [15]. In several simulation experiments we were able to show that the multi-agent model of the cellular automaton reproduces the original behaviour with sufficient accuracy.

3.2 Strategic Layer

The strategic layer extends the basic layer and is responsible for the information assimilation and the decision-making of a driver (Fig. 2). During and before a trip, a road user collects information in many ways, for instance by radio broadcast or variable message signs. If the driver has to select between different travel alternatives he uses the collected information and his experience or attitudes. In most models, perfect rationality and utility maximisation is assumed for such problems. But in the commuter scenario discussed above there is no optimal solution, the process is highly dynamic and it depends on the behaviour of the others.

Additionally, ATISs and other intelligent devices will provide even more information about link travel times, densities, road works, or route guidance in the near future. Thus drivers have to collect even more information and evaluate it with a higher frequency. This clearly indicates that understanding travellers' route choice behaviour is an important consideration for the development and effectiveness of such systems.

There are several different techniques to describe such problems [7,10]. In general, the decision-making process in human beings is based not only on rational elements, but also involves some emotional components that are typically non-rational, like motives or social constraints. As a result, behaviour can also be explained by approaches, which additionally consider beliefs, desires or intentions, the so-called BDI-formalism, which is well-known in the field of multi-agent systems.

Such a formalism for a simple commuter scenario is proposed in [13]. The drivers are represented by their individual mental states. One road user can trust in the information another does not, or only occasionally. Apart from that they have an individual knowledge base and a certain set of plans. A possible plan could be to leave earlier to avoid being late; another one takes the risk and stays in bed longer. The knowledge base contains for instance navigational information: a driver who is familiar with the network topology has more options for his decisions.

It becomes clear that the description of the strategic layer requires very sophisticated methods and that it is crucial for development of intelligent transportation systems. The starting point is the understanding of the human behaviour. New results will be gained from investigations in this field which employ experimental economics [12].

4 Conclusions

This paper discussed intelligent transportation systems, especially ATISs, and their impact on the traffic patterns. Recent traffic control systems try to recommend alternative route, i.e., to change the spatial distribution in the network. A traffic forecast offers the road user a new degree of freedom since he has the opportunity to take advantage of choosing the departure time. This might affect the temporal distribution of traffic in a desirable way.

Nevertheless, all methods of traffic forecast face a fundamental problem: traffic messages can lead to an undesirable feedback (Fig. 1), which is capable of destabilising the traffic state due to the reaction of the road users. We analysed this feedback loop considering a simple commuter scenario and introduced the concept of an anticipatory traffic forecast, which in addition includes human factors in the predictions.

In order to tackle such a demanding task, existing models have to be expanded to account for human behaviour, especially route choice. We propose a multi-agent system where every agent consists of two-layers, namely the tactical and the strategic layer. The tactical layer processes the perceptions of the road user and describes its actions, i.e., acceleration, braking, etc. This layer is modelled by every microscopic traffic flow model, for instance the Nagel-

Schreckenberg cellular automaton. The strategic layer accounts for information assimilation and decision-making of drivers. Such decisions imply social dynamics which might be implemented by a BDI-architecture. In this formalism every driver is represented by a set of mental states, his beliefs, desires, and intentions (BDI). Such a model can be used to improve the quality of traffic forecasts. In the future we try to analyse human factors in intelligent transportation systems, especially human machine interfaces, by means of experimental economics.

Acknowledgement. We would like to thank F. Puppe and R.H. Bordini for useful discussions and sharing insights. This research is sponsored partially by the DLR (German Nat. Aerospace Research Centre) and the CNPq (Brazilian Nat. Council for Research and Technology) within the project SOCIAT.

References

1. J.L. Adler and V.J. Blue, *Toward the design of intelligent traveler information systems*, Transpn. Res. C **6**, 157 (1998).
2. W. Barfield and T.A. Dingus, *Human Factors in Intelligent Transportation Systems* (Lawrence Erlbaum Associates Inc., Mahwah, New Jersey, 1998).
3. ITS International, *Proc. of the 6th World Congress on Intelligent Transport Systems* (ITS World Congress, CD-ROM, Toronto, 1999).
4. H. Gorr, *Die Logik der individuellen Verkehrsmittelwahl* (Focus Verlag, Gießen, 1996).
5. B.S. Kerner, H. Rebhorn, and M. Aleksic, *Forecasting of traffic congestion*, in: *these proceedings.*
6. B. Schürmann, *Application of neural networks for predictive and control purposes*, in: *these proceedings.*
7. J. Bottom, M. Ben-Akiva, M. Bierlaire, I. Chabini, H. Koutsopoulos, and Q. Yang, *Investigation of route guidance generation issues by simulation with DynaMIT*, in: *Proc. of the 14th Int. Symp. on Transp. and Traffic Theory*, A. Ceder, (Ed.), pp. 577–600 (Pergamon, Amsterdam, 1999).
8. W.B. Arthur, *Inductive reasoning and bounded rationality*, Am. Econ. Rev. **84**, 406 (1994).
9. D. Challet and Y.-C. Zhang, *Emergence of cooperation and organization in an evolutionary game*, Physica A **246**, 407–418 (1997).
10. A.L.C. Bazzan, R.H. Bordini, G.K. Andrioti, R.M. Vicari, and J. Wahle, *Wayward agents in a commuting scenario (personalites in the minority game)*, in: *Proc. of the 4th Int. Conf. MultiAgent Systems (ICMAS'2000)*, accepted, IEEE Computer Society.
11. R. Berkemer, *Modal split and social dilemmas*, in: *these proceedings.*
12. R. Nagel, *Unraveling in guessing games: An experimental study*, Am. Econ. Rev. **85**, 1013–1026 (1995).
13. A.L.C. Bazzan, J. Wahle, and F. Klügl, *Agents in traffic modelling – from reactive to social behaviour*, in: *KI-99: Advances in Artificial Intelligence,* W. Burgard, T. Christaller, and A.B. Cremers, (Eds.), (LNAI 1701, Springer, Berlin, 1999).
14. K. Nagel and M. Schreckenberg, *A cellular automaton model for freeway traffic*, J. Phys. I France **2**, 2221 (1992).
15. F. Klügl and F. Puppe, *The multi-agent simulation environment SeSAM*, in: *Simulation in wissensbasierten Systemen* (Universtiät Paderborn, 1998).

Modal Split and Social Dilemmas

R. Berkemer

ITV Denkendorf - Department of Management Research, 73770 Denkendorf, Germany

Abstract. Traffic systems in large cities might be viewed as interdependent decision situations. Natural N-person extensions of familiar games may serve as templates to cover some system relations. An agent based simulation model is proposed. Several decision levels are considered in the model in order to cover aspects of institutional framework.

1 Introduction

Social dilemmas are situations, where there is a contradiction between individual and collective rationality. For instance out of several million car drivers in Tokyo or New York no single individual has a measurable influence on environmental quality of the metropolis. Hence, it is reasonable to neglect air pollution when individuals have a remarkable advantage by using the car (high flexibility, short travel time, etc.). However, it might lead to a collective disaster if everybody neglects environmental issues in the same way.

In this paper a model is proposed, where a large number of agents are confronted with similar rationality traps. The economic concept of external effects might help us to classify some interesting situations which are relevant in metropolitan traffic systems. This is elaborated in Sect. 2. In Sect. 3 several decision levels are introduced. However, for the sake of brevity further discussion will focus on only two levels of decision: a "modal choice level", where agents have to choose between private car and public transport, and an individual "context choice level", where an agent defines its context for the modal choice. Section 4 describes the dynamics for adjusting corresponding agent strategies. The dynamics is closely related to imitation models widely used in the literature on evolutionary game theory [1,2]. However, it is new that two decision levels are covered. Section 5 outlines how the conception might be stepwise exploited in order to gain understanding of critical parameters. Section 6 concludes.

2 External Effects and Social Traps

Suppose for a moment that the entire agent population is divided into two subgroups (car users, and public transport users). The following cases lead to interesting social traps:

1. Members of one subgroup impose negative external effects to the entire population: Particularly the subgroup of car users does so by causing air pollution and noise, which reduces utility of anyone. It is likely that this situation will

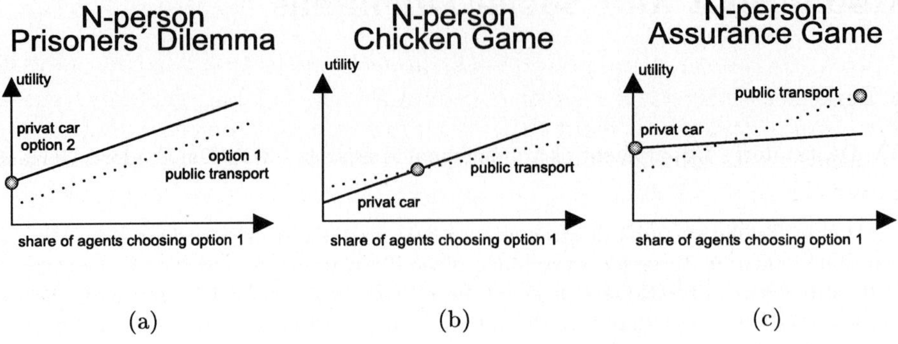

(a) (b) (c)

Fig. 1. Social Trap Situations.

result in a N-person Prisoners Dilemma as illustrated in Fig. 1a. Notice that utility of both subgroups (car users and public transport users) increases with number of agents using public transport. However, the car using option strictly dominates the public transport option. Hence, game theorists will predict the strict Nash equilibrium (indicated by a circle) as the only individually reasonable outcome. No agent should be expected to use public transport. However, anyone using public transport would lead to a better result for all. The situation might be interpreted as one where any agent would advocate in principle traffic calming in the total system, but no agent likes to be restricted in his particular domain.

2. Members of one subgroup impose negative external effects to their own subgroup:

   Again the utility of both subgroups increases with number of agents using public transport. However, there is no dominant strategy any longer. The resulting situation can be labeled as N-person extension of the "Game of Chicken" which is strategically equivalent to the "Hawk-Dove-Game" known from theoretical biology. Special attention might be assigned to such situations when members of one subgroup compete for resources (e.g., car users fighting for parking sites). Game theory now predicts a mixed Nash equilibrium as indicated in Fig. 1b.

3. Even positive external effects may lead to a social trap (see "Assurance Game" in Fig. 1c):

   In this case members of one subgroup (public transport users) might increase the utility of their own subgroup. More people switching to public transport might lead to more stop stations, and shorter cycle times. This may attract even more people which would reinforce this effect. However, there are two stable Nash equilibria and it is not sure whether evolution leads to the more efficient one.

3 Decision Levels and Institutional Framework

In Fig. 2 the interrelations between several decision levels are roughly drafted. Further discussion will restrict on the impact context choice will have on the modal choice level. The context choice is still a decision level where an individual can decide on its own. Essentially individuals choose what distribution of fixed and variable costs will be relevant. One can reduce fixed costs of a private owned car by joining a car sharing project. However, this will lead to higher variable costs when the car has actually to be used. On the other hand one will reduce variable costs of public transport by acquiring a year ticket. For the several

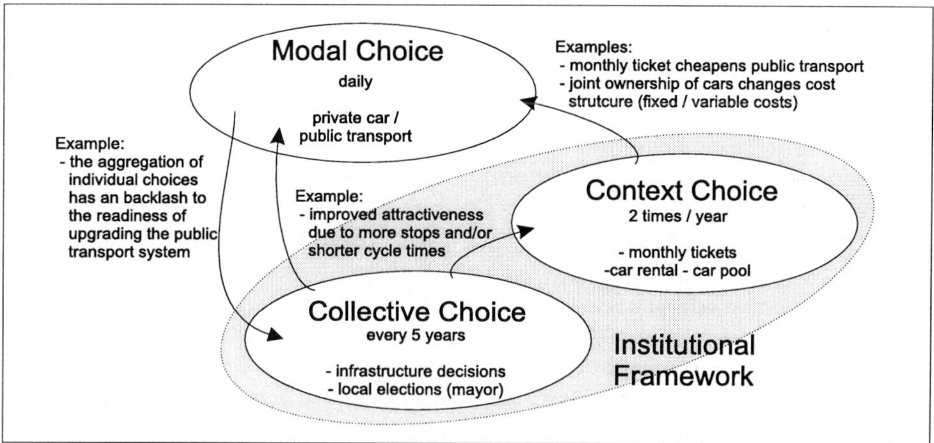

Fig. 2. Three Decision Levels.

decision levels we provide agents with different utility functions. We introduce $u_i(.)$ for the modal choice level and $f_i(.)$ for the context choice level. $u_i(.)$ should only consider variable costs and will be derived from a superposition of the three social trap situations mentioned in Sect. 2. $f_i(.)$ must include also fixed costs (e.g., the insurance of the car).

4 The Dynamics of the Adjustment Process

We employ an agent based simulation model. Consider a population of N agents. Initially each agent is endowed with a strategy for every decision level discussed above.

For a framework simulating agents with multi-level decisions we propose following rules from which an adjustment dynamics should be derived:

- On each decision level there might be two kind of events,
  - applying a strategy,

   - or reviewing a strategy (and eventually adjusting it).
- Rule to connect decision levels,
   - whenever a strategy on decision level l is applied this event will be connected with the event that the strategy on decision level $(l-1)$ is reviewed (but there might be additional reviews on level $(l-1)$).
- There might be two kinds of adaption processes,
   - "naive" adaption - in the evolutionary game theory literature various models are discussed, which require only low rationality from agents. Many models employ imitation behavior [1,3] as we will also do.
   - "sophisticated" adaption, which will require from agents somewhat more deliberation.

Notice that these general rules might be applied to models with an arbitrary number of decision levels. However, restricting on two decision levels and turning back to the modal split model we state following assumptions (first for the modal choice level):

We assume that agent behavior reveals some inertia. In each time period (e.g., once per day) agents usually apply their modal choice strategy (e.g., they are used to travel every morning to work with their private car and so they are expected to do so also next morning). However, from time to time agents review their modal choice strategy.

Formally, we state that in each time period some fraction α_1 of the total population is drawn randomly to do so. When agent i is required to review his strategy he samples another randomly chosen agent j and asks j for her experience (travel time, traffic jam, accidents, etc.). Notice that agent i asks not for her utility but applies his utility function $u_i(.)$ to the experience of agent j. The difference between the actual experienced utility of agent i and the utility i would have experienced, given the modal choice strategy of j will be used as input of a probabilistic switching function. The greater utility difference the more likely agent i will switch to the strategy of agent j. So far the adjustment dynamics on decision level 1 (modal choice). Turning to decision level 2 (context choice) we state following assumptions:

In each time period an agent is required to make a context choice with some probability β. Again we assume behavioral inertia, i.e. a given context choice strategy is just applied. For instance, one has to buy a new monthly ticket for the public transport system, since validity of the old one has just expired. Similarly a minor repair of the own car is done without reflecting about alternatives. However, from time to time agents are assumed to reflect more. The price for the monthly ticket might have increased remarkably. An accident may require a more expensive repair. Such events might cause agents shifting to another context.

Given the event that an agent has to make a context choice with some probability α_2 he is required to make this choice deliberately (i.e. to review his context choice strategy). Thus we can state that in each time period some fraction $(\alpha_2\beta)$ of the total population is drawn randomly in order to review its context choice strategy. When agent i is required to do so he samples another randomly chosen agent j and asks j for her experience (e.g., whether the investment connected with a context decision has been amortized).

He applies his utility function $f_i(.)$ to the experience of agent j. The difference between the actual experienced utility of agent i and the utility agent i would have experienced, given the context choice strategy of j will be used as input of a probabilistic switching function.

Up to now the adjustment process described is very simple. Agents observe other agents and the more promising the observed strategy the more likely there will be imitation. However, recall that there has been also claimed the need for "sophisticated" adaption which will be motivated next.

Suppose that there has been a radical change on the context choice level (e.g., the private car was sold and a year ticket was bought). It is not plausible that the respective agent will thoughtless use his actual experience (related to a completely different context) when reviewing his modal choice strategy next time. Rather we would expect that the change on the context choice level was made with detailed intentions how to behave on the lower level (i.e. the public transport option is intended in the example above).

In order to model such kind of behavior we have to endow agents with the ability to anticipate consequences of a context change at least partly. We formalize this as follows:

- We require that a strategy change on the context level should immediately force a review of the depending modal choice strategy.
- For this "special" review the simple utility difference mechanism is not sufficient. Particularly, the agent should not use his actually experienced utility $u_i(.)$ for comparison. Rather he should use some utility $\tilde{u}_i(.)$.
- We define: $\tilde{u}_i(.) = \lambda u_i^n(.) + (1-\lambda)u_i(.)$ where $u_i^n(.)$ denotes the utility agent i would have experienced if he would have chosen the new context already in the past and $\lambda \in [0,1]$ denotes the "level of sophistication".

Notice that $\tilde{u}_i(.)$ might be interpreted as "anticipated utility". $\lambda = 1$ corresponds to the case where agents are able to predict consequences of context changes accurately. With $\lambda = 0$ there is no sophistication at all.

5 Outlook

For further investigations the following procedure is planned:

- In a first step we will fix both higher decision levels completely. That means there will be no change on the collective choice level and even on the context choice level we will provide all agents with the same decision context, which will never change. Thus there will only occur "naive" adaption on the modal choice level. By introducing further simplifications (identical utility functions, switching probabilities are growing linearly in utility difference, sufficiently large population size N, etc.) one can derive analytical results. We cannot elaborate on this in detail here. However, some hints can be given.

As N gets larger and larger one can apply the strong law of large numbers. On the other hand proceeding to shorter and shorter time periods (which implies $\alpha_1 \to 0$) one can pass from discrete to continuous time. In a nutshell - with appropriate simplifications one finally ends up with the replicatory dynamics (which is well known from evolutionary game theory). Hence, we will be able to predict the results of the stochastic (time discrete) model in advance.

- In a second step we may still fix the context choice level but operate with a heterogeneous population. Agents might be provided with different decision contexts, which will not change. It can be investigated whether agents can efficiently adapt different modal choice strategies to their exogenously given context (i.e. whether agents will build subpopulations).
- In a third step we should allow for endogenous heterogeneity (i.e. agents will also adjust their context choice strategy). By reaching this step one would be able to investigate the central issues the model is designed for. Will agents be able to learn the appropriate modal choice strategy? How relevant is the level of sophistication? Will a variety of different decision contexts facilitate the partial switch to public transport at least for subpopulations?

6 Conclusion

The paper considers several social traps as being relevant in large cities traffic systems. It is emphasized that daily decisions (car or public transport) are made in a specific context. It is up to the individual to define this context. The relationship between context choice and (daily) modal choice is explored. An agent based simulation model is proposed which employs an imitation dynamics for the adjustment process. It can be shown that specific simplifications might enable analytical results. This is extremely valuable concerning internal validation of the model.

References

1. K. Schlag, *Why imitate, and if so, how? Exploring a model of social evolution*, Discussion paper B-296, University of Bonn, 1994.
2. J. Weibull, *Evolutionary Game Theory,* (Cambridge, Mass., 1995).
3. D. Fudenberg and D.K. Levine, *The Theory of Learning in Games,* (Cambridge, Mass., 1998).

Synthesising the Brillouin Information-Thermodynamic Approach with the Social Synergetics Weidlich-Haag Model

J.Z. Hubert

Dept. of Structural Research, H. Niewodniczanski Institute of Nuclear Physics IFJ, Radzikowskiego 152, 31-342 Krakow, Poland

Abstract. The model is an attempt to synthesise the Weidlich-Haag social synergetics probabilistic approach with Brillouin's information-thermodynamics method of reasoning. It proposes mathematical modelling *and* physical explanation of one of the basic human and social phenomena: The need of **change** — change for the sake of change (without visible *external* motivations and reasons; reasons and motivations which in the Weidlich-Haag model are expressed by the traditional concept of utility widely used in economics). The computations make use of the Monte-Carlo method, in which the histories of each individual are followed. The results are discussed in terms of really observed social phenomena. The model can also be regarded as an attempt to "motorise" the elements of a stochastically behaving complex system composed of independent elements with free energy depots. This would extend the approach of other contributors to the TGF 99 conference extending their models of stochastic movement in the geometrical space into motion in the axiological space.

1 Introduction: Aim, Model, and Method

In the social synergetics models discussed here, in front of each individual we have a certain number of options. Each of these has a certain probability p of realisation. A social pattern or structure has been realised if some choices have been made: each individual has chosen a given attitude towards a given *aspect*, i.e., to a given domain of life, of activities etc.
Now, the following questions may be asked:

- When does a pattern have a high probability of realisation? When is a low probability (exceptional) pattern stable, i.e., has a high probability of continuation?
- What thermodynamic conditions warrant such a situation?
- And finally, a question which we shall try to answer explicitly in this paper: How to account in a synergetic model of a society for such basic social phenomena like (cyclic) changes of fashion or changes of artistic styles and maybe even of changes – understood as in Kuhn's generalisations – of currently prevailing paradigms. In short: How to explain **changes within** a system without pointing to a **visible change of the external demands and conditions** in the social and physical environment? What **internal dynamic processes** are responsible?

To try to answer these questions we shall treat the social system as a system in which at the same time (meaning: *not only*) physical and probabilistic laws operate, among others the laws of non-equilibrium thermodynamics and information-thermodynamics.
In our analysis we shall proceed in two steps:

1. Formulate these laws using the concepts and terminology of the synergetic social model (SSM for short — this abbreviation will be used below), as formulated by Weidlich and Haag [1–3].
2. Draw conclusions. In particular try to see how the second law, or rather its generalisation found in Brillouin's Principle [4–6] (stating that information or information negentropy gain must always be paid for by energy dissipation) would affect the behaviour of crucial quantities of the SSM model (like the individual trend parameter, the opinion pressure, the trend influence strength, the affirmation (dissidence) strength and *above all*, a visible and measurable quantity, the attitude distribution among the population).

Thus the main objective and content of this paper may be summarised in two points:

1. Presentation of the main assumptions and of the reasoning sequence implied by the above mentioned synthetic approach. (This will be done in the first part of the paper: sections 2 and 3).
2. Quantitative application of this synthetic approach to a simple model of changes of binary social opinion.

2 Basic Assumptions and Mechanisms Related to Brillouin's Principle

1. Selection and choice of a particular attitude vector (included in any decision process) out of the full set of all possible options means a decrease of the entropy of the system. The decrease of (Shannonian) entropy is denoted by ΔH.
2. If a choice is repeated n times, ΔH is a function of n:

$$\Delta H = f(n) . \tag{1}$$

3. According to Brillouin's Principle this selection process may be realised only at the expense of a thermodynamic negentropy NT. What is the source of this negentropy? It seems that no other source may be pointed to except the subject itself performing the selection process. NT is released in metabolic processes ongoing in the human body and transformed into information (Shannonian) negentropy NI by activities of the human brain.
4. Essential for the whole reasoning is the following point: The efficiency of the transformation process, i.e., both the quantity of thermodynamic negentropy furnished by the metabolic processes and of (obtained in all kinds of non random activities) the information negentropy is, as observed (among others) in everyday experience, *not a constant.*

It depends both on the characteristics of the subject (his/her internal preferences) carrying out a given the selection process *and* on the characteristics of the goal being pursued (which, in turn, conditions the selection process itself). It is as if the "internal transformer" of the negentropy had some preferred direction of transformation. Activities which are in tune with the internal tendencies (preferred directions) increase the transformation power and those which are not decrease it.

Formalising what has been said above we can write a negentropy balance equation:

$$NI - \Delta H = NI_S. \tag{2}$$

The first term on the left hand side NI represents the information negentropy which has been produced (transformed) by an individual and the second one what has been used in the selection and choice processes.

For an option which is "liked" or very welcomed by an individual, arousing his/her joy and creative forces we might expect that it will positively catalyse the negentropy transformation rate and NI will exceed the expenses of negentropy ΔH which, as noted by (1), is (an increasing) function of the number of decision nodes n:

$$NI - \Delta H = NI - f(n) = NI_S > 0. \tag{3}$$

Accordingly for a "disliked" or boring option we might expect that $NI_S < 0$. NI_S is called "stored" negentropy[1].

Connection with the Weidlich-Haag model is made through the (internal) trend parameter Θ. It is assumed that that individual's interest, i.e., the tendency to continue choosing a certain option, is proportional to the **net** amount of negentropy that this choice brings along, thus maximising negentropy transformation rates. Expressing this with the use of the concept of a trend parameter employed in the SSM means that we have

$$\Theta \sim NI - \Delta H = NI - f(n). \tag{4}$$

A conviction that complex systems tend in their development to maximise the intake of free energy from their environment and then, at constant or minimised dissipation, maximise the "production" of information (structural) negentropy, has been held by many authors. Alfred Lotka[2] has written about these maximisation phenomena in economic processes already in the thirties. Odum has applied it to the theory of evolution (to survival competition among different species). Kirkaldy [9] used the thermodynamic extremum principles to predict the final form of **macro**scopic patterns spontaneously arising in **physical systems** transforming far away from the thermodynamic equilibrium, like Benard

[1] The word "stored" cannot be taken too literally. However, NI_S has the same information thermodynamic meaning like the notion of (information) structural negentropy stored in dissipative structures in physical systems.

[2] The same Lotka who was co-author of the Lotka-Volterra equations.

cells, patterns arising in germanium at low temperatures [9] or in pearlite developing in phase transformations in iron [10,11]. Hubert applied the concept of the negentropy transformation and of the negentropy balance equation (2) to the theory of action (praxilogy) and the principle of negentropy maximisation to the theory of creativity [8].

3 Application to the Binary Opinion Formation and Decision Process

It is possible to specify the expression for the **binary opinion formation and binary decision** process, i.e., in front of each individual there are only two possible options. In this case it can be shown that $f(n) = n$.
Let us now assume that the decision moments are equally spaced over the time period t. Then:

$$n = t/\Delta T, \tag{5}$$

(where ΔT is the time between two subsequent decisions). Of course NI is also a function of time $NI = NI(t)$. We shall express it in terms of the quantity denoted as negentropy transformation power P_T. In the case of a linear relationship and taking into account (3) and (5) we get

$$NI_S = (P_T - 1/\Delta T)t + NI_0. \tag{6}$$

where we introduced NI_0 as the amount of information negentropy at disposal of an acting individual at time $t = 0$.

We shall use the explicit form of the transition probability p_{ik}, as proposed in the SSM, for a member of the population to go from opinion "k" to opinion "i" in one unit of time. Taking into account (4) and (6) we obtain:

$$p_{ik} = \nu \ \exp\{-[(P_T - 1/\Delta T)t + NI_0]\}, \tag{7}$$

where ν is a social mobility factor. Note that in the Weidlich-Haag model p_{ik} and p_{ki} are symmetrical, i.e., Θ is for both transition probabilities *the same*, differing only by sign. This comes from the fact that in the transition probabilities *of this model* Θ measures *the difference* of utilities Θ_i and Θ_k of two options (or of a personal preference proportional to the subjective utility of a given option).

In the presented model we are rather trying to describe "**a need to change**". This need is not necessarily connected with a perception of another option to become better or the *objectively* worsening of the quality of the current option. Therefore a person staying at the option "i" is characterised by an internal trend parameter Θ_i expressing his/her *subjective tendency ("pressure")* **to leave** this option or an interest **to stay at it**.

4 Computational Procedure

The history of each individual is followed using Monte-Carlo method computer simulations. The simulation proceeds as follows:

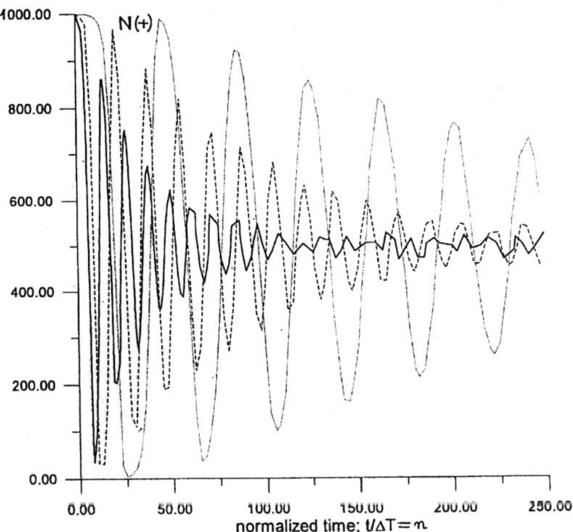

Fig. 1. The vertical axis shows the number of people espousing the option (+). Initially it is the whole population (i.e., 1000 persons in this example). All curves converge to the asymptotic point with $N(+) = N(-) = 500$. These curves differ only by the value of frequency of decision nodes, i.e., of $1/\Delta T$. However for all of them $P_T < (1/\Delta T)$. Shown are the curves for $\Delta T = 0.1$ (solid), $\Delta T = 1.0$ (dashed), and $\Delta T = 10.0$ (finely dashed).

random numbers, between 0 and 1, are drawn at each time step for each individual. If a number is smaller than the mean transition probability given by formula (7) then the individual goes over from the k-th to the i-th level. If it is larger it remains at this level.
After each transition the time t in formula (7) is again set to zero.

This depicts in this model a psychological situation corresponding to replenishment (just after transition) of the initial negentropy stock. (Of course this is so if $(P_T - 1/\Delta T) < 0$, then for $t = 0$ the probability of change is very small[3]). In the new situation, by the force of *stimulation by novelty* an individual regains his "psychic (or moral) energy" to face the new decision situation and to pass over the coming along decision nodes before the next transition takes place.

This is only one possible psychological situation which could be modelled. In other situations the time t needs not to be set to zero after each transition. For example we can model the *dynamics of change* by making NT a function of time and of the choice of the option. NT could also be different for different individuals and it could depend on their interaction (which in Weidlich-Haag model is expressed by the parameter k). This shows many possibilities for adapting this

[3] The negentropy NI_0 is always positive.

model for the information-thermodynamic description of various psychological and social situations happening in real life.

5 Interpretation of the Results and Discussion

Parameter values: Various sets of parameter values have been considered - each supposedly describing a different psychological and social conditions of selection and choice process. The parameters are constant for each set. For all individuals ($N = 1000$), the initial "negentropy stock" was set to $NI_0 = 10$.

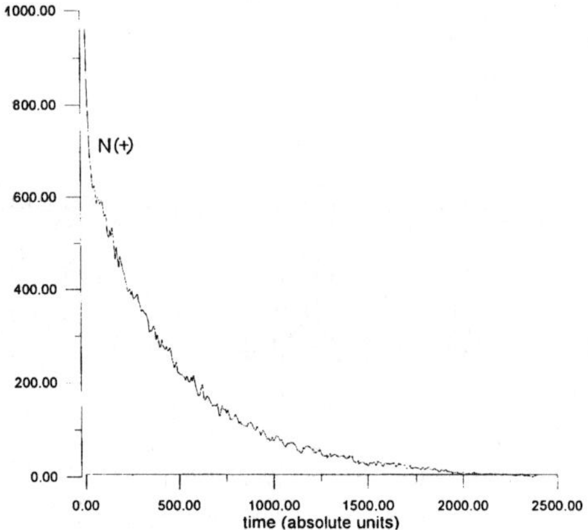

Fig. 2. Here is modelled a situation with two different values of P. For one option it is equal 0.05 for the other it may be any other value bigger than 0.1 (i.e. any other value for which T). $\Delta T = 10$, $P_T(+) = 0.05$, $P_T(-) > \frac{1}{\Delta T}$

The Changing parameters: ΔT: time interval between any two successive "decision nodes", $P(i)$: negentropy transformation rate at level "i", initial division of population between the levels (+) and (−).

Interpretation and individual meaning: The validity of this relationship means that the supply of negentropy is always smaller than its demand (caused by requirements of the selection process). The selection proceeds at the cost of the initial negentropy stock NI_0. When it is exhausted and the argument of the exponential function in (7) reaches 0 the transition probability p_{ik} approaches 1 and an individual may be likely to change its original option. Relatively small P_T, in comparison to $(1/\Delta T)$, means that an individual is not flourishing in any option. Sooner or later, if an occasion presents itself, he/she will change to the opposite option. There is no choice which will be good for him/her.

Social meaning and historical illustration: Comparing the figures it is easily noticed that when ΔT decreases the frequency of oscillations increases. In the social reality the former happens in a more "dense" social milieu with more social contacts, interactions and the increase of the information flow rates. This is exactly what has been going on in the last few hundred years. We can see it if we look at the history of art, of fashion changes or of political options. The pace of changes is markedly quicker now than it was in the middle ages or in antiquity. Technical achievements, like electronic means of communication and mass long distance travelling have dramatically expanded the global information flow and this has caused deep and fast developing social changes and transformations. The latter, in a *very simplified* way, is expressed by the above given information-thermodynamic abstract construct and reasoning.

Figure 2 shows a "non-return" situation. The preference for the option (−) is strong enough to overcome the "call" for the reverse choice at every decision node. Eventually all people leave the option (+) and settle at (−).

References

1. W. Weidlich, "Quantitative Social Science", Physica Scripta **35**, 380-387 (1987).
2. W. Weidlich, G. Haag, "Concepts and Models of a Quantitative Sociology", Springer Series in Sygernetics, Vol.14 (Springer, Berlin, 1983).
3. W. Weidlich, "Physics and Social Science — the Approach of Synergetics", Physics Reports **204** (1991).
4. L. Brillouin, "Science and Information Theory" (Academic Press, New York, 1962).
5. O. Costa de Beauregard, "Le Second Principe de la Science du Temps. Entropie. Information. Irreversibilité" (Aux Editions du Seuil Paris, 1967).
6. I. Prigogine and I. Stengers, "Order out of Chaos: Man's New Dialogue with Nature" (Heinemann, London, 1984).
7. J.Z. Hubert: "Negentropy, action and praxiology: definitions and general systems", (in Polish), "Prakseologia", **3** (59), (1977).
   "Negentropy, action and praxiology: conditions and possibilities - a semi-quantitative analysis", (in Polish), "Prakseologia", **1/2** (61/62), (1977).
   "Negentropy, action and praxiology: freedom and decision", (in Polish), "Prakseologia", **2** (66), (1978).
8. J.Z. Hubert, "Creativity and discovery in scientific thought: in search of an algorithm", D&H **7**, 2, (1980).
9. J. Kirkaldy, J.Z. Hubert, "Steady-state patterns in an exciton gas spinodal decomposition to a Fermi liquid in germanium", J. Phys. C **20**, 1393-1411, 1987.
10. J.Z. Hubert, "Thermodynamic optimisation within stationary macroscopic structures in metals and alloys", Scientific Bulletin of the University of Mining and Metallurgy, **1138**, Metallurgy and Foundry Practice Series 109, (Kraków, 1987).
11. J.Z. Hubert, "Computational applications of the thermodynamic local potential: the case of pattern formatting systems", International Centre for Theoretical Physics Report, Ic/86/255, (1986).

Biology, Internet, Transport Theory

Evolution of Molecular Phenotypes – A Physicist's View of Darwin's Principle

P. Schuster[1,2]

[1] Institut für Theoretische Chemie und Molekulare Strukturbiologie, Universität Wien, Währingerstraße 17, 1090 Wien, Austria
[2] Santa Fe Institute, 1399 Hyde Park Road, Santa Fe, NM 87501, USA

Abstract. The power of Darwinian evolution is based on the dichotomy of genotype and phenotype with the former being the object under variation and the latter constituting the target of selection. Only the simplest case of an evolutionary process, the optimization of RNA molecules *in vitro*, where phenotypes are understood as RNA structures, can be handled explicitly. We derive a model based on differential equations with stochastic terms which includes unfolding of genotypes to yield phenotypes as well as the evaluation of the latter. The relations between genotypes and phenotypes are understood as mappings from sequence space into shape space, the space of molecular structures. Generic properties of this map are derived and analyzed for RNA secondary structures as an example. The optimization of molecular properties in populations is modeled *in silico* through replication and mutation in a flow reactor. The approach towards a predefined structure is monitored and reconstructed in terms of a relay series being an uninterrupted sequence of phenotypes from initial structure to target. Analysis of the molecular shapes in the relay series provides the basis for a novel definition of continuity in evolution. Discontinuities can be identified as major changes in molecular structures.

1 Molecular Phenotypes

Evolution of asexually replicating individuals in the sense of Darwin's principle is characterized by the interplay of two processes which have counteracting influences on heterogeneity of populations: (i) Mutations increase diversity of genotypes, and (ii) selection decreases diversity of phenotypes[1]. In sexually reproducing populations recombination acts as an additional process increasing diversity. It is important to realize that variation and selection operate on different manifestations of the individual, genotype and phenotype, respectively. At a first glance, decoupling, or more precisely uncorrelatedness of the targets for mutation and selection may seem to be a disadvantage: then, a mutation does not occur more frequently because it has a better chance to become selected. Considering the success in non-biological complex optimization problems, however, random variation is well known to be a powerful strategy. Separation of

[1] The genotype is understood as the poly-nucleotide sequence that carries the genetic information of the organism, DNA, or RNA in the case of several families of viruses. The phenotype is the adult organism which enters the reproductive phase and determines thereby fitness commonly understood as the number of progeny transferred into the next generation.

genotype and phenotype occurs trivially in all higher forms of life where the phenotype is an adult multi-cellular organism created through unfolding of the genotype in a manner that reminds of the execution of a computer program. In the case of uni-cellular organisms, procaryotic or eucaryotic, the phenotype comprises cellular metabolism in its full complexity which, in reality, is not yet deducible from the DNA sequence, and thus stays in contrast to the genotype. *In vitro* evolution deals also with optimization in populations of molecules which are capable of replication. In this case the distinction between genotype and phenotype is more subtle.

In his pioneering experiments Sol Spiegelman [2] studied evolution of RNA molecules in the test-tube (Fig. 1). The rate of RNA synthesis increases by orders of magnitude in serial transfer experiments. Spiegelman identified the sequence of nucleotides in an RNA molecule as its genotype and the molecular structure as its phenotype. Genotype and phenotype thus are two different manifestations of the same molecule, known to the biochemist as primary and spatial structure, respectively. Here, we cannot be sure *a priori* that genotype and phenotype are truly distinct features. Considering RNA folding in detail, however, we realize that structure formation is a highly complex process that does not (yet) generally allow to infer structural changes from mutations in the sequence. At best we have to go through a complex algorithm that predicts structure from sequence (Fig. 2). A characteristic of sequence structure relations is that small changes in sequence may but need not have small consequences for the structure, and thus, the sequence-structure map appears to be almost uncorrelated if considered on a (sufficiently) coarse grained level (see Sect. 3).

Different notions of structure imply different models for the molecular phenotype. Examples are: (i) the structure of minimal free energy (mfe) which is formed after long enough time and at sufficiently low temperature, (ii) the mfe structure together with Boltzmann weighted suboptimal conformations in the sense of a partition function, and (iii) kinetic structures or ensembles of structures which take available folding times into account and acknowledge the fact that RNA is produced in the cell through transcription that forms the newly synthesized RNA strand from the 5'-end to the 3'-end. Although it is commonly assumed that small RNA structures form their mfe structures on folding, recent studies by means of a new algorithm resolving the process to formation and cleavage of single base pairs have shown that this is not necessarily true and kinetic structures may play an important role for rather small RNA molecules too [3]. For longer RNA sequences the discrepancy between most stable and kinetically favored structures is well established [4]. Kinetic effects on structures also imply that only sufficiently low lying barriers between the mfe structure and metastable suboptimal conformations are readily surmounted at room temperature. Higher barriers separate valleys of the conformational landscapes and what we observe in experiment are only the subsets of conformations in one particular valley, in other words, those conformations which are accessible within the (temperature dependent) time window of observations. The modified Boltzmann

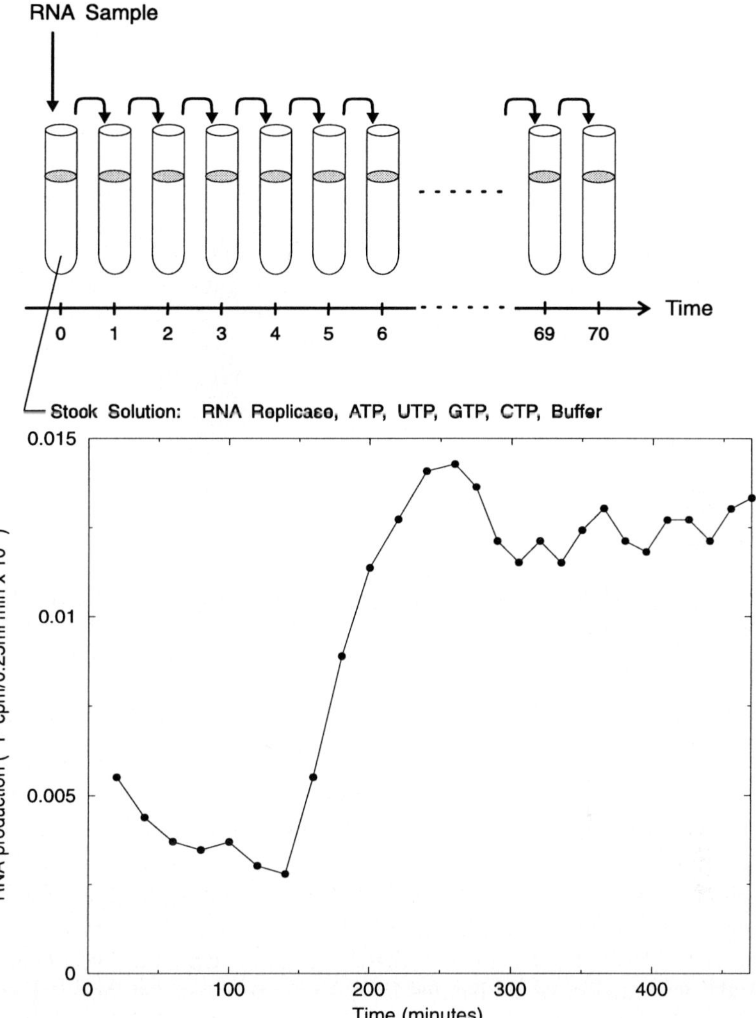

Fig. 1. Stepwise increase in the rate of RNA production. The upper part shows the technique of serial transfer applied to evolution of RNA molecules in the test tube. The material consumed is replaced by transfer of a small sample into a new test tube with fresh stock solution. The stock solution contains an enzyme required for replication, $Q\beta$-replicase, for example, and the activated monomers (ATP, UTP, GTP, and CTP), the building blocks for poly-nucleotide synthesis. The rate of RNA production (lower part) is measured through incorporation of radioactive GTP into the newly synthesized RNA molecules The figure is redrawn from the data in [1].

GCGGAUUUAGCUCAGDDGGGAGAGCMCCAGACUGAAYAUCUGGAGMUCCUGUGTPCGAUCCACAGAAUUCGCACCA

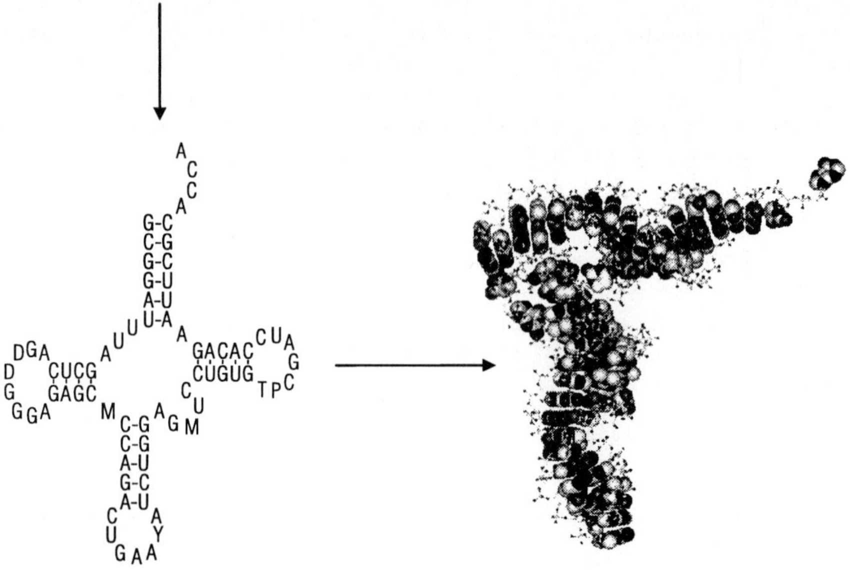

Fig. 2. Folding of RNA sequences into structures. The folding is performed in two steps from the sequence to the secondary structure and from the secondary structure to the full spatial structure. The example shown is the transfer RNA molecule $tRNA^{phe}$. Both steps occur under the condition of minimal free energy (mfe). The secondary structure is commonly defined as a listing of base pairs which is compatible with a planar graph without knots or pseudo-knots.

ensemble corresponding to such a subset is tantamount to an elaborate notion of biopolymer structure.

A relevant feature of Spiegelman's and other evolution experiments *in vitro* is a stepwise increase in the quantity to be optimized. Punctuation is observed even under controlled constant conditions (Fig. 1). Epochal evolution [5] is not restricted to evolution of molecules in the test tube: it has been observed also with bacterial cultures under the constant conditions of a precisely controlled serial transfer experiment [6] as well as in evolution experiments *in silico* mimicking replication and mutation in a chemostat [7–11]. A straightforward but almost trivial interpretation of the phenomenon says that the population waits for some rare event during such quasi-stationary periods. The basic questions apparently concern the nature of the rare mutation and the processes taking place in the search for the infrequent event. We shall try to find an answer which is compatible with the now well established neutral [12] or nearly neutral [13] theory of evolution.

In order to set the stage for a comprehensive theory of evolution we describe one particularly illustrative series of experiments. Bacteria are well suited objects for such studies because generation times can be as short as 20 minutes under optimal conditions. The rate of mutation was determined for many DNA based microbes and, interestingly, was found to have a constant value of about 0.0033 per genome and generation independently of DNA chain length [14]. In order to confront theory with real data we mention serial transfer experiments with *Escherichia coli* bacteria [6,15,16] (see also previous section). Populations of 5×10^8 cells were diluted 1:100 every day and recorded for about three years leading to about 10 000 generations and an average generation time of 3.6 hours. Fitness measured in terms of progeny increased by about 40% during a fast adaptive period over the first 2 000 generations. This increase occurs in steps and not continuously as one might have expected [6]. After the first adaptive period the curve saturates in the remaining 8 000 generations and settles on a plateau at about 1.5 times the initial fitness. More recently the rate of phenotypic evolution as monitored via fitness or cell size was compared with the rate of genomic evolution determined through DNA fingerprinting [16]: phenotypic evolution is fast in the initial phase and slows down during saturation. Evolution of the genotype, on the other hand, behaves differently: it speeds up in the saturation phase. Although the values from two independent experiments differ substantially, it is certain that the rate of genotypic change does not decrease in the same way as phenotypic evolution does.

The mean generation time is the time unit of the rate of evolution and hence it decides whether or not evolution experiments are feasible in accessible times. Higher organisms have generation times from several weeks to more than one decade. Then, time spans required for direct observation of evolutionary phenomena are at least hundreds to thousands of years and thus too long for experiments. At least at present we are thus confined with three experimental systems to study evolution, poly-nucleotide molecules, viroids or viruses, and bacteria.

2 Evolutionary Dynamics

The three scholars of population genetics, Ronald Fisher, John Haldane, and Sewall Wright, built the theory of population genetics and united the previously conflicting issues of Darwinian evolution and Mendelian genetics in an elegant and straightforward way. Evolution is considered as a process on the level of populations, the relevant variables are the frequencies of genes, and the properties of phenotypes enter the model equations as parameters. Such parameters are, among others, life times, litter sizes, and rates of reproduction, all of them contributing to fitness values. This is, at the same time, the basis of success and the most serious limitation of conventional population genetics. Proper choice of parameter values allows to model and analyze typical idealized situations and to study the influences of quantities like, for example, relative fitness, mutation rate, recombination rate or population size on the spreading of genes in populations. Problems arise when it becomes necessary to assign realistic values to the parameters, which are commonly very hard to determine experimentally, or when one aims at studies that deal with phenotypes explicitly. In the latter case we require knowledge on the relation between genotypes and phenotypes in or-

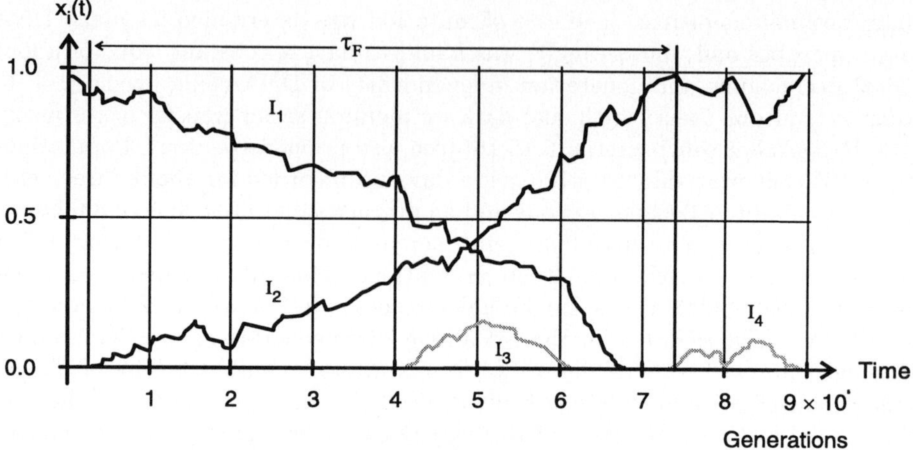

Fig. 3. Evolution of genotype frequencies in asexual populations. The sketch shows typical solution curves representing relative concentration or frequencies $x_i(t)$ of a population modeled by equation (1). New variants are formed by rare mutation events. Depending on replication rates relative to the mean value relative concentrations will increase ($a_i > \bar{a}$), decrease ($a_i < \bar{a}$) or drift randomly ($a_i \approx \bar{a}$) in the neutral case. Stochastic theory shows that fixation of mutants occurs also in neutral evolution: according to Kimura's theory [12] the mean time from the appearance of a mutant, $x_m(0) = 1/N$, until its fixation in the population, $x_m(\tau_F) \approx 1$, is $< \tau_F >= 2N$.

der to be able to derive the consequences of changes in the genomic nucleotide sequence on for the phenotype. Genotype-phenotype maps are highly complex and we shall discuss one particularly simple example in the next Sect. 3.

2.1 Selection Equation

It is straightforward to model selection in populations with asexual replication in precisely the same way as conventional population genetics handles sexual reproduction. Since there is little or no recombination, the appropriate variables are frequencies of different genotypes (I_k) rather than genes: $x_k = [I_k]/\sum_{j=1}^{n}[I_j]$ with $\sum_{j=1}^{n} x_j = 1$, and

$$\frac{dx_k}{dt} = x_k\Big(a_k - \Phi(t)\Big) + \sigma_k(\mathbf{x},t)\,\xi_k(t)\,,\ k = 1,\ldots,n\,, \tag{1}$$

with $\Phi(t) = \bar{a} = \sum_{j=1}^{n} a_j x_j(t)$ and $\sigma_k(\mathbf{x},t)\xi_k(t)$ representing stochastic fluctuations in reproduction corresponding, for example, to a Wiener process with $< \xi(t) >= 0$ and $< \xi(t)\xi(t') >= \delta(t-t')$. The above mentioned parameters are here the replication rate constants a_k. For equal degradation rates or life times these rate constants are tantamount to fitness values ($f_k = a_k$). Mutations are assumed to be rare events, they are not considered explicitly. In a setup with controlled population size N the fluctuations are proportional to $\sqrt{N}$. Typical solution curves are shown in Fig. 3. Genotypes replace each other in the course of evolution. Motoo Kimura [12,17] developed a stochastic version of population genetics that allows to consider the neutral case, $a_1 = \ldots = a_k = \bar{a}$.

The mean rate of evolution in Kimura's model, $\langle k \rangle$, measured as the number of mutant substitutions per generation time can be expressed by

$$\langle k \rangle \;=\; N \cdot v \cdot u(N,s,p) \;=\; N \cdot v \, \frac{1-\exp(-2Nsp)}{1-\exp(-2Ns)} \;, \tag{2}$$

where N is the population size, v the mutation rate per genome and generation and $u(N,s,p)$ the probability of fixation with s is the selective advantage[2] and p is the initial frequency of the mutant. Since every mutant starts inevitably from a single copy we may put $p = 1/N$ and find

$$\langle k \rangle = N \cdot v \frac{1-\exp(-2s)}{1-\exp(-2Ns)} \mathrm{for} p = \frac{1}{N}.$$

In the neutral case the rate of evolution is readily computed to be $\langle k \rangle = v$ and the mean time for the replacement of a given genotype by the next is $< \tau_R >= 1/\langle k \rangle \;= v^{-1}$ generations. For a substantial selective advantage, $s > 1/(2N)$, we find $\langle k \rangle \;= 2Ns \cdot v$ since $u \approx 2s$: the rate of evolution increases linearly with selective advantage s and population size N and a genotype will be replaced by the next after $< \tau_R >= (2Ns \cdot v)^{-1}$ generations (which is shorter than in the neutral case because $s > 1/(2N)$). At still higher values of the selective advantage s the probability of fixation converges to $\lim_{s\to\infty} u(s) \;= 1$ and we have $\lim_{s\to\infty}\langle k \rangle \;= N \cdot v$. Accordingly, the mean rate of evolution for neutral, weakly and strongly advantageous is confined by $v \le \langle k \rangle \le N \cdot v$. We can use the expressions derived for estimates on the times required for the observation of evolutionary phenomena. We should keep in mind, however, that the upper limit of $\langle k \rangle$ is highly unrealistic because it requires to maintain large increases in fitness, which do not occur often under normal conditions, over a sequence of many consecutive mutations (see, however, the initial period of *in silico* evolution of RNA molecules in Sect. 4).

It is also worth considering the mean time for fixation, τ_F, for neutral and advantageous variants. The solution of the deterministic selection equation for two genotypes[3], $x_2 = x$, $x_1 = 1 - x$ and $f_0 = 1$, is readily obtained in analytical form:

$$x(t) \;=\; \frac{x_0}{x_0 \,+\, (1-x_0)\cdot \exp(-s\,t)} \;.$$

From this equation we compute the time τ_F it takes for a variant to grow from a single mutant copy, $x(0) = 1$, to population size, $x\,(\tau_F) = N - 1$: $\tau_F \approx 2 \ln N/s$. For sufficiently large selective advantage, $s > 1/(2N)$ this time is substantially

[2] The selective advantage s is measured additively to the neutral case: The fitness value is $f = f_0(1+s)$ and thus neutrality implies $f = f_0$ or $s = 0$.
[3] Thereby we mean (1) for $n = 2$ and without the stochastic terms.

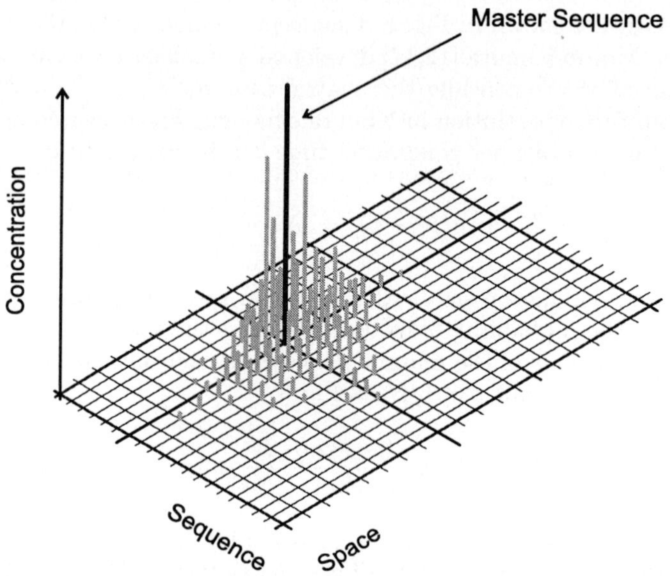

Fig. 4. A quasi-species-type mutant distribution around a master sequence. The quasi-species is an ordered distribution of poly-nucleotide sequences (RNA or DNA) in sequence space $\mathcal{I}^{\ell}_{\kappa}$. A fittest genotype or master sequence I_m being present in highest concentration is surrounded in sequence space by a "cloud" of close relatives. Relatedness of sequences is expressed in error class by the number of mutations which are required to produce them as mutants of the master sequence. In case of point mutations the distance between sequences is the Hamming distance.[8] In precise terms, the quasi-species is defined as the stable stationary solution of (3) [18,19], the mutant distribution is described by the largest eigenvector of the matrix $W = \{W_{ij} = Q_{ij} \cdot a_j ; i, j = 1, \ldots, n\}$ [20–23] (Its diagonal elements are approximations for fitness values, I_k: $f_k \approx W_{kk} = a_k \cdot Q_{kk}$). In reality, such a stationary solution exist only if the error rate of replication is below a maximal value called the error threshold. In this region, i.e. below the often sharply defined mutation rate of error threshold, this eigenvector represents a structured population as shown in the figure. Above the critical error rate the largest eigenvector is (practically) identical with the uniform distribution. The uniform distribution, however, can never be realized in nature or *in vitro* since we have many orders of magnitude more possible nucleic acid sequences (4^{ℓ}) than individuals in the largest populations. The actual behavior is determined by incorrect replication and random drift: the populations migrate through sequence space.

shorter than the mean time of fixation in the neutral case: $\tau_F = 2N$ (Fig. 3). In other words, ${\tau_F}^{-1}$ decreases linearly with s in the deterministic limit $s \to 0$ but the term from neutral evolution guarantees that the reciprocal time of fixation does not fall below ${\tau_F}^{-1} = 1/2N$.

2.2 Molecular Quasi-Species

An extension of conventional population genetics considering evolution as a process in genotype space was proposed by Manfred Eigen in his seminal paper on the theory of the evolution of molecules [18]. His concept can be understood, in essence, as an application of chemical reaction kinetics to molecular evolution. A main issue was to derive the mechanism by which biological information is created[4]. Populations migrate through sequence space as metastable but structured distributions of genotypes. They explore the environment by a variation-selection process and gain information on it. Biological information is laid down in the genotypes being poly-nucleotide sequences, DNA or RNA. The deterministic part of (1) is extended to handle the frequent mutation scenario and yields the replication-mutation equation:

$$\frac{dx_k}{dt} = x_k \Big(Q_{kk} a_k - \Phi(t) \Big) + \sum_{j=1, j \neq k}^{n} Q_{kj} a_j x_j \ , \quad k = 1, \ldots, n \ , \qquad (3)$$

with $Q = \{Q_{ij}; i,j = 1, \ldots, n\}$ being the mutation matrix and $\Phi(t) = \bar{a}$, as before. In essence, the ansatz (3) differs from conventional population genetics as expressed, for example, by (1) in the handling of mutations. The conventional treatment introduces mutation as a stochastic event like a fluctuation in the environment. It is characterized by its probability density, in particular by the expectation value and the variance of the mutation rate. The replication-mutation (3) deals explicitly with mutations and handles error-free and incorrect replication as parallel reactions. Relative frequencies of the corresponding reaction channels are given by the elements of the mutation matrix. In particular, Q_{kj} is the frequency at which I_k is synthesized as an error-copy of I_j. In general, (3) settles down to a stationary state corresponding to a stable or metastable[5] distribution of genotypes. Such distributions of genotypes, called *molecular quasispecies* [19,24], are ordered: in the center there is a most frequent and commonly also fittest genotype (Fig. 4). Quasi-species represent the genetic reservoirs of asexually replicating species, for example viruses and bacteria.

Most fitness landscapes, understood as distributions of fitness values over sequence space, sustain stable quasi-species only for mutation rates below a certain critical value called the *error threshold* [19,24]. At this point a phase transition

[4] We use the symbol I for genotypes or poly-nucleotide sequences (DNA, RNA) as well as $\mathcal{I}$ for the space of genotypes, called sequence space in order to point out that its elements I_k are the carriers of biological information and $\mathcal{I}$ is the corresponding information space. In particular, $\mathcal{I}_\kappa^\ell$ in the sequence space of sequences of chain length ℓ over an alphabet of size κ (**AUGC**: $\kappa = 4$).

[5] There are two reasons why the infinite time solution of (3) may be or become unstable: (i) the steady state can never be reached because the largest population sizes that can be realized are too small to be an approximation to the infinite population limit, and (ii) random drift in the sense of neutral evolution (which is not addressed by deterministic equations) may lead to further onset of selection after epochs of stasis.

like abrupt change in evolutionary dynamics is observed [25] (Exceptions are smooth landscapes which show gradual transitions from quasispecies to uniform distribution). At error frequencies above threshold populations migrate through sequence space in random walk manner and do not approach stationary states (Fig. 4). Stationary mutant distributions were indeed observed and analyzed in the case of RNA evolution *in vitro* [26]. The data reproduce well the predictions derived from (3). The quasispecies concept turned out to be particularly useful for a description of the evolution of viral populations [27,28] and for the development of novel anti-viral strategies [29].

The quasi-species concept (3) provides a solution for the mutation problem but does not yet consider phenotypes explicitly. Basic to models of the mutation matrix Q is the notion of sequence space ($\mathcal{I}_\kappa^\ell$) [30] with the Hamming distance of two sequences I_i and I_j, $d_{ij}^{\,h}$, serving as metric[6]. Restriction to the most frequent class of mutations, which are point mutations causing an exchange of the symbol at one particular position, and the uniform error rate assumption allow to express all elements of Q in terms of only three parameters only, ℓ, q and $d_{ij}^{\,h}$:

$$Q_{ij} \;=\; q^{\ell} \left(\frac{1-q}{q} \right)^{d_{ij}^{\,h}} \;. \tag{4}$$

Herein ℓ is the length of the poly-nucleotide chain, q is the single digit accuracy of replication, implying a uniform error rate of $p = 1 - q$ per digit and replication event, and $d_{ij}^{\,h}$, finally, is the Hamming distance between the two sequences to be inter-converted by mutation. As in the selection (1) the properties of phenotypes enter the kinetic differential equations (3) as parameters. Assignment of parameters is a formidable problem which can be handled only under model assumptions. Considering, for example, a rather short RNA molecule of chain length $\ell = 100$ we are dealing with 1.6×10^{60} different genotypes which may give rise to the same hyper-astronomically large number of phenotypes and thus require information or assumptions on more than 10^{60} fitness values. Commonly the problem is overcome by rather drastic simplifications. As an example we mention the single peak fitness landscape: a replication rate $a = \sigma$ is assigned to the fittest genotype and all other genotypes are assumed to have replication rate $a = 1$ [22]. This ansatz reminds of the mean field approximation often used in physics. Since details of the distribution of fitness values are unknown, one replaces them by a mean value for all genotypes except the fittest one. In this sense, the single peak landscape has been used, for example, to derive analytical expressions for the threshold value of the error rate [18,19,24] (For further work on replication and mutation on model landscapes see [32–38]).

[6] The Hamming distance is a measure of distance of two general strings and was invented in coding theory [31]. It counts the number of positions in which two aligned strings differ.

2.3 Evolution of Genotypes with Phenotypes

The first explicit consideration of phenotypes in a model for molecular evolution was implemented *in silico* to simulate replication and mutation in a flow reactor (Fig. 5) and used to describe optimization of RNA properties [7]. Later on, the relation between genotypes and phenotypes was considered as a mapping from sequence space into phenotype space. To this end we assume a metric phenotype space $\mathcal{S}$ with a hypothetical measure of distance, d_{ij}^{s}, between phenotypes:

$$\psi : \{\mathcal{I}\; d_{ij}^{h}\} \Rightarrow \{\mathcal{S}; d_{ij}^{s}\} \tag{5}$$

or, in other words, $S_k = \psi(I_k)$, a phenotype S_k is uniquely assigned to every genotype I_k. Equation (5) thus describes the unfolding of the genotype to yield a phenotype. Replication rate constants are eventually derived from phenotypes through a mapping f into the real numbers:

$$f : \{\mathcal{S}; d_{ij}^{s}\} \Rightarrow \mathbb{R}^1 \,. \tag{6}$$

The map (6) deals with an evaluation of the phenotype that returns its fitness value. In summary, we obtain the required fitness values from the genotype through the function: $f_k = f(S_k) = f\big(\psi(I_k)\big) = f_k \approx a_k \cdot Q_{kk}$. The mapping $\psi(.)$, in general, cannot be expressed in analytical terms. At best we have algorithms that allow to compute structures from sequences (see next Sect. 3). The situation is not less complex for the derivation of fitness values of phenotypes, but in this case the evaluation is often done by means of simple model functions. For example, $f(.)$ is assumed to be a simple function of the distance between the phenotype and some target to be approached.

Now we are in a position to classify different mappings: (i) $\psi(.)$ maps a discrete vector space, the sequence space $\mathcal{I}$, into another non-scalar discrete (or continuous) space $\mathcal{S}$. We call it a combinatory map [39] since the sequence space $\mathcal{I}$ is derived by a combinatory building principle (see also next Sect. 3). (ii) $f(.)$ maps a discrete or (continuous) non-scalar space $\mathcal{S}$ into the real numbers $\mathbb{R}^1$. It represents an example of a landscape, in particular, it is the fitness landscape assigning a fitness value f_k to a phenotype S_k[7].

Finishing this section we consider environmental influences and indicate how one may generalize to variable environments, $\mathcal{E}(t)$. Both, the unfolding of the genotype as well as the evaluation of the phenotype depend on the environment $\mathcal{E}$. In other words, the same genotype, I_k develops different phenotypes, say S_k, S_k' or S_k'', in different environments, $\mathcal{E}$, $\mathcal{E}'$ or $\mathcal{E}''$. The same phenotype may have different fitness values under different environmental conditions. In principle, the ansatz for evolutionary dynamics presented here can be readily extended to handle situations in variable environments by the introduction of time dependent fitness values:

$$f_k(t) = f\Big(S_k, \mathcal{E}(t)\Big) = f\Big(\psi\big(I_k, \mathcal{E}(t)\big), \mathcal{E}(t)\Big) \,. \tag{7}$$

[7] The expression "landscape" is a generalization of the notion used in common-sense or geography for the representation of a three-dimensional relief on Earth as a mapping from two dimensions (longitude, latitude) into the real numbers (altitude).

Incorporating time dependent fitness values into (1) and (3) we obtain the stochastic differential equation

$$\frac{dx_k}{dt} = x_k \Big(Q_{kk} f_k(t) - \Phi(t) \Big) + \sum_{j=1, j \neq k}^{n} Q_{kj} f_j(t)\, x_j + \sigma_k(\mathbf{x}, t) \xi(t) \ , \quad k = 1, \ldots, N \ , \qquad (8)$$

which we consider as the basic equation describing Darwinian evolution in asexually replicating populations. Needless to say, the mappings (7) encapsulate a great deal of complexity and there is no chance to find simple solutions. They are, nevertheless, suitable to discuss special simplified cases and they provide a reference for computer simulations. Three experimentally accessible realizations of (8) are currently conceivable: (i) evolution of RNA molecules *in vitro*, (ii) life cycles and evolution of viral RNA (or DNA) in host cells, and (iii) metabolism and evolution of bacteria (under constant environmental conditions). In the following two sections we shall present a simplified model of (i) that allows to simulate evolution according to (8) using a realistic algorithm to compute RNA structures from sequences. Virus evolution (ii) can be modeled in principle provided enough data are available on the influence of mutations on viral life cycles. Quantity and quality of these data are rapidly improving now and we can expect full understanding of virus evolution at the molecular level within the next decade. Although (iii) seems to be too complex by far for computer implementations we may expect fast progress in the near future: information on complete DNA sequences is already available in a few cases and many more bacterial genomes will be sequenced soon. The current data are already used in the development of models for the metabolism of procaryotic cells. Still, simulation of bacterial evolution based on such models remains a great challenge for future research.

3 The RNA Model

A straightforward interpretation of serial transfer or flow reactor experiments with RNA identifies the molecular structure with the phenotype [2]. Accordingly, the simplest and the only currently accessible example of a genotype-phenotype map relates RNA sequences with a simplified and coarse-grained version of RNA structure. This simplified RNA structure, which we shall use here, is the so-called secondary structure (Fig. 2). It is explained best as a listing of Watson-Crick (**AU**,**GC**) and **GU** wobble base pairs which is compatible with an unknotted and pseudo-knot-free two-dimensional graph [8]. RNA secondary structures with

[8] RNA secondary structures can be represented by strings written in a short-hand notation using parentheses and dots. Parentheses correspond to bases combined in base pairs, dots represent single bases. The symbols for bases belonging together in pairs are interpreted unambiguously through reading them in the sense of mathematical notation, i.e. from outside to inside. For example, the string of a typical hairpin loop

minimum free energies are readily derived from sequences by means of fast algorithms based on dynamic programming [40–43]. In addition, they are not only a convenient theoretical construct but represent also a relevant and experimentally verified intermediate in the folding of RNA sequences into three-dimensional objects [44]. Moreover, secondary structures are conserved in nature and they were used by biochemists for decades to interpret successfully reactivities and other properties of RNA before three-dimensional structures became available.

Table 1. Various strategies applied to study sequence-structure maps of RNA.

	Method	Advantage	Disadvantage	Ref.
Mathematical model	Random graph theory	Analytical expressions	Limited validity of model assumptions	[39]
Exhaustive folding and enumeration	Folding algorithm and handling of large samples ($> 10^9$ objects)	Exact results	Limited to short chains: **GC**, $\ell \leq 30$ **AUGC**, $\ell \leq 16$	[45,46]
Statistical evaluation	Inverse folding or random walks in sequence space	Applicability to longer sequences	Limited accuracy due to statistics	[47,48]
Simulation of evolutionary dynamics	Chemical kinetics of replication and mutation	Evolutionary relevance	Restriction to small parts of sequence space	[8–11]

Minimum free energy secondary structures of RNA molecules, called here (RNA) shapes for short, were studied in order to discover regularities in sequence to structure mappings [39,45–49]. We applied four strategies listed in Table 1 in order to achieve the goal and obtained the following general results for sequence-structure relations of RNA molecules (with fixed chain length ℓ):

(i) **More sequences than structures.** Estimates on the numbers of acceptable secondary structures through combination of structural elements yields the expression [50]:

$$|\mathcal{S}(\ell)| \leq N_S(\ell) \approx 1.485 \times \ell^{-3/2}(1.849)^\ell ,$$

which is an asymptotic upper limit for the cardinality of the shape space for sequences of chain length ℓ, $|\mathcal{S}(\ell)|$. Exhaustive enumeration of the minimum

reads: $\cdot\cdot((((\cdot\cdot\cdot\cdot))))$. A pseudo-knot occurs when base pairs intercalate, for example in the secondary structure $\cdot\cdot(((\cdot\cdot[[\cdot\cdot\cdot)))\cdot]]\cdot$, where we need two classes of symbols, parentheses and square brackets, for an unambiguous grouping of bases into pairs.

free energy structures (or shapes) formed by binary **GC**-sequences of chain length $\ell = 30$ showed that the actual number of shapes, $|\mathcal{S}_{\mathbf{GC}}(30)| = 218\,820$ amounts to 24% of the acceptable structures, $N_S(30) = 918\,000$. Compared to the cardinalities of sequence spaces, $|\mathcal{I}_{\mathbf{GC}}^{(30)}| = 1.07 \times 10^9$ and $|\mathcal{I}_{\mathbf{AUGC}}^{(30)}| = 1.15 \times 10^{18}$, shape space is smaller than sequence space. In other words we are dealing with many more sequences than shapes or even structures. Thus the mapping from sequence space onto shape space is many to one and not invertible.

(ii) **Few common and many rare shapes**. The distribution of the numbers of sequences forming the same shape, $|S_k|$, is rather broad and strongly biased towards the rare-shape end. Analysis through exhaustive folding [45,46] yielded a clear result independently of chain lengths ℓ and size of alphabet (**AUGC**: $\kappa = 4$; **GC**: $\kappa = 2$): There are relatively few common shapes and many rare ones. In the above mentioned example, **GC**-sequences of chain length $\ell = 30$ ($\mathbf{GC}_{30}$), more than 93% of all sequences fold into common shapes which are made up of only 10.4% of all shapes. An increase in chain length causes these percentages to go up and down, respectively, and in the limit of long chains almost all sequences fold into a vanishingly small fraction of all shapes.

(iii) **Shape space covering**. Sequences forming common shapes are distributed (almost) randomly in sequence space. Accordingly, one need not to search the entire sequence space in order to find a sequence that folds into a given common shape. One can indeed show that it is sufficient to screen a (high-dimensional) sphere around an arbitrarily chosen reference sequence in order to find (with probability one) at least one sequence for every common shape [48]. The radius of this shape space covering sphere, $r_{\mathrm{cov}}(\ell)$, can be estimated straightforwardly [51,52]:

$$r_{\mathrm{cov}}(\ell) = \min\left\{h = 1, 2, \ldots, \ell \,|\, B_h(\ell, \kappa) \geq \frac{\kappa^\ell}{N_s(\ell)}\right\} ,$$

where B_h is the number of sequences contained in a ball of radius h and can be easily obtained from the recursion

$$B_h(\ell, \kappa) = \sum_{i=1}^{h} b_i(\ell, \kappa)\,; \; b_i = b_{i-1} \cdot \frac{(\kappa - 1)(\ell + 1 - i)}{i}\,; \; b_0 = 1 \,.$$

The covering radius is much smaller than the radius of sequence space ($\ell/2$). For example, it amounts to $r_{\mathrm{cov}} = 15$ for **AUGC**-sequences of chain length $\ell = 100$ and thus on has to search only a fraction of sequence space containing a 4.52×10^{-37}-th of all sequences in order to find all common shapes.

(iv) **Common structures form extended neutral networks**. The pre-image of a given shape S_j in sequence space is the set of sequences $M_j = \psi^{-1}(S_j) \doteq \{I_k | \psi(I_k) = S_j\}$. A set of sequences can be converted into a graph $\mathcal{G} = (v[\mathcal{G}], e[\mathcal{G}])$ in sequence space with $v[.]$ and $e[.]$ denoting the vertices and edges, respectively. The *neutral network* $\mathcal{M}_j$ of S_j is constructed by identifying the sequences in M_j with the nodes and drawing edges between all nearest neighbors in $\mathcal{I}$ (these are the pairs of sequences with Hamming distance $d_{ij}^h = 1$):

$$\mathcal{M}_j = \Big(v[\mathcal{M}_j] = \{I_k \,|\, I_k \in M_j\},\\ e[\mathcal{M}_j] = \{(\overline{I_k I_{k'}}) \,|\, I_k, I_{k'} \in M_j \text{ and } d_{k,k'}^h = 1\}\Big).$$

The question, how sequences belonging to a neutral network $\mathcal{M}_j$ are distributed in sequence space was answered by means of random graph theory [39,53]. The central quantity of this approach is the average degree of neutrality of a given network, $\bar{\lambda}(\mathcal{M}_j) = \bar{\lambda}_j$. It is, in other words, the mean fraction of neutral neighbors of sequences belonging to the network: $\bar{\lambda}_j = \sum_{I_k \in \mathcal{M}_j} \lambda_k / |\mathcal{M}_j|$, where λ_k is the number of nearest neighbor sequences of I_k which form shape S_j divided by the total number of nearest neighbors, $\ell \cdot (\kappa - 1)$. Neutral networks show a kind of percolation phenomenon. They are connected and span the entire sequence space if $\bar{\lambda}_j$ exceeds a critical threshold value, whereas they are partitioned into components with one dominating giant part and many small "islands" when $\bar{\lambda}_j$ is below threshold:

$$\mathcal{M}_j \text{ is } \begin{cases} \text{connected:} & \bar{\lambda}_j > (\bar{\lambda})_{\text{cr}} = 1 - \kappa^{-\frac{1}{\kappa-1}} \,, \\ \text{partitioned:} & \bar{\lambda}_j < (\bar{\lambda})_{\text{cr}} = 1 - \kappa^{-\frac{1}{\kappa-1}} \,, \end{cases}$$

where κ is the number of digits in the alphabet of nucleotide bases (**AUGC**: $\kappa = 4$). Connected areas on neutral networks are important in evolution since they define regions in sequence space which are accessible to populations through random drift [8].

The predictions on sequence-structure mappings of RNA shapes made by random graph theory were tested through exhaustive folding of entire sequence spaces [45,46]. In some cases we found deviations from generic behavior and these deviations could be explained by and derived from specific molecular structures. One particularly relevant and illustrative example was observed in the partitioning of neutral networks in the $\mathbf{GC}_{30}$ case. Random graph theory predicts that networks are either connected or their partition contains one largest "giant component". Analysis of the sequence of components ordered with respect to sizes revealed, however, that there also networks with two or four dominant components of equal size, or with three dominant components of size distributions 1:2:1. Most sequences of chain length $\ell = 30$ form shapes with one double-helical region. The four single stranded chains coming out from the stack form a hairpin loop on one side and zero, one or two free ends on the other side. Hairpin loops fall into two groups: (i) three and four membered loops and (ii) loops with five or more single bases. Loops of the former group cannot be shortened by forming an additional base pair at the end of the stack since one and two membered hairpin loops do not occur in real structures[9], In contrast, five membered loops can be

[9] The basis for unacceptability of loops with one and two single bases is high free energy of the corresponding RNA structures.

converted into a base pair and a triloop, six membered loops into a base pair and a tetraloop, etc. Similarly we find at the other side of the stack: (i) no additional base pair can be formed if the number of free ends is zero or one, but (ii) shapes with two free ends allow for elongation of the stack provided the corresponding sequence requirements are fulfilled. Combination of two elements at the two ends of the stack leads to three different classes which form neutral networks with different sequence distributions in sequence space. Shapes with stacks containing two category (i) ends of stacks (class **0**), these are tri- and tetraloops as well as zero or one free ends, form generic neutral networks (connected or one largest component), shapes with one category (i) and one category (ii) end (class **1**) form networks with two largest components, and shapes with two category (ii) ends (class **2**), eventually, form those with three or four largest components. Interpretation of this finding is straightforward: generic neutral networks (class **0**) show a distribution of sequences in sequence space which is close to the binomial distribution (being the distribution of all sequences in sequence space):

$$B(\ell, h, \kappa) = \binom{h}{\ell} (\kappa - 1)^h \Big/ \kappa^\ell ,$$

where h is the Hamming distance between the reference sequence and the error class under consideration (h). For binary sequences ($\kappa = 2$) the distribution $B(\ell, h, 2)$ is symmetric in sequence space and the majority of all sequences is found in the error class $\ell/2$ (or in the error classes $(\ell \pm 1)/2$, respectively). An arbitrarily chosen most frequent binary sequence (50% **G**, 50% **C**) will form a shape of class **1** with a reduced probability because 50% of these sequences have complementary symbols (**G**, **C**) at the end of the stack and will spontaneously form the shape with the additional base pair. The largest probability to form a class **1** shape is therefore displaced from the zone of highest frequency. As a matter of fact we find indeed two components components lying symmetrically with respect to the center of sequence space, one with excess **G**$(+\delta)$ and the other with excess **C**$(-\delta)$. By the same token we explain the occurrence of three or four largest components for shapes of class **2**: there are two labile category (ii) ends and the excess percentages of **G** and **C** are superimposed independently yielding four components of equal size (Displacements: $+2\delta, +\delta - \delta = 0, -\delta + \delta = 0, -2\delta$), two displaced ones and two in the middle. Three components of sizes 1:2:1 originate from a four component system though merging the two central components.

4 Darwinian Evolution *in Silico*

In this section the RNA model will be used as an example of a genotype-phenotype map in computer simulations of evolutionary optimization of RNA shapes and structure related properties. At first the simulation has to be embedded in a physically relevant environment and we choose the flow reactor shown in Fig. 5 as an appropriate device. The chemical reaction mechanism comprises replication and mutation apart from flow terms. It is implemented as a stochastic

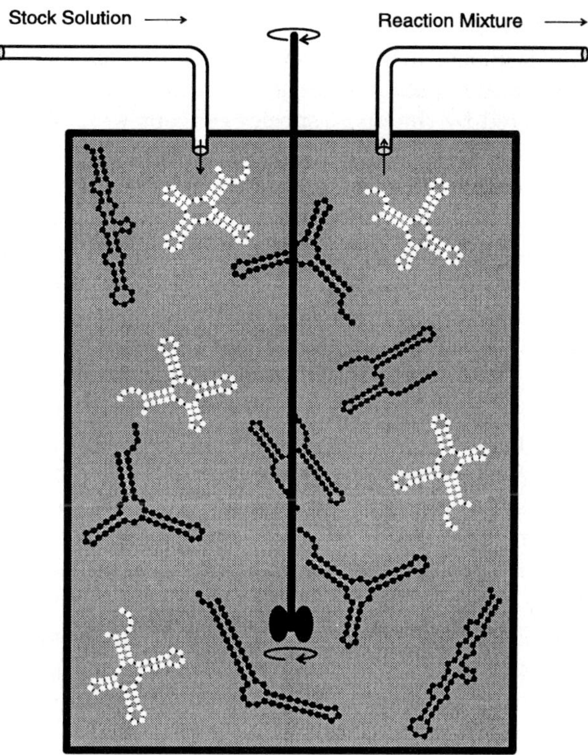

Fig. 5. The flow reactor as a device for RNA structure optimization. RNA molecules with different shapes are produced through replication and mutation. New sequences obtained by mutation are folded into minimum free energy secondary structures. Replication rate constants are computed from structures by means of predefined rules (see text). For example, the replication rate is a function of the distance to a target structure (the white tRNA shown in the reactor). Input parameters of an evolution experiment *in silico* are: the population size N, the chain length ℓ of the RNA molecule, and the single digit replication accuracy q.

process based on the master equation for the reaction mechanism, and individual trajectories are computed by means of an algorithm conceived and analyzed by Daniel Gillespie [54,55]. Under the constraints of the flow reactor the population size fluctuates and fulfills a normal distribution with expectation value N and standard deviation $\sqrt{N}$. The replication rates are determined according to fitness criteria. In previous simulations the kinetic constants were derived from molecular shapes by some predefined and biophysically motivated rules [7,56]. Error-free replication and mutation are parallel reaction channels whose relative frequencies are given by (4). The single digit accuracy of replication, q corresponding to a mutation rate $p = 1 - q$ per site and generation, is an input parameter of the computations. The early computer simulations confirmed three basic features of molecular evolution: (i) population sizes of a few thousand molecules are sufficient for RNA optimization, (ii) stochastic effects dominate in the sense that the sequence of shapes recorded in one particular trajectory were never observed again in subsequent identical simulations[10], and (iii) sharp error thresholds as predicted by the quasi-species concept were observed in computer runs with different mutation rates.

More recently, computer simulations of replication and mutation in the flow reactor were used to show that evolution on the neutral network of a tRNA-structure corresponds to a diffusion process in sequence space where the diffusion coefficient is proportional to the mutation rate [8]. In this simulation as well as in the computer experiments described below, replication rates were assumed to depend on the shape of the molecule independently of the sequence folding into it and thus fulfilling the neutrality condition for structures: $a_k = a(S_j) \, \forall \, I_k \in M_j$. In particular, a function of the kind $a(S_j; S_\tau) = (\alpha + d^{\,s}_{j\tau}/\ell)^{-1}$ was used, where α is some constant, ℓ the chain length of the RNA, and $d^{\,s}_{j\tau}$ the distance between structure S_j and the target structure S_τ. Many measures of distance between structures are conceivable [47], a particularly simple one is the Hamming distance between the short-hand "parentheses representations" (Sect. 3) of the two shapes. General results, however, were found to be largely independent of the specific choices of constants, fitness functions, and distance measures.

Optimization of RNA structures was studied through simulations of the evolution of a population in the flow reactor [10,11]. The approach towards the target structure which happened to be a tRNA clover-leaf occurs in steps: periods of fast decrease in distance to target are interrupted by long quasi-stationary phases or epochs of almost constant average fitness (Fig. 6). The course of the evolutionary optimization process was reconstructed by retrieving the series of phenotypes leading from an initial shape to the target, called the *relay series* of the computer experiment. The relay series is a uniquely defined and uninterrupted sequence of shapes which starts from an initial shape and ends at the target structure. It is retrieved through backtracking, that is in opposite direction from the final shape to the initial structure. The procedure starts by highlighting the final structure which is traced back during its uninterrupted presence until the time of its first appearance in the flow reactor. At this point we search for the parent shape from which it descended by mutation. Now we record time and structure, highlight the parent shape, and repeat the procedure. Recording further backwards yields a series of shapes and times of first appearance which ultimately ends in the initial population[11]. The full relay series of the computer experiment of Fig. 6 is shown in Fig. 7. It contains 42 shapes produced through 41 consecutive transitions (Six characteristic structures along the series are shown on top of Fig. 6.).

---

[10] By two simulation experiments under identical conditions we mean that everything was kept constant except the seeds for random number generators.

[11] It is important to stress two facts about relay series: (i) the same shape may appear two or more times in a given relay series; it was extinct between two consecutive appearances. (ii) A relay series is not a genealogy which is the full recording of a line of genotypes in parent-offspring relation.

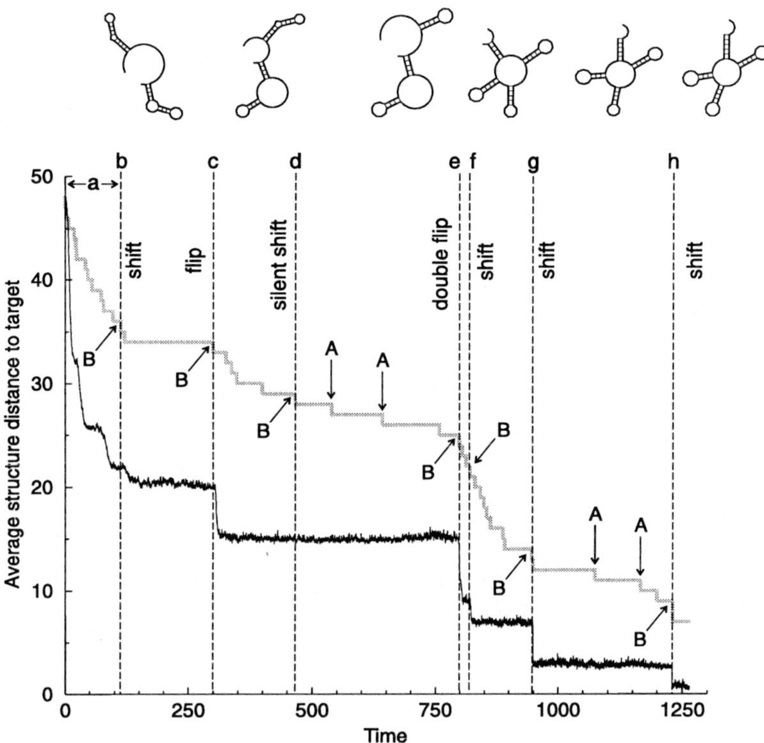

Fig. 6. The recording of an RNA structure optimization experiment in the flow reactor. The computer experiment starts from a homogeneous initial population of about 1000 RNA molecules with an arbitrarily chosen structure and leads to a quasi-species like distribution around the target shape. Fitness expressed as replication rate is computed as a function of the distance between current (S_j) and target structure (S_τ), $d^s_{j\tau}$ (For details see text). The target structure was chosen to be the clover leaf of tRNA^{phe} ($\ell = 76$). The distance to the target structure is averaged over the entire population and plotted against time (black curve). The time scale represents the "real time" of the simulation experiment in arbitrary units. The whole simulation comprises about 1.1×10^7 replications. A mutation rate of $p = 0.001$ per site and replication was applied. From this computer experiment a relay series of 42 shapes (or phenotypes) was reconstructed through backtracking the phenotypes which lead to the target structure (see text and Fig. 7). The six most important shapes are shown at the top of the figure. The relay series is indicated by the step-function (grey) which assigns equal height to every shape. Transitions between phenotypes fall into two classes: (i) continuous (examples marked by **A**) and (ii) discontinuous (**B**). An initial period of about one hundred time units is characterized by fast decrease in the distance to the target (**a**). The individual discontinuous transitions are classified as "shifts", "flips", and "double flips", and marked by **b** to **h**. The "silent shift" at $t \approx 460$ is neutral with respect to distance to target. Discontinuous transitions lead to major changes in RNA shapes which are followed by cascades of minor fitness-improving steps.

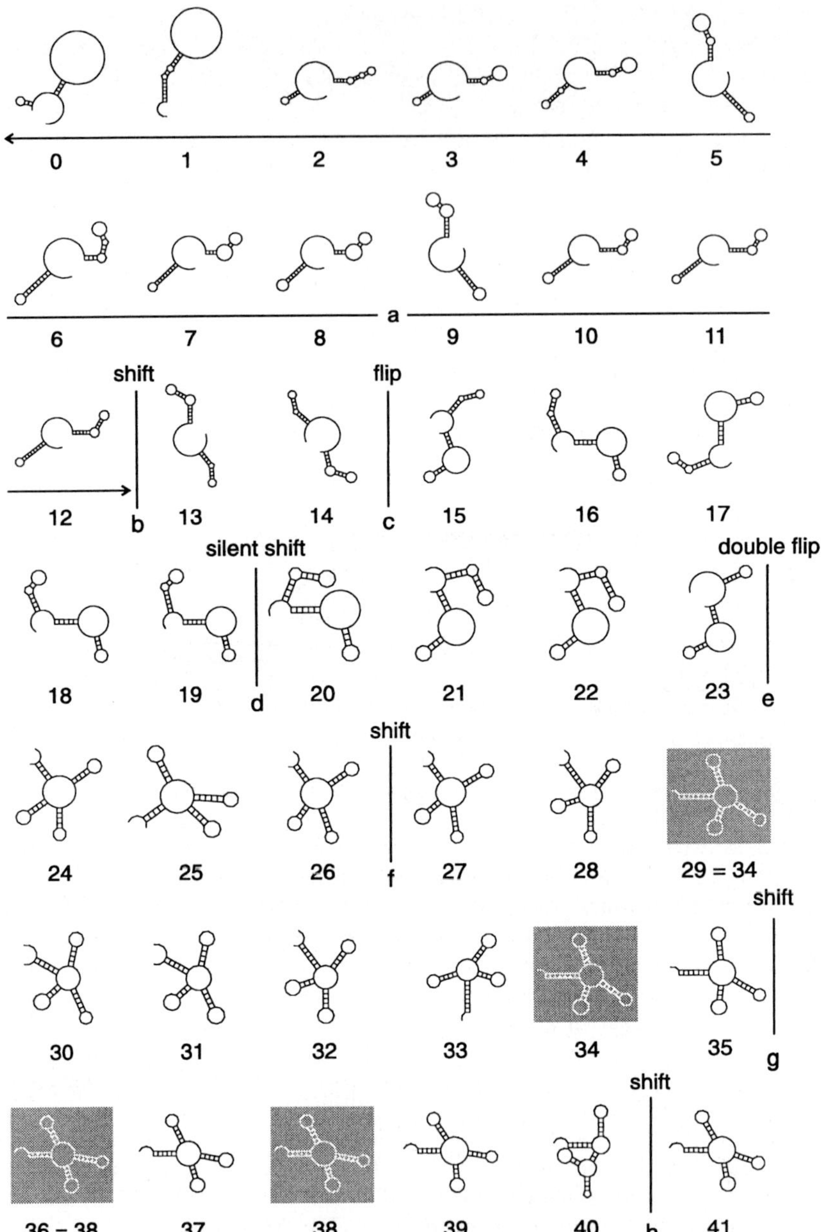

Fig. 7. The relay series of the *in silico* optimization experiment described in Fig. 6. For details see text. It is worth noticing that a given shape may appear twice or more often in a relay series. Examples are the shapes $29 \equiv 34$ and $36 \equiv 38$.

Transitions between two consecutive shapes in the relay series fall into two classes, **A** and **B**. Basis for this classification is the underlying structural change. Class **A** transitions occur frequently on mutation and involve mostly minor changes like closing and opening of a base pairs in the immediate neighborhood of a stack. Class **B** transitions are rare events in the sense that they occur only with special sequences. They lead to major changes in structure. Such major rearrangements involve simultaneous displacement of several base pairs (Different subtypes of class **B** transitions were characterized as "shifts", "flips", and "double flips" depending on the details of the structural change [10].). The majority of rearrangements recorded in RNA optimization experiments with population sizes of a few thousand molecules are class **A** transitions (Four of them are marked in Fig. 6.). Class **B** transitions are less frequent. For example, seven major changes are identified among the 41 transitions of the relay series shown in Fig. 7.

Class **A** and class **B** transitions can be generalized in terms of neighborhood frequencies of neutral networks:

(i) **Continuous transitions (A)**. They represent minor structural changes and lead to structures which are globally frequent in the neighborhood of the neutral network of the initial shape.

(ii) **Discontinuous transitions (B)**. They involve major structural changes leading to globally rare and only locally frequent structures. Accordingly, discontinuous transitions require special sequences that allow major structural changes to occur on single point mutations.

Simulations show an initial period (**a** in Fig. 6) of cascading major and minor transitions followed by a stepwise optimization process with apparent regularities. Each epoch or quasi-stationary phase of evolution ends with a discontinuous transition. Major transitions (**b** to **h**), however, occur only rarely within quiescent periods[12]. Every major transition is followed by a cascade of minor transitions which are accompanied by substantial fitness increase. Then the population approaches the next plateau corresponding to an epoch of neutral evolution at approximately constant fitness. The population undergoes neutral mutations with respect to structure or fitness neutral class **A** transitions until it reaches one of the special sequences from which a fitness-improving discontinuous transition is locally frequent and hence possible. Evolutionary optimization on landscapes with high degree of neutrality proceeds on two time scales: fast periods containing cascades of adaptive changes which are interrupted by long quasi-stationary epochs of neutral evolution during which populations drift randomly on neutral networks until they reach a neighborhood that is suitable for the next major transition.

[12] There is one case (**d**) at time $t \approx 460$ in the computer simulation of Fig. 6. A discontinuous transition is observed inside an epoch. We called it "silent" since it does not change the distance to target and is neutral with respect to fitness.

5 Concluding Remarks

Darwinian evolution is based on chance and necessity which can be both identified and traced down to processes at the molecular level [57]. We introduced here a dynamical concept that starts out from chemical kinetics of replication and mutation but includes the phenotype as in integral part of the model. Central to the model is a genotype-phenotype map which can be used to derive phenotypic properties and, in particular, to calculate the fitness values of the genotypes carried by the phenotype. At present the only tractable case of such a mapping is the sequence-secondary structure map of RNA molecules. The RNA model was used, for example, in computer simulations of evolutionary optimization and provided three results that can be generalized to other, more complex biological systems.

(i) Evolution may show punctuation even under precisely constant environmental conditions like those encountered in a flow reactor. In other words, no external triggers are required for the stepwise course observed with optimization processes. Evolution occurs on two time scales: adaptive processes based on selection are comparatively fast, and random drift on neutral networks is slow. Computer simulations, in this respect, agree with the predictions of the analytical approximations (For details see Sect. 2.1). Both, adaptive evolution and random drift contribute to the success of optimization processes. This finding supports the view on the role of neutral evolution in the development of genetic regulatory networks as proposed by Emile Zuckerkandl [58].

(ii) Accessibility of phenotypes determines the progress of evolution. A notion of nearness between phenotypes has been developed which accounts for the existence of neutral networks [10]. Nearness is defined in a statistical sense with respect to the frequency of occurrence in the one-error neighborhood of the neutral network. A phenotype which is frequently found in this neighborhood can be reached from almost everywhere on the network and we characterized the corresponding transitions as continuous because they occur almost instantaneously. The transitions to other phenotypes are rare. They can occur only at special positions on the network and were denoted as discontinuous. It is straightforward to compare differences between evolution of genotypes and phenotypes in terms of changes per generation. Computer simulations indicate that genotypic evolution may speed up when phenotypic evolution slows down. Precisely the same qualitative result was obtained by careful analysis of long time evolution experiments with bacteria [16].

(iii) The landscape concept, originally introduced as a metaphor by Sewall Wright [59], was put on firm scientific grounds when applied to biopolymers, in particular, to RNA molecules and proteins. It provides the molecular basis of neutrality and neutral evolution which population geneticists could deduce only indirectly from comparisons of sequence data in contemporary organisms. The RNA model is currently based on shapes defined as minimum free energy structures. The concept, however, is very flexible because additional structural features can be taken into account readily. Such features are, for example, the folding process of RNA or suboptimal structures within reach from the ground state at room temperature. We can expect a great variety of new phenomena which wait to be detected in such extended molecular models. These new regularities will help to illustrate and understand otherwise too complex to analyze peculiarities of macroscopic evolution.

Acknowledgement. Financial support of the work presented here was provided by the Austrian *Fonds zur Förderung der wissenschaftlichen Forschung* (Projects P-11065, P-13093, and P-13887), by the *Jubiläumsfonds der Österreichischen Nationalbank* (Project 7813), by the Commission of the European Union (Project PL-970189), and by the Santa Fe Institute.

References

1. D.R. Mills, R.L. Peterson, and S. Spiegelman, Proc. Natl. Acad. Sci. USA **58**, 217–224 (1967).
2. S. Spiegelman, Quart. Rev. Biophys. **4**, 213–253 (1971).
3. C. Flamm, W. Fontana, I.L. Hofacker, and P. Schuster, RNA, (1999).
4. S.R. Morgan and P.G. Higgs, J. Chem. Phys. **105**, 7152–7157 (1996).
5. E. van Nimwegen, *The Statistical Dynamics of Epochal Evolution*, PhD Thesis, (Universiteit Utrecht, Utrecht, NL, 1999).
6. S.F. Elena, V.S. Cooper, and R.E. Lenski, Science **272**, 1802–1804 (1996).
7. W. Fontana and P. Schuster, Biophys. Chem. **26**, 123–147 (1987).
8. M.A. Huynen, P.F. Stadler, and W. Fontana, Proc. Natl. Acad. Sci. USA **93**, 397–401 (1996).
9. E. van Nimwegen, J.P. Crutchfield, and M. Mitchell, Phys. Lett. A **229**, 144–150 (1997).
10. W. Fontana and P. Schuster, J. Theor. Biol. **194**, 491–515 (1998).
11. W. Fontana and P. Schuster, Science **280**, 1451–1455 (1998).
12. M. Kimura, *The Neutral Theory of Molecular Evolution*, (Cambridge University Press, Cambridge, UK, 1983).
13. T. Ohta, Ann. Rev. Ecol. Syst. **23**, 263–286 (1992).
14. J.W. Drake, Proc. Natl. Acad. Sci. USA **88**, 7160–7164 (1991).
15. R.E. Lenski and M. Travisano, Proc. Natl. Acad. Sci. USA **91**, 6808–6814 (1994).
16. D. Papadopoulos, D. Schneider, J. Meier-Eiss, W. Arber, R.E. Lenski, and M. Blot, Proc. Natl. Acad. Sci. USA **96**, 3807–3812 (1999).
17. M. Kimura, Nature **217**, 624–626 (1968).
18. M. Eigen, Naturwissenschaften **58**, 465–523 (1971).
19. M. Eigen and P. Schuster, Naturwissenschaften **64**, 541–565 (1977).
20. C.J. Thompson and J.L. McBride, Math. Biosci. **21**, 127–142 (1974).
21. B.L. Jones, R.H. Enns, and S.S. Rangnekar, Bull. Math. Biol. **38**, 15–28 (1976).
22. J. Swetina and P. Schuster, Biophys. Chem. **16**, 329–345 (1982).
23. M. Nowak and P. Schuster, J. Theor. Biol. **137**, 375–395 (1989).
24. M. Eigen, J. McCaskill, and P. Schuster, Adv. Chem. Phys. **75**, 149–263 (1989).
25. I. Leuthäusser, J. Stat. Phys. **48**, 343–360 (1987).
26. N. Rohde, H. Daum, and C.K. Biebricher, J. Mol. Biol. **249**, 754–762 (1995).
27. E. Domingo, Viral Hepatitis Rev. **2**, 247–261 (1996).
28. E. Domingo and J.J. Holland, Ann. Rev. Microbiol. **51**, 151–178 (1997).

29. E. Domingo, L. Menéndez-Arias, M.E. Quinoñes-Mateu, A. Holguín, M. Gutierrez-Rivas, M.A. Martínez, J. Quer, and J.J. Holland, Prog. Drug. Res. **48**, 99–128 (1997).
30. R.W. Hamming, *Coding and Information Theory,* (Prentice Hall, Englewood Cliffs, NJ, 2nd ed., 1989).
31. R.W. Hamming, Bell Syst. Tech. J. **29**, 147–160 (1950).
32. J.S. McCaskill, J. Chem. Phys. **80**, 5194–5202 (1984).
33. P. Schuster and J. Swetina, Bull. Math. Biol. **50**, 635–660 (1988).
34. P. Tarazona, Phys. Rev. A **45**, 6038–6050 (1992).
35. D. Alves and J.F. Fontanari, Phys. Rev. E **54**, 4048–4053 (1996).
36. D. Alves and J.F. Fontanari, Phys. Rev. E **57**, 7008–7013 (1998).
37. P.R.A. Campos and J.F. Fontanari, Phys. Rev. E **58**, 2664–2667 (1998).
38. P.R.A. Campos and J.F. Fontanari, J. Phys. A **32**, L1–L7 (1999).
39. C. Reidys, P.F. Stadler, and P. Schuster, Bull. Math. Biol. **59**, 339–397 (1997).
40. R. Nussinov and A.B. Jacobson, Proc. Natl. Acad. Sci. USA **77**, 6309–6313 (1980).
41. M. Zuker and P. Stiegler, Nucleic Acids Research **9**, 133–148 (1981).
42. M. Zuker and D. Sankoff, Bull. Math. Biol. **46**, 591–621 (1984).
43. I.L. Hofacker, W. Fontana, P.F. Stadler, L.S. Bonhoeffer, M. Tacker, and P. Schuster, Mh. Chemie **125**, 167–188 (1994).
44. R.T. Batey, R.P. Rambo, and J.A. Doudna, Angew. Chem. Int. Ed. **38**, 2326–2343 (1999).
45. W. Grüner, R. Giegerich, D. Strothmann, C. Reidys, J. Weber, I.L. Hofacker, and P. Schuster, Mh. Chemie **127**, 355–374 (1996).
46. W. Grüner, R. Giegerich, D. Strothmann, C. Reidys, J. Weber, I.L. Hofacker, and P. Schuster, Mh. Chemie **127**, 375–389 (1996).
47. W. Fontana, D.A.M. Konings, P.F. Stadler, and P. Schuster, Biopolymers **33**, 1389–1404 (1993).
48. P. Schuster, W. Fontana, P.F. Stadler, and I.L. Hofacker, Proc. Roy. Soc. Lond. B **255**, 279–284 (1994).
49. W. Fontana, P.F. Stadler, E.G. Bornberg-Bauer, T. Griesmacher, I.L. Hofacker, M. Tacker, P. Tarazona, E.D. Weinberger, and P. Schuster, Phys. Rev. E **47**, 2083–2099 (1993).
50. I.L. Hofacker, P. Schuster, and P.F. Stadler, Discr. Appl. Math. **89**, 177–207 (1998).
51. P. Schuster, J. Biotechnol. **41**, 239–257 (1995).
52. P. Schuster, Physica D **107**, 351–365 (1997).
53. C.M. Reidys, Adv. Appl. Math. **19**, 360–377 (1997).
54. D.T. Gillespie, J. Comp. Phys. **22**, 403–434 (1976).
55. D.T. Gillespie, J. Phys. Chem. **81**, 2340–2361 (1977).
56. W. Fontana, W. Schnabl, and P. Schuster, Phys. Rev. A **40**, 3301–3321 (1989).
57. P. Schuster and W. Fontana, Physica D **133**, 427–452 (1999).
58. E. Zuckerkandl, J. Mol. Evol. **44** (Suppl. 1), S2–S8 (1997).
59. S. Wright, *The roles of mutation, inbreeding, crossbreeding and selection in evolution,* in: *Int. Proc. of the Sixth Int. Congress on Genetics*, D.F. Jones, (Ed.), Vol. 1, pp. 356–366 (1932).

Stochastic Resonance and Brownian Machinery: New Results – New Applications – New Goals

P. Hänggi

Universität Augsburg, Institut für Physik, Universitätstr. 1, 86135 Augsburg, Germany

Abstract. Is it possible to extract energy from random fluctuations and put it to beneficial use? This challenging question has provoked discussions ever since the early days of Brownian motion studies. Generally, noise in dynamical systems is considered a nuisance. But in certain nonlinear systems the presence of noise can in fact enhance the detection of weak signals. This phenomenon, called *Stochastic Resonance* (*SR*) does find useful applications in physical, technological and biomedical contexts [1]. In a second class of systems that are periodic – but which lack reflection symmetry – directed, noise induced transport can take place. The directed motion of particles in periodic potentials requires at least one source of non-equilibrium which must inherit an explicit or inherent statistical asymmetry. Such non-equilibrium systems have become known in the literature under the label of "ratchets" [2]. In both situations the importance of fluctuations is elevated to a level where noise must be viewed as source of order and complexity in its own right.

With this talk I present the basic principles that under-pin the exciting phenomenon of *SR*, and report on a series of novel applications such as its control in terms of harmonic mixing forces [3].

Turning to Brownian machinery we investigate whether microscopic fluctuations – such as Brownian motion – or even the haphazard motion of quantum particles, can cause particles to flow in one direction only. In recent years, this field has been the scene of remarkable activity, fueled mostly by the prospect of potentially high-profile technological and biological (molecular motors) applications. Only recently, though, this concept has been taken from the world of classical thermal Brownian motion to the quantum world [4], where the phenomenon of quantum tunneling presents a new challenge for the characterization of directed motion of electrons and alike. Prominent work of classical and quantum ratchet driven transport is reviewed. The focus will be mainly on *Quantum Brownian Rectifiers* [5], being driven by non-thermal perturbations that are unbiased on average.

Likewise, the rectification of quantum noise also impacts quantum diffusion, which can either be selectively enhanced or suppressed. The discussion of these theoretical results is supplemented by most recent experimental results of rectified electron motion taking place in triangular shaped quantum dot heterostructures, and in a periodic quantum ratchet device composed of asymmetric antidots [6–8]. As a result this new quantum concept can be used to the effect of constructing *shuttles* for electrons that operate on the nanometer scale.

References

1. L. Gammaitoni, P. Hänggi, P. Jung and F. Marchesoni, Rev. Mod. Phys. **70**, 223 (1998).
2. P. Hänggi and R. Bartussek, Lecture Notes in Physics, Vol. **476**, 294, (Springer, Berlin 1996).
   R. D. Astumian, Science **276**, 917 (1997).
   F. Jülicher, A. Ajdari, and J. Prost, Rev. Mod. Phys. **69**, 1269 (1997).
3. L. Gammaitoni, M. Löcher, A. Bulsara, P. Hänggi, J. Neff, K. Wiesenfeld, W. Ditto and M.E. Inchiosa, Phys. Rev. Lett. **82**, 4574 (1999).
4. P. Reimann, M. Grifoni, and P. Hänggi, Phys. Rev. Lett. **79**, 10 (1997).
5. I. Goychuk and P. Hänggi, Phys. Rev. Lett. **81**, 649 (1998).
   I. Goychuk and P. Hänggi, Europhys. Lett. **43**, 503 (1998).
6. For a feature article see P. Hänggi and P. Reimann, Physics World **12**, 21 (1999).
7. H. Linke, W. Zheng, A. Löfgren, H. Xu, P. Omling and P. E. Lindelof, Europhys. Lett. **44**, 341 (1998); **45**, 406 (E).
8. A. Lorke, S. Wimmer, B. Jager, J. P. Kotthaus, W. Wegscheider and M. Bichler, Physica B **312**, 249–251, (1998).

Studies of Bacterial Cooperative Organization

I. Golding, I. Cohen, and E. Ben-Jacob

School of Physics and Astronomy, Raymond and Beverly Sackler Faculty of Exact Sciences, Tel Aviv University, Tel Aviv 69 978, Israel

1 Introduction

During the course of evolution, bacteria have developed sophisticated cooperative behavior and intricate communication capabilities [1–3]. Utilizing these capabilities, bacterial colonies develop complex spatio-temporal patterns in response to adverse growth conditions. It is now understood that the study of cooperative self-organization of bacterial colonies is an exciting new multidisciplinary field of research, necessitating the merger of biological information with the physics of non-equilibrium processes and the mathematics of non-linear dynamics. At this stage, several experimental systems have been identified, and preliminary modeling efforts are making significant progress in providing a framework for the understanding of experimental observations [4–12]. This endeavour is not limited to bacteria alone. Studies have been performed of other types of microorganisms as well, such as amoeba [13] and yeast [14].

Fujikawa and Matsushita [5] reported for the first time that bacterial colonies could grow elaborate branching patterns of the type known from the study of fractal formation in the process of diffusion-limited-aggregation (DLA) [15]. It was shown explicitly that nutrient diffusion was the relevant dynamics responsible for the growth instability. Motivated by these observations, Ben-Jacob *et al.* [16,17,8] conducted new experiments to see how adaptive bacterial colonies could be in the presence of external "pressure". It was seen that the bacteria *Paenibacillus dendritiformis* exhibit branching patterns during growth on a poor substrate. Drawing on the analogy with diffusive patterning in non-living systems [18–21], complex patterns can be expected. The cellular reproduction rate that determines the growth rate of the colony is limited by the level of nutrients available for the cells. The latter is limited by the diffusion of nutrients towards the colony (for low nutrient substrate). Hence, colony growth under certain conditions should be similar to diffusion limited growth in non-living systems as mentioned above [20,21]. The study of diffusive patterning in non-living systems teaches us that the diffusion field drives the system towards decorated (on many length scales) irregular fractal shapes. Indeed, bacterial colonies can develop patterns reminiscent of those observed during growth in non-living systems. But, this is certainly not the end of the story. The colonies exhibit a richer behavior. This, ultimately, is a reflection of the additional levels of complexity involved when the building blocks of the colonies, the bacteria, are themselves living systems. Scientists now start to reveal the cybernetic processes (communication, regulation and control) which are part of the colonial adaptive organization.

How should one approach the modeling of the complex bacterial patterning? With present computational power it is natural to use computer models as a main tool in the study of complex systems. In a "generic model" [8,11,22,23] one tries to elicit, from the experimental observations and the biological knowledge, the generic features and basic principles needed to explain the biological behavior and to include these features in the model. Such modeling, with close comparison to experimental observations, can be used as a research tool to reveal new understanding of the biological systems.

The models employed for the studies of cooperative organization of microorganisms can be grouped into two main categories:

1. Discrete models such as the Communicating Walkers model of Ben-Jacob *et al.* [8] and the bions model of Kessler and Levine [22]. In this approach, the microorganisms (bacteria in the first model and amoebae in the latter) are represented by discrete, random walking entities (walkers and bions, respectively) which can consume nutrients, reproduce, perform random or biased movement, and produce or respond to chemicals. The time evolution of the chemicals is described by reaction-diffusion equations.
2. Continuous reaction-diffusion models [24]. In these models the microorganisms are represented via their 2D density, and a reaction-diffusion equation of this density describes their time evolution. This equation is coupled to the other reaction-diffusion equations for the chemical fields (e.g., nutrients).

2 Research Aim and Methods

Our research aims are to formulate a reaction-diffusion model to describe colonial development, and apply this model (with the necessary extensions and modifications) to study various issues of biological interest: The effects of chemotaxis on the colonial patterns, the expression of mutations, the formation of chiral branches, the formation of vortices, the development of concentric ring patterns, and the ways in which bacteria respond cooperatively to the application of antibiotics.

The reaction-diffusion equations are solved numerically on a two-dimensional tridiagonal lattice, with suitable initial conditions describing the distribution of bacteria and the various chemicals. The numerical simulation can in some cases be supplemented by an analytical study of the equation, for example a Mullins-Sekerka instability analysis of the propagating front [25,26]. The front velocity can sometimes be calculated using the solvability approach [18].

Predictions by the theoretical models are constantly compared with experimental results obtained in our laboratory, in which bacterial colonies are grown and observed macroscopically as well as microscopically. Results of the theoretical study are also used to design new experiments. A comparison is also made with results of the discrete "Communicating Walkers" model, obtained form simulations performed in our group.

Some of these efforts have already yielded interesting results [27–31].

3 A Reaction-Diffusion Model of Branching Growth

In the context of branching growth, reaction-diffusion models have been formulated recently by Cohen [32], Mimura and Matsushita *et al.* [33,34], Kawasaki *et al.* [35], Kitsunezaki [36] and Lacasta *et al.* [37]. The models are all two-dimensional (2D), with $b(\mathbf{x}, t)$ denoting the density of bacteria projected on a 2D plane.

In general, the rate of change of the bacteria density can be described by:

$$\frac{\partial b}{\partial t} = \nabla(D\nabla b) + f(b), \tag{1}$$

where the first term on the LHS describes the random motion ("diffusion") of the bacteria, and the second ("reaction") term describes their growth and death. As mentioned above, this equation is coupled to one or more equations describing the chemical fields in the system. This coupling can come from the dependence of either D or f on other fields (e.g., the nutrient $n(\mathbf{x}, t)$). The equations are then solved on a 2D lattice, with the initial conditions of a bacterial "innoculum" and a constant food level.

In [27] we have investigated several suggested models for bacterial growth, and identified the features which can lead to a branching growth (as opposed to a compact, circular colony). These are mainly:

1. A meta-stable reaction term $f(b)$. That is, $b = 0$ has to be a stable fixed-point [33,34]. This means that in order to initiate bacterial growth, some threshold density b^* must be reached. A related option is to incorporate a cutoff in the reaction term, setting the reaction to zero when the bacterial density is lower than the threshold. Unlike the idea of meta-stability, the cutoff is biologically motivated: It represents the discreteness of the bacteria [26].
2. A non-constant diffusion coefficient. Making D depend on bacterial density, e.g., in the form $D(b) \sim b^k (k > 0)$ (as suggested by Kitsunezaki [36]) can lead to branching growth (Fig. 1), and can be biologically justified as representing the bacterial motion within the layer of lubricant that they produce [29].

4 The Effect of Chemotaxis

Chemotaxis is a bias of movement according to the gradient of a chemical agent. Chemotactic signaling is a chemotactic response to an agent emitted by the bacteria [38]. We have incorporated the effects of chemotaxis toward food and of repulsive chemotactic signaling in the continuous models by introducing a *chemotactic flux* $\boldsymbol{J}_{\text{chem}}$, which is written (for the case of a chemorepellent and a linear diffusion) as [39]:

$$\boldsymbol{J}_{\text{chem}} = -b\chi(r)\nabla r, \tag{2}$$

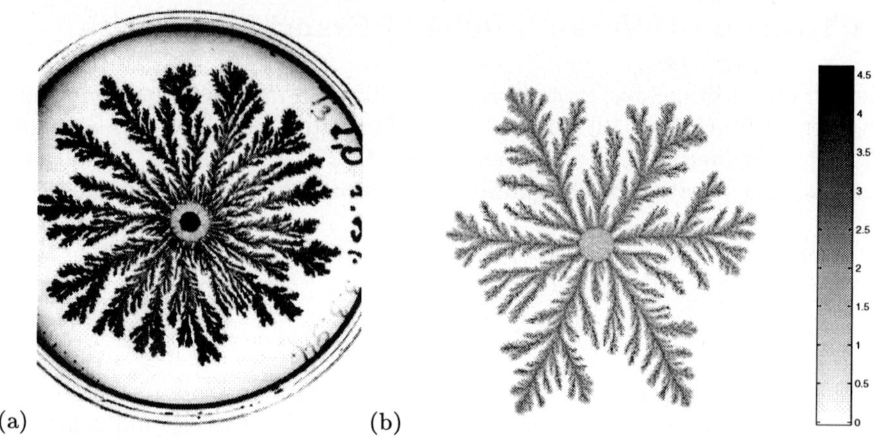

Fig. 1. (a) Typical branching pattern of *P. dendritiformis* var. *dendron* at intermediate agar concentration and peptone level (1 g/l, agar 1.75%). (b) 2D growth pattern of the non-linear diffusion model.

where $r(\mathbf{x}, t)$ is the concentration of chemorepellent and $\chi(r)$ is the chemotactic sensitivity to the repellent. In addition, we write an equation describing the diffusion and the production and decomposition of the chemorepellent.

When chemotaxis is included in the non-linear diffusion model, the results fit our predictions and compare well with those of the discrete "Communicating Walkers" model, as well as with experimental observations [27].

5 Sector Formation in Expanding Colonies

There is a well known observed (but rarely studied) phenomenon of bursts of new sectors of mutants during the growth of bacterial colonies [40,41], see Fig. 2a. If the mutants have the same growth dynamics as the "normal", wild-type, bacteria they will usually go unnoticed (unless some property such as coloring distinguishes them). If, however, the mutants have different growth dynamics, a distinguished sector with a different growth pattern might indicate their presence.

In [31], we generalize the non-linear diffusion model to study mutants. This is done by introducing two fields, for the densities of the wild-type bacteria ("type 1") and the mutants ("type 2"), and allowing some probability of transition from wild-type to mutants.

Following our experimental observations and the numerical studies we are able to appreciate what factors – geometrical, regulatory and others – favor the segregation of the mutant population. These factors include the expansion of the colony, the branching of the pattern, the effect of chemotaxis and advantages possessed by the individual mutants. Fig. 2b shows an example of our numerical results.

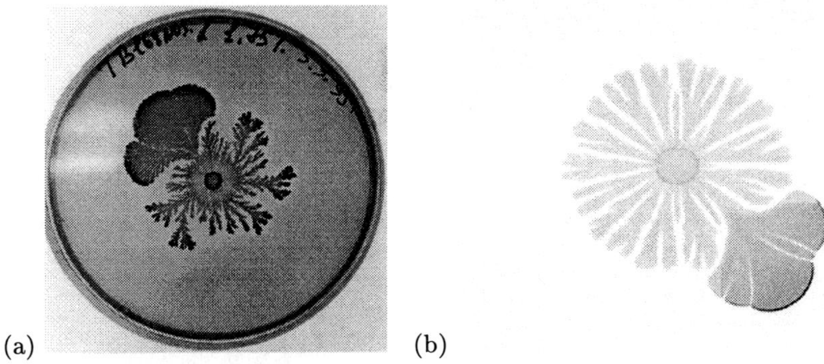

(a) (b)

Fig. 2. (a) Emerging sectors in branching colonies of *P. dendritiformis* , obtained at 1 g/l peptone and 1.75% agar. (b) Mutant with a higher sensitivity to repulsive chemotactic signaling, in a branching colony: Numerical simulation of the continuous model. The mutant erupts in a fan-like sector from the colony of wild-type bacteria.

6 Weak Chirality in *P. Dendritiformis*

Colonies of *P. dendritiformis* var. *dendron* grown on hard substrate exhibit branching patterns with a global twist with the same handedness (Fig. 3a). Similar observations during growth of other bacterial strains have also been reported [42,7]. We refer to such growth patterns as having weak chirality.

Ben-Jacob *et al.* [43] proposed that the high viscosity of the "lubrication" fluid during growth on a hard surface limits the rotation of bacteria while tumbling. They further assumed that the rotation should be relative to a specified direction, and used gradient of a chemotaxis signaling field (specifically, the long-range repellent chemotaxis) as a specific direction. It was shown in [43] that inclusion of the above features in the Communicating Walkers model indeed leads to a weak chirality which is highly reminiscent of the observed one.

In [30] we obtain weak chirality in the reaction-diffusion model by modifying the chemotactic mechanism and causing it to twist: We alter the expression for the chemotactic flux $\boldsymbol{J}_{\text{chem}}$ so that it is not oriented with the chemical gradient (∇R) anymore. Instead it is oriented with a rotated vector $\hat{\mathbf{R}}(\theta)\nabla R$, where $\hat{\mathbf{R}}(\theta)$ is the two-dimensional rotation operator and θ is the rotation angle. The effect of rotating the repulsive chemotaxis, as depicted in Fig. 3b, is to make the pattern chiral, with the degree of chirality determined by the rotation angle θ.

7 Bacterial Response to Antibiotics

Bacterial resistance to antibiotics is one of the major problems facing medicine today [44,45]. This resistance is in many cases obtained by a cooperative process, for example metabolic inactivation of the drug done outside the cells – a significant reduction in the drug's concentration can be done only with high

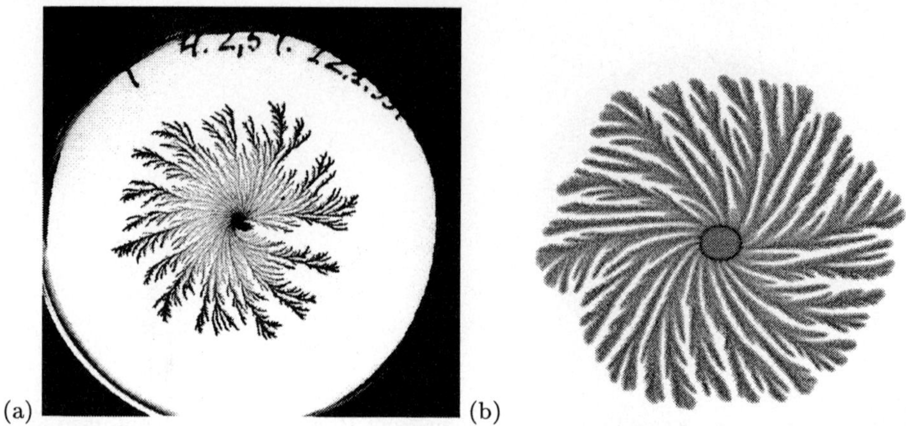

Fig. 3. (a) Weak chirality exhibited by the *P. dendritiformis* var. *dendron* during growth on 4 g/l peptone and 2.5% agar concentration. (b) Growth patterns of the non-linear diffusion model with a "squinting" repulsive chemotactic signaling, leading to weak chirality.

concentration of the in-activating agent which requires a dense population of bacteria. It is also known that resistance genes are distributed, in the form of plasmids, among the bacteria [46,47].

Our group is currently conducting an experimental study of the effect of various antibiotics on bacterial colonies. We intend to model these effects in the framework of a reaction-diffusion model. Our preliminary results show that while in some cases the observations can be explained within our existing models, e.g., the bacteria seem to apply repulsive chemotactic signaling, or simply seem to increase their death rate, there are , however, many cases in which the results cannot be explained that simply, and we plan to extend our models. Possible extensions include:

1. Explicitly including the antibiotics as an additional chemical field in the model, which is decomposed by the bacteria, and diffuses in the petri dish. Additional fields might also be necessary, such as enzymes produced by the bacteria for deactivating the antibiotics.
2. Describing the genetic communication between bacteria – how plasmids are distributed from one bacteria to the others around it, enabling them to resist the antibiotics.

8 Modeling the Chiral and Vortex Morphotypes

Ben-Jacob *et al.* [17,43,48] have isolated two new variants. The first, *P. dendritiformis* var. *chiralis* (also denoted $\mathcal{C}$ morphotype), is characterized by a strong twist (of a specific handedness) of the branches of it's colonies (Fig. 4). They have

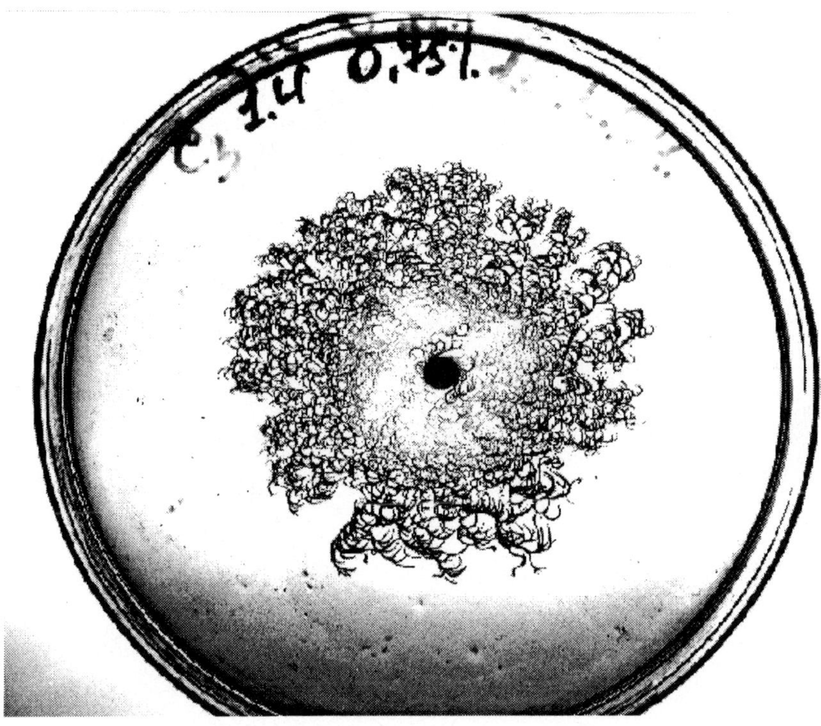

Fig. 4. Patterns exhibited by the $\mathcal{C}$ morphotype for 1.4 g/l peptone level and 0.75% agar concentration.

suggested that the bacterial flagella, acting as a singular perturbation, leads to the observed chirality. The mechanism seems to involve a cell-cell co-alignment which limits the average rotation of the bacteria.

A typical pattern exhibited by the second variant, *P. vortex* (also denoted $\mathcal{V}$ morphotype), is shown in Fig. 5. Each branch is produced by a leading droplet of cells and emits side-branches, each with his own leading droplet. In many cases the branches have a well defined global twist. Each leading droplet consists of several to millions of cells that rotate around a common center (hence, the name vortex).

In order to model these morphotypes in a reaction-diffusion framework, the random "diffusion" of the bacteria must be replaced by a much more complex "swarming" behavior, in which bacteria are treated as a "bio-fluid", for which a Navier-Stokes-like equation is written, describing the time evolution of the bacterial velocity field. We are currently pursuing this direction.

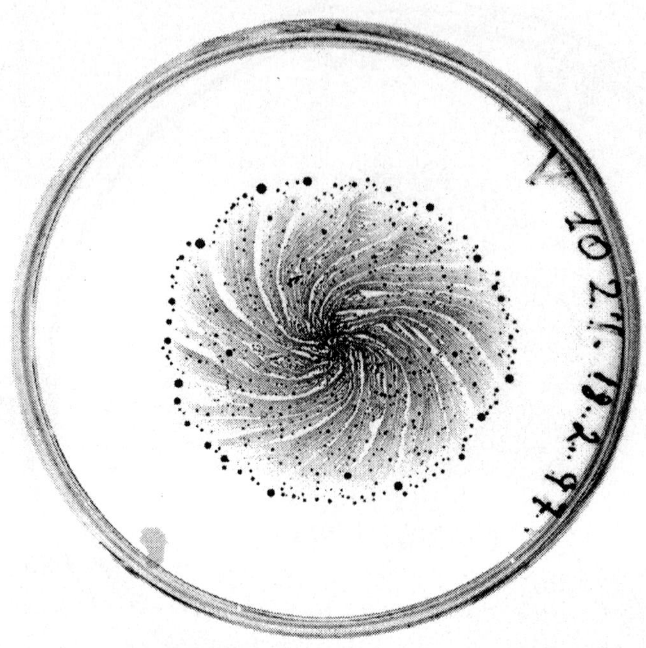

Fig. 5. Patterns exhibited by the $\mathcal{V}$ morphotype for 10 g/l peptone level and 2% agar concentration.

9 Modeling the Formation of Concentric-Rings Patterns

Under conditions of high agar concentration and intermediate to high peptone levels, *P. dendritiformis* exhibits patterns of concentric rings (Fig. 6). Similar patterns have also been observed in other species [49]. Observations of *Bacillus subtilis* reveal that colonial dynamics is periodic, with repeating cycles of fast expansion followed by reproduction and differentiation. Several researchers have tried to understand and model this process, which incorporates a high degree of synchronization [34,12,37]. The models proposed thus-far seem to be lacking, since they either fail to capture the correct dynamics of the colony (only the ring pattern), or in other cases include some global phenomenological parameter which enables them to bypass the problem of synchronization.

We are currently investigating this problem, and we believe that chemical communication might have a role in the synchronization of the different growth phases across the colony.

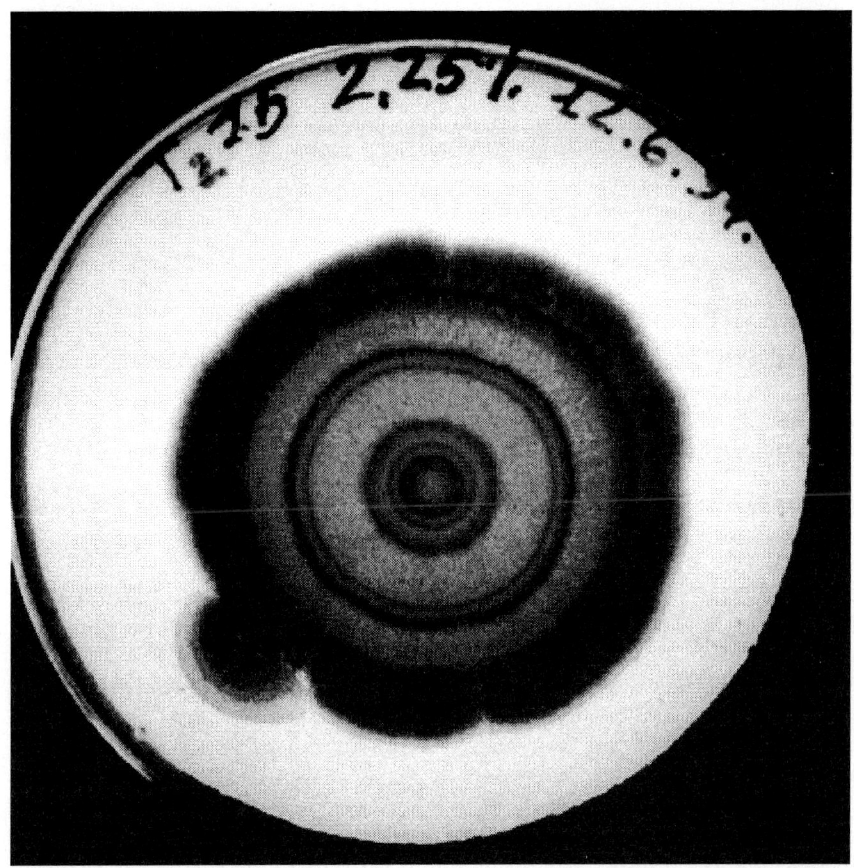

Fig. 6. Concentric rings of *P. dendritiformis* in a compact growth for 15 g/l peptone level and 2.25% agar concentration.

References

1. J.A. Shapiro, *Bacteria as multicellular organisms*, Sci. Am. **258**, 62–69 (1988).
2. E. Ben-Jacob, H. Levine, and I. Cohen, *Cooperative self-organization of microorganisms*, to appear in: Adv. Phys.
3. E. Ben-Jacob and H. Levine, *The artistry of microorganisms*, Sci. Am. **279**, 82–87 (1998).
4. N.H. Mendelson and B. Salhi, *Patterns of reporter gene expression in the phase diagram of Bacillus subtilis colony forms,* J. Bacteriol. **178**, 1980–1989 (1996).
5. H. Fujikawa and M. Matsushita, *Fractal growth of Bacillus subtilis on agar plates,* J. Phys. Soc. Jap. **58**, 3875–3878 (1989).
6. T.J. Pedley and J.O. Kessler, *Bioconvection*, Sci. Prog. **76**, 105–123 (1989).
7. T. Matsuyama, R.M. Harshey, and M. Matsushita, *Self-similar colony morphogenesis by bacteria as the experimental model of fractal growth by a cell population,* Fractals **1**, 302–311 (1993).

8. E. Ben-Jacob, O. Shochet, A. Tenenbaum, I. Cohen, A. Czirók, and T. Vicsek, *Generic modeling of cooperative growth patterns in bacterial colonies*, Nature **368**, 46–49 (1994).
9. E. Ben-Jacob, I. Cohen, O. Shochet, I. Aronson, H. Levine, and L. Tsimering, *Complex bacterial patterns*, Nature **373**, 566–567 (1995).
10. D.E. Woodward, R. Tyson, M.R. Myerscough, J.D. Murray, E.O. Budrene, and H.C. Berg, *Spatio-temporal patterns generated by salmonella typhimurium*, Biophys. J. **68**, 2181–2189 (1995).
11. J.O. Kessler and M.F. Wojciechowski, *Collective behavior and dynamics of swimming bacteria*, in: *Bacteria as Mullticellular Organisms*, J.A. Shapiro and M. Dworkin, (Eds.), pp. 417–450 (Oxford University Press Inc., New York, 1997).
12. S.E. Esipov and J.A. Shapiro, *Kinetic model of proteus mirabilis swarm colony development*, J. Math. Biol. **36**, 249–268 (1998).
13. H. Levine, I. Aranson, L. Tsimring, and T.V. Truong, *Positive genetic feedback governs camp spiral wave formation in Dictyostelium*, Proc. Nat. Acad. Sci. USA **93**, 6382–6 (1996).
14. E. Boschke and Th. Bley, *Growth patterns of yeast colonies depending on nutrient supply,* Acta Biotechnol. **18**, 17–27 (1998).
15. T.A. Witten and L.M. Sander, *Diffusion-limited aggregation, a kinetic critical phenomenon,* Phys. Rev. Lett. **47**, 1400 (1981).
16. E. Ben-Jacob, H. Shmueli, O. Shochet, and A. Tenenbaum, *Adaptive self-organization during growth of bacterial colonies,* Physica A **187**, 378–424 (1992).
17. E. Ben-Jacob, A. Tenenbaum, O. Shochet, and O. Avidan, *Holotransformations of bacterial colonies and genome cybernetics,* Physica A **202**, 1–47 (1994).
18. D.A. Kessler, J. Koplik, and H. Levine, *Pattern selection in fingered growth phenomena,* Adv. Phys. **37**, 255 (1988).
19. J.S. Langer, *Dendrites, viscous fingering, and the theory of pattern formation,* Science **243**, 1150–1154 (1989).
20. E. Ben-Jacob and P. Garik, *The formation of patterns in non-equilibrium growth,* Nature **343**, 523–530 (1990).
21. E. Ben-Jacob, *From snowflake formation to the growth of bacterial colonies,* Part I: *Diffusive patterning in non-living systems,* Contemp. Phys. **34**, 247–273 (1993).
22. D.A. Kessler and H. Levine, *Pattern formation in dictyostelium via the dynamics of cooperative biological entities,* Phys. Rev. E **48**, 4801–4804 (1993).
23. M.Y. Azbel, *Survival-extinction transition in bacteria growth,* Europhys. Lett. **22**, 311–316 (1993).
24. H. Parnas and L. Segel, *A computer simulation of pulsatile aggregation in Dictyostelium discoideum,* J. Theor. Biol. **71**, 185–207 (1978).
25. W.W. Mullins and R.F. Sekerka, *Stability of a planar interface during solidification of a dilute binary alloy,* Appl. Phys. **35**, 444 (1964).
26. D.A. Kessler and H. Levine, *Fluctuation-induced diffusive instabilities,* Nature **394**, 556–558 (1998).
27. I. Golding, Y. Kozlovsky, I. Cohen, and E. Ben-Jacob, *Studies of bacterial branching growth using reaction-diffusion models of colonial development,* Physica A **260**, 510–554 (1998).
28. I. Cohen, I. Golding, Y. Kozlovsky, and E. Ben-Jacob, *Continuous and discrete models of cooperation in complex bacterial colonies*, Fractals **7**, 235–247 (1999).
29. Y. Kozlovsky, I. Cohen, I. Golding, and E. Ben-Jacob, *Lubricating bacteria model for branching growth of bacterial colonies*, Phys. Rev. E **59**, 7025–7035 (1999).

30. E. Ben-Jacob, I. Cohen, I. Golding, and Y. Kozlovsky, *Modeling branching and chiral colonial patterning of lubricating bacteria*, in: *Proceedings of 1998 IMA workshop: Pattern Formation and Morphogenesis*, in press, (Springer, Berlin, 1999).
31. I. Golding, I. Cohen, and E. Ben-Jacob, *Studies of sector formation in expanding bacterial colonies*, in press, Europhys. Lett.
32. I. Cohen, *Mathematical Modeling and Analysis of Pattern Formation and Colonial Organization in Bacterial Colonies*, MSc Thesis, (Tel-Aviv University, Israel, 1997).
33. M. Mimura, H. Sakaguchi, and M. Matsushita, *A reaction-diffusion approach to bacterial colony formation*, preprint, (1997).
34. M. Matsushita, J. Wakita, H. Itoh, I. Rafols, T. Matsuyama, H. Sakaguchi, and M. Mimura, *Interface growth and pattern formation in bacterial colonies,* Physica A **249**, 517–524 (1998).
35. K. Kawasaki, A. Mochizuki, M. Matsushita, T. Umeda, and N. Shigesada, *Modeling spatio-temporal patterns created by bacillus-subtilis,* J. Theor. Biol. **188**, 177–185 (1997).
36. S. Kitsunezaki, *Interface dynamics for bacterial colony formation,* J. Phys. Soc. Jpn. **66**, 1544–1550 (1997).
37. A.M. Lacasta, I.R. Cantalapiedra, C.E. Auguet, A. Peñaranda, and L. Ramírez-Piscina, *Modeling of spatiotemporal patterns in bacterial colonies*, in press, Phys. Rev. E.
38. E.O. Budrene and H.C. Berg, *Dynamics of formation of symmetrical patterns by chemotactic bacteria,* Nature **376**, 49–53 (1995).
39. J.D. Murray, *Mathematical Biology,* (Springer, Berlin, 1989).
40. J.A. Shapiro and D. Trubatch, *Sequential events in bacterial colony morphogenesis,* Physica D **49**, 214–223 (1991).
41. A. Grondin, H.C. Jarell, and L.R. Berube, *Discontinuous expansion linked to sector formation in pseudomonas aeruginosa colonies*, Archives of Microbiol. **172**, 59–62 (1999).
42. T. Matsuyama and M. Matsushita, *Fractal morphogenesis by a bacterial cell population,* Crit. Rev. Microbiol. **19**, 117–135 (1993).
43. E. Ben-Jacob, I. Cohen, O. Shochet, A. Czirók, and T. Vicsek, *Cooperative formation of chiral patterns during growth of bacterial colonies,* Phys. Rev. Lett. **75**, 2899–2902 (1995).
44. C.F. Amáblie-Cueva, M. Cárdenas-García, and M. Ludgar, *Antibiotic resistance,* Am. Sci. **83**, 320–329 (1995).
45. World Health Organization, *Report of The Director-General,* in: *The World Health Report 1996: Fighting Disease, Fostering Development,* (Geneva, 1996).
46. S.B. Levy, *The challenge of antibiotic resistance,* Sci. Am., (1998).
47. R.V. Miller, *Bacterial gene swapping in nature,* Sci. Am. **278**, (1998).
48. E. Ben-Jacob, I. Cohen, A. Czirók, T. Vicsek, and D.L. Gutnick, *Chemomodulation of cellular movement and collective formation of vortices by swarming bacteria and colonial development,* Physica A **238**, 181–197 (1997).
49. I. Rafols, *Formation of concentric rings in bacterial colonies*, MSc Thesis, (Chuo University, Japan, 1998).

Collective Motion and Optimal Self-Organisation in Self-Driven Systems

T. Vicsek[1,2], A. Czirók[1,2], and D. Helbing[1,2]

[1] Department of Biological Physics, Eötvös University, Budapest, Pázmány Péter Sétány 1A, 1117 Budapest, Hungary
[2] Collegium Budapest – Institute for Advanced Study, 1014 Budapest, Hungary

Abstract. We discuss simulations of flocking and the principle of optimal self-organisation during self-driven motion of many similar objects. In addition to driving and interaction the role of fluctuations is taken into account as well. In our models, the particles corresponding to organisms locally interact with their neighbours by choosing at each time step a velocity depending on the directions of motion of them. Our numerical studies of flocking indicate the existence of *new types of transitions*. As a function of the control parameters both disordered and long-range ordered phases can be observed, and the corresponding phase space domains are separated by singular "critical lines". In particular, we demonstrate both numerically and analytically that there is a disordered-to-ordered-motion transition at a finite noise level even in one dimension. We also present computational and analytical results indicating that driven systems with repulsive interactions tend to reach an optimal state corresponding to minimal interaction. This *extremal principle* is expected to be relevant for a class of biological and social systems involving driven interacting entities.

In this paper we address the question whether there are some global, perhaps universal features of collective motion. Such behaviour takes place when many organisms are simultaneously moving and parameters like the level of *perturbations* or the mean *distance* between the individuals is changed.

The collective motion of various organisms (flocking of birds, for example) is a very common phenomenon. In addition to the aesthetic aspects, studies on collective motion can have interesting applications as well: a better understanding of the swimming patterns of large schools of fish can be useful in the context of large scale fishing strategies, or modelling the motion of a crowd of people can help urban designers.

1 Phase Transitions as a Function of the Level of Noise in the Collective Motion of Self-Propelled Particles

The motion of organisms is usually controlled by interactions with other organisms in their neighbourhood and randomness also plays an important role. Recently, several models have been proposed to capture the main features of the collective motion of organisms [1] such as schools of fish, herds of quadrupeds, flocks of birds [2], groups of bacteria, ants [3] or pedestrians. A typical feature of the models leading to collective motion are the assumption that the particles (organisms): i) are driven (propelled) with some constant or average velocity and ii) tend to follow each other. In [4] a model of *self-propelled particles* (SPP) was introduced which displayed interesting novel transitions of collective motion as a function of the magnitude of the perturbations and the density of the particles. In this model the fluctuations are assumed to be uncorrelated and have a zero average.

The models we shall discuss are transport-related, non-equilibrium analogues of the ferromagnetic models [5]. The analogy is as follows: the Hamiltonian tending to align the spins in the same direction in the case of equilibrium ferromagnets is replaced by the rule of aligning the direction of motion of particles, and the amplitude of the random perturbations can be considered to be proportional to the temperature. From a hydrodynamical point of view, in the SPP systems the momentum of the particles is *not* conserved. Thus, the flow field emerging in these models can considerably differ from the usual behaviour of fluids.

It appears that self-propelled particles display qualitatively different behaviours as a function of the dimension of the space they are embedded into. Correspondingly, we shall treat these cases separately, starting with the highly non-trivial one-dimensional case.

1.1 Phase Transition in One Dimension

In 1d the particles cannot get around each other and some of the features of the dynamics present in higher dimensions are not found. On the other hand, motion in 1d implies new interesting aspects, e.g., groups of the particles have to be able to change their direction for the opposite in an organised manner. The model should be such that it takes into account the specific crowding effects typical for 1d (the particles can slow down before changing direction and dense regions may be built up of momentarily oppositely moving particles). In a way the system studied below can be considered as a model of organisms (people, for example), moving in a narrow channel. Imagine that a fire alarm goes on, the tunnel is dark, smoky, everyone is extremely excited. People are both trying to follow the others in order to escape together and behave in an erratic manner due to smoke and excitement.

Thus, we consider N off-lattice particles along a line of length L. The particles are characterised by their coordinate x_i and dimensionless velocity u_i which are updated as

$$x_i(t + \Delta t) = x_i(t) + v_0 u_i(t) \Delta t, \tag{1}$$

$$u_i(t + \Delta t) = G\Big(\langle u(t) \rangle_{S(i)}\Big) + \xi_i, \tag{2}$$

where ξ_i is a random number drawn with a uniform probability from the interval $[-\eta/2, \eta/2]$. The local average velocity $\langle u \rangle_{S(i)}$ for the ith particle is calculated over the particles located in the interval $[x_i - \Delta, x_i + \Delta]$, where we fix $\Delta = 1$. The function G *tends to set the velocity both, close to the average of the neighbours and to a prescribed value* v_0: $G(u) > u$ for $u < 1$ and $G(u) < u$ for $u > 1$. In the numerical simulations [6] one of the simplest choices for G was implemented as

$$G(u) = \begin{cases} (u+1)/2 & \text{for } u > 0, \\ (u-1)/2 & \text{for } u < 0, \end{cases} \tag{3}$$

and random initial and periodic boundary conditions were applied. According to our simulations, for not too large, but finite perturbations the system reaches an ordered state characterised by a spontaneous broken symmetry and clustering of the particles (cf. Fig. 1, left). In contrast, for large noise, the system remains in a disordered state. The transition is continuous, with an order parameter behaving, as a function of the overall density ϱ, according to a power law with an exponent $\beta = 0.60 \pm 0.05$. This exponent is different from both, the the mean-field value 1/2 and the value $\beta = 0.42 \pm 0.03$ found in 2d.

To understand the phase transitions observed in the various SPP models, efforts have been made to set up a continuum theory in terms of $v(x,t)$ and $\rho(x,t)$, representing the coarse-grained velocity and density fields, respectively. The first approach [7] has been made by Toner and Tu for $d > 1$. They were able to treat the problem analytically and show the existence of an ordered phase in 2d, but their theory does not allow an ordered phase in 1*d*.

However, as we have shown, there exist SPP systems in *one* dimension which exhibit an ordered phase for low noise level. This finding prompts the need of constructing an alternative continuum model for 1d [6]:

$$\partial_t u = f(u) + \mu^2 \partial_x^2 u + \alpha \frac{(\partial_x u)(\partial_x \rho)}{\rho} + \zeta, \tag{4}$$

$$\partial_t \rho = -v_0 \partial_x (\rho u) + D \partial_x^2 \rho. \tag{5}$$

Here, $u(x,t)$ is the coarse-grained dimensionless velocity field, $f(u)$ is an anti-symmetric function with $f(u) > 0$ for $0 < u < 1$ and $f(u) < 0$ for $u > 1$, and the noise term satisfies $\overline{\zeta} = 0$, and $\overline{\zeta^2} = \sigma^2/\rho\tau^2$. These equations are different both from the equilibrium field theories and from the non-equilibrium system defined by Toner and Tu. The main difference comes from the nature of the coupling term $(\partial_x u)(\partial_x \rho)/\rho$. This term can be derived, but here we only present a plausible interpretation for its origin.

The main point is that one would like to have a term in the equation for the velocity which would result in the slowing down (and eventually in the "turning back") of the particles under the influence of a larger number of particles moving oppositely. When two groups of particles move in the opposite direction, the density locally increases and the velocity decreases at the point they meet. Let us consider a particular case, when particles move from left to right and the velocity is locally decreasing while the density is increasing as x increases (particles are moving towards a "wall" formed between two oppositely moving groups). The term $(\partial_x u)$ is less, the term $(\partial_x \rho)$ is larger than zero in this case. Together they have a negative sign resulting in the slowing down of the local velocity. This is a consequence of the fact that there are more slower particles (in a given neighbourhood) in the forward direction than faster particles coming from behind, so the average interaction experienced by a particle in the point x slows it down (the particle is trying to take the average velocity of its neighbours and simultaneously move with a preferred velocity). Thus, the $(\partial_x u)(\partial_x \rho)/\rho$ term does what we expect from it.

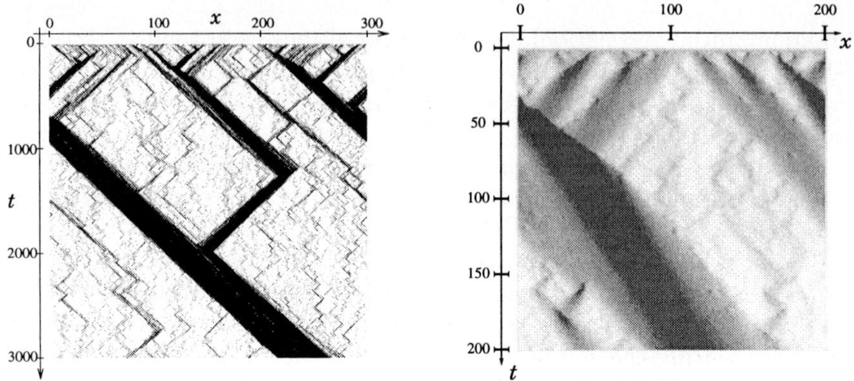

Fig. 1. Left: Result of a microsimulation of the one-dimensional SPP model given by (1) to (3). Right: Solution of the continuum equations (4) and (5) for collective motion in one-dimension. In this space-time (horizontal-vertical) plot the regions of higher densities are indicated with darker shades.

The above continuum equations can be investigated by both stability analysis and numerical integration. Here, we demonstrate our numerical results in the form of a space-time plot of the density $\rho(x,t)$ (Fig. 1, right). A darker band going from left to right for increasing time corresponds to a group of particles moving in an orderly fashion from left to right. The initial conditions are random. Correspondingly, there are many small groups moving in different directions. In the course of time, the groups merge (aggregate) and sooner or later join into a single large group moving in one direction which is determined in a non-trivial way by the initial conditions.

1.2 Flocking in Two Dimensions

In this case a simpler model (introduced in [4]) can be used to describe the dynamics. It consists of particles moving on a plane with periodic boundary condition. The particles are characterised by their (off-lattice) location $\boldsymbol{x}_i$ and velocity $\boldsymbol{v}_i$ pointing into the direction ϑ_i. To account for the self-propelled nature of the particles, the magnitude of the velocity is fixed to v_0. A simple local interaction is defined in the model: at each time step, a given particle assumes the *average direction of motion* of the particles in its local neighbourhood $S(i)$, but with some uncertainty ξ:

$$\vartheta_i(t + \Delta t) = \langle \vartheta(t) \rangle_{S(i)} + \xi. \tag{6}$$

Here, the noise ξ is a random variable with a uniform distribution in the interval $[-\eta/2, \eta/2]$. The locations of the particles are updated as

$$\boldsymbol{x}_i(t + \Delta t) = \boldsymbol{x}_i(t) + v_0 \boldsymbol{v}_i(t) \Delta t, \tag{7}$$

with $|\boldsymbol{v}_i| = 1$.

We have studied the scaling properties of this model in details and the results are available in [8].

In related simulations [9–11], the boundary conditions were changed to *reflective circular walls* to gain insight into the importance of the boundary conditions in the SPP models. To avoid singular behaviour at the boundaries, i.e., the aggregation of particles in a narrow zone, a *short range "hard-core" repulsion* was also incorporated into the model: e.g., in the simulations described in [9], if the distance $d_{ij} = |\boldsymbol{x}_j - \boldsymbol{x}_i|$ between two particles i and j was smaller than a certain ϵ^* value they repelled each other, and the direction of their motion was given by the following expression instead of (6):

$$\vartheta_i(t + \Delta t) = \Phi\Big(- \sum_{j \neq i, d_{ij} < \epsilon^*} \mathbf{N}\big(\boldsymbol{x}_j(t) - \boldsymbol{x}_i(t)\big)\Big). \tag{8}$$

Here, $\mathbf{N}(\boldsymbol{u})$ denotes $\boldsymbol{u}/|\boldsymbol{u}|$, and $\Phi(\boldsymbol{z})$ is the polar angle of the vector $\boldsymbol{z} = (|\boldsymbol{z}| \cos \Phi, |\boldsymbol{z}| \sin \Phi)$.

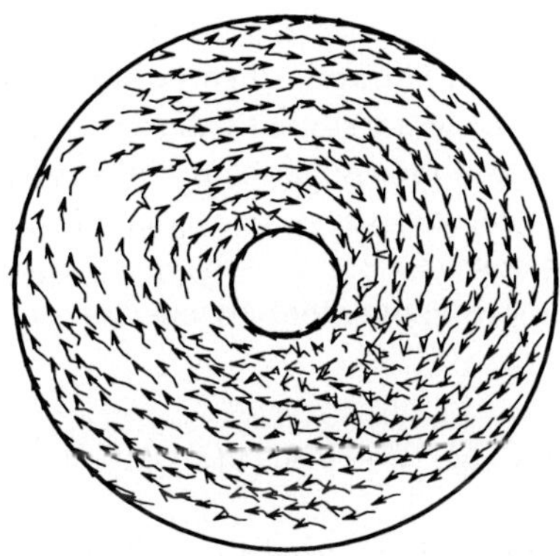

Fig. 2. A possible stationary state of the two-dimensional SSP model with reflective boundary conditions. In such simulations rotation of the particles develops (see Fig. 2) in the high-density, low-noise regime. The direction of the rotation is selected by spontaneous symmetry breaking. Thus, both clockwise and anti-clockwise spinning "vortices" can emerge. (After [9].)

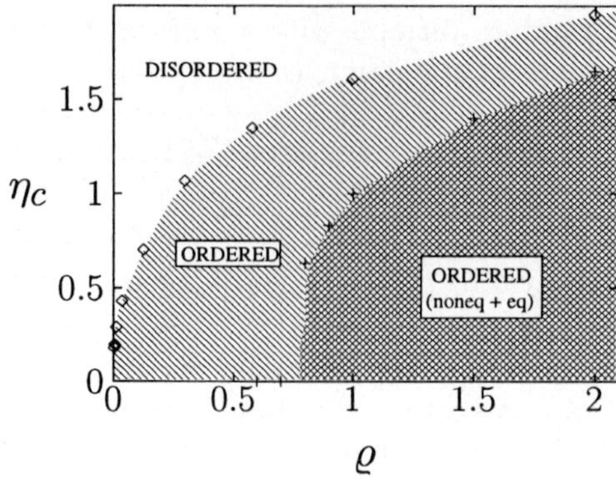

Fig. 3. Phase diagram of the 3d SPP model and the corresponding ferromagnetic system. The diamonds show our numerical estimates for the critical noise for a given density for the SPP model, and the crosses show the same for the static case. The SPP system becomes ordered in the whole region below the curved line connecting the diamonds, while in the static case the ordered region extends only down to the percolation threshold $\rho \simeq 0.8$. (After [12].)

1.3 Phase Diagram of the Three-dimensional Case

In two dimensions, an effective long range interaction is built up, because the migrating particles have a high chance to get close to each other and interact, in contrast to the situation in three dimensions (where, as is well-known, random trajectories do not meet). The less frequent interaction in 3d disfavours ordering. On the other hand, in three dimensions even conventional ferromagnets order. Thus, it is interesting to see how these two competing features determine the behaviour of 3d SPP systems.

For the 3d case it is more convenient to use the equations in the following form [12]

$$\boldsymbol{v}_i(t+\Delta t) = \mathbf{N}\Big(\mathbf{N}(\langle \boldsymbol{v}(t)\rangle_{S(i)}) + \boldsymbol{\xi}\Big), \tag{9}$$

where the noise $\boldsymbol{\xi}$ is uniformly distributed in a sphere of radius η. The positions of the particles are updated as in (7).

The simulations were started from a disordered configuration. $S(i)$ was assumed to be a sphere of unit radius centred around the position of the ith particle. After some relaxation time, a steady state emerged. Our numerical results suggest the existence of a continuous kinetic phase transition as $L \to \infty$.

Next we discuss the role of the overall density ϱ. We observed that the long-range ordered phase was present for any ϱ, but for a fixed value of η vanished with decreasing ϱ. To demonstrate how much this behaviour is different from that of diluted ferromagnets, we have also determined the location of the transition to an ordered state for $v_0 = 0$. In this limit our model reduces to an equilibrium system of randomly distributed "spins" with a ferromagnetic-like interaction. This system is analogous to the three-dimensional diluted Heisenberg model, and we find a major difference between this *static* and the self-propelled case: In the static case the system *does not order* for densities below a critical value of about 0.8 (see, e.g., [13]) which in the units we are using corresponds to the percolation threshold of randomly distributed spheres in 3d.

This situation is demonstrated in the phase diagram shown in Fig. 3. Here, the diamonds show our numerical estimates for the critical noise at a given density for the SPP model and the crosses show the same for the static case. The SPP system becomes ordered in the whole region below the curved line connecting the diamonds, while in the static case, the ordered region extends only down to $\rho \simeq 0.8$.

2 Self-Optimisation in Systems of Driven Particles

One of the main features of the above models was that the particles (individual organisms) had the tendency to align their direction of motion with their neighbours. This is a common situation in the living world, however, there are some other relevant systems in which their individual elements ("entities") have conflicting interests. A simple and relevant example corresponds to pedestrians heading into opposite directions along a corridor.

A central question arising in the context of such systems is whether their stationary state is optimal for the individuals or, due to the conflicting interests built into the models, the long-time behaviour itself remains far from optimal (there remains a higher-then-optimal permanent level of collisions or "conflicts" in the system). In the case of pedestrians, we find that they spontaneously organise their motion into lanes minimising the amount of interactions in the system [14] (Fig. 4).

Extremal principles are fundamental in our interpretation of phenomena in nature. One of the best known examples is the second law of thermodynamics [15–17], governing most physical and chemical systems and stating the continuous increase of entropy ("disorder") in closed systems. Most systems in our natural environment, however, are open, which is true for driven physical systems, but even more for biological, economic, and social systems. As a consequence, these systems are often characterised by self-organised structures [16–20], which calls for principles that apply on time scales shorter or comparable to the life spans of these systems.

Recent simulations point to the possible existence of optimality principles in certain kinds of driven multi-particle or multi-agent systems. As examples we mention, apart from lane formation in pedestrian crowds [21],

1. particle-size segregation in sheared granular media [22],
2. the self-organisation of coherent motion in a mixture of cars and lorries [23], and
3. the evolution of trail systems [20].

Fig. 4. Formation of lanes of uniform walking directions in crowds of oppositely moving pedestrians. Black circles represent pedestrians walking from left to right, white ones represent pedestrians walking into the opposite direction. The simulation assumed periodic boundary conditions, but we could also use walls on both sides or randomly feed pedestrians into the left and right boundaries of the simulation area, without destroying the effect of lane formation (see the Java simulation applets available at `http://www.theo2.physik.uni-stuttgart.de/helbing/`).

In these systems, the respective interacting entities (pedestrians, driver-vehicle units, or particles) have to coordinate each other in order to reach a system state which is "favourable" to them. It was conjectured [20,23] that the resulting system states are optimal in some sense, but there are many open questions to be addressed:

1. Is there really a quantity which is optimised by the self-organising system in the course of time?
2. If yes, which quantity is it? Is there a systematic way to derive it?
3. What are the conditions for the existence of such a quantity?
4. Is there any systematic connection between self-organisation and optimisation?
5. If optimal self-organised systems exist at all, are they exceptional or quite common?

To illustrate the non-trivial aspects of optimal self-organisation, let us consider the dynamics of pedestrian crowds. We focus on a system of oppositely moving pedestrians in a corridor, for which lane formation has been observed in empirical studies [24].

By $\boldsymbol{x}_\alpha(t)$ we denote the position of pedestrian α at time t, by $\boldsymbol{v}_\alpha(t) = d\boldsymbol{x}_\alpha(t)/dt$ his/her actual velocity, by v_0 the absolute value of his/her desired velocity, and by $\boldsymbol{e}_\alpha$ the desired direction of motion. Then, the equation mimicing pedestrian motion reads

$$\boldsymbol{v}_\alpha(t) = v_0 \boldsymbol{e}_\alpha + \sum_{\beta(\neq\alpha)} \boldsymbol{f}_{\alpha\beta}\big(d_{\alpha\beta}(t)\big) \tag{10}$$

(in the over-damped limit). $\boldsymbol{f}_{\alpha\beta}$ represents repulsive interactions between pedestrians α and β, which were assumed to decrease monotonically with their distance $d_{\alpha\beta}(t) = \|\boldsymbol{x}_\alpha(t) - \boldsymbol{x}_\beta(t)\|$. We will not specify the exact form of $\boldsymbol{f}_{\alpha\beta}$, since it turns out to be quite irrelevant for the kind of phenomena we want to describe, here, even if they are velocity-dependent.

For two opposite desired walking directions, simulations of the above model reproduce the observed formation of lanes of uniform directions of motion (Fig. 4). It is clear that lane formation will maximise the average velocity in the respective desired walking direction and, therefore, the quantity

$$E(t) = \frac{\langle\langle \boldsymbol{v}_\alpha \cdot \boldsymbol{e}_\alpha \rangle_\alpha \rangle_t}{v_0} = \frac{1}{N} \sum_{\alpha=1}^{N} \frac{1}{T} \int\limits_{t-T/2}^{t+T/2} dt' \, \frac{\boldsymbol{v}_\alpha(t') \cdot \boldsymbol{e}_\alpha}{v_0} \le 1 \, , \tag{11}$$

which is a measure of the "efficiency" or "success" of motion. (Here, $\langle\langle . \rangle_\alpha \rangle_t$ denotes the average over the pedestrians and over time, and $T > 0$ is a suitable time window.) Moreover, optimisation of efficiency immediately implies that the system minimises the quantity

$$\left\langle\left\langle - \sum_{\beta(\ne\alpha)} \boldsymbol{f}_{\alpha\beta} \cdot \boldsymbol{e}_\alpha \right\rangle_\alpha \right\rangle_t = v_0 - \langle\langle \boldsymbol{v}_\alpha \cdot \boldsymbol{e}_\alpha \rangle_\alpha \rangle_t = v_0(1 - E) \, , \tag{12}$$

i.e., the average interaction intensity opposite to the respective desired direction of motion.

Hence, even without the use of difficult mathematics we could show that *the system minimises the interaction strengths* of the pedestrians, if it shows segregation into lanes of uniform directions of motion. Note, however, that lane formation is not a trivial effect of this model, but eventually arises only due to the smaller relative velocity and interaction rate that pedestrians with the same walking direction have [14]. In more detail, the mechanism of lane formation can be understood as follows: Pedestrians moving in a mixed crowd or moving against the stream will have frequent and strong interactions. In each interaction, the encountering pedestrians move a little aside in order to pass each other. This sidewards movement tends to separate oppositely moving pedestrians. Moreover, once the pedestrians move in uniform lanes, they will have very rare and weak interactions. Hence, the tendency to break up existing lanes is negligible. Furthermore, the most stable configuration corresponds to a state with a *minimal interaction rate*. Therefore, lane formation and minimal interaction rate are two sides of the same medal. Nevertheless, lane formation does not occur in all driven repulsive systems.

To give an analytical description of this segregation phenomenon, we will set up continuum equations for the considered system (for details see [14]), but at the same time, we will generalise the problem to arbitrary kinds of populations a that are composed of identical moving entities α. In the above example, the different populations correspond to two opposite desired directions of walking. Assuming a homogeneous distribution of pedestrians in walking direction, we can reduce the problem to the investigation of the one-dimensional dynamics perpendicular to it. (Imagine a projection of pedestrian dynamics on a cross section of the walkway). The distribution of the N_a entities of population a over the locations x of this one-dimensional space will be represented by the densities $\rho_a(x,t) \ge 0$.

Assuming conservation of the number

$$N_a = \int dx\, \rho_a(x,t) \tag{13}$$

of entities in each population a, we obtain the continuity equations [15]

$$\frac{\partial \rho_a(x,t)}{\partial t} + \frac{\partial}{\partial x}\Big[\rho_a(x,t)V_a(x,t)\Big] = 0\,. \tag{14}$$

Here, $V_a(x,t)$ is the average velocity of entities of population a which, in the one-dimensional case, can be represented by the spatial derivative

$$V_a(x,t) = c\,\frac{\partial S_a(x,t)}{\partial x} \tag{15}$$

of some function S_a of the densities ρ_b. For higher-dimensional spaces, (15) becomes a so-called potential condition. A prefactor $c \neq 1$ like $c(x,t) = [1 - \sum_a \rho_a(x,t)/\rho_{\max}]$ allows to limit the density to the maximum value $\rho_{\max}$, reflecting that motion is slowed down in crowded areas.

In linear approximation, we have

$$S_a(x,t) = S_a^0 + \sum_b S_{ab}\,\rho_b(x,t)\,, \tag{16}$$

but the constants S_a^0 do not matter at all. The function $S_a(x,t)$ may be interpreted as the "(expected) success" per unit time for an entity of population a at location x, as it is plausible that the entities move into the direction of the greatest increase of success, corresponding to formula (15). Positive S_{ab} belong to profitable or attractive interactions between populations a and b, whereas competitive or repulsive interactions correspond to negative S_{ab}. If an entity of kind a interacts with entities of kind b at a rate ν_{ab} and the associated outcome of the interaction can be quantified by some "payoff" P_{ab}, we have the relation $S_{ab} = \nu_{ab}P_{ab}$.

Now, we will proof that, for symmetric interactions characterised by $S_{ba} = S_{ab}$, the overall success

$$S(t) = \sum_a \int dx\, \rho_a(x,t) S_a(x,t) \tag{17}$$

is a so-called Lyapunov function which monotonically increases in the course of time, just like some thermodynamic non-equilibrium potentials [25]. By deriving (17) with respect to t, using (16), and properly interchanging indices a and b, we eventually obtain $dS(t)/dt = \sum_a \int dx\, \partial\rho_a(x,t)/\partial t \sum_b (S_{ab} + S_{ba})\rho_b(x,t)$. By inserting (14) and (15), and applying (16), we get $dS(t)/dt = -2\sum_a \int dx\, \big[S_a(x,t) -$

$S_a^0] \, \partial/\partial x \big[\rho_a(x,t) \, c \, \partial S_a(x,t)/\partial x\big]$. Making use of partial integration (for spatially periodic systems), we finally arrive at

$$\frac{dS(t)}{dt} = 2 \sum_a \int dx \; c\rho_a(x,t) \left[\frac{\partial S_a(x,t)}{\partial x} \right]^2 \geq 0 \, . \tag{18}$$

This result establishes self-optimisation for symmetrical interactions and can be easily transferred to discrete spaces (Fig. 5). In case of slightly asymmetric interactions, small non-linear contributions to (16) or small diffusion, relation (18) will still be a good approximation, i.e., the system will behave close to optimal. This is exemplified by mixed freeway traffic [23].

According to (18), the stationary solution $\rho_a^{\rm st}(x)$ is characterised by

$$c\rho_a^{\rm st}(x) = 0 \qquad \text{or} \qquad q \frac{\partial S_a^{\rm st}(x)}{\partial x} = \sum_b S_{ab} \frac{\partial \rho_b^{\rm st}(x)}{\partial x} = 0 \tag{19}$$

for all a, which is fulfilled by homogeneous or step-wise constant solutions. For the case of two species, a linear stability analysis shows that the homogeneous solution (Fig. 5a) is unstable if

$$\rho_a^{\rm hom} S_{aa} + \rho_b^{\rm hom} S_{bb} > 0 \quad \text{or} \quad S_{ab} S_{ba} > S_{aa} S_{bb} \, , \tag{20}$$

where $\rho_a^{\rm hom} = N_a/I$ denotes the homogeneous density and I the spatial extension of the system. On condition (20), the stable stationary solution corresponds to a self-organised, non-homogeneous state (complete segregation). If the populations interact in a symmetrical way, the underlying self-organisation process is related to self-optimisation (see (18)). Even more interesting, segregation is most likely to occur for symmetric interactions, which can be seen as follows: Introducing $\overline{S} = (S_{ab} + S_{ba})/2$ and $\Delta S = (S_{ab} - S_{ba})/2$, we have $S_{ab} = (\overline{S} + \Delta S)$, $S_{ba} = (\overline{S} - \Delta S)$, and the latter condition in (20) becomes $(\overline{S} + \Delta S)(\overline{S} - \Delta S) = \overline{S}^2 - (\Delta S)^2 > S_{aa} S_{bb}$. Since the interaction strengths $|S_{aa}|$, $|S_{bb}|$, $|S_{ab}|$, $|S_{ba}|$, and $|\overline{S}|$ will normally have the same order of magnitude, this condition for self-organisation can only be fulfilled for small $|\Delta S|$, i.e. $S_{ab} \approx S_{ba}$. *Hence, if there is self-organisation in the considered class of systems, it is likely to come with optimality,* at least approximately.

For the above reasons, one may speak of "optimal self-organisation" [14]. A good example is uni-directional multi-lane traffic of cars and lorries [23], which develops a coherent state with significantly reduced interactions (lane-changing rates) only in a small density range, where the interactions between cars and lorries become sufficiently symmetric.

In our pedestrian example, we find optimal self-organisation, since the symmetry condition is satisfied exactly and the latter condition of (20) is fulfilled (because the interaction rate ν_{ab} of oppositely moving pedestrians is much higher than for pedestrians with the same desired walking direction [14]). Notice that the spatial regions occupied by one population need not be connected (Figs. 5b-d), and that the corresponding configuration may correspond to a *relative* optimum as in Figs. 5c and 5d. If $S_{aa} < 0$ for all a, the distributions $\rho_a^{\rm st}(x)$ tend

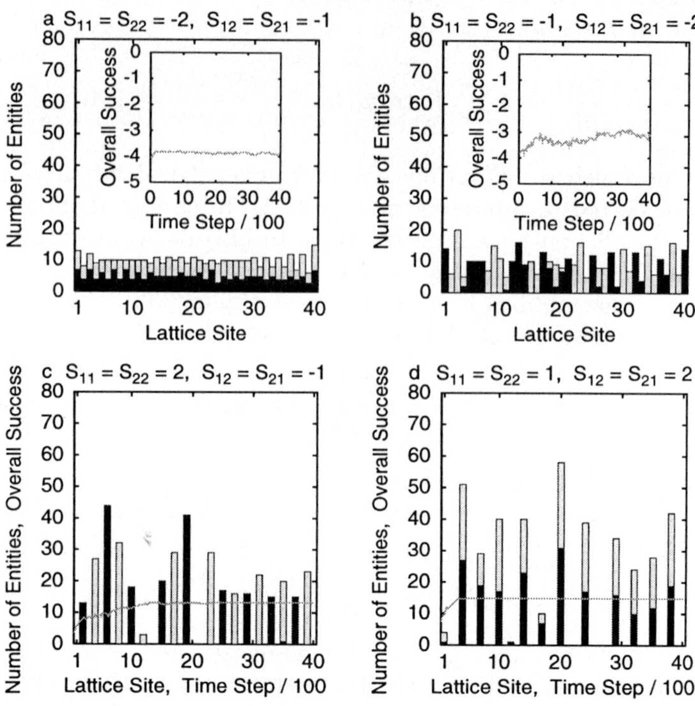

Fig. 5. Illustration of the various forms of self-optimisation for two different populations: **a** Homogeneous distribution in space, **b** segregation of populations without agglomeration, **c** attractive agglomeration, **d** repulsive agglomeration. Cases **b** to **d** are examples of "optimal self-organisation", since the finally evolving optimal states are self-organised, non-homogeneous states.

The above results were obtained with a one-dimensional, discrete version of the game-dynamical model defined by equations (14) to (16). We assumed a periodic lattice with I lattice sites $x \in \{1, \dots, I\}$ and two populations $a \in \{1, 2\}$ with a total of $N = N_1 + N_2 \gg I$ entities (here: $I = 40$ and $N_1 = N_2 = 200$). Furthermore, we applied the following update steps: 1. Calculate the successes $S_a(x,t) = S_a^0 + \sum_b S_{ab} n_x^b(t)/I$, where $n_x^b(t) = \rho_b(x,t)I$ represents the number of entities of population b at site x. 2. For each entity α, determine a random number ξ_α that is uniformly distributed in the interval $[0, S_{\max}]$ with a large constant $S_{\max}$ (here: $S_{\max} = 20$). 3. Move entity α belonging to population a from site x to site $x+1$, if $c(x+1,t)[S_a(x+1,t) - S_a(x-1,t)] > \xi_\alpha$, but to site $x-1$, if $c(x-1,t)[S_a(x-1,t) - S_a(x+1,t)] > \xi_\alpha$. Our simulations started with a random initial distribution of the entities and assumed $c(x,t) = 1$. We applied a random sequential update rule, but a parallel update yields qualitatively the same results. The above figures show the numbers n_x^1 and $n_x = (n_x^1 + n_x^2)$ of entities as a function of the lattice site x at time $t = 4000$ and the evolution of the overall success $S(t)$ as a function of time t. The fluctuations around the monotonic increase of $S(t)$ are caused by the fluctuations ξ_α and the random sequential update.

to be flat, as in the case of lane formation by repulsive pedestrian interactions (Fig. 5b). Instead, we have agglomeration (local clustering), if $S_{aa} > 0$ for all a (Figs. 5c and 5d). The case of Fig. 5c allows to understand the conjectured optimality of trail systems which, based on attractive interactions, result by a bundling of trails ending at different destinations [20]. Figure 5d describes the segregation of populations with repulsive interactions ("ghetto formation").

We call attention to the fact that there is a class of living systems (for which we have given some realistic examples) to which existing methods, notions, and principles from thermodynamics can be successfully applied, if they are generalised in a suitable way. In particular, we have proven that a wide class of driven systems, which can be represented as a game between symmetrically interacting populations, approach a stationary state characterised by maximal overall success and minimal dissipation. In other words, as individual entities are trying to optimise their *own* success, a class of systems tends to reach a state with the highest *global* success, which is not trivial at all. The precondition for this is symmetrical interactions, which are very common in nature and most likely to be approximately found in a wide class of self-organising systems, as we have shown.

Acknowledgement. D.H. is grateful to the DFG for financial support by a Heisenberg scholarship. This work was in part supported by OTKA F019299 and FKFP 0203/1997. The authors also want to thank Illés J. Farkas for his collaboration in producing Fig. 4 and Robert Axelrod, Eshel Ben-Jacob, and Martin Treiber for valuable comments.

References

1. J.K. Parrish and L. Edelstein-Keshlet, Science **284**, 99 (1999).
2. C.W. Reynolds, Computer Graphics **21**, 25 (1987).
3. E.M. Rauch, M.M. Millonas, and D.R. Chialvo, Phys. Lett. A **207**, 185 (1995).
4. T. Vicsek, A. Czirók, E. Ben-Jacob, I. Cohen, and O. Shochet, Phys. Rev. Lett. **75**, 1226 (1995).
5. R.B. Stinchcombe, in: *Phase Transitions and Critical Phenomena,* C. Domb and J. Lebowitz, (Eds.), Vol. 7, (Academic Press, New York, 1983).
6. A. Czirók, A.-L. Barabási, and T. Vicsek, Phys. Rev. Lett. **82**, 209 (1999).
7. J. Toner and Y. Tu, Phys. Rev. Lett. **75**, 4326 (1995).
8. A. Czirók, H.E. Stanley, and T. Vicsek, J. Phys. A **30**, 1375 (1997).
9. A. Czirók, E. Ben-Jacob, I. Cohen, and T. Vicsek, Phys. Rev. E **54**, 1791 (1996).
10. Y.L. Duparcmeur, H.J. Herrmann, and J.P. Troadec, J. Phys. I France **5**, 1119 (1995).
11. J. Hemmingsson, J. Phys. A **28**, 4245 (1995).
12. A. Czirók, M. Vicsek, and T. Vicsek, Physica A **264**, 299 (1999).
13. J. Kertész, D. Stauffer, and A. Coniglio, in: *Percolation Structures and Processes,* G.D. Deutcher, R. Zallen, and J. Adler, (Eds.), p. 121 (Adam Hilger, 1983).
14. D. Helbing, and T. Vicsek, New J. of Physics **1**, No. 13, (1999); (see `http://www.njp.org/`).

15. J. Keizer, *Statistical Thermodynamics of Nonequilibrium Processes,* (Springer, New York, 1987).
16. H. Haken, *Information and Self-Organization,* (Springer, Berlin, 1988).
17. G. Nicolis and I. Prigogine, *Self-Organization in Nonequilibrium Systems. From Dissipative Structures to Order through Fluctuations,* (Wiley, New York, 1977).
18. T. Vicsek, *Fractal Growth Phenomena,* (World Scientific, Singapore, 1992).
19. D. Helbing, *Quantitative Sociodynamics,* (Kluwer, Dordrecht, 1995).
20. D. Helbing, J. Keltsch, and P. Molnár, Nature **388**, 47 (1997).
21. D. Helbing and P. Molnár, Phys. Rev. E **51**, 4282 (1995).
22. S.B. Santra, S. Schwarzer, and H. Herrmann, Phys. Rev. E **54**, 5066 (1996).
23. D. Helbing and B.A. Huberman, Nature **396**, 738 (1998).
24. D. Helbing, *Verkehrsdynamik,* (Springer, Berlin, 1997).
25. R. Graham and T. Tél, in: *Stochastic Processes and Their Applications*, S. Albeverio, P. Blanchard, and L. Streit (Eds.), p. 153 (Kluwer, Dordrecht, 1990).

Active Brownian Particles with Internal Energy Depot

F. Schweitzer

GMD Institute for Autonomous Intelligent Systems, 53754 Sankt Augustin, Germany

Abstract. Active motion relies on the supply of energy. In order to turn passive into active motion, we need to consider mechanisms of energy take-up, storage and conversion. A suitable approach which considers both the energetic and stochastic aspects of active motion is provided by the model of active Brownian particles. For a supercritical supply of energy these particles are able to move in a "high velocity" or active mode, which results in deviation from the Maxwellian velocity distribution. We investigate different types of complex motion of active Brownian particles moving in external potentials. Among the examples are the occurrence of stochastic limit cycles, transitions between Brownian and directed motion, the "uphill" motion against the direction of an external force, or the establishment of positive or negative net currents in a ratchet potential, dependent on energy supply and stochastic influences.

1 Passive vs. Active Motion

The motion of a "simple" Brownian particle is due to fluctuations of the surrounding medium, i.e., the result of random impacts of the molecules or atoms of the liquid or gas, the particle is immersed in. This type of *undirected* motion would be rather considered as *passive* motion, simply because the Brownian particle does not play an active part in this motion. Passive motion can be also directed, if it is driven, e.g., by convection, currents or by external fields.

Active motion, on the other hand relies on the supply of energy. Already in physico-chemical systems a *self-driven motion* of particles can be observed [1]. On the biological level, *active* self-driven motion can be found on different scales, ranging from cells [2] or simple microorganisms up to higher organisms, such as bird or fish. Last, but not least, also human movement can be described as active motion [3], as well as the motion of cars. All these types of active motion occur under energy consumption and energy conversion and may also involve processes of energy storage.

Recent investigations on *interacting* self-driven entities [4–6] focus on collective effects, such as the formation of swarms or crowds, rather than on the origin of the entities' velocity; i.e., it is usually postulated that the entities move with a certain non-zero velocity. In order to describe both the *random* aspects and the *energetic* aspects of active motion, we have introduced a model of *active Brownian particles* [7–11]. These are Brownian particles with the ability to take up energy from the environment, to store it in an internal depot and to convert internal energy to perform different activities, such as metabolism or motion. Possible changes of the environment or signal-response behavior are neglected here.

In order to turn passive (Brownian) motion into active motion, we need to consider mechanisms of energy take-up. A very simple mechanism is the pumping of energy by *space-dependent friction* [7] which in a certain spatial range can be also negative. Inside this area the Brownian particle, instead of loosing energy because of dissipative processes, is pumped with energy, which in turn increases its velocity. While such an approach will be able to model the spatially inhomogeneous supply of energy, it has the drawback not to consider processes of storage and conversion of energy. In fact, with only a space-dependent friction, the Brownian particle is instantaneously accelerated or slowed down, whereas, e.g., biological entities or cars have the capability to stretch their supply of energy over a certain time interval.

In order to develop a more realistic model of active motion, we have considered an *internal energy depot* for the Brownian particles [10,11], which allows to store the taken-up energy in the internal depot, from where it can be converted, e.g., into kinetic energy, namely for the acceleration of motion. Additionally, the internal dissipation of energy, due to storage and conversion (or metabolism in a biological context) can be considered.

With these extensions, the Brownian particle becomes in fact a *Brownian motor* [12–14], which is fueled somewhere and then uses the stored energy with a certain efficiency [11] to move forward, also against external forces. Provided a supercritical supply of energy, we find that the motion of active Brownian particles in the two-dimensional space can become rather complex as shown in the sections below.

2 Pumping from an Internal Energy Depot

The motion of simple Brownian particles in a space-dependent potential $U(\boldsymbol{r})$ can be described by the Langevin equation:

$$\dot{\boldsymbol{r}} = \boldsymbol{v}\,;\;\; m\,\dot{\boldsymbol{v}} = -\gamma_0 \boldsymbol{v} - \nabla U(\boldsymbol{r}) + \mathcal{F}(t) \tag{1}$$

where γ_0 is the friction coefficient of the particle at position $\boldsymbol{r}$, moving with velocity $\boldsymbol{v}$. $\mathcal{F}(t)$ is a stochastic force with strength D and a δ-correlated time dependence

$$\langle \mathcal{F}(t) \rangle = 0\,;\;\; \langle \mathcal{F}(t)\mathcal{F}(t') \rangle = 2D\,\delta(t-t'). \tag{2}$$

Using the fluctuation-dissipation theorem, we assume that the loss of energy resulting from friction, and the gain of energy resulting from the stochastic force, are compensated in the average, and D can be expressed as $D = k_B T \gamma_0$, where T is the temperature and k_B is the Boltzmann constant.

In addition to the dynamics described above, the Brownian particles considered here are active in the sense that they are able to take up energy from the environment, which can be stored in an *internal depot*, *e*. $q(\boldsymbol{r})$ shall be the space-dependent flux of energy into the depot. The internal energy can be converted into kinetic energy with a rate $d(\boldsymbol{v})$ which should be a function of the actual velocity of the particle. Further, we consider internal dissipation, which is assumed to be proportional to the depot energy, c being the rate of energy loss. The resulting balance equation for the internal energy depot e of an active Brownian particle is then given by:

$$\frac{d}{dt}e(t) = q(\boldsymbol{r}) - c\, e(t) - d(\boldsymbol{v})\, e(t). \tag{3}$$

A simple ansatz for $d(\boldsymbol{v})$ reads $d(\boldsymbol{v}) = d_2 v^2$, with $d_2 > 0$. The energy conversion results in an additional acceleration of the Brownian particle in the direction of movement. Hence, the equation of motion has to consider an additional driving force, $d_2 e(t)\boldsymbol{v}$. In [10, 11], we have postulated a stochastic equation for pumped Brownian particles, which is consistent with the Langevin equation (1):

$$\dot{\boldsymbol{r}} = \boldsymbol{v}\,;\ m\dot{\boldsymbol{v}} + \gamma_0 \boldsymbol{v} + \nabla U(\boldsymbol{r}) = d_2 e(t)\boldsymbol{v} + \mathcal{F}(t) \tag{4}$$

From now on, $m = 1$ is used. The Langevin equation (4) can be rewritten in the known form (1) by introducing a *velocity-dependent friction coefficient*:

$$\gamma(\boldsymbol{v}) = \gamma_0 - d_2\, e(t) \tag{5}$$

Here, the value of $\gamma(\boldsymbol{v})$ changes dependent on the value of the internal energy depot, which itself is a function of the velocity. If the term $d_2\, e(t)$ exceeds the "normal" friction γ_0 the velocity dependent friction coefficient can be negative which means the active particle's motion is pumped with energy. In order to get an estimate of the range of energy pumping, we assume a constant influx of energy into the internal depot $q(\boldsymbol{r}) = q_0$, and a fast relaxation of the internal energy depot, (3), which reads in an adiabatic approximation:

$$e_0 = \frac{q_0}{c + d_2 \boldsymbol{v}^2} \tag{6}$$

The quasi-stationary value e_0 can be used to approximate the velocity-dependent friction coefficient $\gamma(\boldsymbol{v})$ (5):

$$\gamma(\boldsymbol{v}) = \gamma_0 - \frac{q_0\, d_2}{c + d_2 \boldsymbol{v}^2} \tag{7}$$

which is plotted in Fig. 1.

Dependent on the parameters γ_0, d_2, q_0, c the friction function (7) may have a zero, where the friction is just compensated by the energy supply. It reads in the considered case:

$$\boldsymbol{v}_0^2 = \frac{q_0}{\gamma_0} - \frac{c}{d_2} \tag{8}$$

We see that for $\boldsymbol{v} < \boldsymbol{v}_0$, i.e., in the range of small velocities pumping due to negative friction occurs, as an additional source of energy for the Brownian particle. Hence, slow particles are accelerated, while the motion of fast particles is damped.

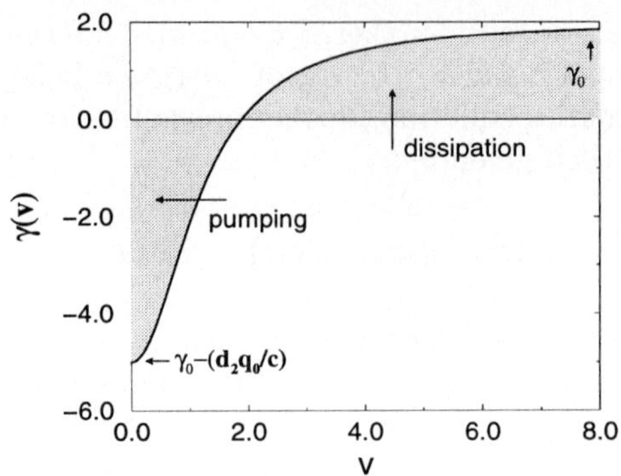

Fig. 1. Velocity-dependent friction coefficient $\gamma(\boldsymbol{v})$ (7) vs. velocity $\boldsymbol{v}$. The velocity ranges for "pumping" ($\gamma(\boldsymbol{v}) < 0$) and "dissipation" ($\gamma(\boldsymbol{v}) > 0$) are indicated. Parameters: $q_0 = 10$; $c = 1.0$; $\gamma_0 = 20$, $d_2 = 10$. [16]

Due to the pumping mechanism discussed here, the conservation of energy clearly does not hold for the particle, i.e., we now have a non-equilibrium, canonical-dissipative system instead of an equilibrium canonical system. This should result in deviations from the known Maxwellian velocity distribution.

We restrict the further discussion to the two-dimensional space $\boldsymbol{r} = \{x_1, x_2\}$. Then the stationary velocities v_0, (8), where the friction is just compensated by the energy supply, define a cylinder, $v_1^2 + v_2^2 = \boldsymbol{v}_0^2$, in the four-dimensional state space $\{x_1, x_2, v_1, v_2\}$ which attracts all deterministic trajectories of the dynamic system [16]. The probability density for the velocity $P(\boldsymbol{v}, t)$ can be described by a Fokker-Planck equation, which reads for the friction function (7), and in the absence of an external potential, i.e., $U(x_1, x_2) \equiv 0$:

$$\frac{\partial P(\boldsymbol{v},t)}{\partial t} = \frac{\partial}{\partial \boldsymbol{v}} \left[\left(\gamma_0 - \frac{d_2 q_0}{c + d_2\, \boldsymbol{v}^2} \right) \boldsymbol{v}\, P(\boldsymbol{v},t) + D\, \frac{\partial P(\boldsymbol{v})}{\partial v} \right] \tag{9}$$

The stationary solution of (9) yields:

$$P^0(\boldsymbol{v}) = C \left(1 + \frac{d_2 \boldsymbol{v}^2}{c} \right)^{\frac{q_0}{2D}} \exp\left(-\frac{\gamma_0}{2D}\, \boldsymbol{v}^2 \right), \tag{10}$$

where C results from the normalization condition. Compared to the Maxwellian velocity distribution of "simple" Brownian particles, a new prefactor appears now in (10) which results from the internal energy depot. In the range of small values of $\boldsymbol{v}^2$, the prefactor can be expressed by a power series truncated after the first order, and (10) reads then:

$$P^0(\boldsymbol{v}) \sim \exp\left[-\frac{\gamma_0}{2D} \left(1 - \frac{q_0 d_2}{c\gamma_0} \right) \boldsymbol{v}^2 + \cdots \right]. \tag{11}$$

In (11), the sign of the expression in the exponent depends significantly on the parameters which describe the balance of the energy depot. For a subcritical pumping of energy, $q_0 d_2 < c\gamma_0$, the expression in the exponent is negative and an *unimodal velocity distribution* results, centered around the maximum $\boldsymbol{v}_0 = 0$. This corresponds to the *Maxwellian velocity distribution.* However, for supercritical pumping,

$$q_0 d_2 > c\gamma_0 \tag{12}$$

the exponent in (11) becomes positive, and a *crater-like velocity distribution* results, which indicates strong deviations from the Maxwell distribution (Fig. 2). The maxima of $P^0(\boldsymbol{v})$ correspond to the solutions for $\boldsymbol{v}_0^2$, (8).

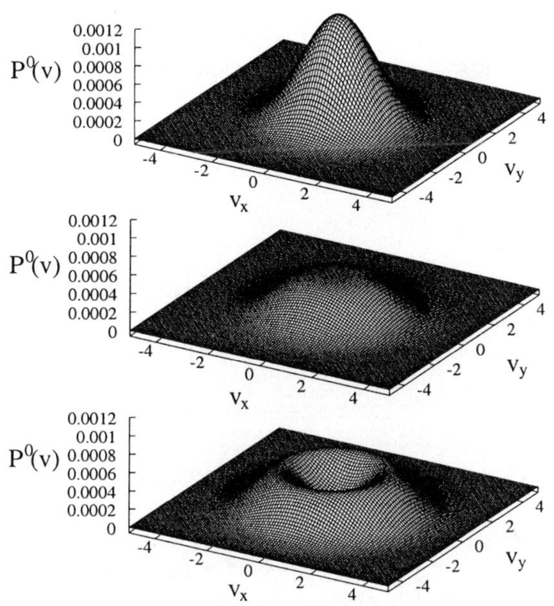

Fig. 2. Normalized stationary solution $P^0(\boldsymbol{v})$ (10), for $d_2 = 0.07$ (top), $d_2 = 0.2$ (middle) and $d_2 = 0.7$ (bottom). Other parameters: $\gamma_0 = 2$, $D = 2$, $c = 1$, $q_0 = 10$. Note that $d_2 = 0.2$ is the bifurcation point for the given set of parameters [16]

3 Motion in a Parabolic Potential

In the following we discuss the motion of a Brownian particle with an internal energy depot in a two-dimensional parabolic potential:

$$U(x_1, x_2) = \frac{a}{2}(x_1^2 + x_2^2). \tag{13}$$

This potential originates a force directed to the minimum of the potential, however, the random force in (4) keeps the particle moving, even without the take-up of energy (Fig. 3). If we consider a take-up of energy which is constant in space, $q(x_1, x_2) = q_0$, Fig. 3 demonstrates that the particle reaches out farther, moving on a stochastic limit cycle, if some critical conditions are satisfied. In [11] it was shown that for the case of the harmonic potential these critical conditions coincide with those obtained for $U(x_1, x_2) \equiv 0$, i.e., (12) has to be fulfilled again.

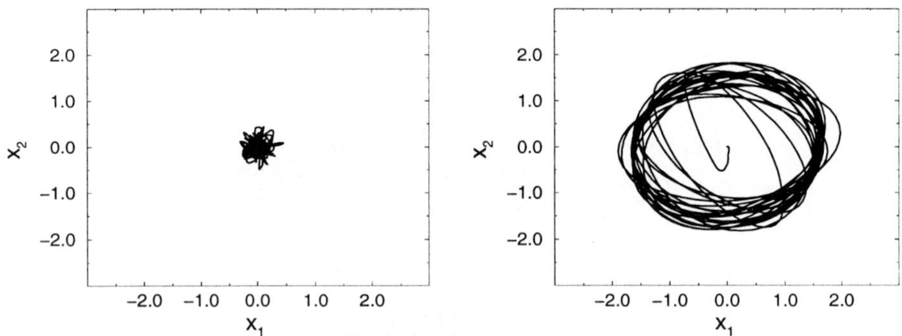

Fig. 3. Stochastic motion of an active Brownian particle in a parabolic potential. (left) $q = 0$ (simple Brownian motion), (right) $q_0 = 1.0$ (other parameters: $\gamma_0 = 0.2$ $d_2 = 1.0$, $c = 0.1$, $D = 0.01$, $a = 2$, initial conditions: $(x_1, x_2) = (0, 0)$, $(v_1, v_2) = (0, 0)$, $e(0) = 0$) [10].

If we consider that the energy influx is not constant, but space dependent, for example

$$q(x_1, x_2) = \begin{cases} q_0 \,\text{if } \left[(x_1 - b_1)^2 + (x_2 - b_2)^2\right] \leq R^2 \\ 0 \;\text{ else} \end{cases} \tag{14}$$

then the internal depot of the active Brownian particles can be refilled only in a restricted area. Equation (14) assumes that the supply area (*energy source*) is modeled as a circle, the center being different from the minimum of the potential. Noteworthy, the active particle is *not* attracted by the energy source due to long-range attraction forces. In the beginning, the internal energy depot of the particle is empty and active motion is not possible. So, the particle may hit the supply area because of the action of the stochastic force. But once the energy depot is filled up, it increases the particles mobility, as presented in Fig. 4. Most likely, the motion into the energy area becomes accelerated, therefore an oscillating movement between the energy source and the potential minimum occurs after an initial period of stabilization.

Interestingly, the oscillating motion breaks down after a certain time. Then the active particle, with an empty internal depot, moves again like a simple Brownian particle, until a new cycle starts, as indicated in Fig. 4. This way the particle motion is of intermittent type. We found that the trajectories eventually

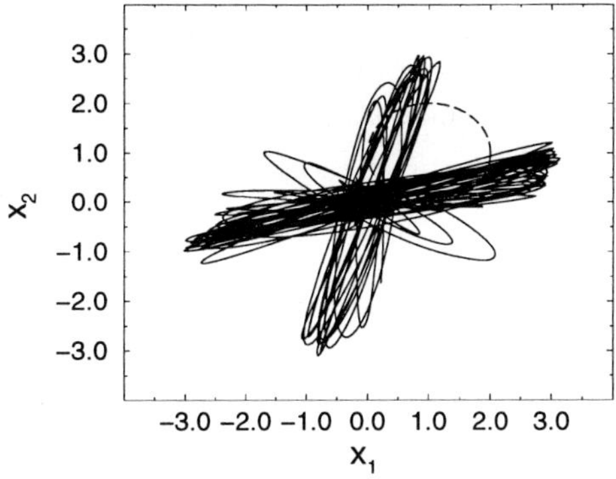

Fig. 4. Trajectories in the x_1, x_2 space for the stochastic motion of an active Brownian particle in a parabolic potential. The circle (coordinates (1,1), radius $R = 1$) indicates the area of energy supply (14). Parameters: $d_2 = 1$, $q_0 = 10$, $\gamma_0 = 0.2$, $c = 0.01$, $D = 0.01$, $a = 2$, initial conditions: $(x_1, x_2) = (0, 0)$, $(v_1, v_2) = (0, 0)$, $e(0) = 0$ [10].

cover the whole area inside certain boundaries, however during an oscillation period the direction is most likely kept.

We may also consider that active motion occurs in more complex landscapes which typically not only contain localized areas of energy supply, but also *obstacles.* We have discussed such a case while assuming a hard-core like obstacle where the particle is simply reflected at the boundary if it hits the obstacle. Considering a continuous supply of energy but only a deterministic motion, we found a chaotic motion of the active particle in the phase space $\Gamma = \{x_1, x_2, v_1, v_2, e\}$ [10]. Hence, we concluded that for the motion of Brownian particles with energy depots reflecting obstacles have an effect similar to stochastic influences (external noise).

4 Motion in a Linear/Periodic Potential

In the following, we restrict the discussion to the *one-dimensional case*, i.e., the space coordinate is given by x. For the potential, we assume a linear function, $U(x) = ax$, hence the resulting force is a constant: $\boldsymbol{F} = -\boldsymbol{\nabla} U = -a$. Further the flux of energy into the internal depot of the particle is assumed as constant, $q(x) = q_0$. Then, the dynamics for the pumped Brownian motion is described by the following set of equations:

$$\begin{aligned} \dot{\boldsymbol{v}} &= -\big(\gamma_0 - d_2 e(t)\big)\boldsymbol{v} + \boldsymbol{F} + \sqrt{2D}\boldsymbol{\xi}(t) \\ \dot{e} &= q_0 - ce - d_2 \boldsymbol{v}^2 e. \end{aligned} \tag{15}$$

In the deterministic case, the stationary solutions of (15) obtained from $\dot{\boldsymbol{v}} = 0$ and $\dot{e} = 0$ lead to a cubic polynom for the velocity $\boldsymbol{v}_0$:

$$d_2\gamma_0 \boldsymbol{v}_0^3 - d_2 \boldsymbol{F} v_0^2 - (q_0 d_2 - c\gamma_0)\boldsymbol{v}_0 - c\boldsymbol{F} = 0. \tag{16}$$

Here, $\boldsymbol{v}_0^{n+1}$ is defined as a vector $|v_0|^n \boldsymbol{v}_0$. Depending on the value of F and in particular on the sign of the term $(q_0 d_2 - c\gamma_0)$, (16) has either one or three real solutions for the stationary velocity, $\boldsymbol{v}_0$. This is also shown in the bifurcation diagram, Fig. 5.

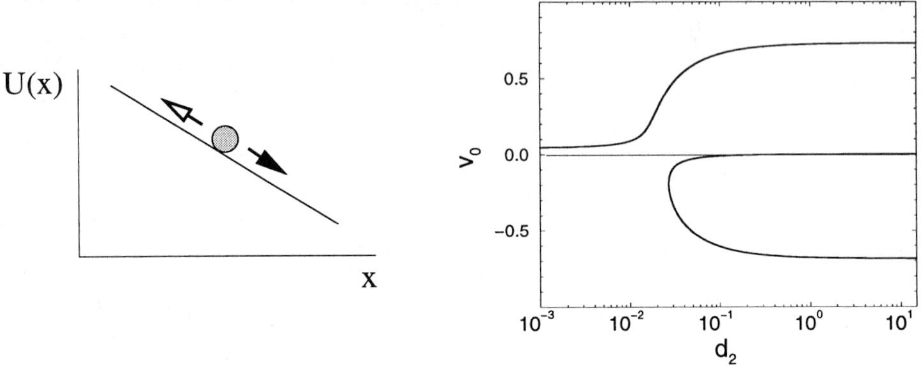

Fig. 5. (right) Sketch of the one-dimensional motion of the particle in the presence of a constant force $\boldsymbol{F} = -\boldsymbol{\nabla} U(x) = \text{const}$. Provided a supercritical amount of energy from the depot, the particle might be able to move "uphill", i.e., against the direction of the force. (left) Stationary velocities $\boldsymbol{v}_0$ (16) vs. conversion rate d_2. Above a critical value of d_2, a negative stationary velocity indicates the possibility to move against the direction of the force. Parameters: $\boldsymbol{F} = +7/8$, $q_0 = 10$, $\gamma_0 = 20$, $c = 0.01$ [17].

The always existing solution expresses a direct response to the force in the form: $\boldsymbol{v}_0 \sim \boldsymbol{F}$, i.e., it results from the analytic continuation of Stokes' law, $\boldsymbol{v}_0 = \boldsymbol{F}/\gamma_0$, which is valid for $d_2 = 0$. We denote this solution as the "normal" mode of motion, since the velocity $\boldsymbol{v}$ has the same direction as the force $\boldsymbol{F}$ resulting from the external potential $U(x)$.

As long as the supply from the energy depot is small, we will also name the normal mode as the *passive mode*, because the particle is simply driven by the external force. More interesting is the case of three stationary velocities, $\boldsymbol{v}_0$, which significantly depends on the (supercritical) influence of the energy depot. In this case the particle will be able to move in a "high velocity" or *active mode* of motion. Fig. 5 shows that the former passive normal mode, which holds for subcritical energetic conditions, is transformed into an active normal mode, where the particle moves into the same direction, but with a much higher velocity. *Additionally*, in the active mode a new high-velocity motion *against* the direction of the force $\boldsymbol{F}$ becomes possible, which corresponds to an "uphill" motion.

It is obvious that the particle's motion "downhill" is stable, but the same does not necessarily apply for the possible solution of an "uphill" motion. In [17], we have investigated the necessary conditions for such a motion in a linear potential. For the *deterministic* case, we found the following critical condition for a possible "uphill" motion of the pumped Brownian particle:

$$d_2^{crit} = \frac{F^4}{8q_0^3}\left(1+\sqrt{1+\frac{4\gamma_0 q_0}{F^2}}\right)^3 . \tag{17}$$

In the limit of negligible internal dissipation $c \to 0$ (17) describes how much power has to be supplied by the internal energy depot to allow a *stable uphill motion* of the particle.

This result will be used to explain the motion of an *ensemble* of N active Brownian particles in a piecewise linear, asymmetric potential (Fig. 6), which is known as a *ratchet potential* [14,15].

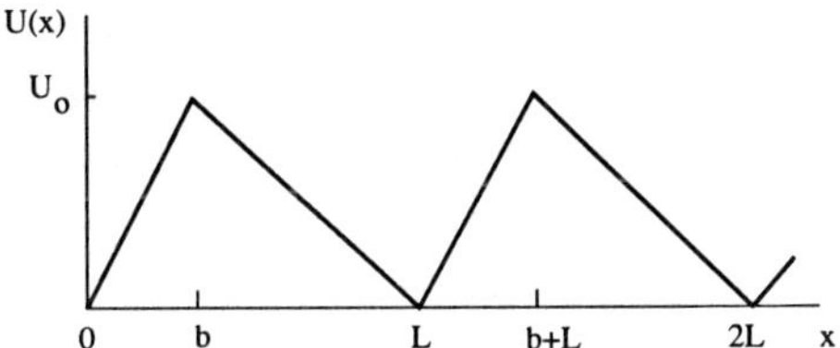

Fig. 6. Sketch of the ratchet potential $U(x)$. For the computer simulations, the following values are used: b=4, L=12, $U_0 = 7$ in arbitrary units.

Figure 7 shows the *net current* expressed by the mean velocity $\langle v \rangle$, dependent on the conversion rate, d_2. For the *deterministic motion* ($D = 0$), we see the existence of *two different critical values* for the parameter d_2. For values of d_2 near zero and less than d_2^{crit1}, there is no net current at all. This is due to the subcritical supply of energy from the internal depot, which does not allow an uphill motion on any flank of the potential. With an increasing value of d_2, we see the occurrence of a negative net current at d_2^{crit1}. That means, the energy depot provides enough energy for the uphill motion along the flank with the lower slope. For $d_2^{crit1} \leq d_2 \leq d_2^{crit2}$, a stable motion of the particles up and down the flank with the lower slope is possible, but not for the steeper slope. Only for $d_2 > d_2^{crit2}$, the energy depot also supplies enough energy for the particles to climb up the steeper slope, consequently a periodic motion of the particles into the positive direction becomes possible now.

For the computer simulations, we have assumed that the start locations of the particles are equally distributed over the first period of the potential, $\{0, L\}$ and their initial velocity is zero. Hence, with respect to Fig. 6 more particle start into the positive direction, which eventually results in a larger positive net current for $d_2 > d_2^{crit2}$ [18]. If we insert the two different values for F (Fig. 6) into the critical condition (17), we find for the critical values $d_2^{crit1} = 1.03$ and $d_2^{crit1} = 11.3$, which agrees with the onset of the negative and the positive current in the deterministic computer simulations, Fig. 7.

In the deterministic case, the particles will keep their direction determined by the initial conditions provided the energy supply allows them to move "up-

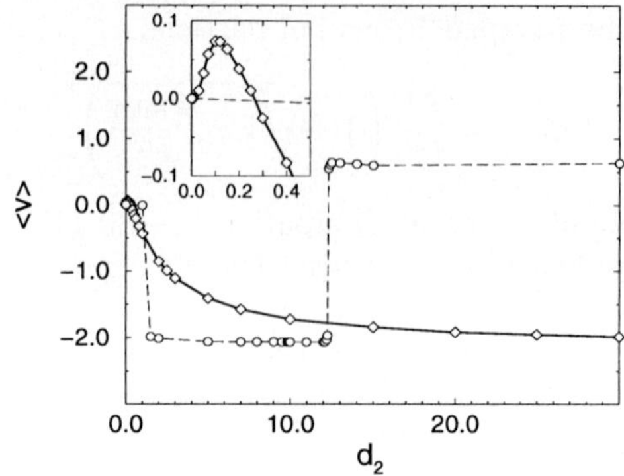

Fig. 7. Average velocity $\langle v \rangle$ vs. conversion parameter d_2. The data points are obtained from simulations of 10.000 particles with arbitrary initial positions in the first period of the ratchet potential. (◇) stochastic case ($D = 0.1$), (∘) deterministic case ($D = 0$). Parameters: $q_0 = 1.0$, $\gamma_0 = 0.2$, $c = 0.1$ [18].

hill". In the stochastic case, however, the initial conditions will be "forgotten" after a short time, hence due to stochastic influences the particle's "upill" motion along the steeper flank will soon turn into a "downhill" motion. In [18], we have investigated the two-dimensional separatrix plane, which separates the motion into positive and negative directions in the three-dimensional phase space, $\{x, v, e\}$. We found that, if a particle moves into the positive direction, most of the time the trajectory is very close to the separatrix. That means it will be rather susceptible for small perturbations, i.e., even small fluctuations might be able to destabilize the motion into the positive direction. The motion into negative direction, on the other hand, is not susceptible in the same manner, since the respective trajectory remains in a considerable distance from the separatrix or comes close to the separatrix only for a very short time.

Thus, the stochastic fluctuations reveal the instability of the motion into the positive direction, i.e., the "uphill" motion along the *steeper* slope. Hence, in the stochastic case the net current is always negative. In addition, we find a *very small* positive net current in the range of small d_2 (cf. the insert in Fig. 7), because the fluctuations allow some particles to escape the potential barriers.

In order to investigate how much the strength D of the stochastic force may influence the magnitude of the net current into the negative direction, we have varied D for a fixed conversion parameter $d_2 = 1.0$. As Fig. 7 indicates, for this setup there will be only a negligible net current $\langle v \rangle \approx 0$ in the deterministic case ($D = 0$), but a remarkable net current $\langle v \rangle = -0.43$ in the stochastic case for $D = 0.1$. As Fig. 8 shows, there is a *critical strength* of the stochastic force, $D^{crit}(d_2 = 1.0) \simeq 10^{-4}$, where an onset of the net current can be observed, while for $D < D^{crit}$ no net current occurs. On the other hand, there is also an *optimal strength* of the stochastic force, D^{opt}, where the amount of the net current, $|\langle v \rangle|$, reaches a *maximum*. An increase of the stochastic force above D^{opt}

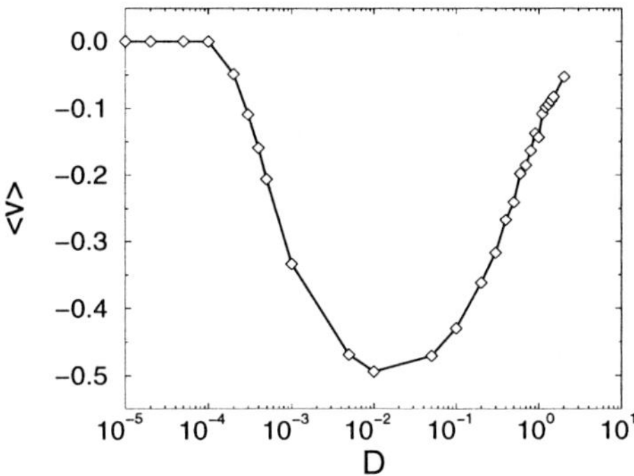

Fig. 8. Average velocity $\langle v \rangle$ vs. strength of the stochastic force D. The data points are obtained from simulations of 10.000 particles with a fixed conversion parameter $d_2 = 1.0$, for the other parameters see Fig. 7 [18].

will only increase the *randomness* of the particle's motion, hence the net current decreases again. In conclusion, this sensitive dependence on the stochastic force may be used to adjust a *maximum net current* for the particles movement in the ratchet potential.

Acknowledgement. The author would like to thank W. Ebeling (Berlin) and B. Tilch (Stuttgart) for collaboration.

References

1. A.S. Mikhailov and D. Meinköhn, *Self-Motion in Physico-Chemical Systems Far from Thermal Equilibrium*, in: *Stochastic Dynamics,* L. Schimansky-Geier and T. Pöschel, (Eds.), pp. 334-345 (Springer, Berlin, 1997).
2. H. Gruler and A. de Boisfleury-Chevance, *Directed Cell Movement and Cluster Formation: Physical Principles*, J. de Physique I France **4**, 1085-1105 (1994).
3. D. Helbing and P. Molnár, *Social force model for pedestrian dynamics*, Phys. Rev. E **51/5**, 4282-4286 (1995).
4. T. Vicsek, A. Czirok, E. Ben-Jacob, I. Cohen, and O. Shochet, *Novel Type of Phase Transition in a System of Self-Driven Particles*, Phys. Rev. Lett. **75**, 1226-1229 (1995).
5. E.V. Albano, *Self-organized collective displacements of self-driven individuals*, Phys. Rev. Lett. **77**, 2129-2132 (1996).
6. D. Helbing and T. Vicsek, *Optimal Self-Organization,* New J. Physics **1**, 13.1-13.17 (1999).
7. O. Steuernagel, W. Ebeling, and V. Calenbuhr, *An Elementary Model for Directed Active Motion*, Chaos, Solitons & Fractals **4**, 1917-1930 (1994).
8. L. Schimansky-Geier, M. Mieth, H. Rose, and H. Malchow, *Structure Formation by Active Brownian Particles*, Physics Lett. A **207**, 140-146 (1995).

9. F. Schweitzer, *Active Brownian Particles: Artificial Agents in Physics*, in: *Stochastic Dynamics,* L. Schimansky-Geier and T. Pöschel, (Eds.), pp. 339-352 (Springer, Berlin, 1997).
10. F. Schweitzer, W. Ebeling, and B. Tilch, *Complex Motion of Brownian Particles with Energy Depots*, Phys. Rev. Lett. **80**, 5044-5047 (1998).
11. W. Ebeling, F. Schweitzer, and B. Tilch, *Active Brownian Particles with Energy Depots Modelling Animal Mobility*, BioSystems **49**, 17-29 (1999).
12. M.O. Magnasco, *Brownian Combustion Engines*, in: *Fluctuations and Order: The New Synthesis*, M. Millonas, (Ed.), pp. 307-320 (Springer, New York, 1996).
13. F. Jülicher and J. Prost, *Cooperative Molecular Motors*, Phys. Rev. Lett. **75/13**, 2618-2621 (1995).
14. J. Luczka, R. Bartussek, and P. Hänggi, *White Noise Induced Transport in Periodic Structures*, Europhys. Lett. **31**, 431 (1995).
15. J. Rousselet, L. Salome, A. Ajdari, and J. Prost, *Directional motion of Brownian particles induced by a periodic asymmetric potential*, Nature **370**, 446-448 (1994).
16. U. Erdmann, W. Ebeling, L. Schimansky-Geier, and F. Schweitzer, *Brownian Particles far from Equilibrium*, Europ. Phys. J. B, in press, 1999.
17. F. Schweitzer, B. Tilch, and W. Ebeling, *Uphill Motion of Active Brownian Particles in Piecewise Linear Potentials*, Europ. Phys. J. B, in press, 1999.
18. B. Tilch, F. Schweitzer, and W. Ebeling, *Directed Motion of Brownian Particles with Internal Energy Depot*, Physica A **273**, 294-314 (1999).

Associative Memory of a Pulse-Coupled Noisy Neural Network with Delays: The Lighthouse Model

H. Haken

Institute for Theoretical Physics 1, Center of Synergetics, University of Stuttgart, Germany

Abstract. We start from the basic equations of a pulse-coupled neural network with arbitrary couplings ("synaptic strengths") between its elements. The axonal pulses are described by means of a phase, whose rotation speed depends on the dendritic inputs ("lighthouse model"). We include the effects of noise by means of fluctuating forces. We also allow for delays between the neurons. The introduction of time-averaged axonal pulse rates ω_ℓ allows us to convert the original, highly nonlinear and stochastic equations into rather simple equations for ω_ℓ that can be solved directly. The solutions can be interpreted as action of an associate memory.

1 Introduction

A good deal of the contributions to this book are devoted to the collective behavior of active members – or agents – that interact within a group. Examples are provided by pedestrians or cars, by users of the Internet, or shareholders in financial markets. (For references consult these proceedings.) W. Weidlich's field of Sociodynamics [1] provides us with a wealth of further examples. An outstanding example of the cooperation between agents is provided by the human brain, where myriads of neurons cooperate so to produce perception, speech, motor control, a.s.o. In my paper I will present a model of a neural net [2] that is rather realistic and can be treated explicitly. As I have shown elsewhere [3], it allows for the treatment of a number of experimentally observed phenomena, such as phase locking between different neurons or the impact of noise. Here I will show how such a network functions as associative memory.

2 What is Associative Memory?

According to a generally accepted view (e.g. Kohonen [4]), such a memory serves the completion of data if an incomplete set is provided to a system. A telephone book is a simple example. Our network model solves this problem by acting as a dynamical system.

3 The Basic Equations

First, I briefly remind the reader of some basic facts of real neurons. A neuron can be conceived as an element that receives inputs from other neurons, processes them and then sends out outputs to other neurons. The scheme is shown in Fig. 1.

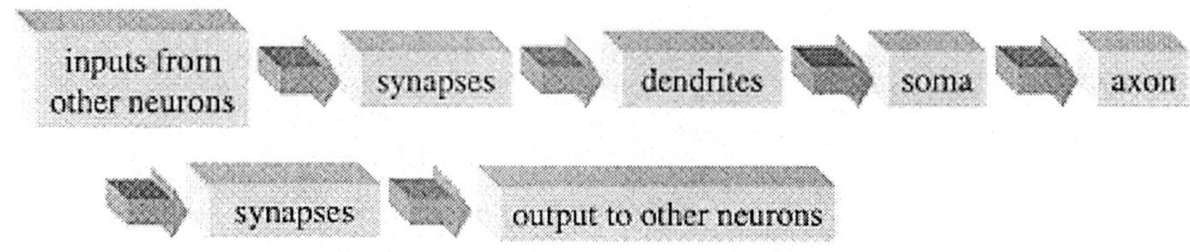

Fig. 1. Block diagram of neuronal connections and activities.

The soma of a neuron receives its inputs through dendrites in the form of electric currents. It then sums up over these inputs, which might be excitatory or inhibitory, and once a certain threshold is reached it fires. This means, it sends out short pulses along its axon, which then splits up and connects with the dendrites of other neurons via synapses. Once the electric signal carried by the pulse reaches a synapse, it causes vesicles to open and to release neurotransmitters that diffuse across the synaptic cliff and eventually cause currents in the dendrites of other neurons. The messages sent out from a neuron via the axon are pulse-coded , i.e., the signal is encoded by means of time intervals between the individual pulses. Since I derived the basic equations elsewhere [2], I shall start from them directly.

The generation of a dendritic current $\psi_m(t)$ of dendrite m is described by

$$\dot{\psi}_m(t) = \sum_k a_{mk} P_k(t - \tau_{mk}) - \gamma \psi_m(t) + F_{\psi,m}(t). \tag{1}$$

The coefficients a_{mk} are called synaptic strengths, $P_k(t - \tau)$ is a pulse of axon k with delay time τ, γ dendritic damping, $F_{\psi,m}(t)$ a stochastic force, caused for instance by random opening of vesicles. By means of a periodic, sharply peaked function f, P_k is expressed by a phase ϕ_k

$$P_k(t) = f(\phi_k(t)). \tag{2}$$

The phase ϕ_k is subject to the equation

$$\dot{\phi}_k(t) = S\left(\sum_m c_{km} \psi_m(t - \tau'_{km}) + p_{\text{ext},k}(t - \tau''_{km}) - \Theta_k\right) + F_{\phi,k}(t). \tag{3}$$

$S(X)$ is a sigmoid function, i.e., equal to zero for $X < 0$, quasi-linear for $0 \leq X \leq X_M$, and equal to a constant S_M. $p_{\text{ext},k}$ is an external signal, Θ_k a threshold, $F_{\phi,k}$ a fluctuating force.

In the following I assume that the neural net operates in the (practically) linear regime of S so that S in (3) can be replaced by its argument. It is a simple matter to eliminate ψ_m from (1) and (3), whereby we obtain

$$\begin{aligned}\ddot{\phi}_k(t) + \gamma\dot{\phi}_k(t) &= \sum_{\ell,m} c_{km} a_{m\ell} P_\ell\left(t - \tau'_{km} - \tau_{m\ell}\right) \\ &+ \gamma \sum_m p_{\text{ext},k}\left(t - \tau''_{km}\right) - \gamma\Theta_k + \gamma F_{\phi,k}(t) \\ &+ \sum_m c_{km} F_{\psi,m}\left(t - \tau'_{km}\right) \\ &+ \sum_m \dot{p}_{\text{ext},k}\left(t - \tau''_{km}\right) + \dot{F}_{\phi,k}(t),\end{aligned} \tag{4}$$

which can be cast into the simpler form (where we replace k by j)

$$\ddot{\phi}_j(t) + \gamma\dot{\phi}_j(t) = \sum_{\ell,m} A_{j\ell;m} P_\ell\left(t - \tau_{j\ell m}\right) C_j(t) + \hat{F}_j(t), \tag{5}$$

where $A_{j\ell;m} = c_{jm} a_{m\ell}$, $\tau_{j\ell m} = \tau'_{jm} + \tau_{m\ell}$,

$$C_j(t) = \gamma \sum_m p_{\text{ext},j}\left(t - \tau''_{jm}\right) - \gamma\Theta_j + \sum_m \dot{p}_{\text{ext},j}\left(t - \tau''_{km}\right) \tag{6}$$

$$\hat{F}_j(t) = \gamma F_{\phi,j}(t) + \sum_m c_{jm} F_{\psi,m}\left(t - \tau'_{jm}\right) + \dot{F}_{\phi,j}(t). \tag{7}$$

We will use the explicit form

$$P_\ell(t) = f\left(\phi_\ell(t)\right) = \dot{\phi}_\ell(t) \sum_n \delta(\phi_\ell(t) - 2\pi n). \tag{8}$$

The initial conditions in (5) are $\dot{\phi}_j(0) = \phi_j(0) = 0$.

4 The Steady State

In spite of the highly nonlinear character of (5) due to (8) and the delay times, we can derive an exact relationship between the pulse-rates of the axons j, where we use time-averaged pulse rates. We shall assume that $C_j(t)$ is practically constant over the averaging interval T and that T is large enough to cover the delay times τ. We integrate (5) over the time interval T and divide both sides by T

$$\begin{aligned}&T^{-1}\left(\dot{\phi}_j(T + t_0) - \dot{\phi}_j(t_0) + \gamma\phi_j(T + t_0) - \gamma\phi_j(t_0)\right) \\ &= T^{-1} \sum_{\ell,m} A_{j\ell;m} \int_{t_0}^{t_0+T} P_\ell(t - \tau_{j\ell m}) dt + T^{-1} \int_{t_0}^{t_0+T} C_j(t) dt + T^{-1} \int_{t_0}^{t_0+T} \hat{F}_j(t) dt.\end{aligned} \tag{9}$$

We discuss the individual terms in the mathematical limit $T \to \infty$. (For all practical purposes, T can be chosen finite but large enough.) We obtain

$$T^{-1}(\dot{\phi}(T+t_0) - \dot{\phi}(t_0)) \to 0, \tag{10}$$

because the phase-velocity is finite. Since $\frac{1}{2\pi}(\phi_j(T+t_0) - \phi_j(t_0))$ is the number of rotations of the phase-angle ϕ_j,

$$\frac{1}{2\pi} T^{-1}(\phi_j(T+t_0) - \phi_j(T)) = \omega_j \tag{11}$$

is the pulse rate. Since

$$\int_{t_0}^{t_0+T} P_\ell(t-\tau)dt \tag{12}$$

contains the δ-peaks according to (8), we obtain

$$(12) = \int_{t_0}^{t_0+T} \sum_n \delta\left(\phi_\ell(t-\tau) - 2\pi n\right) \dot{\phi}_\ell(t-\tau)dt \tag{13}$$

$$= \int_{t_0-\tau}^{t_0+T-\tau} \sum_n \delta\left(\phi_\ell(t) - 2\pi n\right) \dot{\phi}_\ell(t)dt \tag{14}$$

$$= \int_{\phi_\ell(t_0-\tau)}^{\phi_\ell(t_0+T-\tau)} \sum_n \delta(\phi_\ell - 2\pi n)d\phi_\ell \tag{15}$$

= number of pulses of axon ℓ in interval $[\phi_\ell(t_0-\tau), \phi_\ell(t_0+T-\tau)]$.

Thus, when we divide (12) by T and assume steady state conditions, we obtain the pulse rate

$$T^{-1} \int_{t_0}^{t_0+T} P_\ell(t-\tau)d\tau = \omega_\ell. \tag{16}$$

For slowly varying or constant $C_j(t) \approx C_j$, we obtain

$$T^{-1} \int_{t_0}^{t_0+T} C_j(t)dt = C_j(t_0). \tag{17}$$

Because we assume that the fluctuating forces possess an ensemble average and under the further assumption that the time average equals the ensemble average, we find

$$T^{-1} \int_{t_0}^{t_0+T} \hat{F}_j(t)dt = 0. \tag{18}$$

Lumping the results (10), (11), (16), (17), (18) together and dividing all terms by $(2\pi\gamma)$, we obtain the *linear* equations

$$\omega_j = \sum_{\ell,m} A_{j\ell;m}(2\pi\gamma)^{-1}\omega_\ell + c_j(t), \tag{19}$$

where $c_j(t) = (2\pi\gamma)^{-1}C_j(t)$.

The equations (19) can be solved under the usual conditions. Depending on the coupling coefficients $A_{j\ell} = \sum_m A_{j\ell;m}$, even those ω_ℓ may become nonzero, for which $c_\ell = 0$. On the other hand, only those solutions are allowed for which $\omega_\ell \geq 0$ for all ℓ. This imposes limitations on c_ℓ and $A_{\ell\ell'}$.

5 Associative Memory and Pattern Filter

The results of the preceding section can be used to discuss the action of our neural network. To this end, we introduce the vectors $\boldsymbol{\omega}$ and $\mathbf{c}$ by means of their transposed vectors

$$\boldsymbol{\omega}^T = (\omega_1, ..., \omega_N), \quad \mathbf{c}^T = (c_1, ..., c_N). \tag{20}$$

When we use prototype vectors $\mathbf{c_u} = \mathbf{v_u}$ according to (19), to each of them a frequency pattern $\boldsymbol{\omega}_u$ of axonal firing rates is attached $\boldsymbol{\omega}_u = (1 - \tilde{A})^{-1}\mathbf{v_u}$, where $\tilde{A}$ is a matrix with elements $A_{j\ell}/(2\pi\gamma)$. We assume that $(1 - \tilde{A})^{-1}$ exists, and that the dimension of the space spanned by $\mathbf{v}_u, u = 1, ..., n$ is equal to the dimension of the $\boldsymbol{\omega}$-space. If a smaller number of prototype vectors $\mathbf{v}_u$ is given, we may define further ad-hoc vectors so that the original requirement is fulfilled. Let us choose a specific form

$$(1 - \tilde{A})^{-1} = \sum_u \Lambda_u \mathbf{v_u} \cdot \mathbf{v_u^+}, \tag{21}$$

where $\mathbf{v}_u^+$ is defined by $(\mathbf{v}_{u'}^+ \mathbf{v}_u) = \delta_{uu'}$. Equation (21) can be interpreted in two ways: For a given matrix $\tilde{A}$, the vectors $\mathbf{v}_u(\mathbf{v}_u^+)$ are the right (left) eigenvectors of it with eigenvalues $\Lambda_u = (1 - \lambda_u)^{-1}$. In the field of neurocomputers, the r.h.s. of (21) is known as learning matrix with attention parameters Λ_u [5] so that the r.h.s. of (21) defines $\tilde{A}$. When we apply $(1 - \tilde{A})^{-1}$ to an arbitrary input

$$\mathbf{q} = \sum_{\mathbf{u}} \mathbf{v_u} \left(\mathbf{v_u^+ q}\right), \tag{22}$$

it is converted into a neuronal firing pattern

$$\boldsymbol{\omega} = (1 - A)^{-1}\mathbf{q} = \sum_{\mathbf{u}} \Lambda_\mathbf{u} \mathbf{v_u} \left(\mathbf{v_u^+ q}\right) \tag{23}$$

that may act in a two-fold way.

1. Depending on A_u, only specific patterns are amplified ("pattern filter")
2. A firing pattern $\boldsymbol{\omega}$ belonging to $\mathbf{v_u}$ appears, even if $\mathbf{v_u}$ is only partly contained in $\mathbf{q}$ ("associative memory"). Note that only those $\mathbf{q}$'s are admitted in the frame of the network discussed here for which no negative components of $\boldsymbol{\omega}$ occur. Otherwise the input must be filtered correspondingly.

In order to elucidate these results, let us specialize our results to two coupled neurons, where

$$A_{11} = A_{22} = 0, \quad A_{12} = A_{21} = A.$$

We then obtain

$$\omega_1 = 2\pi \frac{(c_1 2\pi + c_2 A/\gamma)}{4\pi^2 - A^2/\gamma^2}, \tag{24}$$

$$\omega_2 = 2\pi \frac{(c_1 A/\gamma + c_2 2\pi)}{4\pi^2 - A^2/\gamma^2}. \tag{25}$$

Their difference and sum are particularly simple

$$\omega_2 - \omega_1 = \frac{c_2 - c_1}{1 + A/(2\pi\gamma)}, \tag{26}$$

$$\omega_1 + \omega_2 = \frac{c_1 + c_2}{1 - A/(2\pi\gamma)}. \tag{27}$$

These results exhibit a number of remarkable features of the coupled neurons: According to (26) their frequency sum, i.e., their activity is enhanced by positive coupling A. Simultaneously, according to (26) some frequency pulling occurs. According to (24), neuron 1 becomes active even for vanishing or negative c_1 (provided $| c_1 2\pi | < c_2 A/\gamma$), if neuron 2 is activated by c_2. This has an important application to the interpretation of the perception of Kanisza figures (Fig. 2), and more generally, to associative memory.

6 The Strong Coupling Limit

Our solution of the basic equations (1) – (3) relies on the assumption that the system operates in the linear range of S. This requires a not too strong coupling as expressed by the coefficients $A_{j\ell;m}$ or a not too small damping γ. If these conditions are violated, an instability occurs bringing the net into a state, where saturation of S has to be taken into account [3]. Interestingly enough, this problem can be also treated. For simplicity, we drop the fluctuating forces. We first solve (1), where we even allow for m-dependent damping constants γ_m. Assuming the initial condition $\psi_m(0) = 0$, we obtain

$$\psi_m(t) = \int_0^t \exp\left(-\gamma_m(t-\sigma)\right) \sum_k a_{mk} P_k(\sigma - \tau_{mk}) d\sigma. \tag{28}$$

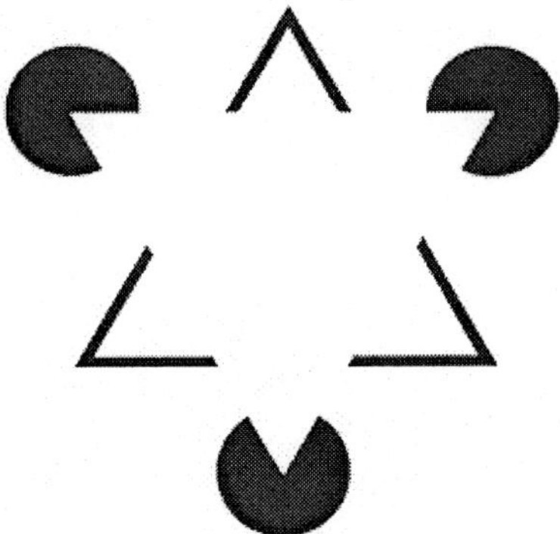

Fig. 2. The Kanisza figure as an example for the action of associative memory. In spite of the fact that only few indications, such as corners, are given, our brain recognizes a full triangle.

Inserting this in (3) yields

$$\dot{\phi}_j(t) = S\left(\sum_m \sum_\ell c_{jm} a_{m\ell} [...]_{jm\ell} + \sum_m p_{\text{ext},j}(t - \tau''_{jm}) - \Theta_j\right), \qquad (29)$$

where

$$[...]_{jm\ell} = \int_0^{t-\tau'_{jm}} \exp\left(-\gamma_m(t - \tau'_{jm} - \sigma)\right) P_\ell(\sigma - \tau_{m\ell}) d\sigma. \qquad (30)$$

We note that $\dot{\phi}/2\pi$ is equal to the pulse rate ω_ℓ. In order to proceed further, we put the delay times $\tau'_{jm} = 0$ and assume p_{ext} time-independent. Using similar arguments as above in Sect. 4, we may approximate (30) by ω_ℓ/γ. Thus, we obtain the equations

$$\omega_j = \frac{1}{2\pi} S\left(\sum_{m\ell} A_{j\ell;m}\omega_\ell + \frac{1}{\gamma} C_j\right), \qquad (31)$$

which quite obviously generalizes (19) to the nonlinear case. A systematic discussion of the case in which S saturates is beyond the scope of my paper. May it suffice to illustrate the great variety of solutions for two neurons, where we consider $c_1 > 0, c_1 = 0$ and $A_{11} = A_{22} = 0, \quad A_{12} = A_{21} = A \geq 0$. We introduce a threshold Θ_1 so that

$$S(X) = S_M = \text{const. for } X \geq \Theta_1. \qquad (32)$$

We start with

1. $A = 0$. Then $\omega_1 > 0, \omega_2 = 0$, i.e., neuron 1 fires while neuron 2 is silent.
2. $A > 0$ so that

$$A/\gamma + c_1 \leq \Theta_1, \quad A/\gamma \cdot S_M < \Theta_1. \tag{33}$$

Then

$$\omega_1 = \frac{1}{2\pi} S_M, \quad \omega_2 = \frac{1}{2\pi} A/\gamma \cdot \omega_2. \tag{34}$$

3. $A > 0$ so that

$$A/\gamma + c_1 \geq \Theta_1, \quad A/\gamma \cdot S_M \geq \Theta_1. \tag{35}$$

Then

$$\omega_1 = \frac{1}{2\pi} S_M, \quad \omega_2 = \frac{1}{2\pi} S_M. \tag{36}$$

The cases 2. and 3. show the action of an associative memory: in addition to neuron 1, neuron 2 fires in spite of the absence of a direct external signal. The case 2. is probably more important, because the firing rate ω_2 depends on the synaptic strength A. On the other hand, case 3., in a way, leads us back to the fundamental work of McCullough and Pitts [6]. If we identify $\omega_j = 0$ with neuron j "inactive" (no) and $\omega_j > 0$ with neuron j "active" (yes), the network acts as a logical network.

References

1. W. Weidlich, *Sociodynamics,* in press.
2. H. Haken, *What can synergetics contribute to the understanding of brain functioning,* in: *Analysis of Neurophysiological Brain Functioning,* C. Uhl, (Ed.), pp. 7-40 (Springer, Berlin, 1999).
3. H. Haken, *Phase locking and noise in the lighthouse model of a neural net with delay,* in: *Stochaos: Stochastic and Chaotic Dynamics in the Lakes,* D.S. Broomhead, E.A. Luchinskaya, P.V.E. McClintock, and T. Mullin, (Eds.), in press, (American Institute of Physics, Woodbury, New York).
4. T. Kohonen, *Associative Memory - A System Theoretical Approach,* (Springer, Berlin, 1987).
5. H. Haken, *Synergetic Computers and Cognition,* (Springer, Berlin, 1991).
6. W. McCulloch and W. Pitts, *A logical calculus of the ideas immanent in nervous activity,* Bull. Math. Biophysics **5**, 115-133 (1943).

Application of Neural Networks for Predictive and Control Purposes

B. Schürmann

Siemens AG, Corporate Technology, Information and Communications, 81730 München, Germany

Abstract. For a company to stay competitive, providing "intelligent" application solutions, services and products to its customers is indispensable. For example, forthcoming telematics applications require technologies like adaptive control, model building by learning, and focusing attention on relevant features. Neural Networks offer appropriate architectures for these purposes. This contribution gives an overview on activities at Siemens Corporate Research using Neural Networks for prediction and control purposes.

1 Introduction

The activities of the Siemens business groups cover a variety of fields, examples being traffic, telecommunications, medical care, control, energy, and finance. As a central R&D division, Siemens Corporate Technology has to provide innovative solutions in order to keep the company competitive. One of its core technologies is Neural Computation. In this contribution, research activities and applications are presented which are or might become relevant for telematics.

The department of Neural Computation covers know-how on a wide spectrum of technologies, among them modeling, control, optimization, prognosis, decision support, cooperation, distribution and autonomy. This know-how is developed and extended further by means of long-term research projects. Subjects of these are, among others, information extraction from large databases, scalability of KDD-algorithms[1], reinforcement learning, learning causal nets, distributed learning systems, learning algorithms for networks of spiking neurons and hierarchical integration of neural modules.

In Sect. 2, a short introduction to standard feedforward neural networks is given. Forecast applications for traffic and finance are discussed in Sect. 3. Reinforcement learning is explained in the next section followed by an application to traffic light control. In Sect. 5, spiking neurons, a potential future direction of neuroscience, are presented. Finally, in the outlook we discuss the next generation of artificial neural systems which make use of research results obtained in the neurosciences. The relevance for telematics is outlined.

[1] KDD = Knowledge Discovery in large Databases.

2 Neural Networks - From Biology to Mathematics

The human brain is one of the most complex structures in nature with probably the most advanced features; simply speaking, it provides us with "intelligence". However, so far we only partly understand how the brain works. It is common opinion that the connections between neurons, the basic units of the brain, represent all the information which has been acquired by the brain. Neurons receive action potentials, i.e. electrical pulses, from other neurons; once their activation exceeds a certain threshold, they are firing a pulse themselves. This principle mechanism can be further simplified assuming that only the average firing rates carry relevant information. This leads to a picture as shown in Fig. 1d. Connecting such mathematical neurons using specific topologies, yields artificial neural networks as shown in Fig. 1c.

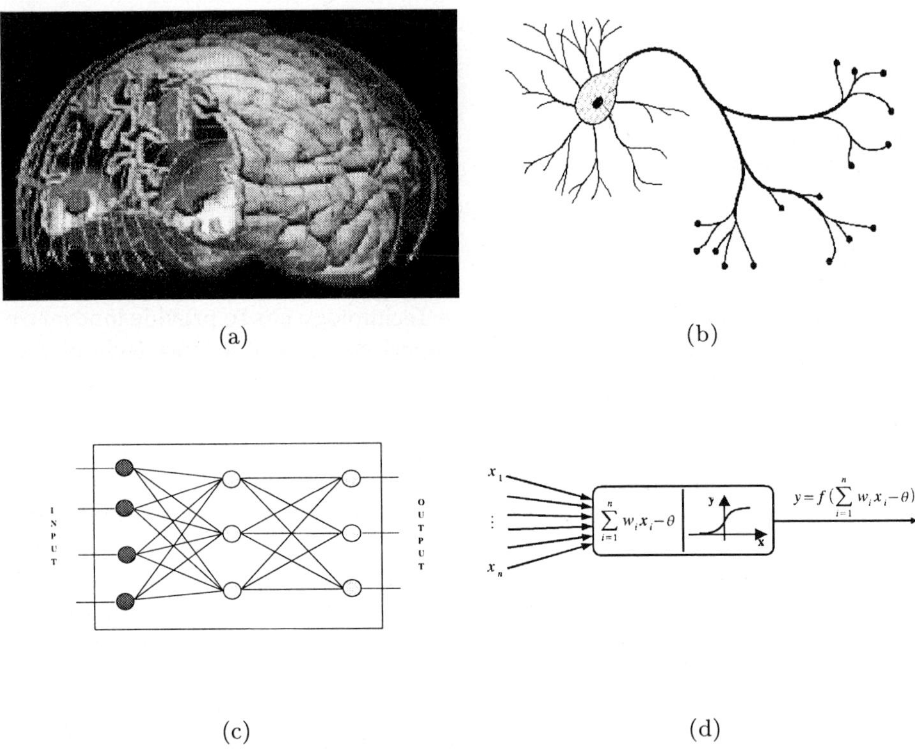

Fig. 1. The human brain (a) and its smallest functional unit, the neuron (b), versus an artificial neural network (c) consisting of artificial "neurons" (d), its basic building blocks.

Feedforward neural networks are characterised by topologies where outputs of neurons of a certain layer are connected to neurons of the next layer. These topologies have been proven to be able to approximate any continuous real-valued function. Recurrent neural networks allow for connections of neurons to previous layers, i.e. the output of such a network can be input for the next processing step.

In the huqman brain, learning is the process of adapting connections between neurons by enforcing or weakening these. In artificial neural networks, learning means adapting the weights w_i and creating or destroying connections. A standard procedure for the adaptation of weights is error backpropagation in supervised learning. Here, the output of a neural network is compared with a target output to be learned; the gradient of the error with respect to a particular weight is used to change this weight such that the error decreases. Variants of learning procedures and a presentation of the state-of-the-art can be found in [1].

3 A Feedforward Forecast Topology and Its Applications

Very important applications of feedforward neural networks are forecasts. In order to use these networks one has to describe the forecast as a mapping of present and possibly past input data to future data. Figure 2 shows a topology which we have developed for such purposes.

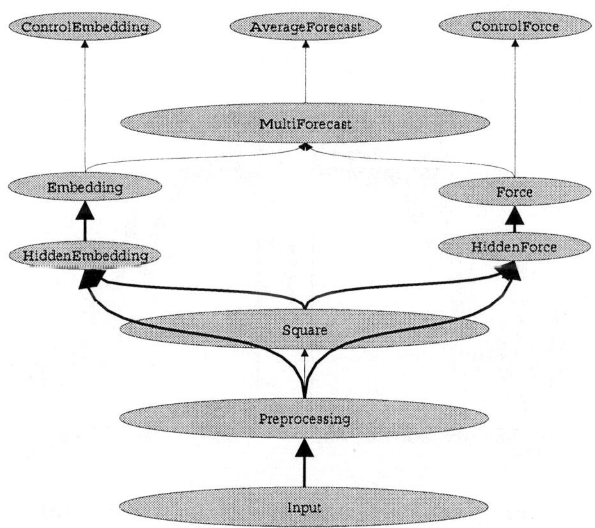

Fig. 2. The topology of a feedforward neural network for forecast purposes. Only the bold connections can be modified by learning.

A preprocessing layer following the input layer is trained to scale the inputs and - due to the saturation of the neuron's activation function - to eliminate outliers. These preprocessed inputs and their squares are fed into two hidden layers where the relevant information is built. The following layers just produce various information for the learning procedure. They are linear units providing temporal differences and curvatures as well as temporal averages. All these quantities are compared to corresponding ones derived from the training data. These layers are responsible for a smooth and not too volatile output of the neural network.

3.1 Traffic Forecast

Short term traffic forecast is useful in several respects: It might be part of an information system warning drivers approaching jams, or it might be part of a predictive control system. This raises the question of the quality achievable for short-time traffic forecasts.

For this purpose we took data collected from 6 subsequent induction loops located on highway A9 north of Munich. The forecast is performed by a feedforward neural network whose inputs are fed with present and past data from the 6 induction loops. The outputs of the neural network should provide a 5-minutes forecast of the traffic state at an intermediate detector position. The choice of inputs and the forecast horizon take into account the range of propagation velocities of perturbations as measured in real traffic [2]. The topology of the neural network shown in Fig. 2 is employed. The resulting accuracy of the forecast is about 10 km/h for the velocity, which is in the range of the uncertainties in the measurements (cf. [2–6] for further references to our research activities in traffic dynamics).

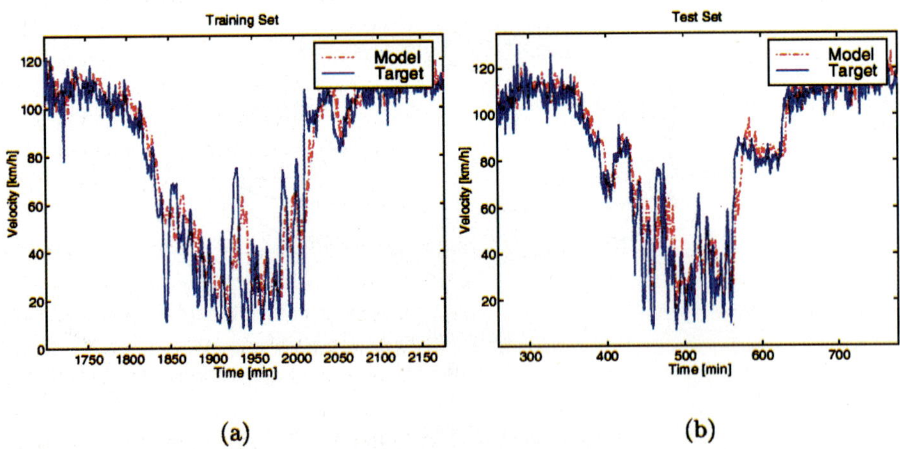

Fig. 3. Measurements of the speed during rush hours. (a) Comparison with neural network output for training data set. (b) Comparison with neural network output for test data set (another day).

3.2 Strategic Portfolio Management

We implemented a strategic portfolio management system in collaboration with one of the five largest assurance companies in the world and with a Siemens business group. This project was sponsored by the EU. The task was to allocate funds across three different investment opportunities, -stocks, bonds, cash-, over the seven largest economies (Spain, France, Germany, Italy, Japan, United Kingdom and USA). There were allocation constraints dependent on the market capitalization of the 21 possible assets.

Our approach to solve this problem can be regarded as a neural network realization of the "Global Portfolio Optimization" of Black and Litterman [7] which combines the mean-variance approach of Markowitz [8] with the capital asset pricing model (CAPM) of Sharpe and Lintner [9,10]. Black and Litterman claim that their method avoids the unrealistic solutions the Markowitz optimization often produces in practice (large short positions, or "corner" solutions) and releases the investor from the difficult task of formulating views of the returns for all assets. Their approach uses excess returns of the CAPM equilibrium to define a neutral benchmark portfolio which is typically well balanced. The investor may deviate slightly from this portfolio by a different weighting of such assets with specific expectations. In contrast to Black and Litterman, our approach is based on excess return forecasts estimated by neural networks. These predictions are transformed to allocation decisions which obey the allocation constraints given by the investor, and which implicitly control the risk exposure of the portfolio by parameter-optimizing over time. The allocation constraints correspond to the neutral views of Black and Litterman, and they may be identical to them or differ due to institutional or legal reasons.

Our approach to construct profitable portfolios while obeying allocation constraints and controling its risk exposure consists of the following steps (see also [11]):

1. Identification of forecast models $f_1, \cdots, f_k$ for all k assets. These are based on the specific neural network architecture described in [1] assuming an economic decison model. The network includes an outlayer detection, an integated MLP-RBF[2] approximation and various penalty functions to avoid too volatile outputs.
2. Integration of all models $f_1, \cdots, f_k$ in a single model with k outputs. These are the basis of the successive allocation scheme.
3. Computation of an allocation strategy a_i using the estimated excess returns. The so-called market share constraints force the strategy to obey that the allocations a_i may deviate from a given benchmark portfolio only within predefined intervals.
4. Optimization of the parameters of this allocation strategy computes profitable portfolios and also implicitly controls the risk exposure by attenuating the influence of uncertain forecasts.

[2] MLP = Multilayer-perceptron; RBF = Radial basis function.

Our approach yields an improvement of 5% in profit in comparison to the benchmark portfolio.

4 Reinforcement Learning

4.1 Markov Decision Processes

The term "Reinforcement Learning" denotes a class of learning algorithms where an agent learns to interact with its environment so as to maximize a given performance criterion. Reinforcement Learning is best analyzed and understood within the framework of discrete time Markov Decision Processes (MDP, see Fig. 4) which consist of a state space S, an action space A, a transition probability function $p : S \times S \times A \to [0,1]$, and a reward function $r : S \times A \to \mathbf{R}$. At each time step $t = 0, 1 \ldots$ the agent observes the actual state $x_t \in S$ of the environment and applies an action (or control) $a_t \in A$. Then, the system changes its state according to the probability function p, i.e. $p(x_t, y, a_t)$ denotes the probability for state $y \in S$ being the successor state x_{t+1}. Furthermore, the agent receives a scalar reward of amount $r(x_t, a_t)$.

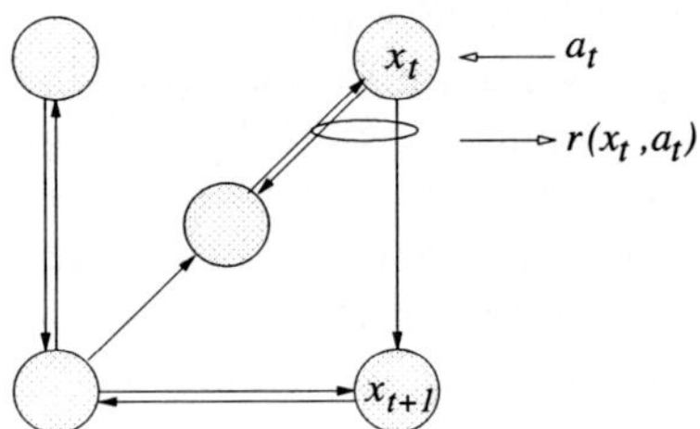

Fig. 4. Markov Decision Process

The goal of the agent is to learn a control policy π, i.e. a mapping $\pi : S \to A$ from states to actions so as to maximize the discounted cumulated reward

$$E\left(\sum_{t=0}^{\infty} \gamma^t r(x_t, \pi(x_t))\right) \tag{1}$$

where $\gamma \in [0, 1[$ denotes a discount factor and $E(\cdot)$ denotes the expectation with respect to the future states of the stochastic process.

4.2 Q-Learning

Various reinforcement learning algorithms have been proposed (see [12] and the references therein). Here, we briefly describe Q-learning – a simulation based method where the agent learns to *evaluate* state-action pairs in order to optimize its policy. This is done using a parameterized real-valued function $\tilde{Q}(x, a; w)$ which may be implemented by neural networks, feature mappings or other architectures. The idea is to adapt the parameters w such that $\tilde{Q}(x, a; w)$ approximates the cumulated reward the agent obtains if it starts at state x, applies action a and follows an optimal policy thereafter. The Q-learning update rule is:

$$\begin{aligned} w \; &:= \; w + \sigma d_t \nabla_w \tilde{Q}(x_{t-1}, a_{t-1}; w), \\ &\text{with } \; d_t = r(x_{t-1}, a_{t-1}) + \gamma \max_{a \in A} \tilde{Q}(x_t, a; w) - \tilde{Q}(x_{t-1}, a_{t-1}; w) \end{aligned} \tag{2}$$

where σ denotes a small learning rate. After learning, a near-optimal policy $\tilde{\pi}$ can be obtained from the approximate Q-function $\tilde{Q}(x, a; w)$ through the maximization process.

$$\tilde{\pi}(x) = \arg\max_{a \in A} \tilde{Q}(x, a; w).$$

We applied variants of Q-learning to the traffic light control problem (see next section) and to the admission control and routing problem in telecommunication networks [13].

4.3 Optimization of Traffic Lights

One particularly challenging application of reinforcement learning is traffic light control. In particular, we investigate the selforganisation of subsequent traffic lights to form so-called green waves. For simplicity we consider two subsequent intersections controlled by traffic lights connected by one-way-streets (Fig. 5). In addition, the flows into the network and the routing are taken to be constant. We are looking for a relation between state information (from sensor signals) and action (switching traffic lights or not) such that the expectation of the discounted long-term reward

$$\text{CostToGo} = \sum_{\tau=0}^{\infty} \gamma^{\tau} r(t + \tau) \tag{3}$$

is optimized. The immediate reward $r(t)$ is equal to the number of cars passing the intersection, and $\gamma < 0$ is a discount factor.

We find an increase and a final saturation of the long-term reward accompanied by a reduction of its variance (see figure 6). The final solution corresponds to a green wave.

The learning procedure described above has been extended to account for nonstationary environments like variable flows or routing [14].

5 Spiking Networks

Feedforward and recurrent neural networks may be used for parametrically extracting the underlying structure of dynamical systems. In the case of feedforward networks, each observable of the dynamical system should be embedded

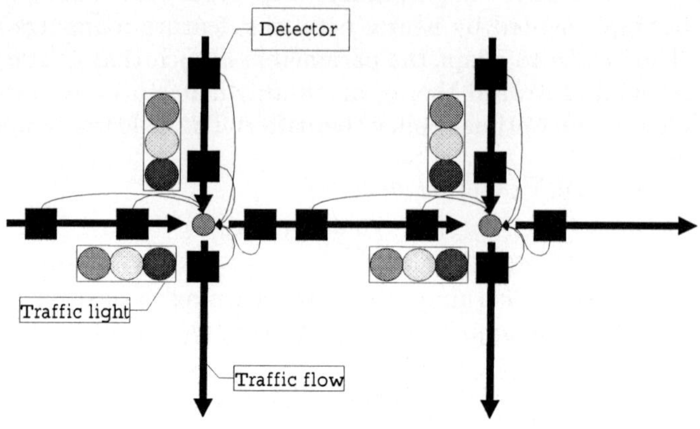

Fig. 5. Topology of traffic light controlled intersections.

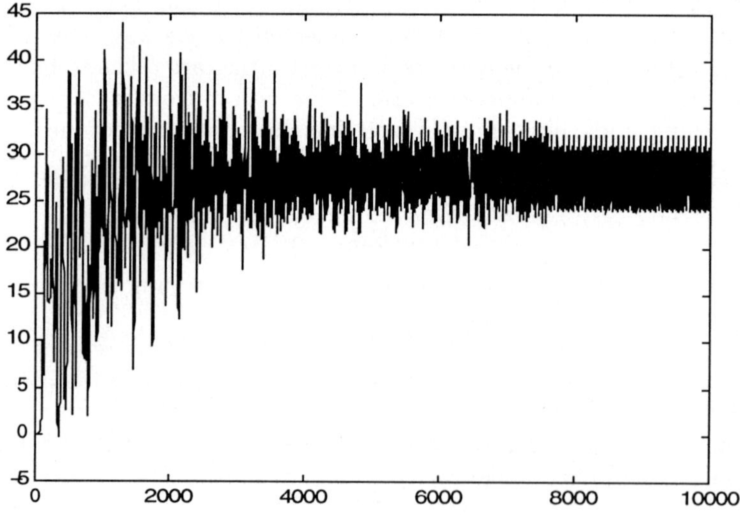

Fig. 6. Discounted long term reward as a function of time.

explicitly in a vector in order to catch the nonlinear structure which relates the past points with the future (see upper part of Fig. 7). A disadvantage of this method is that the exact embedding dimension of each variable and the embedding lag are not known and should therefore be estimated by heuristic methods. This is due to the fact that the encoding of the underlying dynamical structure is static (value of the static hidden neurons) and therefore the time should be explicitly "simulated" by the embedding.

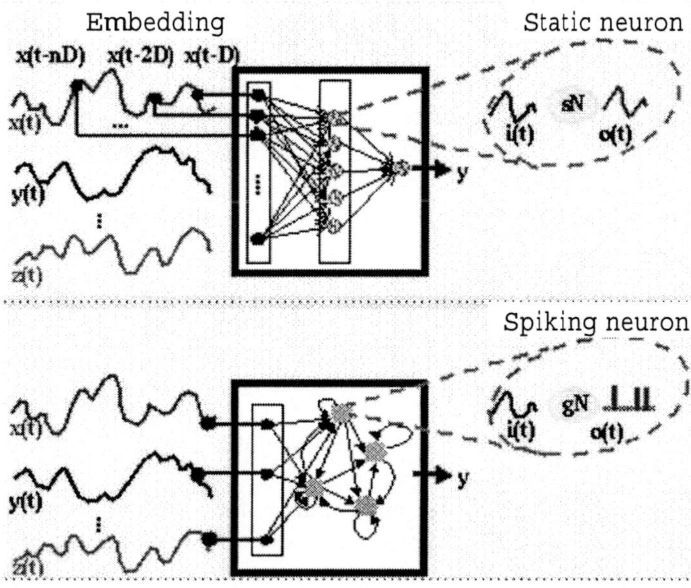

Fig. 7. Feedforward and recurrent neural networks (top) versus networks of spiking neurons (bottom).

In the case of recurrent neural networks such kind of embedding is not required because the internal coding can now catch the dynamics due to the internal recurrences of the architecture. The problem now is that the internal temporal coding is also a continuous variable and therefore apparently we have gained nothing, i.e. we merely have transformed a high-entropy continuous dynamical system in an another high-entropy continuous dynamical system. In fact, in both cases, i.e. for the observables and for the internal coding, we eventually have to describe at each time a density function.

In conclusion, it seems that for an efficient encoding of a dynamical system we need two ingredients: The first one is the ability of performing a dynamical coding (like recurrent networks), and the second one is a compression mechanism, i.e. a more efficient coding. Networks of spiking neurons offer these two mathematical ingredients. The promising mathematical convenience of using spiking networks for the processing of high-dimensional dynamical signals is the fact that the integrate-and-fire trained network can be viewed as an effective mechanism for comprising a high-entropy continuous process into a high-entropy point process (cf. [15]). The internal encoding scheme by spiking networks is dynamic, so that it is able to catch the underlying dynamics without explicit embedding of the observed variables, and at the same time the coding is given by a finite number of point processes (namely one for each spiking neuron) instead of by a continuous stochastic time process. In fact, if the output is a finite number of point processes, we can express the transmitted information in the framework of information theory, because each point process can be thought of as an output variable of the communication channel. Our learning principle can perhaps be used for developing algorithms for achieving this goal. If the internal coding would be given by continuous-time stochastic processes, the outputs are an infinite number of stochastic variables, namely one at each time. Further theoretical and experimental investigations on this principle are required for unveiling the process of understanding the way the brain encodes and processes information.

6 Outlook

There is a fair chance that in the next decade, genuinely intelligent systems will be developed. Computational neuroscience, a newly established interdisciplinary field of research, provides a suitable environment for such endeavors. Computational neuroscience is intended to take into account several levels of abstraction by constraining the mechanistic models by neuropsychological (macroscopic functional level), neurophysiological (system level) and neurobiological (microscopic level of networks and neurons) facts (see figure 8).

We believe that a computational perspective offers an approach that allows a detailed study of these problems. For example, we adopt a computational neuroscience perspective in order to analyze the attentional enhancement of the spatial resolution of the area containing the objects of interest (cf. [16]). Such kind of model consists of several modules with feedforward and feedback interconnections describing the mutual links between different areas of the visual cortex. Each module analyses the visual input with different spatial resolution and can be thought of as a hierarchical predictor at a given level of resolution. The attention control decides in which local regions the spatial resolution should be enhanced in a serial fashion. In this sense, the scene is first analyzed at a coarse resolution level and the focus of attention enhances iteratively the resolution at the location of an object until the object is identified.

In the near future, new telematics services will arise. They will provide more and more customer-specific information which is extracted from huge databases. The extraction of information has to be done in an intelligent way, for example context-sensitive. Such features precisely characterize neurocognitive systems.

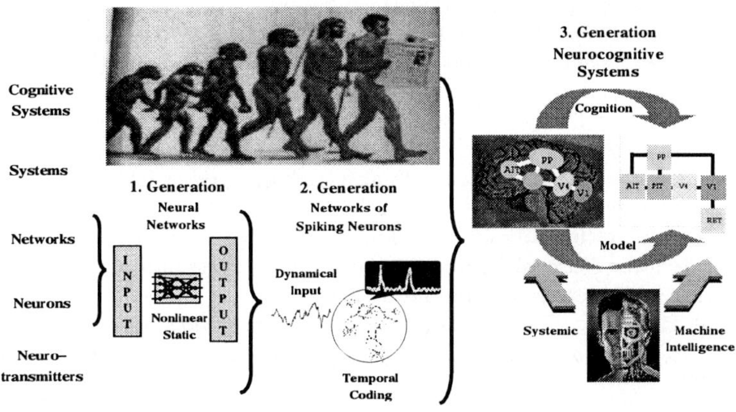

Fig. 8. Generations of neural networks.

Acknowledgement. I wish to thank my colleagues from the Department of Neural Computation, and in particular Rudi Sollacher, for their support during the preparation of this contribution.

References

1. R. Neuneier and H.G. Zimmermann, *How to Train Neural Networks*, in: *Neural Networks: Tricks of the Trade*, G.B. Orr and K.-R. Müller, (Eds.), pp. 373–423 (Springer, Berlin, 1998).
2. R. Sollacher and P. Konhäuser, *SANDY – Nichtlineare Dynamik im Straßenverkehr*, in: *Statusseminar Technische Anwendungen von Erkenntnissen der Nichtlinearen Dynamik*, pp. 103–112 (VDI Technologiezentrum Physikalische Technologien, Düsseldorf, 1999).
3. R. Sollacher, Ch. Stutz, and H. Lenz, *Prognose und Steuerung von Verkehr*, in: *Statusseminar Technische Anwendungen von Erkenntnissen der Nichtlinearen Dynamik*, pp. 177–180 (VDI Technologiezentrum Physikalische Technologien, Düsseldorf, 1999).
4. R. Sollacher and H. Lenz, *Nonlinear Control of Stop-and-Go Traffic*, in: These proceedings.
5. C. Wagner, C. Hoffmann, R. Sollacher, J. Wagenhuber, and B. Schürmann, *Second-Order Continuum Traffic Flow Model*, Phys. Rev. E **54**, 5073–5085 (1996).
6. H. Lenz, Ch. Wagner, and R. Sollacher, *Multi-anticipative car-following model*, Eur. Phys. J. B **7**, 331–335 (1999).
7. F. Black and R. Litterman, *Global Portfolio Optimization*, Fin. Analysts J., Sep., (1992).

8. H.M. Markowitz, *Portfolio Selection*, J. of Finance **7**, 77–91 (1952).
9. W.F. Sharpe, *Capital Asset Prices: A Theory of Market Equilibrium Under Conditions of Risk*, J. of Finance, Sep., (1964).
10. J. Lintner, *The Valuation of Risk Assets and the Selection of Risky Investments in Stock Portfolios and Capital Budgets*, Rev. of Eco. and Stat., Feb., (1965).
11. H.G. Zimmermann and R. Neuneier, *Active Portfolio-Management with Neural Networks*, in: *Proc. of Comp. Finance CF'1999*, A.S. Weigend and B. LeBaron, (Eds.), (Springer, 1999).
12. D. Bertsekas and J.N. Tsitsiklis, *Neuro-Dynamic Programming*, (Athena Scientific, Melmont, Massachusetts, 1996)
13. P. Marbach, O. Mihatsch, and J.N. Tsitsiklis, *Call Admission Control and Routing in Integrated Services Networks Using Neuro-Dynamic Programming*, to appear in IEEE J. on Selected Areas in Communic., (2000).
14. M. Appl and R. Palm, *Fuzzy Q-Learning in Nonstationary Environments*, in: *Proc. of the 7th Eur. Congress on Intelligent Techniques& Soft Computing (EUFIT'99)*, (1999).
15. G. Deco and B. Schürmann, *Spatiotemporal Coding in the Cortex: Information Flow-based Learning in spiking Neural Networks*, Neural Comp. **11**, 919–934 (1999).
16. G. Deco and B. Schürmann, *A Neuro-Cognitive Visual System for Object Recognition based on Interactive Attentional Top-Down Hypothesis Testing*, submitted, Perception, (1999).

Optimizing Traffic in Virtual and Real Space

D. Helbing[1,2,3], B.A. Huberman[1], and S.M. Maurer[1]

[1] Xerox PARC, 3333 Coyote Hill Road, Palo Alto, CA 94304, USA
[2] II. Institute of Theoretical Physics, University of Stuttgart, Pfaffenwaldring 57/III, 70550 Stuttgart, Germany
[3] Collegium Budapest – Institute for Advanced Study, 1014 Budapest, Hungary

Abstract. We show how optimization methods from economics known as portfolio strategies can be used for minimizing down-load times in the Internet and travel times in freeway traffic. While for Internet traffic, there is an optimal restart frequency for requesting data, freeway traffic can be optimized by a small percentage of vehicles coming from on-ramps. Interestingly, the portfolio strategies can decrease the average waiting or travel times, respectively, as well as their standard deviation ("risk"). In general, portfolio strategies are applicable to systems, in which the distribution of the quantity to be optimized is broad.

1 Virtual Traffic in the World Wide Web

Anyone who has browsed the World Wide Web has probably discovered the following strategy: whenever a web page takes too long to appear, it is useful to press the reload button. Very often, the web page then appears instantly. This motivates the implementation of a similar but automated strategy for the frequent "web crawls" that many Internet search engines depend on. In order to ensure up-to-date indexes, it is important to perform these crawls quickly. More generally, from an electronic commerce perspective, it is also very valuable to optimize the speed and the variance in the speed of transactions, automated or not, especially when the cost of performing those transactions is taken into account. Again, restart strategies may provide measurable benefits for the user.

The histogram in Figure 1 shows the variance associated with the down-load times for the text on the main page of over 40,000 web sites. Based on such observations, an economics-based strategy has recently been proposed for quantitatively managing the time of executing electronic transactions [1]. It exploits an analogy with the modern theory of financial portfolio management by associating cost with the time it takes to complete the transaction and taking into account the "risk" given by the standard deviation of that time. Before, such a strategy has already been successfully applied to the numerical solution of hard computational problems [2].

In modern portfolio theory, risk averse investors may prefer to hold assets from which they expect a lower return if they are compensated for the lower return with a lower level of risk exposure. Furthermore, it is a non-trivial result of portfolio theory that simple diversification can yield portfolios of assets which have higher expected return *as well as* lower risk. In the case of latencies (waiting times) on the Internet, thinking of different restart strategies is analogous to asset

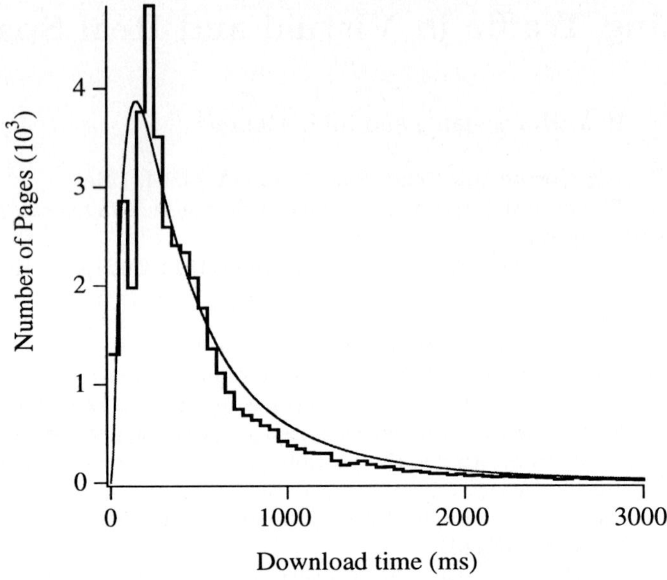

Fig. 1. Latency distribution of the down-load times of the `index.html` file on the main page of over forty thousand web sites, with a fit to a log-normal distribution, see formula (5). The parameters are $\sigma = 0.99$ and $\mu = 5.97$.

diversification: there is an efficient trade-off between the average time a request will take and the standard deviation of that time ("risk").

Consider a situation in which data have been requested but not received (down-loaded) for some time. This time can be very long in cases where the latency distribution has a long tail. One is then faced with the choice to either continue to wait for the data, to send out another request or, if the network protocols allow, to cancel the original request before sending out another. For simplicity, we consider the case in which it is possible to cancel the original request before sending out another one after waiting for a time period of duration τ. If $p(t)$ denotes the probability distribution for the down-load time without restart, the probability $P(t)$ that a page has been successfully down-loaded in time less than t is given by

$$P(t) = \begin{cases} p(t) & \text{if } t \leq \tau \,, \\ \left[1 - \int_0^\tau dt\, p(t)\right] P(t - \tau) & \text{if } t > \tau \,. \end{cases} \tag{1}$$

As a consequence, the resulting average latency $\langle t \rangle$ and the risk σ in down-loading a page are given by

$$\langle t \rangle = \int_0^\infty dt\, t P(t) \tag{2}$$

and

$$\sigma^2 = \langle (t - \langle t \rangle)^2 \rangle = \langle t^2 \rangle - \langle t \rangle^2 . \tag{3}$$

If we allow an infinite number of restarts, the recurrence relation (1) can be solved in terms of the partial moments $M_n(\tau) = \int_0^\tau dt\, t^n P(t)$:

$$\begin{aligned} \langle t \rangle &= \frac{1}{M_0}\Big[M_1 + \tau(1 - M_0)\Big] , \\ \langle t^2 \rangle &= \frac{1}{M_0}\left\{M_2 + \tau(1 - M_0)\left[2\frac{M_1}{M_0} + \tau\left(\frac{2}{M_0} - 1\right)\right]\right\} . \end{aligned} \tag{4}$$

In the case of a log-normal distribution

$$p(t) = \frac{1}{\sqrt{2\pi} x \sigma} \exp\left(-\frac{(\log x - \mu)^2}{2\,\sigma^2}\right) , \tag{5}$$

$\langle t \rangle$ and $\langle t^2 \rangle$ can be expressed in terms of the error function:

$$M_n(\tau) = \frac{1}{2}\exp\left(\frac{\sigma^2 n^2}{2} + \mu n\right)\left[1 + \mathrm{erf}\left(\frac{\log\tau - \mu}{\sigma\sqrt{2}} - \frac{\sigma n}{\sqrt{2}}\right)\right] . \tag{6}$$

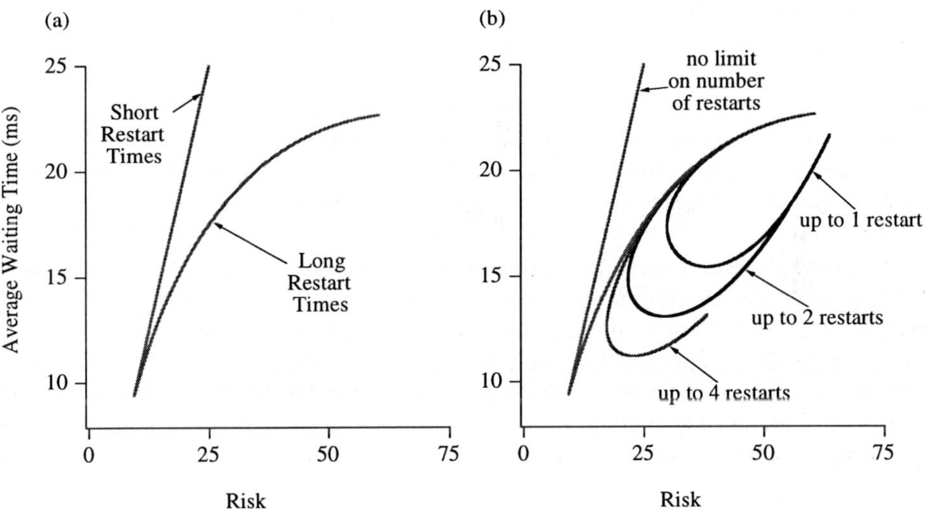

Fig. 2. (a) Expected average latency versus risk calculated for a log-normal distribution with $\mu = 2$ and $\sigma = 1.5$. The curve is parameterized over a range of restart times τ. (b) Family of curves obtained when we limit the maximum number of allowed restarts.

The resulting $\langle t \rangle$-versus-σ curve is shown in Fig. 2a. As can be seen, the curve has a cusp point that represents the restart time τ that is preferable to all others. No strategy exists in this case with a lower expected waiting time or with a lower variance. The location of the cusp can be translated into the optimum value of the restart time to be used to reload the page.

There are many variations to the restart strategy described above. In particular, in Fig. 2b, we show the family of curves obtained from the same distribution used in Fig. 2a, but with a restriction on the maximum number of restarts allowed in each transaction. Even a few restarts yield an improvement.

Clearly, in a network without any kind of usage-based pricing, sending many identical data requests, to begin with, would be the best strategy as long as we do not overwhelm the target computer. On the other hand, everyone can reason in exactly the same way, resulting in congestion levels that would render the network useless. This paradoxical situation, sometimes called a social dilemma, arises often in the consideration of "public goods" such as natural resources and the provision of services which require voluntary cooperation [3]. This explains much of the current interest in determining the details of an appropriate pricing scheme for the Internet, since users do consume Internet bandwidth greedily when down-loading large multimedia files for example, without consideration of the congestion caused by such activity.

Note that the histogram in Fig. 1 represents the variance in the down-load time between *different* sites, whereas a successful restart strategy depends on a variance in the down-load times for the *same* document on the *same* site. For this reason, we cannot use the histogram in Fig. 1 to predict the effectiveness of the restart strategy, but need to apply the similarly looking distribution of the respective Internet site. While a spread in the average down-load times of pages from different sites reduces the gains that can be made using a common restart strategy, it is possible to take advantage of geography and the time of day to fine tune and improve the strategy's performance. As a last resort, it is possible to fine tune the restart strategy on a site per site basis.

As a final caution, we point out that with current client-server implementations, multiple restarts are detrimental and very inefficient since every duplicated request will enter the server's queue and are processed separately until the server realizes that the client is not listening to the reply. This is an important issue for a practical implementation, and we neglect it here: our main assumption is that the restart strategy only affects the congestion by modifying the perceived latencies. This is only true if the restart strategy is implemented in an efficient and coordinated way on both the client and server side.

2 Real Traffic on Freeways with Ramps

The recent study of the properties of "synchronized" congested highway traffic [4] has generated a strong interest in the rich spectrum of phenomena occurring close to on-ramps [5,6]. In this connection, a particularly relevant problem is that of choosing an optimal injection strategy of vehicles into the highway. While there exist a number of *heuristic* approaches to optimizing vehicle injection into freeways by on-ramp controls, the results are still not satisfactory.

What is needed is a strategy that is flexible enough to adapt in real time to the transient flow characteristics of road traffic while leading to minimal travel times for all vehicles on the highway.

Our study presents a solution to this problem that explicitly exploits the naturally occurring fluctuations of traffic flow in order to enter the freeway at optimal times. This method leads to a more homogeneous traffic flow and a reduction of inefficient stop-and-go motions.

In contrast to conventional methods, the basic performance criterion behind this technique is *not* the traffic volume, the optimization of which usually drives the system closer to the instability point of traffic flow and, hence, reduces the reliability of travel time predictions [7]. Instead, we will focus on the optimization of the travel time distribution itself, which is a global measure of the overall dynamics on the whole freeway stretch. It allows the evaluation of both the expected (average) travel time of vehicles and their variance, where a high value of the variance indicates a small reliability of the expected travel time when it comes to the prediction of individual arrival times.

Both the average and the variance of travel times are influenced by the inflow of vehicles entering the freeway over an on-ramp. From these two quantities one can again construct a relation between the average payoff (the negative mean value of travel times) and the risk (their standard deviation). The optimal strategy will then correspond to the point in the curve that yields the lowest risk at a high average payoff. In the following, we will show that the variance of travel times has a minimum for on-ramp flows that are different from zero, but only in the congested traffic regime (which shows that the effect is not trivial at all). This finding implies that traffic flow can be optimized by choosing the appropriate vehicle injection rate into the freeway. Hence, in order to reach well predictable and small average travel times at high flows in the overall system, it makes sense to temporarily hold back vehicles by a suitable on-ramp control based on a traffic-dependent stop light [8]. At intersections of freeways, this may require additional buffer lanes [9].

In order to obtain the travel time distribution of vehicles on a highway, we simulated traffic flow via a discretized and noisy version of the optimal velocity model by Bando *et al.* [10], which describes the empirical known features of traffic flows quite well [11]. Moreover, we extended the simulation to several lanes with lane-changing maneuvers and different vehicle types (fast cars and slow trucks) [12,13]. For lane changes, we assumed symmetrical ("American") rules, i.e. vehicles could equally overtake on the left-hand or on the right-hand lane. Lane changing maneuvers were performed, when an *incentive criterion* and a *safety criterion* were satisfied [14]. The incentive criterion was fulfilled, when a vehicle could go faster on the neighboring lane, while the safety criterion required that lane changing would not produce any accident (i.e., there had to be a sufficiently large gap on the neighboring lane) [12,13].

In addition to a two-lane stretch of length $L = 10$ km, we simulated an on-ramp section of length 1 km with a third lane that could not be used by vehicles from the main road. However, vehicles entered the beginning of the on-ramp lane at a specified injection rate. Injected vehicles tried to change from the on-ramp to the main road as fast as possible, i.e. they cared only about the safety criterion, but not about the incentive criterion. The end of the on-ramp was treated like a resting vehicle, so that any vehicle that approached it had to stop, but it changed to the destination lane as soon as it found a sufficiently large gap. If the on-ramp was completely occupied by vehicles waiting to enter the main road, the injected vehicles formed a queue and entered the on-ramp as soon as possible. After injected vehicles had completed the 10 kilometer long two-lane measurement stretch, they were automatically removed from the freeway [13].

Our simulations were carried out for a circular road. After the overall density was selected, vehicles were homogeneously distributed over the road at the beginning, with the same densities on both lanes of the main road. The experiments started with uniform distances among the vehicles and their associated optimal velocities. The vehicle type was determined randomly after specifying the percentages r of cars ($\geq 90\%$) and $(1-r)$ of trucks ($\leq 10\%$). Notice that the effects discussed in the following should be more pronounced for increased $r(1-r)$, since lane-changing rates seem to be larger and traffic flow more unstable, then.

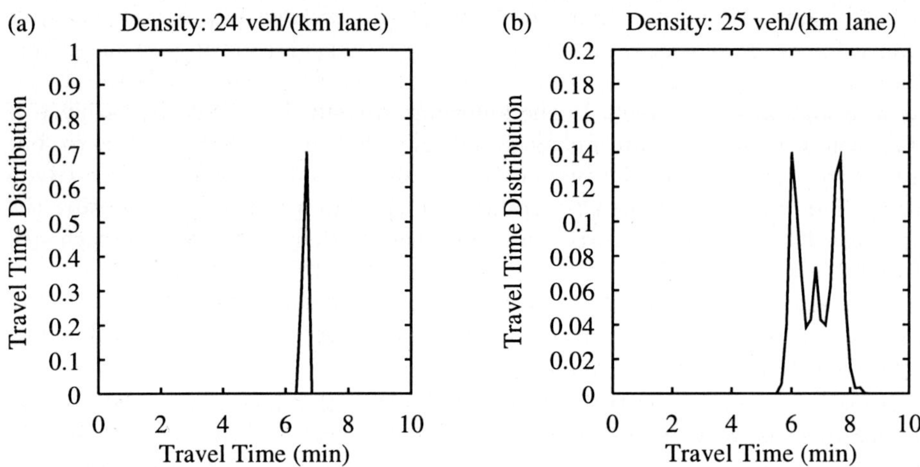

Fig. 3. Examples of travel time distributions for mixed traffic composed of a majority of cars and a minority of trucks. (a) A narrow distribution results for stable traffic flow. (b) For unstable traffic flow, the distribution is broad. (In the simulations underlying the above results, no vehicles were injected to the main road over the on-ramp.)

We determined the travel times of all vehicles by calculating the difference in the times at which they passed the beginning and the end of the 10 kilometer long two-lane section. For mixed traffic composed of a high percentage of cars and a small percentage of slower trucks, we found narrow travel time distributions at small vehicle densities, where traffic flow was stable, while for unstable traffic flow at medium densities, the travel time distributions were broad (Fig. 3).

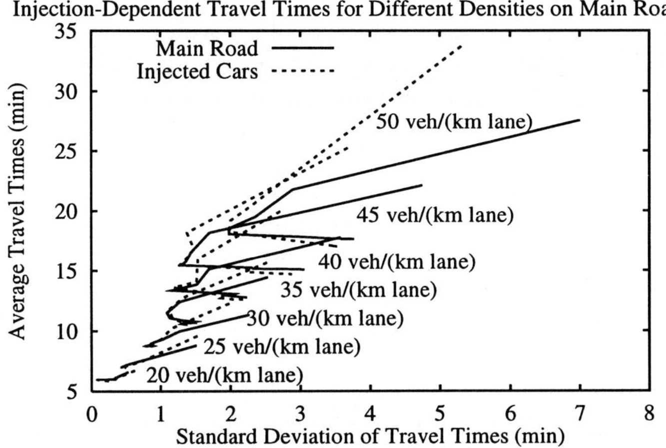

Fig. 4. Average and standard deviation of the travel times of vehicles on the main road and from the ramp as a function of the injection rate $Q_{\text{rmp}} = 1/(n \text{ s})$ with $n = 2^k$ and $k \in \{2, 3, \ldots, 10\}$ for various vehicle densities on the main road (measured without injection). With increasing injection rate, the average travel times are growing due to the higher resulting vehicle density on the freeway. However, at high enough vehicle densities, the standard deviation shows minima for medium interaction rates. Injected vehicles require longer travel times, since they cause a more crowded destination lane. (After [13].)

If we plot the average of travel times as a function of their standard deviation (Fig. 4), we obtain curves parameterized by the injection rate of vehicles into the road and find the following: 1. With growing injection rate

$$Q_{\text{rmp}} = \frac{1}{n \text{ s}}, \tag{7}$$

the travel time increases monotonically. This is because of the increased density caused by injection of vehicles into the freeway. 2. The average travel time of *injected* vehicles is higher, but their standard deviation lower than for the vehicles circling on the main road. This is due to the fact that vehicle injection produces a higher density on the lane adjacent to the on-ramp, which leads to smaller velocities. The difference between the travel time distributions of injected vehicles and those on the main road decreases with the length L of the simulated road, since lane-changes tend to equilibrate densities between lanes.

In addition, the standard deviation of the travel times has a *minimum* for *finite* injection rates, as entering vehicles tend to fill existing gaps and thus homogenize traffic flow. This minimum is optimal in the sense that there is no other value of the injection rate that can produce travel with smaller variance. In particular, gap-filling behavior mitigates inefficient stop-and-go traffic at medium densities. Above a density of 45 vehicles per kilometer and lane on the main road (measured without injection), the minimum of the travel times' standard deviation occurs for $n \approx 60$. The reduction of the average travel time by smaller injection rates is less than the increase of their standard deviation. This result suggests that, in order to generate predictable and reliable arrival times, one should operate traffic at medium injection rates. For the case of 40 vehicles per kilometer and lane, the minimum of the standard deviation of travel times is located at $n \approx 30$, while for 35 vehicles per kilometer and lane, it is at $n \approx 15$. Below about 25 vehicles per kilometer and lane, vehicle injection does not reduce the standard deviation of travel times, since the travel time distribution is narrow anyway. At these densities, traffic flow is stable and homogeneous, so that no inefficient stop-and-go traffic exists and therefore no large gaps can be filled [12].

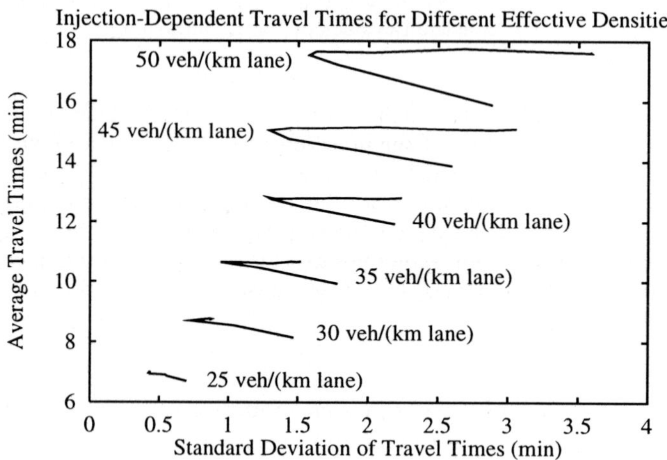

Fig. 5. As Fig. 4, but as a function of the resulting *effective* density ρ_{eff} on the freeway rather than the density ρ_{main} on the main road without injection. We find shorter travel times at *high* injection rates because of the homogenization of traffic. The standard deviation of travel times is varying stronger than the average travel time, which indicates that medium injection rates are the optimal choice at high vehicle densities. At small vehicles densities, the standard deviation does not show a minimum, since traffic flow is homogeneous anyway. (After [13].)

The curves displayed in Fig. 4 correspond to a given density ρ_{main} on the main road *without* injection of vehicles. The effective density ρ_{eff} on the freeway *resulting* from the injection of vehicles can be approximated by

$$\rho_{\text{eff}} = \rho_{\text{main}} + \frac{N_{\text{inj}}}{IL} , \tag{8}$$

where $I = 2$ lanes, $L = 10$ km. N_{inj} is the average number of injected vehicles present on the main road and can be written as

$$N_{\text{inj}} = N_{\text{tot}} \frac{\mathcal{T}_{\text{inj}}}{\mathcal{T}_{\text{tot}} - \mathcal{T}_{\text{inj}}}, \tag{9}$$

where $N_{\text{tot}} = 1000$ is the total number of injected vehicles during the simulation runs, $\mathcal{T}_{\text{inj}}$ is their average travel time, and $\mathcal{T}_{\text{tot}}$ the time interval needed by all $N_{\text{tot}} = 1000$ vehicles to complete their trip. We point out that, in addition to these measurements, we used two other methods of density measurement which yielded similar results.

In contrast to Fig. 4, we also computed the dependence of the travel time characteristics on the resulting *effective* densities of vehicles. Fig. 5 shows the average of the travel times for vehicles in the main road as a function of their standard deviation. Once again, we observe a minimum of the standard deviation of travel times at high vehicle densities and medium injection rates. However, this time, an increase of the injection rate *reduces* the average travel times!

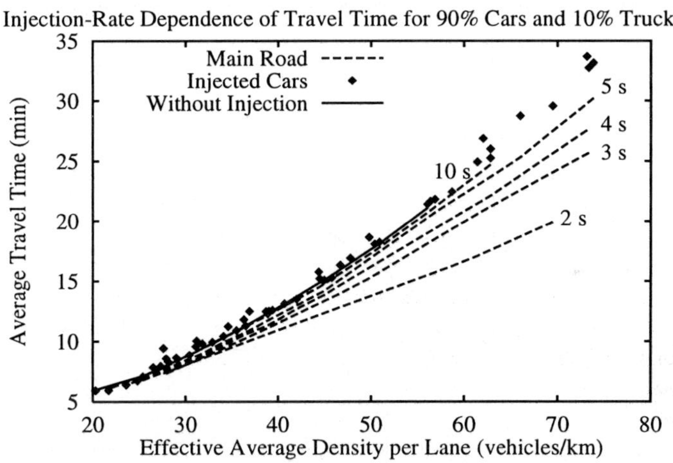

Fig. 6. Average travel times of vehicles on the main road as a function of the resulting effective vehicle density on the freeway for various injection rates $Q_{\text{rmp}} = 1/(n \text{ s})$ ($n \in \{2, 3, 4, 5, 10\}$). Obviously, the travel times are reduced by vehicle injection, which comes from a reduction of inefficient stop-and-go traffic. In the limit of high injection rates, one observes a linear dependence of average travel times on effective density, which is in agreement with analytical results [13]. We mention that the travel times of *injected* vehicles did not depend on the injection rate. However, when we checked what happens if the vehicles on the main road try to change to the left lane along the on-ramp in order to give way to entering vehicles (as they do in many European countries), we found that both, injected vehicles and the vehicles on the main road, profited from this behavior [13].

Figure 6 investigates the surprising reduction of the average travel times in more detail. While the injected cars experienced travel times that agreed with the case of no injection, the vehicles on the main road clearly profited from vehicle injection, if the effective density was the same. This means that, for given ρ_{eff}, one can actually increase the average velocity $V_{\text{main}} = L/T_{\text{main}}$ of vehicles by injecting vehicles at a considerable rate without affecting their travel times. This result is due to the increased degree of homogeneity caused by entering vehicles that fill gaps on the main road, which mitigates the less efficient stop-and-go traffic.

We point out that the injection-based reduction of travel times on the main road at a given effective density ρ_{eff} is related with a higher proportion

$$P = 1 - \frac{\rho_{\text{main}}}{\rho_{\text{eff}}} \tag{10}$$

of injected vehicles, which implies a reduction in the number of vehicles circling on the main road. The relation between the injection rate and the percentage of injected vehicles is roughly linear (Fig. 7).

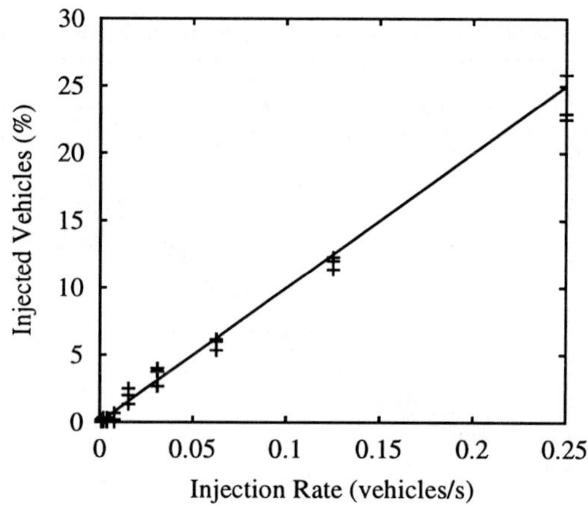

Fig. 7. Proportion of injected vehicles as a function of the injection rate Q_{rmp}.

The dependencies of the average travel times and their standard deviation on the proportion of injected vehicles are depicted in Fig. 8. We find that the decrease in the average travel times is minor, while a significant reduction of the variance of travel times can be achieved by less than 5% of injected vehicles.

3 Summary and Conclusions

Portfolio strategies can be successfully applied to systems in which the distribution of the quantity to be optimized is broad. This is the case for the down-load

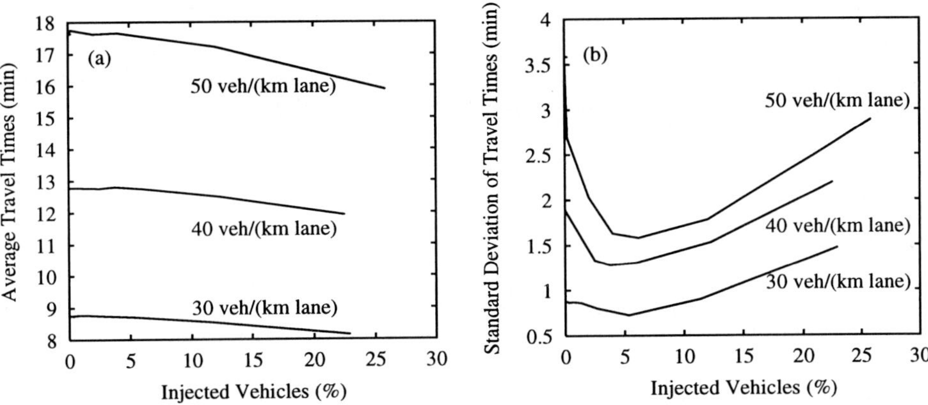

Fig. 8. (a) Average travel times and (b) their standard deviation as a function of the proportion of injected vehicles.

times in the World Wide Web as well as the travel time distributions in congested traffic flow. In this article, we showed how to reduce the average waiting times as well as their standard deviation ("risk") by suitable injection strategies. In the case of the World Wide Web, it is possible to enforce smaller average waiting times by restarting a data request when the data were not received after a certain time interval τ. This deforms the waiting time distribution towards smaller values, which automatically reduces the variance as well. Since the long tail of the waiting time distribution comes from the intermittent ("bursty") behavior of Internet traffic, restart strategies partially manage to "calm down" these "Internet storms" by withdrawing requests in busy periods and restarting them later on. Similarly, the injection of vehicles to a freeway over an on-ramp can homogenize inefficient stop-and-go traffic by filling large gaps. In other words: The strategy exploits the naturally occurring fluctuations of traffic flow in order to allow the entry of new vehicles to the freeway at optimal times. In this way, the variation of travel times can be considerably reduced, which is favorable for more reliable travel time predictions. Moreover, at a given effective density, the average travel time decreases with increasing injection rate, i.e., with an increasing percentage of injected vehicles on the main road.

Acknowledgement. D.H. wants to thank the German Research Foundation (DFG) for financial support (Heisenberg scholarship He 2789/1-1). S.M. thanks for financial support by the Hertz Foundation.

References

1. R.M. Lukose and B.A. Huberman, *A methodology for managing risk in electronic transactions over the internet*, in press, Netnomics, (2000).
2. B.A. Huberman, R.M.Lukose, and T. Hogg, *An economics approach to hard computational problems,* Science **275**, 51 (1997).
3. R. Hardin, *Collective Action,* (Johns Hopkins University Press, 1982).
4. B.S. Kerner and H. Rehborn, *Experimental properties of phase transitions in traffic flow,* Phys. Rev. Lett. **79**, 4030 (1997); B.S. Kerner and H. Rehborn, *Experimental properties of complexity in traffic flow,* Phys. Rev. E **53**, R4275 (1996).
5. H.Y. Lee, H.-W. Lee, and D. Kim, *Origin of synchronized traffic flow on highways and its dynamic phase transition,* Phys. Rev. Lett. **81**, 1130 (1998).
6. D. Helbing and M. Treiber, *Gas-kinetic-based traffic model explaining observed hysteretic phase transition,* Phys. Rev. Lett. **81**, 3042 (1998); D. Helbing and M. Treiber, *Jams, waves, and clusters*, Science **282**, 2001 (1998); D. Helbing, A. Hennecke, and M. Treiber, *Phase diagram of traffic states in the presence of inhomogenities,* Phys. Rev. Lett. **82**, 4360 (1999).
7. K. Nagel and S. Rasmussen, *Traffic at the edge of chaos,* in: *Artificial Life IV*, R.A. Brooks and P. Maes, (Eds.), (MIT Press, Cambridge, MA, 1994).
8. D. Helbing, *New simulation models for traffic optimization,* in: *Proc. of the Workshop "Verkehrsplanung und -simulation"*, V. Claus, D. Helbing, and H.J. Herrmann, (Eds.), (Informatik Verbund Stuttgart, University of Stuttgart, 1999).
9. P.H.L. Bovy, (Ed.), *Motorway Traffic Flow Analysis,* (Delft University Press, Delft, 1998).
10. M. Bando, K. Hasebe, K. Nakanishi, A. Nakayama, A. Shibata, and Y. Sugiyama, *Phenomenological study of dynamical model of traffic flow,* J. Phys. I France **5**, 1389 (1995).
11. D. Helbing and M. Schreckenberg, *Cellular automata simulating experimental properties of traffic flows,* Phys. Rev. E **59**, R2505 (1999).
12. D. Helbing and B.A. Huberman, *Coherent moving states in highway traffic,* Nature **396**, 738 (1998).
13. B.A. Huberman and D. Helbing, *Economics-based optimization of unstable flows,* Europhys. Lett. **47**, 196 (1999).
14. K. Nagel, D.E. Wolf, P. Wagner, and P. Simon, *Two-lane traffic rules for cellular automata: A systematic approach,* Phys. Rev. E **58**, 1425 (1998).

Is Boltzmann's Equation Physically Insufficient?

S. Grossmann

Fachbereich Physik, Philipps-Universität Marburg, Germany

Abstract. Boltzmann's equation describes the dynamics of the one particle phase space distribution density including two particle interactions. These drive the system towards equilibrium. The physical properties of this equilibrium turn out to be those of an ideal gas, giving the leading order of the equation of state, the pressure, etc, only. No viral corrections originating from the two particle interactions appear, as one finds them in equilibrium statistical mechanics. This means the Boltzmann dynamics is insufficient to imply the *proper* equilibrium. It has to be upgraded, clearly, in its so called flow terms, which contribute only in order n (particle density) corresponding to a free motion between collisions. If properly derived, the equation of motion contains also order n^2 contributions in the flow terms, weighted by the real part of the forward scattering amplitude. Then the equilibrium limit of the dynamics coincides with the findings of equilibrium statistical physics. Various implications can be identified. It is pointed out, that the corresponding upgrade has to be checked carefully in interdisciplinary modeling, as is used, e.g., in traffic flow or granular systems descriptions.

Foreword

I am very moved to have the great honor and pleasure to contribute a lecture to this challenging conference in honor of Professor Wolfgang Weidlich's retirement. He has contributed to the one particle dynamics, in particular to its master equation, with so many ideas, transgressing the disciplinary fences, always creating a stimulating, open minded, friendly atmosphere, never losing his humor and smiling, stimulating his many pupils to impressive new work.

1 Introduction

The Boltzmann equation, given in 1872, is a basic issue in physics, but also in other fields as, e.g., traffic flow. It originates from considering the first order moment or mean value of the density operator equation, properly truncated and neglecting fluctuations. It describes the temporal behavior of

$$f(\vec{x}, \vec{p}, t) = \left\langle \sum_{a=1}^{N} \delta(\vec{x} - \vec{x}_a(t))\, \delta(\vec{p} - \vec{p}_a(t)) \right\rangle , \tag{1}$$

the one particle distribution function in an N-particle system. Many observable properties of key interest rely on the knowledge of $f(\vec{x}, \vec{p}, t)$. This explains the ongoing interest and numerous renaissances of the Boltzmann description, be it for gases and liquids, be it in semiconductor physics, in cellular automata for hydrodynamic flow, for traffic dynamics and further interdisciplinary work.

$$\partial_t f = \vec{v} \cdot \vec{\partial}_x f + m \vec{b} \cdot \vec{\partial}_p f = \partial_c f. \tag{2}$$

Here $\vec{v} = \vec{p}/m$ is the particle velocity, mass m, momentum $\vec{p}$, and $m\ \vec{b} = -\vec{\partial}_x U$ describes the particle acceleration by external forcing. $\partial_c f$ denotes the collision term

$$\partial_c f = \int\int\int \sigma(pp_1 \leftarrow p' p_1') \left[f' f_1' - f f_1 \right] d^3p' d^3p_1' d^3p_1 . \quad (3)$$

The range of application of (2) is characterized by low density n. To be more precise, the interaction range a has to be much less than the mean free path l_F of the particle, i.e., $\mathrm{a} \ll l_F$. With R. Clausius' 1858 expression for $l_F = 1/\left(\sqrt{2} n \sigma_{tot}\right)$ ($\approx 67nm$ in air) the low density criterion can be phrased

$$n\,\mathrm{a}^3 \ll 1. \quad (4)$$

For the time behavior this means, collisions are like flashes, their duration t_c being much less than the time of change of f, denoted as Δt.

The physical status of the Boltzmann equation (2) is to describe an *almost perfect gas,* collecting contributions of orders n **and** n^2, originating from the gas particles' kinetic and their two-particle potential energy.

The Boltzmann equation (2) describes transport phenomena, the approach to equilibrium, increasing entropy (the famous H-theorem (Eta-theorem, 1872)) and, of course, also equilibrium itself. The transport equations are the continuity equation (conservation of mass), the Navier-Stokes equation (conservation of momentum), and the energy conservation. In particular for the particle flow one gets $n\vec{u}$, where $\vec{u}$ is the mean local velocity. The momentum flow, equal to the pressure tensor, is

$$\begin{aligned} \Pi_{ij}\left(\vec{x},t\right) = \int p_i \frac{p_j}{m} f\left(\vec{x},\vec{p},t\right) d^3p &= \varrho\left(u_i u_j + \langle \delta v_i \delta v_j \rangle\right) \\ &= \varrho\left(u_i u_j + \delta_{ij} \varkappa T/m\right). \end{aligned} \quad (5)$$

Thus the pressure P is $P = n\varkappa T$, the formula for the *ideal* gas. This is a big disappointment. Where is the *almost* of the almost ideal gas, where are the n^2 terms in the equation of state? The approach to equilibrium needed the two particle interactions, being of order n^2, therefore one has to expect n^2-terms also for the equilibrium properties. This indeed is provided in equilibrium statistical mechanics:

$$P = n\varkappa T\left(1 + nB\right), \quad (6)$$

the *viral* coefficient B being given in terms of the two particle interaction $w(r)$, classically as $B = \frac{1}{2}\int d^3r \left(1 - e^{-\beta w(r)}\right)$. In this respect the Boltzmann equation clearly is physically inconsistent. The n^2-dynamics fails to meet the n^2-statics. How come and how to cure?

2 Free-Flow is Quasi-Free Only

The idea to improve the Boltzmann equation is to ask, if the particle motion between collisions is really free. Does the flying particle really miss "to see the others" between two collisions? At least for quantum objects one can argue *no, the objects are wave packets, which are extended and can take notice of other particles.* Then $\vec{v}$ is the group velocity $\partial\omega(\vec{k})/\partial\vec{k}$ or $\partial\varepsilon(\vec{p})/\partial\vec{p}$. If the single particle energy is purely kinetic, $\varepsilon = \vec{p}^2/2m$, one obtains $\partial\epsilon/\partial\vec{p} = \vec{v}$. If an external potential $U(\vec{x})$ contributes, it is $\varepsilon = \vec{p}^2/2m + U(\vec{x})$. And if other particles are present one expects an additional energy $V(\vec{x}, \vec{p}, t)$, which should be $\propto f(\vec{x}, \vec{p}, t)$, indicating how many other particles are on the stage. Then $\varepsilon = \varepsilon(\vec{x}, \vec{p}, t) = \vec{p}^2/2m + U(\vec{x}) + V(\vec{x}, \vec{p}, t)$. The group velocity as well as the wave packet's acceleration are then depending on f and thus on position and momentum, $\vec{v}(\vec{x}, \vec{p}, t)$ and $\vec{b}(\vec{x}, \vec{p}, t)$ being of the type $A_1 + \int A_2 f$. Since these terms contribute to the hydrodynamic conservation equations, they affect the pressure tensor and contribute to it in order n^2, thus can lead to viral contributions.

This is the essence of the argument. Details follow now.

3 The Upgrade of Boltzmann's Equation Derived

The argument of the previous section, that the free inter collisional flight is to be corrected by terms $\propto f$ in $\vec{v}$ and $\vec{b}$ (i.e., additional terms of order n^2 in the Boltzmann equation (2) show up) can be made precise by careful re-derivation of (2) from the quantum mechanical hierarchy [1]. The N-particle density operator $\varrho_t^{(N)}$ satisfies the von Neumann equation

$$i\hbar\partial_t\varrho_t^{(N)} = \left[H^{(N)}, \varrho_t^{(N)}\right]. \tag{7}$$

Only a sketch of the derivation of the corrected, consistent Boltzmann equation is indicated now. Take partial traces to N=1,2 and Wigner transform, e.g.,

$$f^{(1)}(\vec{r}_1, \vec{p}_1, t) = (2/2\pi\hbar)^2 \int dy_1 \exp\left(\frac{2i}{\hbar}\vec{y}_1 \cdot \vec{p}_1\right) \left\langle \vec{r}_1 - \vec{y}_1 \left|\varrho_t^{(1)}\right| \vec{r}_1 + \vec{y}_1 \right\rangle .$$

One arrives at

$$\left(\partial_t + \frac{\vec{p}_1}{m} \cdot \vec{\partial}_{r_1}\right) f^{(1)}(\vec{r}_1, \vec{p}_1, t) = \int d^3r_2 d^3p_2 O_{12} f^{(2)}(\vec{r}_1, \vec{p}_1; \vec{r}_2, \vec{p}_2; t) .$$

The operator O_{12} is proportional to the two particle interaction $w(r_1 - r_2)$. Correspondingly, the two particle density operator satisfies

$$i\hbar\partial_t\varrho_t^{(2)}(1,2) = \left[H^{(2)}(1,2), \varrho_t^{(2)}(1,2)\right] + Tr_{(3)}\left[W(1,3) + W(2,3), \varrho^{(3)}(1,2,3)\right]. \tag{8}$$

Skip now the three particle contributions $\propto \varrho^{(3)}$. From the equation for the two particle density operator $\varrho^{(2)}$ one can first eliminate $O_{12}f^{(2)}$ on the *rhs* and get

$$rhs = \int d^3r_2\, d^3p_2 \left(\partial_t + \frac{\vec{P}}{M}\cdot\vec{\partial}_R + \frac{\vec{p}}{\mu}\partial_{\vec{r}}\right) f^{(2)}, \tag{9}$$

written in center of mass coordinates and their relative coordinate brethren. One can then easily integrate (formally) the two particle equation to find

$$\varrho_t^{(2)}(1,2) = \exp\left[-\frac{i}{\hbar}H^{(2)}(t-t_0)\right]\varrho_{t_0}^{(2)}\exp\left[+\frac{i}{\hbar}H^{(2)}(t-t_0)\right].$$

This expression has to describe at t_0 two particles some distance apart intending to collide. We therefore use conveniently center-of-mass position and momentum $\vec{R},\vec{P}$ as well as the corresponding variables $\vec{r},\vec{p}$ for the relative motion. The eigenstates of $H^{(2)}$ are the product states of the *free* center-of-mass motion and the *scattering* states for the relative motion, $|\vec{P},\vec{p}^{(-)}\rangle$. The corresponding energy is $\vec{P}^2/2M + \vec{p}^2/2\mu$. The scattering state has to be calculated from the time independent Schrödinger equation

$$\left[-\frac{\hbar^2}{2\mu}\nabla_r^2 + \frac{1}{2}\left(W(\vec{r}) + W(-\vec{r})\right)\right]\langle\vec{r}|\vec{p}\rangle^{(-)} = \frac{\vec{p}^2}{2\mu}\langle\vec{r}|\vec{p}\rangle^{(-)}, \tag{10}$$

together with the boundary condition “incoming plane wave plus outgoing spherical wave”.

The *initial* state $\varrho_{t_0}^{(2)}$ is factorized according to the conventional molecular chaos hypothesis. It describes the two particles before collision, thus spatially separated and with different momenta,

$$\begin{aligned}\varrho_{t_0}^{(2)} \propto f^{(1)}&\left(\vec{r}_1 + \frac{\vec{r}+\vec{x}}{2}, \vec{p}_1 + \vec{p} + \vec{k}, t\right)\\ f^{(1)}&\left(\vec{r}_1 + \frac{\vec{r}-\vec{x}}{2}, \vec{p}_1 + \vec{p} - \vec{k}, t\right).\end{aligned} \tag{11}$$

Now these single particle distribution functions are expanded with respect to the space and momentum differences, because on the molecular scale they vary smoothly. This leads to terms of type $\vec{\partial}_x^0, \vec{\partial}_x; \vec{\partial}_p^0, \vec{\partial}_p$. The zeroth order contribution $\vec{\partial}_x^0, \vec{\partial}_p^0$ lead to the conventional flow and the conventional collision terms, respectively. This is the well known, common procedure.

But there are first order derivatives $\vec{\partial}_x, \vec{\partial}_p$ in addition. They are both of order n^2 due the product $f\cdot f$, and can be reformulated as $\vec{\partial}_x f\cdot\vec{\partial}_p\varepsilon$ and $\vec{\partial}_p f\cdot(-\vec{\partial}_x\varepsilon)$, respectively, with an ε of the type discussed in Sect. 2. More precisely, on obtains

$$\varepsilon(\vec{x},\vec{p},t) = \frac{\vec{p}^2}{2m} + U(\vec{x}) + \int F(\vec{p},\vec{q})\, f(\vec{x},\vec{q},t)\, d^3q. \tag{12}$$

The kernel $F(\vec{p},\vec{q})$ comes as the (real part of the) **forward scattering amplitude**

$$F(\vec{p},\vec{q}) = h^3 \Re\left(T\left(\frac{\vec{q}-\vec{p}}{2},\frac{\vec{q}-\vec{p}}{2}\right)\right). \tag{13}$$

The T-matrix $T\left(\vec{p}',\vec{p}\right) = \langle\vec{p}'|W|\vec{p}\rangle^{(-)}$ is the nontrivial part of the S-matrix, $\mathsf{T}=\mathsf{S}-1$. On the energy shell T is related to the scattering amplitude [1]

$$T\left(\vec{p}',\vec{p}\right) = \langle\vec{p}'|W|\vec{p}\rangle^{(-)} = -\frac{1}{2\pi h\mu}\widehat{f}\,(\Theta,E_p). \tag{14}$$

$\widehat{f}$ has the dimension of a length, its (absolute) square is the differential cross section of the two particle scattering. Thus $[T]$ =energy/momentum^3 and consequently $[F]$ =energy·length^3. With these dimensions the integral on F in (12) indeed has the dimension "energy", because $f d^3q$ is the particle density with $\left[f d^3 q\right] = 1/\text{length}^3$.

Thus everything is well defined and has a clearcut prescription how to be calculated. Because there is also a classical collision theory (cf. papers by W. Thirring and Narnhofer [2]), one has to expect the completely analogous result for classical systems.

To sum up, the flow terms in Boltzmann's equation (2) instead of $\vec{v}\cdot\vec{\partial}_x f + \vec{b}\cdot\vec{\partial}_p f$ have to be substituted by

$$\vec{\partial}_p\varepsilon\cdot\vec{\partial}_x f - \vec{\partial}_x\varepsilon\cdot\vec{\partial}_p f. \tag{15}$$

The additional term in ε of (12) is $\propto f$, thus the flow terms are upgraded by order n^2 contributions.

4 Virial Corrected Equilibrium

The upgraded Boltzmann equation (2) with (15) on the *lhs* has among various other consistent features two particularly noteworthy ones. First, its equilibrium value for the pressure has the viral contribution of order n^2 in agreement with the direct results of equilibrium statistical mechanics, which starts from the partition function. Second, it allows for a new mode, so called 2nd sound, cf. Sect. 5.

If $F(\vec{p},\vec{q})$ depends on $\vec{q}-\vec{p}$, one can show $\langle\partial\varepsilon/\partial\vec{p}\rangle = \langle\vec{p}/m\rangle = \vec{u}$ and the continuity equation holds as before. The momentum collisional invariant leads to the momentum conservation equation as before, $\partial_t\varrho u_i + \partial_i\Pi_{ij} = 0$, but with a redefined pressure tensor

$$\Pi_{ij} = \varrho u_i u_j + \varrho\left\langle\left(\frac{\partial\varepsilon}{\partial p_i}-u_i\right)\left(\frac{p_j}{m}-u_j\right)\right\rangle + \delta_{ij}\frac{n^2}{2}\langle F(p,q)\rangle. \tag{16}$$

[1] We denote the scattering amplitude by $\widehat{f}\,(\Theta)$ in order to avoid a mix-up with $f(\vec{r},\vec{p},t)$, the distribution function.

Π_{ij} can be proven to be symmetric, if ε depends on $|\vec{p}|$ only, i.e., if rotational invariance holds. The equilibrium pressure reads

$$P = n\varkappa T + \frac{1}{2} n^2 \langle F \rangle_E . \tag{17}$$

Comparing with (6) the viral coefficient as obtained from the dynamical approach to equilibrium thus is

$$B = \frac{1}{2\varkappa T} \frac{\int\int d^3p d^3q f_E(p) f_E(q) F(p,q)}{\int\int d^3p d^3q f_E(p) f_E(q)} . \tag{18}$$

B thus is the equilibrium average (with f_E) of the scattering amplitude

$$B(T) = -\frac{8\sqrt{\pi}\hbar^2}{(m\varkappa T)^{5/2}} \int_0^\infty dk k^2 e^{-k^2/m\varkappa T} \left[\widehat{f}(0,k) + \widehat{f}^*(0,k) \right] . \tag{19}$$

It is precisely this what one obtains from quantum equilibrium statistics [3].

Let me add that there are also viral corrections to other equilibrium properties, obtained by the upgraded Boltzmann equation. These include the energy density $\frac{3}{2} n\varkappa T \left(1 - \frac{2}{3} nT dB/dT\right)$, the entropy density

$$-\varkappa n \left[\ln\left(n\lambda_T^3\right) - \frac{5}{2} + n\left(B + T\frac{dB}{dT}\right) \right] ,$$

the velocity of sound $c^2 = \partial P / \partial \varrho|_{S/N}$, etc. Also the transport coefficients viscosity, diffusivity, and heat conductivity enjoy viral corrections (which should not be mixed up with the Enskog terms, having even another sign).

For instance,

$$\eta = \eta_0 (1 + n\delta_\eta), \text{ with } \delta_\eta \approx -0.85B. \tag{20}$$

Recently, molecular dynamics calculations also confirmed the corrections to the flow terms [4].

5 New Mode in the Upgraded Boltzmann Equation

The three hydrodynamic conservation equations (mass, momentum, energy) describe sound and diffusive modes. These modes are perturbations of the local equilibrium. In case of sound that means, $\omega t_F \ll 1$, the sound period must be larger than the time between collisions. Because the sound velocity $c_1 = \sqrt{\partial P / \partial \varrho|_{S/N}}$ is of the order of $\sqrt{\varkappa T}$, from $\omega = c_1 k$ it follows that $k l_F \ll 1$ or the sound wave length should exceed the mean free path.

Small amplitude, wave like, *fast* perturbations do not leave time to establish local equilibrium by collisions. Therefore, if $\omega t_F \gg 1$, the collisions are less important, one has to consider distortions of total equilibrium. These can only come from the flow terms, because due to ω^{-1} very small, $\partial_c f$ can be omitted. Such wave like excitations of total equilibrium are called *0th sound.* In low temperature condensed matter physics the 0th sound has first been studied by Yu.L. Klimontovich, W.P. Silin (1952) and by L.D. Landau (1957).

0th sound did not play a role for Boltzmann gases, because no high frequency perturbations of total equilibrium are possible in the Boltzmann equation (2). In contrast, this is quite different for the upgraded Boltzmann equation! The new flow terms very well allow for 0th sound, as shall be shown now.

We ask for solutions $f = f_E + f_1$, where f_E denotes the total (Maxwellian) equilibrium and the small disturbance f_1 represents a wave solution $f_1 \propto \exp(i\vec{k}\cdot\vec{x} - i\omega t)$. The collision term does not contribute. The linearized equation for f_1 then is

$$\partial_t f_1 + \vec{\partial}_p \varepsilon\,(f_E) \cdot \vec{\partial}_x f_1 - \vec{\partial}_x \varepsilon\,(f_1) \cdot \vec{\partial}_p f_E = 0. \tag{21}$$

(Note that the missing terms $\vec{\partial}_x f_E = 0 = \vec{\partial}_x \varepsilon\,(f_E)$ vanish.)

For wave like f_1 this equation of motion implies the k-ω -dispersion relation

$$1 + 2\varkappa T B \int d^3p \frac{\vec{k}\cdot\vec{\partial}_{\vec{p}} f_E}{\omega - \frac{\vec{p}}{m}\cdot\vec{k}} = 0. \tag{22}$$

Apparently, a solution can only exist if $B \neq 0$, i.e., if the forward scattering amplitude F indeed contributes to ε. In the conventional Boltzmann equation this term $\propto F$ is missing, therefore it has no 0th sound modes. In the upgraded equation, 0th sound as a new excitation is possible.

To solve the dispersion relation (22) one can proceed with the ansatz $\omega = Ck$. Then C is determined as $C = c_0\,(1 - i\delta)$. The 0th sound velocity can be written as a multiple of the 1st sound velocity $c_1 = [(5/3)\,\varkappa T/m]^{1/2}$,

$$c_0 = \sqrt{\frac{6}{5}\frac{\pi}{2}} \frac{1}{\sqrt{-\ln\left(2\sqrt{\pi} n B\right)}} c_1. \tag{23}$$

The attenuation is

$$\delta = \frac{2}{\pi} \ln \frac{1}{2\sqrt{\pi} n B}. \tag{24}$$

The 0th sound mode exists as long as the condition

$$0 < 2\sqrt{\pi} n B < 1 \tag{25}$$

is satisfied. If $B \to 0$, an infinite attenuation prohibits this new mode; also negative B has this consequence. 0th sound thus is possible only if the two particle interaction is predominantly repulsive, $B > 0$. For air, under normal atmospheric conditions one gets $c_0 \approx 0.74 c_1$ and $\delta \approx 3.44$, i.e., strong attenuation.

6 Summary and Remarks

A physically consistent Boltzmann equation has been derived and its implications have been discussed. It leads to an equilibrium state in agreement with conventional equilibrium statistical physics. It has upgraded flow terms and can be written as

$$\partial_t f + \overrightarrow{\partial}_p \varepsilon \cdot \overrightarrow{\partial}_x f - \overrightarrow{\partial}_x \varepsilon \cdot \overrightarrow{\partial}_p f = \partial_c f. \tag{26}$$

The single particle energy consists of the kinetic energy, the external potential (if applicable), and the energy of the surrounding other particles $\propto f(x,p,t)$ and $\propto F(p,q)$, the forward scattering amplitude, more precisely, its real part.

$$\varepsilon(\overrightarrow{x}, \overrightarrow{p}, t) = \frac{1}{2m}\overrightarrow{p}^2 + U(\overrightarrow{x}) + \int F(\overrightarrow{p}, \overrightarrow{q}) f(\overrightarrow{x}, \overrightarrow{q}, t). \tag{27}$$

The collisional invariants are mass (m), momentum ($\overrightarrow{p}$), and energy (ε). The equilibrium properties enjoy viral contributions, and a new sound mode is possible, perturbing the total, spatially homogeneous equilibrium. $F(\overrightarrow{p}, \overrightarrow{q})$ has to be calculated as the real part of the well known scattering amplitude at zero scattering angle, as the *forward scattering amplitude*,

$$F \propto \widehat{f}(0; E_{p-q}) + \widehat{f}^*(0; E_{p-q}). \tag{28}$$

By the optical theorem, representing the unitarity of the S-Matrix, and by the dispersion relations, implied by its causality, the real part of $\widehat{f}(0;E)$ can be expressed in terms of an integral over the total cross sections $\sigma_{tot}(E')$. The Born approximation, incidentally, does not contribute, because its forward contribution does not depend on energy. If bound states are just coming into existence they may give rise to resonant contributions. It is textbook wisdom [5] that the optical theorem states

$$\Im\left(\widehat{f}(0;E)\right) = \frac{k}{4\pi}\sigma_{tot}(E), \tag{29}$$

and the dispersion relation reads as a principle value (P) integral,

$$\Re\left(\widehat{f}(0;E)\right) = \widehat{f}_{Born} + \frac{1}{\pi} P \int \frac{\Im\left(\widehat{f}\left(0;E'\right)\right) dE'}{E' - E} + \sum_n \frac{dn}{E_n - E}. \tag{30}$$

The Born approximation $\widehat{f}_{Born} = -\left(\mu/2\pi\hbar^2\right) \int d^3r W(r)$ is a constant.

Before concluding I first wish to emphasize that in contrast to the present case of an example from physics, namely on the dynamics of a real gas, the derivation of such an upgraded Boltzmann equation in other applications may not always be traced back to a well founded equation as, e.g., von Neumann's equation for the density matrix operator. In particular, in interdisciplinary applications one models the equation of motion by thinking about the system's behavior and not by rigorous derivation. But one should carefully check also then, if not also in those other cases a contribution $\sim n^2$ has to be added in the flow terms on the *lhs* of the model-Boltzmann equation. F then is nothing but a model kernel, but has serious nonlinear implications, allowing for instance new, unexpected solutions.

Second, I apologize that a talk cannot give sufficient references. The interested reader can, nevertheless, find more details and hints to other work in the few added references. Historically the first indication of the necessity to upgrade the flow terms for real-gas dynamics has been obtained in the derivation of a master equation for localized momentum states [6]. Later work contributed to the viral corrections in equilibrium states and in the upgraded transport theory [7,8]. The present presentation adds minor aspects.

References

1. K. Bärwinkel and S. Grossmann, Z. Phys. **198**, 277 (1967).
2. W. Thirring, *Lehrbuch der Mathematischen Physik*, (Springer, Wien, 1977).
3. E. Beth and G.E. Uhlenbeck, Physica **4**, 915 (1937). Strictly speaking, the equilibrium statistical result has an additional term $\propto \widehat{f}$-derivative. A corresponding one has been skipped when deriving the upgraded Boltzmann equation. It generally is less important.
4. F.J. Alexander, A.L. Garcia, and B. Alder, *A consistent Boltzmann algorithm*, preprint, 1995.
5. L.D. Landau and E.M. Lifschitz, *Quantum mechanics*, (Pergamon Press, Oxford, 1962).
6. S. Grossmann, Physica **30**, 779 (1964).
7. S. Grossmann, Il Nuovo Cimento (X) **37**, 698 (1965).
8. S. Grossmann, Z. Naturforsch. **20a**, 861 (1965).

Fractional Evolution Equations and Irreversibility

R. Hilfer[1,2]

[1] ICA-1, Universität Stuttgart, Pfaffenwaldring 27, 70569 Stuttgart, Germany
[2] Institut für Physik, Universität Mainz, 55099 Mainz, Germany

Abstract. The paper reviews a general theory predicting the general importance of fractional evolution equations. Fractional time evolutions are shown to arise from a microscopic time evolution in a certain long time scaling limit. Fractional time evolutions are generally irreversible. The infinitesimal generators of fractional time evolutions are fractional time derivatives. Evolution equations containing fractional time derivatives are proposed for physical, economical and traffic applications. Regular non-fractional time evolutions emerge as special cases from the results. Also for these regular time evolutions it is found that macroscopic irreversibility arises in the scaling limit.

1 Introduction

An evolution equation for the time evolution of physical and other systems is typically of the form [1]

$$\frac{\mathrm{d}}{\mathrm{d}t} f(t) = \mathrm{B} f(t) \tag{1}$$

where $t \in \mathbb{R}$ denotes time and B is an operator on a Banach space $(B, \|\cdot\|)$. Depending on the initial data $f(0) = f_0$ for the state or observable f of the system at time $t = 0$ the problem is to find $f(t)$ at later times $t > 0$.

Many examples of (1) arise not only in physics but also in the social or economical sciences (see other articles in this book and also [2]). All evolution equations pertain to phenomena on a characteristic microscopic or macroscopic time scale. If the time scale is changed then the form of an evolution equation will usually also change. Of course this micro-macro transition affects the variables f, or the operator B, or both. Recently, it was found that not only f and B, but also the generator $\mathrm{d}/\mathrm{d}t$ of the time evolution in (1) may change in a transition from microscopic to macroscopic of time scales. Expressed somewhat imprecisely the result states that the infinitesimal generator $\mathrm{d}/\mathrm{d}t$ of time evolution may be changed into a fractional derivative "$\mathrm{d}^\alpha/\mathrm{d}t^\alpha$" of order α with $0 < \alpha \leq 1$. My objective in this paper is to discuss the origin of this result.

Derivates of fractional order exhibit algebraic scaling properties with non-integer exponents. Extending classical evolution equations to fractional evolution equations therefore introduces dynamic scaling in a natural way. It provides also a more flexible class of solutions for the comparison with empirical data.

Given the generality and mathematical universality of the appearance of fractional derivatives it is tempting to propose fractional evolution equations also for economical, social or traffic dynamics. Let me point out that this was done implicitly already in [2] where continuous time random walk (CTRW) models were introduced for modeling stochastic processes in the economical sciences. One of the input functions for a CTRW model is the so called waiting time density $\psi(t)$ that describes the distribution of time intervals between random events (jumps). Random events may correspond to hopping (jumps) of particles in physics, to purchases or business transactions in economical applications, or to delays due to a traffic jam in a traffic model. Exploiting the relationship between continuous time random walks and generalized master equations it was shown in [3] that CTRW-models to a fractional master equation whenever $\psi(t)$ is a generalized Mittag-Leffler function. Many applications of fractional derivatives in the economical sciences are therefore obtained simply by combining the results from [2] and [3].

A concrete example from the economical sciences is the model for the probability distribution $f(z,t)$ of cumulative sales z after time t introduced in [2, p. 138]. Combined with the results from [3,4] the model from [2] is equivalent to a fractional diffusion equation of the form

$$\mathrm{D}_t^\alpha f(z,t) = B\frac{\partial^2}{\partial z^2} f(z,t) \tag{2}$$

where D_t^α denotes a suitably defined fractional derivative operator of order α and B is a constant. [1] A precise formulation of such a fractional diffusion equation and its exact solution are given in [6]. Exact solutions for a whole class of fractional derivative operators are also given in [7]. Fractional equations could also be postulated for traffic flow. An example would be the fractional differential equation

$$\mathrm{D}_t^\alpha f(z,t) = A\frac{\partial}{\partial z} f(z,t) + B\frac{\partial^2}{\partial z^2} f(z,t) \tag{3}$$

where now $f(z,t)$ denotes the probability distribution for the number z of cars in a traffic jam at time t, and where A, B are phenomenological parameters. For $\alpha = 1$ this equation reduces to an equation proposed in [8]. Another example would be a fractional Lighthill-Whitham model defined by the equations [9]

$$\frac{\partial}{\partial t} f_1(z,t) = -\frac{\partial(f_1 f_2)}{\partial z} \tag{4a}$$

$$\mathrm{D}_t^\alpha f_2(z,t) = A - Bf_2 \tag{4b}$$

where A and B are phenomenological parameters. Here $f_1(z,t)$ is a macroscopic density of vehicles at position z at time t, and $f_2(z,t)$ is their average velocity. An exact solution for the fractional relaxation equation (4b) is given in [7].

My purpose in this article is not to discuss specific applications of fractional calculus (see [5] for other examples), but rather to investigate the reasons for

[1] Not all definitions of fractional derivatives are suitable. For information about "suitable" definitions of fractional derivatives see [5].

the importance of fractional time derivatives from a more fundamental point of view. The results presented here provide a general justification for the usage of fractional time derivatives in model equations pertaining to macroscopic time scales [7]. Fractional time derivatives appear universally because they arise as attractive fixed points in a scaling limit. This is similar to the thermodynamic limit in the theory of phase transitions, and to the concept of universality in that context. The appropriate scaling limit is introduced as a long time limit describing a change between microscopic and macroscopic time scales. The results also contain a mathematical mechanism for the emergence of macroscopic irreversibility from microscopic reversibility that does not depend on traditional [10,11] phase space arguments.

2 Semi-groups and Long Time Limit

Consider a general time evolution $\mathrm{T}(t)$ in physics, economics, or other sciences. An example of what is meant by $\mathrm{T}(t)$ is given by the left hand side of (1). In that case it is identified as a simple translation, defined as

$$\mathcal{T}(t)f(s) = f(s-t), \tag{5}$$

because the infinitesimal generator $-\mathrm{d}/\mathrm{d}t$ of $\mathcal{T}(t)$ appears on the left hand side of eq. (1). The subsequent considerations concern the possible operators that may appear on the left hand side of an evolution equation.

A time evolution may be characterized generally as a pair $(\{\mathrm{T}_\tau(t) : 0 \le t < \infty\}, (B_\tau, \|\cdot\|))$. Here $\mathrm{T}_\tau(t) = \mathrm{T}(t\tau)$ is a semi-group of operators $\{\mathrm{T}(t) : 0 \le t < \infty\}$ mapping a Banach space $(B_\tau(\mathbb{R}), \|\cdot\|)$ of functions $f_\tau(s) = f(s\tau)$ on $\mathbb{R}$ to itself. The elements of the Banach space represent the observables or states of the system. The argument $t \ge 0$ of $\mathrm{T}_\tau(t)$ represents a time duration, the argument $s \in \mathbb{R}$ of $f_\tau(s)$ a time instant. The index $\tau > 0$ indicates the units (or scale) of time. Below, τ will again be frequently suppressed to simplify the notation. The elements $f_\tau(s) = f(s\tau) \in B_\tau$ represent observables or the state of a physical system as function of the time coordinate $s \in \mathbb{R}$. The semi-group conditions require

$$\mathrm{T}_\tau(t_1)\mathrm{T}_\tau(t_2)f_\tau(t_0) = \mathrm{T}_\tau(t_1+t_2)f_\tau(t_0) \tag{6a}$$

$$\mathrm{T}_\tau(0)f_\tau(t_0) = f_\tau(t_0) \tag{6b}$$

for $t_1, t_2 > 0$, $t_0 \in \mathbb{R}$ and $f_\tau \in B_\tau$. The first condition defining the composition law of the semi-group may be viewed as representing the unlimited divisibility of time. The second condition is the unit element of the semi-group.

Reproducibility of experiments requires homogeneity with respect to time translations. The postulate of homogeneity assumes that $\mathrm{T}(t)$ commutes with translations, i.e.

$$\mathcal{T}(t_1)\mathrm{T}(t_2)f(t_0) = \mathrm{T}(t_2)\mathcal{T}(t_1)f(t_0) \tag{7}$$

for all $t_2 > 0$ and $t_0, t_1 \in \mathbb{R}$. This postulate allows to shift the origin of time and it reflects the basic symmetry of time translation invariance for scientific laws.

A time evolution $\mathrm{T}(t)$ should also be causal in the sense that the function $g(t_0) = (\mathrm{T}(t)f)(t_0)$ should depend only on values of $f(s)$ for $s < t_0$.

Finally, the time evolution $\mathrm{T}(t)$ is assumed to be strongly continuous in t by demanding

$$\lim_{t \to 0} ||\mathrm{T}(t)f - f|| = 0 \tag{8}$$

for all $f \in B$. It is also assumed to preserve boundedness in the sense that for all t and for $f \in L^p(\mathbb{R})$ the assumption $0 \leq f \leq 1$ almost everywhere implies $0 \leq \mathrm{T}(t)f \leq 1$ almost everywhere. For simplicity it will also be assumed that the operators $\mathrm{T}(t)$ are linear. For the general case see [7].

Let $L^p(\mathbb{R}^n)$ denote the Lebesgue spaces of p-th power integrable functions, and let $\mathcal{S}$ denote the Schwartz space of test functions for tempered distributions [12]. It is well known that all bounded linear operators on $L^p(\mathbb{R}^n)$ commuting with translations (i.e. fulfilling (7)) are of convolution type [12]. More precisely, suppose the operator $\mathrm{T} : L^p(\mathbb{R}^n) \to L^q(\mathbb{R}^n)$, $1 \leq p, q, \leq \infty$ is linear, bounded and commutes with translations. Then there exists a unique tempered distribution μ such that $\mathrm{T}f = \mu * f$ for all $f \in \mathcal{S}$. For $p = q = 1$ the tempered distributions in this theorem are finite Borel measures. If the measure is bounded and positive this means that the operator T can be viewed as a weighted averaging operator. In the following the case $n = 1$ will be of interest. A positive bounded measure μ on $\mathbb{R}$ is uniquely determined by its distribution function $\widetilde{\mu} : \mathbb{R} \to [0, 1]$ defined by

$$\widetilde{\mu}(x) = \frac{\mu(] - \infty, x[)}{\mu(\mathbb{R})}. \tag{9}$$

The tilde will again be omitted to simplify the notation.

Now, let $\mathrm{T}(t)$ be a strongly continuous time evolution as defined above fulfilling the conditions of homogeneity and causality, and being such that $f \in L^p(\mathbb{R})$ and $0 \leq f \leq 1$ almost everywhere implies $0 \leq \mathrm{T}f \leq 1$ almost everywhere. Then $\mathrm{T}(t)$ corresponds uniquely to a convolution semi-group of measures μ_t through the formula

$$\mathrm{T}(t)f(s) = (\mu_t * f)(s) = \int_{-\infty}^{\infty} f(s - s')\mathrm{d}\mu_t(s') \tag{10}$$

with $\operatorname{supp} \mu_t \subset \mathbb{R}_+$ for all $t \geq 0$. Here a convolution semi-group of measures on $\mathbb{R}$ is a family $\{\mu_t : t > 0\}$ of positive bounded measures on $\mathbb{R}$ with the properties that

$$\mu_t(\mathbb{R}) \leq 1 \quad \text{for } t > 0, \tag{11a}$$

$$\mu_{t+s} = \mu_t * \mu_s \quad \text{for } t, s > 0, \tag{11b}$$

$$\delta = \lim_{t \to 0} \mu_t \tag{11c}$$

where the last equality follows from the Mellin transform

$$\int_0^\infty x^{s-1} H_{01}^{10}\left(x \,\middle|\, \begin{matrix}(-,-)\\(0,B_1)\end{matrix}\right) \mathrm{d}x = \Gamma(B_1 s) \tag{20}$$

from the definition (23). With (19), (15), (17) and the convolution theorem one finds by Laplace inversion

$$\begin{aligned}
\mathrm{A}_\alpha \overline{f}(\overline{s}) &= \lim_{\overline{t}\to 0} \frac{1}{2\pi i} \int_{\eta-i\infty}^{\eta+i\infty} e^{\overline{s}\overline{u}} \left(\frac{e^{-\overline{t}\overline{u}^\alpha}-1}{\overline{t}}\right) \overline{f}(\overline{u})\,\mathrm{d}\overline{u} \\
&= \frac{1}{2\pi i} \int_{\eta-i\infty}^{\eta+i\infty} e^{\overline{s}\overline{u}} \lim_{\overline{t}\to 0} \left(\frac{e^{-\overline{t}\overline{u}^\alpha}-1}{\overline{t}}\right) \overline{f}(\overline{u})\,\mathrm{d}\overline{u} \\
&= -\frac{1}{2\pi i} \int_{\eta-i\infty}^{\eta+i\infty} e^{\overline{s}\overline{u}} \overline{u}^\alpha \overline{f}(\overline{u})\,\mathrm{d}\overline{u}.
\end{aligned} \tag{21}$$

This formal result can be made rigorous [25]. The infinitesimal generator A_α of the macroscopic time evolutions $\overline{\mathrm{T}}_\alpha(\overline{t})$ is therefore related to the infinitesimal generator $\mathrm{A} = -\mathrm{d}/\mathrm{d}\overline{t}$ of $\overline{\mathfrak{T}}_{\overline{t}}$ through

$$\begin{aligned}
\mathrm{A}_\alpha \overline{f}(\overline{s}) = -(-\mathrm{A})^\alpha \overline{f}(\overline{s}) &= -\frac{1}{\Gamma(-\alpha)} \int_0^\infty \frac{\overline{f}(\overline{s}-y) - \overline{f}(\overline{s})}{y^{\alpha+1}}\,\mathrm{d}y \\
&= -\frac{1}{\Gamma(-\alpha)} \int_0^\infty y^{-\alpha-1} (\overline{\mathfrak{T}}_y - \mathbf{1}) \overline{f}(\overline{s})\,\mathrm{d}y \\
&= -\mathrm{D}^\alpha \overline{f}(\overline{s})
\end{aligned} \tag{22}$$

showing that A_α is the fractional power of the derivative $\mathrm{d}/\mathrm{d}\overline{t}$. The last equality defines the fractional derivative of order α, denoted as D^α, through the Marchaud-Hadamard-Balakrishnan algorithm [23,25].

5 Conclusion

The preceding results demonstrate that fractional time derivatives may arise from a suitable scaling limit as the infinitesimal generators of time evolutions on macroscopic time scales. The order α of the derivative is restricted to the unit interval, and its value is determined by the microscopic time evolution. Physically, the order α is a quantitative measure for the decay of the temporal correlations or history dependence in the microscopic time evolution. For the most frequent case $\alpha = 1$ the results show that macroscopic irreversibility of regular evolution equations such as (1) may be the viewed as a general consequence of a long time scaling limit.

Appendix: Definition and Properties of H-Functions

The H-function of order $(m, n, p, q) \in \mathbb{N}^4$ and parameters $A_i \in \mathbb{R}_+ (i = 1, \dots, p)$, $B_i \in \mathbb{R}_+ (i = 1, \dots, q)$, $a_i \in \mathbb{C} (i = 1, \dots, p)$, and $b_i \in \mathbb{C} (i = 1, \dots, q)$ is defined for $z \in \mathbb{C}, z \neq 0$ by the contour integral [26–30]

$$H^{m,n}_{p,q}\left(z \,\middle|\, \begin{matrix} (a_1, A_1), \dots, (a_p, A_p) \\ (b_1, B_1), \dots, (b_q, B_q) \end{matrix}\right) = \frac{1}{2\pi i} \int\limits_{\mathfrak{L}} \eta(s) z^{-s} \, \mathrm{d}s \tag{23}$$

where the integrand is

$$\eta(s) = \frac{\prod\limits_{i=1}^{m} \Gamma(b_i + B_i s) \prod\limits_{i=1}^{n} \Gamma(1 - a_i - A_i s)}{\prod\limits_{i=n+1}^{p} \Gamma(a_i + A_i s) \prod\limits_{i=m+1}^{q} \Gamma(1 - b_i - B_i s)} . \tag{24}$$

In (23) $z^{-s} = \exp\{-s \log|z| - i \arg z\}$ and $\arg z$ is not necessarily the principal value. The integers m, n, p, q must satisfy

$$0 \leq m \leq q, \qquad 0 \leq n \leq p \tag{25}$$

and empty products are interpreted as being unity. The parameters are restricted by the condition

$$\mathbb{P}_a \cap \mathbb{P}_b = \emptyset \tag{26}$$

where

$$\mathbb{P}_a = \{\text{poles of } \Gamma(1 - a_i - A_i s)\} = \left\{\frac{1 - a_i + k}{A_i} \in \mathbb{C} : i = 1, \dots, n; k \in \mathbb{N}_0\right\}$$

$$\mathbb{P}_b = \{\text{poles of } \Gamma(b_i + B_i s)\} = \left\{\frac{-b_i - k}{B_i} \in \mathbb{C} : i = 1, \dots, m; k \in \mathbb{N}_0\right\} \tag{27}$$

are the poles of the numerator in (24). The integral converges if one of the following conditions holds [30]

$$\mathfrak{L} = \mathfrak{L}(c - i\infty, c + i\infty; \mathbb{P}_a, \mathbb{P}_b); \quad |\arg z| < C\pi/2; \quad C > 0 \tag{28a}$$

$$\mathfrak{L} = \mathfrak{L}(c - i\infty, c + i\infty; \mathbb{P}_a, \mathbb{P}_b); \quad |\arg z| = C\pi/2; \quad C \geq 0; \quad cD < -\operatorname{Re} F \tag{28b}$$

$$\mathfrak{L} = \mathfrak{L}(-\infty + i\gamma_1, -\infty + i\gamma_2; \mathbb{P}_a, \mathbb{P}_b); \quad D > 0; \quad 0 < |z| < \infty \tag{29a}$$

$$\mathfrak{L} = \mathfrak{L}(-\infty + i\gamma_1, -\infty + i\gamma_2; \mathbb{P}_a, \mathbb{P}_b); \quad D = 0; \quad 0 < |z| < E^{-1} \tag{29b}$$

$$\mathfrak{L} = \mathfrak{L}(-\infty + i\gamma_1, -\infty + i\gamma_2; \mathbb{P}_a, \mathbb{P}_b); \quad D = 0; \quad |z| = E^{-1}; C \geq 0; \operatorname{Re} F < 0 \tag{29c}$$

$$\mathfrak{L} = \mathfrak{L}(\infty + i\gamma_1, \infty + i\gamma_2; \mathbb{P}_a, \mathbb{P}_b); \quad D < 0; \quad 0 < |z| < \infty \tag{30a}$$

$$\mathfrak{L} = \mathfrak{L}(\infty + i\gamma_1, \infty + i\gamma_2; \mathbb{P}_a, \mathbb{P}_b); \quad D = 0; \quad |z| > E^{-1} \tag{30b}$$

$$\mathfrak{L} = \mathfrak{L}(\infty + i\gamma_1, \infty + i\gamma_2; \mathbb{P}_a, \mathbb{P}_b); \quad D = 0; \quad |z| = E^{-1}; C \geq 0; \operatorname{Re} F < 0 \tag{30c}$$

where $\gamma_1 < \gamma_2$. Here $\mathfrak{L}(z_1, z_2; \mathbb{G}_1, \mathbb{G}_2)$ denotes a contour in the complex plane starting at z_1 and ending at z_2 and separating the points in $\mathbb{G}_1$ from those in $\mathbb{G}_2$, and the notation

$$C = \sum_{i=1}^{n} A_i - \sum_{i=n+1}^{p} A_i + \sum_{i=1}^{m} B_i - \sum_{i=m+1}^{q} B_i \tag{31}$$

$$D = \sum_{i=1}^{q} B_i - \sum_{i=1}^{p} A_i \tag{32}$$

$$E = \prod_{i=1}^{p} A_i^{A_i} \prod_{i=1}^{q} B_i^{-B_i} \tag{33}$$

$$F = \sum_{i=1}^{q} b_i - \sum_{i=1}^{p} a_j + (p-q)/2 + 1 \tag{34}$$

was employed. The H-functions are analytic for $z \neq 0$ and multi-valued (single valued on the Riemann surface of $\log z$).

The following properties of a general H-function are used in the text. First, the order reduction formula

$$H^{m,n}_{p,q}\left(z \,\middle|\, \begin{matrix} (a_1, A_1), (a_2, A_2) \ldots, (a_p, A_p) \\ (b_1, B_1), (b_2, B_2) \ldots, (b_{q-1}, B_{q-1})(a_1, A_1) \end{matrix}\right)$$
$$= H^{m,n-1}_{p-1,q-1}\left(z \,\middle|\, \begin{matrix} (a_2, A_2), \ldots, (a_p, A_p) \\ (b_1, B_1), \ldots, (b_{q-1}, B_{q-1}) \end{matrix}\right) \tag{35}$$

holds for $n \geq 1$ and $q > m$, and similarly

$$H^{m,n}_{p,q}\left(z \,\middle|\, \begin{matrix} (a_1, A_1), (a_2, A_2) \ldots, (a_{p-1}, A_{p-1})(b_1, B_1) \\ (b_1, B_1), (b_2, B_2) \ldots, (b_q, B_q) \end{matrix}\right)$$
$$= H^{m-1,n}_{p-1,q-1}\left(z \,\middle|\, \begin{matrix} (a_1, A_1), \ldots, (a_{p-1}, A_{p-1}) \\ (b_2, B_2), \ldots, (b_q, B_q) \end{matrix}\right) \tag{36}$$

for $m \geq 1$ and $p > n$. A change of variables in (23) shows

$$H^{m,n}_{p,q}\left(z \,\middle|\, \begin{matrix} (a_1, A_1), \ldots, (a_p, A_p) \\ (b_1, B_1), \ldots, (b_q, B_q) \end{matrix}\right)$$
$$= H^{n,m}_{q,p}\left(\frac{1}{z} \,\middle|\, \begin{matrix} (1-b_1, B_1), \ldots, (1-b_q, B_q) \\ (1-a_1, A_1), \ldots, (1-a_p, A_p) \end{matrix}\right) \tag{37}$$

which allows to transform an H-function with $D > 0$ and $\arg z$ to one with $D < 0$ and $\arg(1/z)$. For $\gamma > 0$

$$\frac{1}{\gamma} H_{p,q}^{m,n}\left(z \,\middle|\, \begin{matrix}(a_1,A_1),\ldots,(a_p,A_p)\\(b_1,B_1),\ldots,(b_q,B_q)\end{matrix}\right) = H_{p,q}^{m,n}\left(z^{\gamma} \,\middle|\, \begin{matrix}(a_1,\gamma A_1),\ldots,(a_p,\gamma A_p)\\(b_1,\gamma B_1),\ldots,(b_q,\gamma B_q)\end{matrix}\right) \tag{38}$$

while for $\gamma \in \mathbb{R}$

$$z^{\gamma} H_{p,q}^{m,n}\left(z \,\middle|\, \begin{matrix}(a_1,A_1),\ldots,(a_p,A_p)\\(b_1,B_1),\ldots,(b_q,B_q)\end{matrix}\right) = H_{p,q}^{m,n}\left(z \,\middle|\, \begin{matrix}(a_1+\gamma A_1,A_1),\ldots,(a_p+\gamma A_p,A_p)\\(b_1+\gamma B_1,B_1),\ldots,(b_q+\gamma B_q,B_q)\end{matrix}\right) \tag{39}$$

holds. Finally, the Laplace transform of an H-function is

$$\begin{aligned}\mathcal{L}\left\{H_{p,q}^{m,n}(z)\right\}(u) &= \int_0^{\infty} e^{-ux} H_{p,q}^{m,n}\left(x \,\middle|\, \begin{matrix}(a_1,A_1),\ldots,(a_p,A_p)\\(b_1,B_1),\ldots,(b_q,B_q)\end{matrix}\right) \mathrm{d}x \\ &= H_{q,p+1}^{n+1,m}\left(u \,\middle|\, \begin{matrix}(1-b_1-B_1,B_1),\ldots,(1-b_q-B_q,B_q)\\(0,1)(1-a_1-A_1,A_1),\ldots,(1-a_p-A_p,A_p)\end{matrix}\right) \\ &= \frac{1}{u} H_{p+1,q}^{m,n+1}\left(\frac{1}{u} \,\middle|\, \begin{matrix}(0,1)(a_1,A_1),\ldots,(a_p,A_p)\\(b_1,B_1),\ldots,(b_q,B_q)\end{matrix}\right)\end{aligned} \tag{40}$$

for $\operatorname{Re} s > 0$, $C > 0$, $|\arg z| < \frac{1}{2}C\pi$ and $\min_{1\le j\le m} \operatorname{Re}(b_j/B_j) > -1$.

References

1. In classical mechanics the states are points in phase space, the observables are functions on phase space, and the operator B is specified by a vector field and Poisson brackets. In quantum mechanics (with finitely many degrees of freedom) the states correspond to rays in a Hilbert space, the observables to operators on this space, and the operator B to the Hamiltonian. In field theories the states are normalized positive functionals on an operator algebra of observables, and then B becomes a derivation on the algebra of observables. The equations (1) need not be first order in time. An example is the initial-value problem for the wave equation for $g(t,x)$

$$\frac{\partial^2 g}{\partial t^2} = c^2 \frac{\partial^2 g}{\partial x^2}$$

in one dimension. It can be recast into the form of (1) by introducing a second variable h and defining

$$f = \begin{pmatrix} g \\ h \end{pmatrix} \quad \text{and} \quad \mathrm{B} = \begin{pmatrix} 0 & 1 \\ 1 & 0 \end{pmatrix} c \frac{\partial}{\partial x}$$

2. R. Hilfer, *Stochastische Modelle für die betriebliche Planung,* (GBI – Verlag, München, 1985).
3. R. Hilfer and L. Anton, *Fractional master equations and fractal time random walks,* Phys. Rev. E, Rapid Commun. **51**, 848 (1995).
4. R. Hilfer, *On fractional diffusion and its relation with continuous time random walks,* in: *Anomalous Diffusion: From Basis to Applications*, A. Pekalski, R. Kutner, and K. Sznajd-Weron, (Eds.), p. 77 (Springer, 1999).
5. R. Hilfer, *Applications of Fractional Calculus in Physics,* (World Scientific Publ. Co., Singapore, 2000).
6. R. Hilfer, *Exact solutions for a class of fractal time random walks,* Fractals **3**, 211 (1995).
7. R. Hilfer, *Fractional time evolution,* in: *Applications of Fractional Calculus in Physics*, R. Hilfer, (Ed.), p. 87 (World Scientific, Singapore, 2000).
8. M. Paczuski and K. Nagel, *Self-organized criticality and $1/f$ noise in traffic,* in: *Traffic and Granular Flow*, D.E. Wolf, M. Schreckenberg, and A. Bachem, (Eds.), p. 73 (World Scientific, Singapore, 1996).
9. D. Helbing, *Traffic modeling by means of physical concepts,* in: *Traffic and Granular Flow*, D.E. Wolf, M. Schreckenberg, and A. Bachem, (Eds.), p. 87 (World Scientific, Singapore, 1996).
10. J.L. Lebowitz, *Macroscopic laws, microscopic dynamics, time's arrow and Boltzmann's entropy,* Physica A **194**, 1 (1993).
11. J. Lebowitz, *Statistical mechanics: A selective review of two central issues,* Rev. Mod. Phys. **71**, S346 (1999).
12. E. Stein and G. Weiss, *Introduction to Fourier Analysis on Euclidean Spaces,* (Princeton University Press, Princeton, 1971).
13. R. Hilfer, *Foundations of fractional dynamics,* Fractals **3**, 549 (1995).
14. W. Feller, *An Introduction to Probability Theory and Its Applications*, Vol. II, (Wiley, New York, 1971).
15. E. Seneta, *Regularly Varying Functions*, (Springer Verlag, Berlin, 1976).
16. R. Hilfer, *Classification theory for an equilibrium phase transitions,* Phys. Rev. E **48**, 2466 (1993).
17. R. Hilfer, *On a new class of phase transitions,* in: *Random Magnetism and High-Temperature Superconductivity*, W.P. Beyermann, N.L. Huang-Liu, and D.E. MacLaughlin, (Eds.), p. 85 (World Scientific Publ. Co, Singapore, 1994).
18. R. Hilfer, *Fractional dynamics, irreversibility and ergodicity breaking,* Chaos, Solitons & Fractals **5**, 1475 (1995).
19. R. Hilfer, *An extension of the dynamical foundation for the statistical equilibrium concept,* Physica A **221**, 89 (1995).
20. S. Bochner, *Harmonic Analysis and the Theory of Probability,* (University of California Press, Berkeley, 1955).
21. K. Yosida, *Functional Analysis,* (Springer, Berlin, 1965).
22. C. Berg and G. Forst, *Potential Theory on Locally Compact Abelian Groups,* (Springer, Berlin, 1975).
23. A.V. Balakrishnan, *Fractional powers of closed operators and the semigroups generated by them,* Pacific J. Math. **10**, 419 (1960).
24. U. Westphal, *Ein Kalkül für gebrochene Potenzen infinitesimaler Erzeuger von Halbgruppen und Gruppen von Operatoren,* Compos. Math. **22**, 67 (1970).
25. U. Westphal, *Fractional powers of infinitesimal generators of semigroups,* in: *Applications of Fractional Calculus in Physics*, R. Hilfer, (Ed.), (World Scientific, Singapore, 2000)

26. C. Fox, *The G and H functions as symmetrical Fourier kernels,* Trans. Am. Math. Soc. **98**, 395 (1961).
27. B.L.J. Braaksma, *Asymptotic expansions and anlytic continuations for a class of Barnes-integrals,* Compos. Math. **15**, 239 (1964).
28. A.M. Mathai and R.K. Saxena, *The H-function with Applications in Statistics and Other Disciplines,* (Wiley, New Delhi, 1978).
29. H.M. Srivastava, K.C. Gupta, and S.P. Goyal, *The H-functions of One and Two Variables with Applications,* (South Asian Publishers, New Delhi, 1982).
30. A.P. Prudnikov, Yu.A. Brychkov, and O.I. Marichev, *Integrals and Series*, Vol. 3, (Gordon and Breach, New York, 1990).

Dynamical Theory of Steady State Selection in Open Driven Diffusive Systems

G.M. Schütz *

Institut für Festkörperforschung, Forschungszentrum Jülich, 52425 Jülich, Germany

Abstract. The stationary states of one-dimensional driven diffusive systems, connected to boundary reservoirs with fixed particle density are shown to be selected by an extremal principle for the macroscopic current. Given the current one obtains the exact first- and second-order non-equilibrium phase transition lines for the bulk density as a function of the boundary densities. The basic dynamical mechanism behind the extremal principle is an intriguing generic interplay between the motion of shocks and localized perturbations.

1 Steady State Selection in Open Driven Systems

Imagine a driven particle system – it may be *any* system such as ribosomes moving along a m-RNA, ions diffusing in a narrow channel, or even cars proceeding on a long road – where classical objects move with preference in one direction and which is coupled at its two ends to external reservoirs. Such a system with open boundaries where particles can enter and leave will settle into an non-equilibrium steady state characterized by some bulk density and the corresponding particle current. Which stationary bulk density will the system assume as a function the boundary densities?

At first glance this appears to be an ill-posed question as undoubtedly the answer to this problem of steady state selection (first considered in general terms by Krug [4]) depends on the system under investigation. However, guided by the insights gained from the exact solution [1] of a prototypical toy model for driven diffusion, the totally asymmetric simple exclusion process (TASEP), we have developed a dynamical theory of boundary-induced phase transitions for homogeneous one-species driven particle systems with a single conserved density [2,3]. The phase diagram for the bulk density is governed by an extremal principle for the current – *irrespective of the local dynamics.* As a function of non-universal effective right and left boundary densities $\rho_{\mathrm{R,L}}$ the steady-state current j is given by the macroscopic current-density relation

$$j = \begin{cases} \max\limits_{\rho\in[\rho_{\mathrm{R}},\rho_{\mathrm{L}}]} j(\rho) & \text{for } \rho_{\mathrm{L}} > \rho_{\mathrm{R}} \\ \min\limits_{\rho\in[\rho_{\mathrm{L}},\rho_{\mathrm{R}}]} j(\rho) & \text{for } \rho_{\mathrm{L}} < \rho_{\mathrm{R}} . \end{cases} \qquad (1)$$

* Much of this paper reviews work done in collaboration with E. Domany [1], A.B. Kolomeisky, E.B. Kolomeisky, J.P. Straley [2] and V. Popkov [3].

The structure of the phase diagram which exhibits a variety of first- and second-order non-equilibrium transitions is determined by the number of extrema of the current. The microscopic details of the system enter only in so far as they determine the functional form of the current $j(\rho)$ and the effective boundary densities which depend on the true boundary densities through the details of the coupling mechanism. At the second-order phase transition lines the intrinsically non-universal properties of the boundary layer are dominated by a universal power law decay of the density profile to its bulk value [4,5]. The extremal principle (1) and hence the various phases and the nature of the transitions can be understood dynamically by the interplay of local fluctuations (which lead to an overfeeding effect [1,2]) with the branching and coalescence of shocks [2,3] (see below).

2 Exact Solution of the TASEP with Open Boundaries

In the simplest non-trivial case the current is a convex function with a local extrema. This is realized in the TASEP where particles hop on a 1-d lattice with constant rate to the right, provided the neighboring lattice site is vacant. Otherwise an attempted move is rejected. In a finite, open system particles are injected with rate α at the left boundary site 1. At the right boundary site L particles leave the system with rate β. This corresponds to a coupling to reservoirs of constant densities $\rho_{\mathrm{L}} = \alpha = \rho_{\mathrm{L}}^{\mathrm{true}}$ and $\rho_{\mathrm{R}} = 1 - \beta = \rho_{\mathrm{R}}^{\mathrm{true}}$. This model was first introduced in 1968 to describe the kinetics of protein synthesis [6,7] and has since then obtained paradigmatic status as an interesting and exactly solvable model for a many-body system out of equilibrium [8,9]. It even serves as a very simple toy model for traffic flow [10,11], exhibiting shocks (traffic jams) and a current-density relation

$$j = \rho(1 - \rho) \tag{2}$$

which has a single maximum of the current as observed in real traffic [11], albeit at a different density ρ^*.

Employing probabilistic tools, Liggett [12] obtained a recursion relation for the stationary distribution of the TASEP in system size L from which he could extract the bulk density as a function of α and β and hence the phase diagram. However, in order to understand the physical origin of the phase transitions, one has to study the local density profile. From the steady-state recursion for the 2^L configurational probabilities one obtains closed recursions for the unnormalized string weights

$$Y_{L,k} \propto \langle \underset{k \ldots L}{\ldots 00000} \rangle \tag{3}$$

for the probability of finding the last $L - k + 1$ sites empty and

$$X^p_{L,k} \propto \langle \underset{p \ldots k \ldots L}{\ldots 1 \ldots 00000} \rangle \tag{4}$$

resp. [13]. These quantities satisfy

$$Y_{L,k} = Y_{L,k-1} + \alpha\beta Y_{L-1,k} \quad \text{for } 1 < k < L+1 \tag{5}$$
$$Y_{L,1} = \beta Y_{L-1,1} \quad \text{with } Y_{0,1} = 1 \tag{6}$$
$$X^p_{L,k} = X^p_{L,k-1} + \alpha\beta X^p_{L-1,k} \quad \text{for } p+1 < k < L+1 \tag{7}$$
$$X^p_{L,L+1} = X^p_{L,L} + \alpha X^p_{L-1,L} \quad \text{with } X^p_{L,p+1} = \alpha\beta Y_{L-1,p+1}. \tag{8}$$

For general α, β these recursion relations are not straightforward to solve with standard approaches. However, solving explicitly for small system size, guessing the pattern behind the expressions and verifying the guess by cross-checking with the recursion relations yields the bulk density ρ as well as the density profile $t_p = \rho_p - \rho_{p-1}$ for $1 \le p \le L$ [1]. One finds in agreement with the extremal principle (1)

$$\rho = \begin{cases} \alpha = \rho_{\mathrm{L}} & \text{for } \beta < \alpha < 1/2 \\ \beta = 1 - \rho_{\mathrm{R}} & \text{for } \alpha < \beta < 1/2 \\ 1/2 = \rho^* & \text{for } \alpha, \beta > 1/2. \end{cases} \tag{9}$$

and

$$t_p = A(\alpha, \beta)\Phi_p(\alpha)\Phi_{L-p}(\beta) \tag{10}$$

with

$$\Phi_n(x) = \frac{1-2x}{2[x(1-x)]^n} + 2\binom{2n}{n}\, {}_2\mathrm{F}_1(1, n+1/2, 1/2; (1-2x)^2. \tag{11}$$

In an alternative matrix product approach [14] the same exact results were obtained.

The phase diagram (Fig. 1) has three phases, a maximal-current phase C in the domain $\rho_{\mathrm{L}} > \rho^*, \rho_{\mathrm{R}} < \rho^*$ (bulk density $\rho = \rho^*$), with second-order transitions to the low-density ($\rho = \rho_{\mathrm{L}} < \rho^*$) and the high-density phase ($\rho = 1 - \rho_{\mathrm{R}} > \rho^*$) resp. These phases are separated by a first-order transition along the line $j_{\mathrm{L}} = j_{\mathrm{R}}$ in the domain $\rho_{\mathrm{L}} < \rho^*, \rho_{\mathrm{R}} > \rho^*$. The phase diagram of the TASEP is generic for all systems with a single maximum in the current-density relation in that the phase transition lines are given by the same relations in terms of the current. For a system with two maxima of the current the phase diagram consists of seven distinct phases, including two maximal current phases with bulk densities corresponding to the respective maxima of the current and a minimal current phase in a regime defined by

$$j(\rho_{\mathrm{R}}), j(\rho_{\mathrm{L}}) > j(\rho_{\min}); \quad \rho_{\mathrm{L}} < \rho_{\min} < \rho_{\mathrm{R}}. \tag{12}$$

Somewhat contrary to intuition the system organizes itself into a state with bulk density ρ_{bulk} corresponding to the local minimum of the current even though both boundaries support a higher current [3].

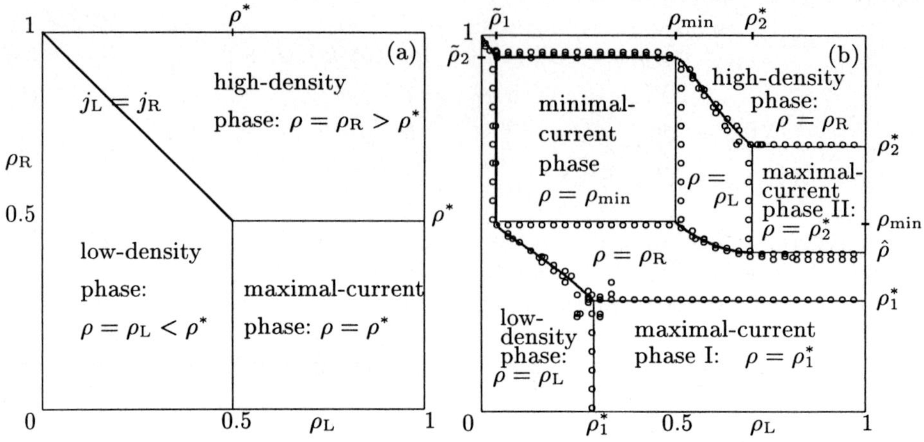

Fig. 1. Exact phase diagram of the TASEP (a) and of a lattice gas model with two maxima at $\rho^*_{1,2}$ and a minimum at $\rho_{min} = 0.5$ in the current density relation (b). Full (bold) lines indicate phase transitions of second (first) order. Circles show the results of Monte-Carlo simulations of a system with 150 sites [3].

3 Shocks and Overfeeding

To understand the origin of the extremal principle for the current we first note that in the absence of detailed balance stationary behavior cannot be understood in terms of a free energy, but has to be derived from the system dynamics. Following Kolomeisky et al. [2] there are two basic dynamical phenomena to consider. A density wave, i.e., a localized perturbation in a stationary region of background density ρ (Fig. 2), spreads out in the course of time, but keeps a constant center-of-mass velocity v_c. Under mild assumptions on the nature of the steady state this collective velocity is given by the derivative

$$v_c = \frac{d}{d\rho} j \tag{13}$$

of the current [9].

In a driven system with non-linear current-density relation one observes shocks, i.e., abrupt changes in the local density (Fig. 2). Shocks are stable collective excitations which travel with a mean velocity v_s determined by the boundary densities of the shock. Irrespective of the specific system mass conservation yields

$$v_s = \frac{j_L - j_R}{\rho_L - \rho_R}. \tag{14}$$

In equilibrium phenomena a domain wall is a localized region where the order parameter interpolates between degenerate ground states. In driven non-

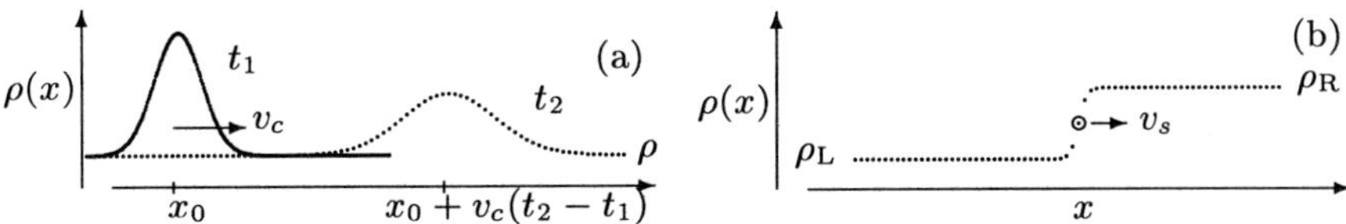

Fig. 2. (a) Diffusive spreading of a density perturbation in the steady state at two times $t_2 > t_1$. The collective velocity describes the motion of the center of mass of the perturbation. (b) Motion of a shock. To the left (right) of the domain wall particles are distributed homogeneously with an average density ρ_L (ρ_R).

equilibrium systems a shock is like a domain wall, separating two possible *stationary* states of the system. A first-order transition between these states takes place when the shock velocity changes sign. Since the shock velocity is determined by boundary conditions a first-order (discontinuous) transition in the bulk density may be induced by boundary effects. This picture is well-supported by the linear increase of the local density (10) exactly at the first-order transition which corresponds to an average over all shock positions resulting from an unbiased random walk.

The factorized form of the density profile (10) suggests to study the two boundaries separately from each other. An overfeeding occurs when the collective velocity with which variations of the boundary density penetrate into the bulk changes sign. Injected particles create a localized increase of the density which cause back-moving density waves for boundary densities where the current has negative slope. These waves block further injection attempts and thus prevent a change in the stationary bulk current. The result is a second-order (continuous) phase transition in the bulk density.

These velocities and the underlying single-shock picture are sufficient to understand the phase diagram of systems with a single maximum in the current. If the current has a local minimum a single shock may branch into two distinct shocks, moving away from each other. This phenomenon follows from the stability criterion [9]

$$v_c(\rho_\mathrm{L}) > v_s(\rho_\mathrm{L}, \rho_\mathrm{R}) > v_c(\rho_\mathrm{R}) \tag{15}$$

for a single shock which one obtains by considering the flow of fluctuations in the neighborhood of the shock. This behavior which explains the emergence of the minimal current phase can be checked numerically (Fig. 3).

With these observations the dynamical origin of all phase transition lines can be understood by considering the time evolution of judiciously chosen shock initial states. Because of ergodicity, the steady state does not depend on the initial conditions and a specific choice involves no loss of generality. Thus one constructs the complete phase diagram and obtains the extremal principle (1).

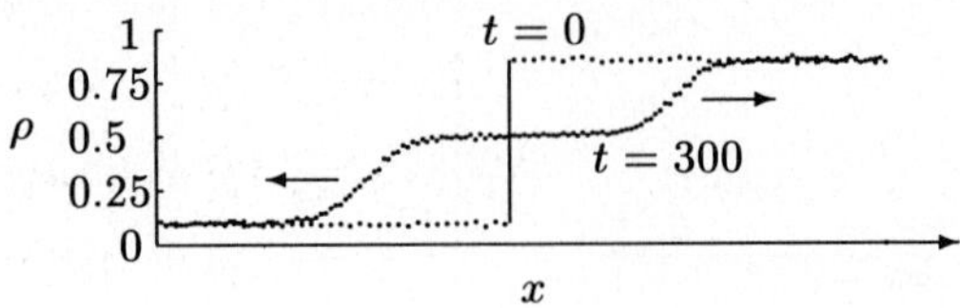

Fig. 3. Monte-Carlo simulation of the particle density distribution in a lattice gas in the initial state (one shock) and after 300 Monte-Carlo sweeps, showing branching into two shocks. 3000 histories are averaged over [3].

4 Conclusions

The interplay of density fluctuations and shock diffusion, coalescence and branching resp. determines the steady state selection of driven diffusive systems with a single conserved density and leads to the extremal principle (1). In the absence of bulk phase transitions we expect this scenario to remain valid also in higher dimensions. A surprising phenomenon is the occurrence of a self-organized minimal current phase. Since little reference is made to the precise nature of the dynamics we argue that the phase diagram is generic and hence knowledge of the macroscopic current-density relation of a given physical system is sufficient to calculate the exact non-equilibrium phase transition lines. Experimental evidence for the first-order transitions has been found in the kinetics of biopolymerization as well as more directly in traffic flow data close to an on-ramp [15,16].

References

1. G. Schütz and E. Domany, J. Stat. Phys. **72**, 277 (1993).
2. A.B. Kolomeisky, G.M. Schütz, E.B. Kolomeisky, and J.P. Straley, J. Phys. A **31**, 6911 (1998).
3. V. Popkov and G.M. Schütz, Europhys. Lett. **48**, 257 (1999).
4. J. Krug, Phys. Rev. Lett. **67**, 1882 (1991).
5. K. Oerding and H.K. Janssen, Phys. Rev. E **58**, 1446 (1998).
6. J.T. MacDonald, J.H. Gibbs, and A.C. Pipkin, Biopolymers **6**, 1 (1968).
7. G.M. Schütz, Int. J. Mod. Phys. B **11**, 197 (1997).
8. V. Privman, (Ed.), *Nonequilibrium Statistical Mechanics in One Dimension,* (Cambridge University Press, Cambridge, 1997).
9. G.M. Schütz, in: *Phase Transitions and Critical Phenomena*, C. Domb and J. Lebowitz (Eds.), to appear, (Academic Press, London).
10. T. Nagatani, J. Phys. A **28**, 7079 (1995).
11. D. Chowdhury, A. Schadschneider, and L. Santen, to appear, Phys. Rep.
12. T.M. Liggett, Trans. Amer. Math. Soc. **179**, 433 (1975).
13. B. Derrida, E. Domany, and D. Mukamel, J. Stat. Phys. **69**, 667 (1992).
14. B. Derrida, M.R. Evans, V. Hakim, and V. Pasquier, J. Phys. A **26**, 1493 (1993).
15. L. Neubert, L. Santen, A. Schadschneider, and M. Schreckenberg, Phys. Rev E **60**, 6480 (1999).
16. V. Popkov, L. Santen, A. Schadschneider, and G.M. Schütz, preprint.

Application of a Deterministic Scheme for the Boltzmann Equation in Modelling Shock Wave Focusing

P. Kowalczyk, T. Płatkowski, and W. Waluś

Inst. of Applied Mathematics and Mechanics; Dept. of Mathematics, Informatics, and Mechanics; Warsaw University, Banacha 2, 02–097 Warszawa, Poland

Abstract. A recently developed accelerated deterministic method of approximation of the Boltzmann collision operator is applied in numerical modelling of the process of shock wave focusing in a rarefied noble gas. The results are compared with the results obtained previously for the same problem with the Boltzmann collision operator evaluated by the Monte Carlo quadrature.

1 Introduction

The central mathematical object in the kinetic theory of rarefied gases is the Boltzmann equation (see [3]) for the distribution function $f = f(t, \boldsymbol{x}, \boldsymbol{v})$

$$\frac{\partial f}{\partial t} + \boldsymbol{v}\frac{\partial f}{\partial \boldsymbol{x}} = J(f, f). \tag{1}$$

In (1) $t \geq 0$ denotes time, $\boldsymbol{x} \in \Omega \subset \mathbb{R}^3$ position, and $\boldsymbol{v} = [v_x, v_y, v_z] \in \mathbb{R}^3$ velocity. The collision operator $J(f, f)$ is defined as follows

$$J(f, f)(\boldsymbol{v}) = \int_{\mathbb{R}^3} \int_{\mathbf{S}^2} B(q, \boldsymbol{n}) \left(f(\boldsymbol{v}')f(\boldsymbol{w}') - f(\boldsymbol{w})f(\boldsymbol{v})\right) d\boldsymbol{n} d\boldsymbol{w}, \tag{2}$$

where $\boldsymbol{n} \in \mathbf{S}^2$, $q = |\boldsymbol{v} - \boldsymbol{w}|$, and $\boldsymbol{v}' = \frac{1}{2}(\boldsymbol{v} + \boldsymbol{w}) + \frac{1}{2}q\boldsymbol{n}$, $\boldsymbol{w}' = \frac{1}{2}(\boldsymbol{v} + \boldsymbol{w}) - \frac{1}{2}q\boldsymbol{n}$ are the post-collision velocities. In the case of the hard sphere model with diameter d, which we assume in our calculations, the kernel of the collision operator $B(q, \boldsymbol{n}) = d^2 q$.

A challenging problem in establishing an efficient numerical method of solving the Boltzmann equation is approximation of the collision operator $J(f, f)$. There are two main approaches to this problem. The first one is based on Monte Carlo quadratures, see [1] and references therein. The Monte Carlo quadrature is relatively efficient when used in numerical solvers for the Boltzmann equation, however possesses intrinsic errors due to the stochastic nature of this approximation. The second, deterministic approach consists in evaluating conservatively the collision operator applying the so-called *discrete velocity approximation*, e.g., [2,8,9] and references therein. Being free of the stochastic errors, the method has large (at least quadratic) computational cost, see [2]. Thus, there is need for modifications of the discrete velocity approximation that lower the computational cost of the method.

2 Discrete velocity approximation of the collision term and acceleration procedure

Let $\Delta v\mathbf{Z}^3$ be a discrete regular lattice in the velocity space $\mathbb{R}^3$ (Δv denotes the step of the lattice). For fixed $\boldsymbol{v}_i \in \Delta v\mathbf{Z}^3$ the collision operator (2) can be approximated in the following way (see [9])

$$J(f,f)(\boldsymbol{v}_i) \approx \sum_{\boldsymbol{v}_j \in \boldsymbol{v}_i + \Delta v\mathbf{Z}^3} \sum_{(k,l)} \Gamma_{ij}^{kl}(f_k f_l - f_i f_j), \tag{3}$$

where $f_i = f(\boldsymbol{v}_i)$. In (3) the first sum $\sum_{\boldsymbol{v}_j}$ replaces the integration over $\mathbb{R}^3$ with respect to $\boldsymbol{w}$ in $J(f,f)$. The second sum $\sum_{(k,l)}$ replaces the integration over $\mathbf{S}^2$, and denotes the summation over all k and l such that $\boldsymbol{v}_k$ and $\boldsymbol{v}_l$ constitute post-collision velocities for particles with pre-collision velocities $\boldsymbol{v}_i$ and $\boldsymbol{v}_j$. The coefficients Γ_{ij}^{kl} in (3) are defined as follows

$$\Gamma_{ij}^{kl} = \frac{4\pi}{r_{ij}}(\Delta v)^3 B\left(|\boldsymbol{v}_i - \boldsymbol{v}_j|, \frac{\boldsymbol{v}_k - \boldsymbol{v}_l}{|\boldsymbol{v}_k - \boldsymbol{v}_l|}\right),$$

where r_{ij} denotes the number of the post-collision pairs $\boldsymbol{v}_k$ and $\boldsymbol{v}_l$ on the velocity lattice for given $\boldsymbol{v}_i$ and $\boldsymbol{v}_j$.

Below we describe briefly a particular method of calculating the discrete velocity approximation (3) of the collision operator. This method suits well for the acceleration procedure which we shall propose at the end of this Section.

First we replace $\Delta v\mathbf{Z}^3$ in (3) by its twice coarser sub-lattice $2\Delta v\mathbf{Z}^3$ and balance the sum by the multiplicative factor 8. The use of $2\Delta v\mathbf{Z}^3$ guarantees that the centre $\boldsymbol{c}_{ij} = \frac{1}{2}(\boldsymbol{v}_i + \boldsymbol{v}_j)$ of the sphere spanned by $\boldsymbol{v}_i$ and $\boldsymbol{v}_j$ with diameter $|\boldsymbol{v}_j - \boldsymbol{v}_i|$ belongs to $\Delta v\mathbf{Z}^3$. The summation over $2\Delta v\mathbf{Z}^3$ is performed in a manner reminiscent of the integration in $\mathbb{R}^3$ using the spherical coordinates. If $V(m)$ denotes the set of discrete velocities $\boldsymbol{v} \in \Delta v\mathbf{Z}^3$ lying on the sphere of radius $\sqrt{m}\Delta v$ centred at 0 for $m = 1, 2, \ldots$ (note that $V(m)$ does not depend on Δv), we can write

$$\sum_{\boldsymbol{v}_j \in \boldsymbol{v}_i + \Delta v\mathbf{Z}^3} \ldots \approx 8 \sum_{\boldsymbol{v}_j \in \boldsymbol{v}_i + 2\Delta v\mathbf{Z}^3} \ldots \approx 8 \sum_{m=1}^{\infty} \sum_{\boldsymbol{v}_j \in \boldsymbol{v}_i + V(4m)} \ldots . \tag{4}$$

Note that we must use $V(4m)$ in (4) for $\boldsymbol{c}_{ij}$ be a lattice point. Moreover, recall that all the discrete post-collision pairs of velocities $\boldsymbol{v}_k$ and $\boldsymbol{v}_l$ for the pre-collision velocities $\boldsymbol{v}_i$, $\boldsymbol{v}_j$ lie on $\boldsymbol{c}_{ij} + V(m)$. Since $V(4m) = 2V(m)$ (see [9]), in carrying out the summations we can only use $V(m)$.

Taking into account the above considerations and the the fact in actual computations we use a finite (cubic) discrete velocity lattice $\mathbf{L}_N = \{\boldsymbol{v}_1, \ldots, \boldsymbol{v}_N\}$, we obtain

$$\sum_{\boldsymbol{v}_j \in \boldsymbol{v}_i + \Delta v\mathbf{Z}^3} \sum_{(k,l)} \ldots \approx 8 \sum_{m=1}^{m_{\max}} \sum_{\boldsymbol{v}_j \in \boldsymbol{v}_i + 2V(m)} \sum_{\boldsymbol{v}_k \in \boldsymbol{c}_{ij} + V(m)} \ldots, \tag{5}$$

where the integer $m_{\max}$ is such that $2\sqrt{m_{\max}}\Delta v$ is the diameter of $\mathbf{L}_N$.

Acceleration procedure for the calculation of the collision term (3) consists in limiting the number of radii of the spheres $V(m)$ taken in (5). Let us choose $1 \leq m_s \leq m_{\max}$ and split the sum over m into two parts

$$\sum_{m=1}^{m_{\max}} \cdots = \sum_{m=1}^{m_s} \cdots + \sum_{m=m_s+1}^{m_{\max}} \cdots . \tag{6}$$

Our earlier findings reported in [6] show that for judiciously chosen m_s we can simply omit the second sum in (6) without losing much of the accuracy. This acceleration procedure reduces the computational cost of the above numerical method by order of magnitude, which is important in calculations of complex multidimensional flows. A detailed exposition of this discrete velocity approximation and the acceleration procedure is given in [6].

3 Numerical Modelling of the Shock Wave Focusing

The focusing of the shock wave was studied by many researches, see [4,7] and references therein. Our numerical study was motivated by the experimental results on the shock wave focusing obtained by Walenta [7] in a low-density shock tube.

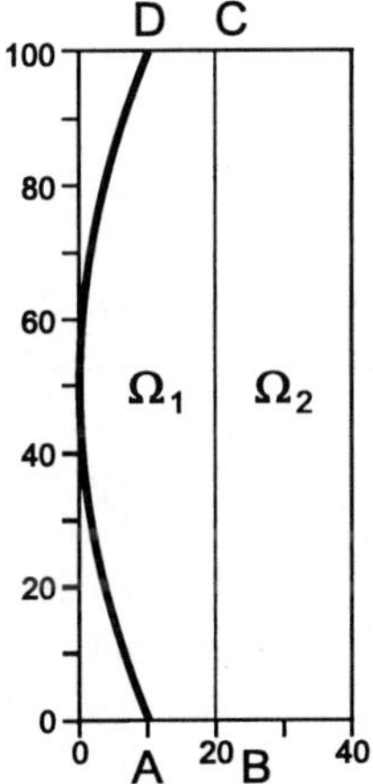

Fig. 1. The domain Ω.

To model the shock reflection and focusing we considered the planar 2-D flow in the domain Ω, which corresponded to the physical region occupied by the gas (see Figure 1). Initially Ω contained an impermeable surface BC which divided Ω into two parts Ω_1 and Ω_2. In each part of Ω the gas was in equilibrium, described by the relevant Maxwellians with parameters satisfying the Rankine-Hugoniot relations. At $t = 0$ the surface was removed, allowing the development of the shock wave. The interaction of the gas with the reflecting surface (denoted by AD in Figure 1) was modelled by the Maxwell boundary condition: $f(t,\boldsymbol{x},\boldsymbol{v}) = (1-\alpha)f(t,\boldsymbol{x},\boldsymbol{v}_r)+\alpha f_M(t,\boldsymbol{x},\boldsymbol{v})$, for $\boldsymbol{x} \in \partial\Omega$ and $\boldsymbol{v}\cdot\boldsymbol{n}(\boldsymbol{x}) > 0$, where $\boldsymbol{n}(\boldsymbol{x})$ denoted the inner unit normal to the boundary $\partial\Omega$ at the point $\boldsymbol{x}$. In the boundary condition $\boldsymbol{v}_r = \boldsymbol{v} - 2(\boldsymbol{n}(\boldsymbol{x}) \cdot \boldsymbol{v})\boldsymbol{n}(\boldsymbol{x})$ and $f_M(t,\boldsymbol{x},\boldsymbol{v}) = \frac{\rho_{in}(t,\boldsymbol{x})}{2\pi(T_w)^2} \exp\left(-\frac{\boldsymbol{v}^2}{2RT_w}\right)$, with T_w being the temperature of the reflector, and $\rho_{in}(t,\boldsymbol{x}) = \int_{\boldsymbol{n}(\boldsymbol{x})\cdot\boldsymbol{v}'<0} |\boldsymbol{n}(\boldsymbol{x})\cdot \boldsymbol{v}'| f(t,\boldsymbol{x},\boldsymbol{v}')d\boldsymbol{v}'$. α denoted the accommodation coefficient. At the other parts of the boundary, except the right-hand side of the region, the gas was specularly reflected.

In order to obtain subsequent pictures of the shock wave focusing, we solved numerically the Boltzmann equation (1) with the initial and boundary conditions described above. The numerical scheme was based on the splitting procedure, which decoupled the Boltzmann equation on each time step into the free flow equation and the spatially homogeneous Boltzmann equation. The former, accompanied with the relevant boundary conditions, was solved using the second order finite volume scheme for the hyperbolic advection equation. The latter was solved using the explicit Euler scheme in which the collision operator was approximated by the method described in Section 2. The macroscopic parameters were normalised with respect to their unperturbed values. More details of the setup of the numerical calculations are given in [5].

4 Numerical results

Figures 2 and 3 show the shock reflected from the parabolic reflector for $M = 2.8$ and $\alpha = 0.3$. The iso-lines of the pressure in the computational domain are shown at the instant of time, which corresponds to the highest pressure in the focus area. Figure 2 presents results obtained by the accelerated deterministic algorithm with $m_s = 5, 7, 9$ respectively (if the algorithm was not accelerated, we would have $m_s = m_{\max} = 48$). Figure 3 shows results obtained by a numerical scheme with the Monte Carlo (MC) quadrature for the collision operator. The number of MC points in the quadrature formula was 20 and 200 respectively. Although the qualitative pictures of the focused shock are similar for both methods, the thickness of the shock is larger and the iso-lines are smoother for the shock obtained by the accelerated deterministic method than for shock computed with the MC quadrature. Note that the iso-lines for $m_s = 7$ and $m_s = 9$ are similar, whereas for $m_s = 5$ and $m_s = 7$ they show noticeable discrepancies, confirming the observation of [6] that the truncated radius cannot be too small in order to obtain plausible results. In Figure 4 we show the pressure distributions along the axis of the shock tube at various instants of time. Note that the shock obtained by the deterministic method is slower than the shock calculated by the method with the MC quadrature. This observation should be compared with the finding of [7], that the reflected shock in the experiment is slower than the shock obtained

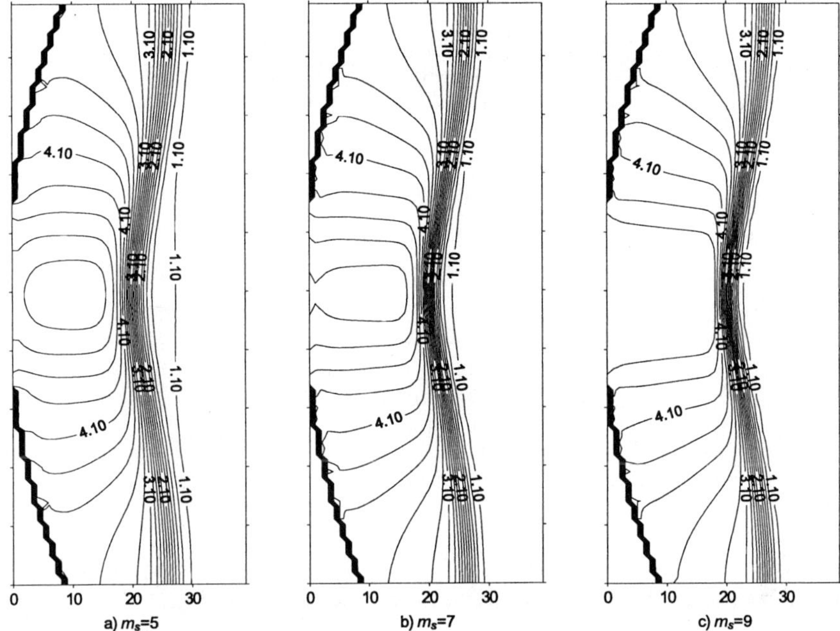

Fig. 2. The pressure iso-lines for $M = 2.8$, $\alpha = 0.3$; deterministic method.

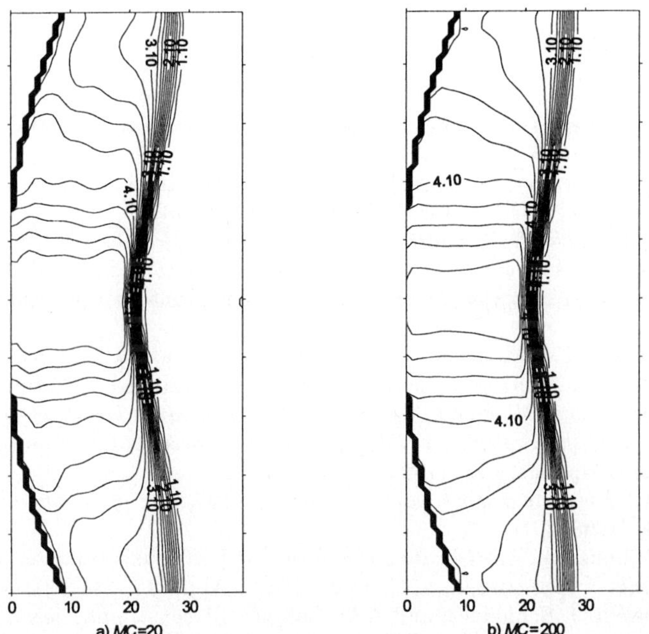

Fig. 3. The pressure iso-lines for $M = 2.8$, $\alpha = 0.3$; method with MC quadrature.

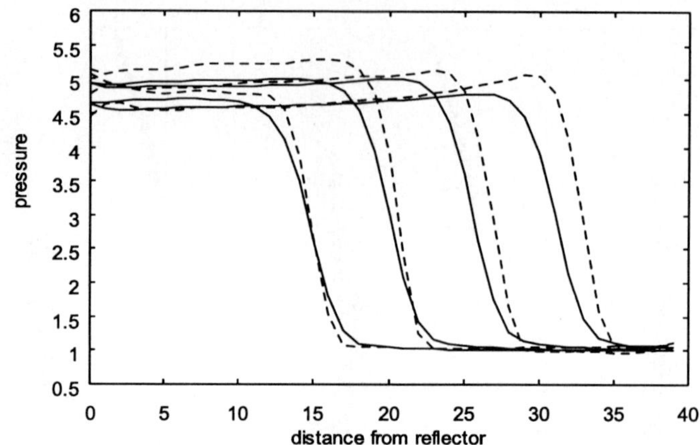

Fig. 4. The pressure distribution for $M = 2.8$, $\alpha = 0.3$. Solid lines – accelerated deterministic method, dashed lines – method with the MC quadrature.

in the previous numerical studies – in the DSMC calculations reported in [7] and in the calculations in which the MC quadrature was used [5].

Acknowledgement. This work was supported by the Polish Government Grant No. 2 P03A 007 17.

References

1. V.V. Aristov and F.G. Tcheremissine, *Direct numerical solutions of the kinetic Boltzmann equation*, (Comp. Center of Russ. Acad. of Sci., Moscow, 1992).
2. C. Buet, *A discrete-velocity scheme for the Boltzmann operator of rarefied gas dynamics*, Trans. Theory Stat. Phys. **25**, 33–60 (1996).
3. C. Cercignani, *The Boltzmann Equation and its Applications*, (Springer, 1988).
4. H. Grönig, *Shock Wave Focusing Phenomena*, in: *15th Shock Waves and Shock Tubes*, Bershader and Hanson (Eds.), pp. 43–56 (Stanford University Press, Stanford, 1986).
5. P. Kowalczyk, T. Płatkowski , and W. Waluś, *Focusing of a Shock Wave in a Rarefied Gas: A Numerical Study, ShockWaves*, submitted, (1999).
6. T. Płatkowski and W. Waluś, *An Acceleration Procedure for Discrete Velocity Approximation of the Boltzmann Collision Operator*, accepted, Computers and Mathematics with Applications, (1999).
7. Z. Walenta, *Focusing a shock wave: Microscopic structure of the phenomenon*, Arch. Mech. **50**, (1998).
8. F.G. Tcheremissine, *Conservative evaluation of Boltzmann collision integral in discrete ordinates approximation,* Comput. Math. Appl. **35**, 215–221 (1998).
9. A. Palczewski, J. Schneider, and A.V. Bobylev, *A consistency result for discrete - velocity model of the Boltzmann equation*, SIAM J. Numer. Anal. **34**, 1865–1883 (1997).

Nonlinear Waves and Moving Clusters on Rings

U. Erdmann, J. Dunkel, and W. Ebeling

Institut für Physik, Humboldt-Universität zu Berlin, Invalidenstraße 110,
10115 Berlin, Germany

Abstract. The dynamics of a ring of masses including dissipative forces (passive and active friction) and Toda interactions between the masses are investigated. The characteristic attractor structure and the influence of noise by coupling to a heat bath are studied. The system may be driven from the thermodynamic equilibrium to far from equilibrium states by including negative friction. We show, that over-critical pumping with free energy may lead to a partition of the phase space into attractor regions corresponding to several types of collective motions including uniform rotations, one- and multiple soliton excitations and relative oscillations. With Lennard-Jones like interaction potentials the particles form clusters moving along the ring.

1 Introduction

Models of active Brownian particles were recently used for modeling several new types of complex motion [1–4]. Chemical interactions of active Brownian particles were studied in detail by SCHIMANSKY-GEIER and coworkers [5,6]. Here we will develop a model of active Brownian particles with conservative (repulsive) interactions under the influence of white noise. We will study one-dimensional systems of N active Brownian particles with repulsive interactions, which form a ring. These models are closely related to recent studies of Toda rings [7,8]. In these studies we investigated Toda rings with noise and dissipation. In our previous studies of Toda rings no active element was included. A first approach to investigate active Toda rings with the aim to model dissipative solitons was given recently [3]. In the mentioned work it was shown theoretically and by simulations that in active Toda rings running excitations may be generated. Here we discuss possible applications to traffic problems.

2 Nonlinear Brownian Dynamics on Rings

The model investigated here is a one-dimensional model of coupled masses on a ring with the total length L which obey the following equation of motion:

$$\ddot{x}_i = W(x_{i+1}, x_i, x_{i-1}) + R(v_i) + Z(t), \qquad (1)$$

where $R(v_i)$ is a nonlinear friction term, $W(x_{i+1}, x_i, x_{i-1})$ describes the interaction between the neighbors and $Z = \sqrt{2D}\xi(t)$ is a stochastic force described by Gaussian white noise.

In order to model active motion the friction function consists of an additional term which can be understood as an external energy supply

$$R(v) = -\gamma(v)\, v = -\gamma_0 \left(1 - \frac{1 + d_* V^2}{1 + d_* v^2}\right) v. \tag{2}$$

This kind of friction, which is plotted in Fig. 1a, proposed in recent work [2,9], includes both an active and a passive term, where γ_0 is the usual passive friction. We see that in the range of small velocities pumping due to negative friction occurs, as an additional source of energy for the Brownian particle. The parameters V, d_* characterize the pumping. At large velocities the friction function converges to γ_0, at small velocities with $v < V$ the friction function is negative. Hence, slow particles are accelerated, while the motion of fast particles is damped. The friction function $R(v)$ leads to a kind of "potential" which depends on v,

$$V_R(v) = \frac{1}{2}\, \gamma_0 v^2 - \frac{1}{2}\, \gamma_0 \left[V^2 \ln(1 + d_* v^2) - \frac{1}{d_*}\, \ln(1 + d_* V^2)\right]. \tag{3}$$

what is shown in Fig. 1b. In the stationary case the particles are supposed to move with a velocity which is similar to the minima of the potential ($\pm V$). Slow particles follow the gradient of this "potential" from this viewpoint. The

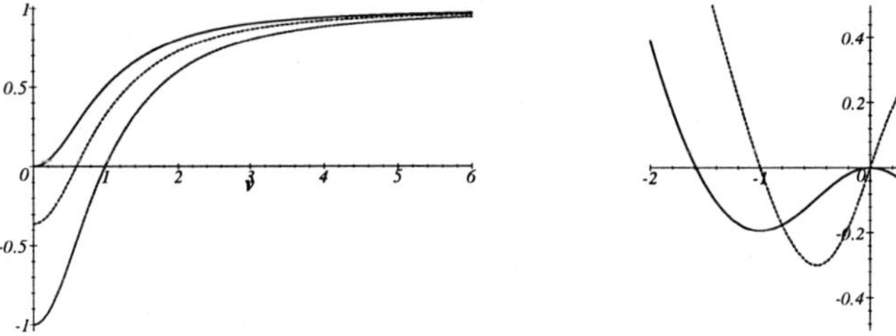

Fig. 1. (a) For different strengths V of the energy supply, $\gamma(v)$ as in (2) shows different qualitative behavior. $V = 0$ (solid line): The system leads to the equilibrium case $v = 0$ for every particle. For non-vanishing positive V (dashed lines) friction becomes negative $\Longrightarrow$ transition to a non-equilibrium case. (b) $R(v)$ (dashed) and the corresponding friction potential (solid). d_*, V and γ_0 set to 1.0 here.

interaction between the masses is modeled by Toda springs first:

$$U_T(r) = \frac{a}{b}\, \exp^{-b\,(r-\sigma)} + a\,(r - \sigma) - \frac{a}{b}, \tag{4}$$

with $r = x_i - x_{i-1}$. Closely related to the Toda potential is the exponential potential.

$$U_E(r) = \frac{a}{b} \exp^{-br} . \tag{5}$$

Both potentials ((4), (5)) lead to the dynamics for a system of many particles with periodic boundary conditions (Fig. 4a) The main parameter is b which describes the strength of the non-linearity of the interacting force.

3 Simulation of the Stochastic Dynamics

In the deterministic case the equations of motion and periodic boundary conditions for a ring (length L) of N particles are given by

$$\begin{aligned} \dot{x}_i &= v_i \\ \dot{v}_i &= -\gamma(v_i)\, v_i + A\,\{\exp[-b(x_i - x_{i-1})] - \exp[-b(x_{i+1} - x_i)]\} \\ i &= 1, 2, \ldots, N \qquad x_{N+1} = x_1 + L \qquad x_0 = x_N - L \end{aligned} \tag{6}$$

The amplitude A is $A = \exp(b)/b$. The investigated system is equivalent to a ring of N Toda springs with equilibrium length $\sigma = 1.0$ and frequency $\omega = 1.0$. Initially, we place the particles equally distributed in distances $\rho = L/N$.

There exist $N + 1$ stable attractors for a system of N particles on a ring (Fig. 2). To every attractor belongs a specific mean of certain physical variables (e.g. mean energy, mean momentum per particle). The stationary state, which is attracted depends on the initial conditions. The attractors vary according to there size.

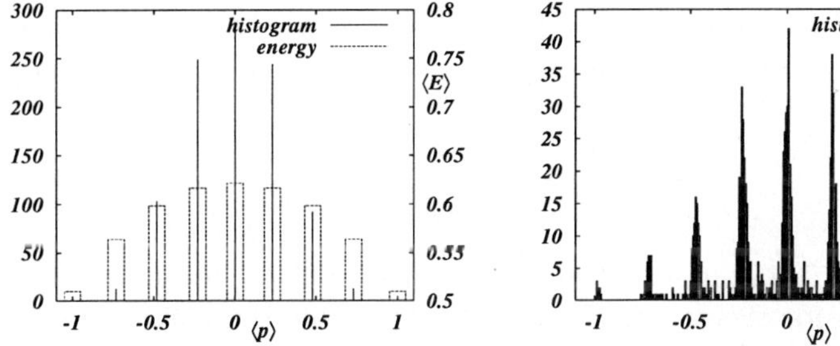

Fig. 2. (a) Attractors for $N = 8$, $b = 10.0$ and $\rho = 1.0$: frequency (thin lines) of appearance per 1000 runs, characteristic temporal mean of impulse (horizontal) and energy (vertical); (b) With non-vanishing stochastic forces the discrete attractor deviate an smear out with increasing noise strength (adapted from [10]).

The outermost lines in Fig. 2 correspond to the rotation of the ring ($\langle v \rangle_t = \pm V$) which means that all particles move equally distributed into the same direction (Fig. 3a). The second lines in Fig. 2 represent a soliton-like attractor

(Fig. 3b). There is always one particle moving in opposite direction to the others. In between two impacts the velocity of the particles increase up to V. We observe clockwise and counterclockwise cnoidal waves. The most frequently visited attractor is the one with $\langle v \rangle_t = 0$ (exists only for even N). Neighboring particles oscillate opposingly ($v_i = -v_{i-1} = v$ and $-V < v < V$) around the equilibrium distance (Fig. 3c).

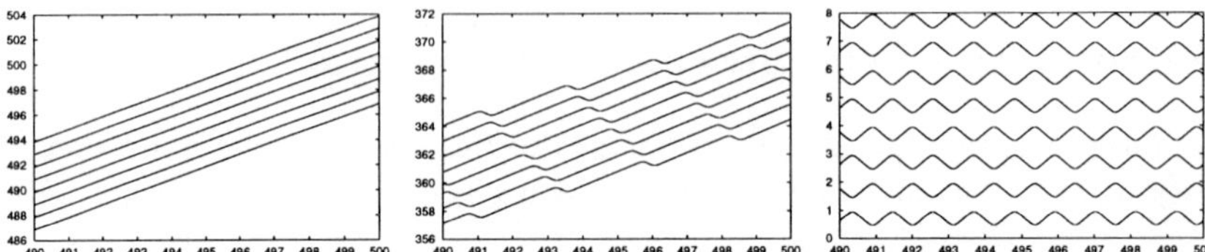

Fig. 3. Time evolution of the coordinates $x(t)$ of the particles for (a) the rotating case, (b) the one soliton case and (c) the oscillating case $\langle v \rangle_t = 0$; same parameters as in Fig. 2. Every line corresponds to the evolution of on particle between $t = 490 \ldots 500$.

4 Cluster Formation

In the previous sections, the kind of interaction between the particles resulted in a mono-stable potential for every particle. Either the particles move constantly in one direction with a equilibrium distance one to another or they do not move and oscillate around the equilibrium coordinate. Now we introduce an interaction (7) which shows, apart from the finite range, similar behavior as the Lennard-Jones potential:

$$U_{LJ}(r) = C \, \ln\left(\frac{\sigma}{r}\right) \, \exp^{-br} \tag{7}$$

Compared to the Toda potential the Lennard-Jones like interaction aims to an bistable potential for every particle (Fig. 4). Therefore the particles are able to form stable clusters additionally. Now the same attractor structure as discussed above can be observed in the cluster (Fig. 5). Reminding that the case $\langle v \rangle = 0$ is the most probable it becomes obvious that the system forms stationary clusters in most of the cases. A moving cluster can be observed if one or more particles within the same cluster are in a state corresponding to the outermost attractor of Fig. 2a. In this case the coupling of the particles can push the cluster in the direction of the moving particle (Fig. 5 between $t = 250 \ldots 300$). Now we add a relatively small external force. Equation (1) becomes:

$$\ddot{x}_i = W(x_{i+1}, x_i, x_{i-1}) + R(v_i) + K + Z(t). \tag{8}$$

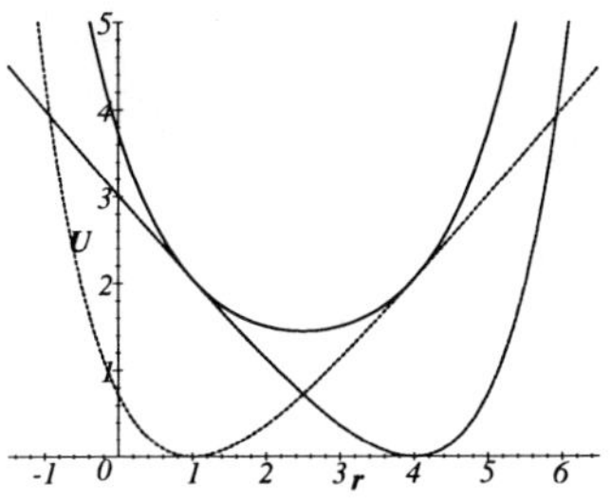

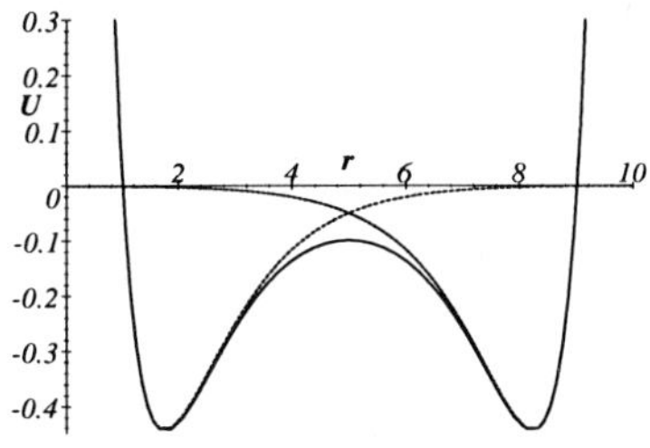

Fig. 4. (a) The resulting potential for particle laying between two other particles connected to them via Toda-springs. The potential has a cosh-shape; (b) In the opposite to (a) the resulting Lennard-Jones type potential shows bistable behavior.

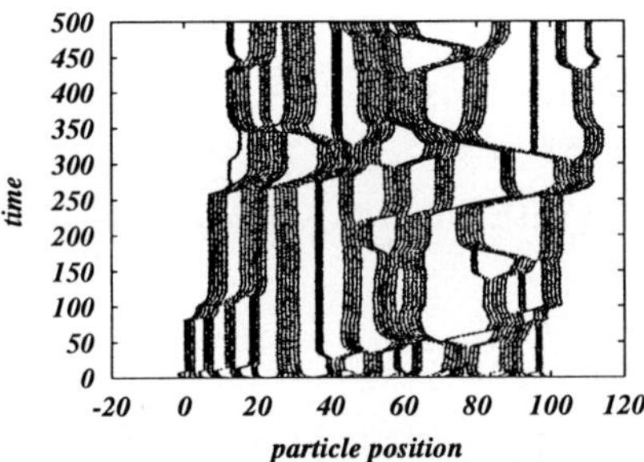

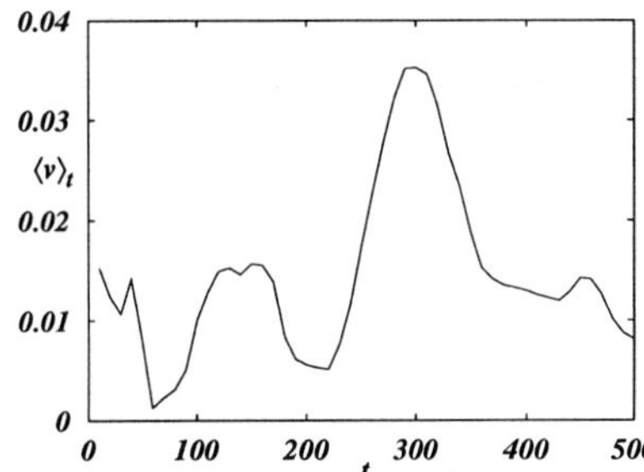

Fig. 5. $N = 50$ particles with Lennard-Jones like interaction without noise (a) Evolution of the position of the particles; (b) Temporal mean of velocity. Parameters: $L = N\, r_0$ with $r_0 = 2.0$ (equilibrium distance), $b = 10.0$; all other parameters set equal to 1.0.

This external field leads to a symmetry breaking. Apart form the chosen attractor basin the particles will "feel" the gradient of the field permanently. The diagrams in Fig. 6 remain us to a behavior as it can be observed in single lane traffic situations with out obstacles. Depending on the noise level one can find increasing probability of clusters (jams) with increasing noise.

5 Conclusions

In this paper we investigated a system of interacting active Brownian particles. The study of these type of systems is of central importance to the further development of the theory of complex processes. Of special interest are clustering phenomena and phase transitions in non-equilibrium. We have shown that several dynamical properties of our model remind dynamic models of vehicular

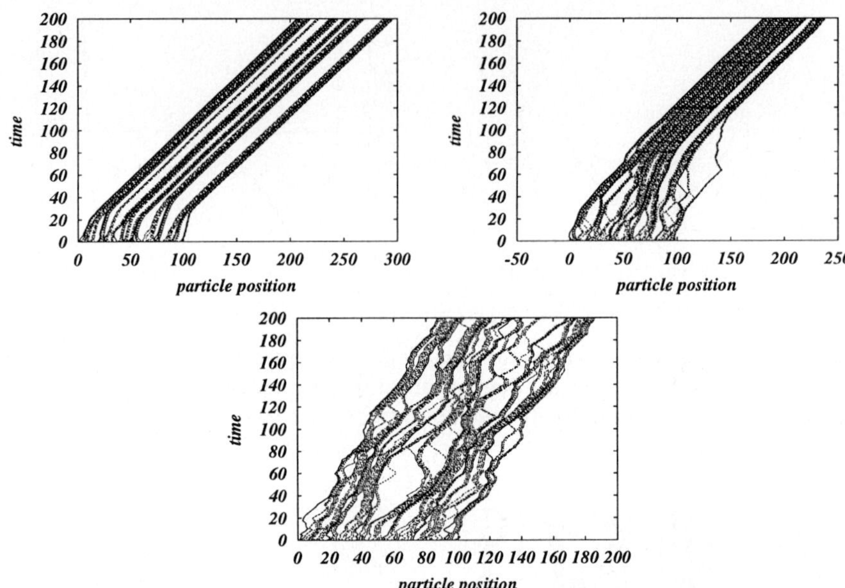

Fig. 6. Evolution of the position of the particles with noise: (a) $D = 0.0$, (b) $D = 0.1$, (c) $D = 0.2$. Other parameters as in Fig. 5.

traffic. Further investigations might help to understand the dynamics of complex transport processes even better especially from the viewpoint of interacting Brownian particles.

References

1. M. Schienbein and H. Gruler, Bull. Math. Biology **55**, 585 (1993).
2. F. Schweitzer, W. Ebeling, and B. Tilch, Phys. Rev. Lett. **80**, 5044 (1998).
3. V. Makarov, W. Ebeling, and M. Velarde, in press, Int. J. Bifurc. & Chaos, (1999).
4. U. Erdmann, W. Ebeling, F. Schweitzer, and L. Schimansky-Geier, submitted, Eur. Phys. J. B, (1999).
5. L. Schimansky-Geier, M. Mieth, H. Rosé, and H. Malchow, Physica A **207**, 140 (1995).
6. U. Erdmann, Interj. of Complex Systems **114**, (1999).
7. W. Ebeling and M. Jenssen, SPIE **3726**, 112 (1999).
8. W. Ebeling and M. Jenssen, in press, Physica D, (1999).
9. W. Ebeling, F. Schweitzer, and B. Tilch, BioSystems **49**, 17 (1999).
10. W. Ebeling, U. Erdmann, J. Dunkel, and M. Jenssen, in press, J. Stat. Phys., (1999).

Freezing by Heating in a Pedestrian Model

D. Helbing[1,2], I.J. Farkas[1], and T. Vicsek[1,2]

[1] Department of Biological Physics, Eötvös University, Pázmány Péter Sétány 1A, 1117 Budapest, Hungary
[2] Collegium Budapest – Institute for Advanced Study, 1014 Budapest, Hungary

Abstract. We investigate a simple model corresponding to pedestrians walking in opposite directions and interacting via a repulsive potential. The pedestrians move off-lattice in a periodic corridor and are subject to random forces as well. We show that this model – which can be considered as a continuum version of some driven diffusive systems – exhibits a paradoxical, new kind of transition called here "freezing by heating". The most interesting feature of the transition is that a *crystallized state* with a higher total energy is obtained from a fluid state *by increasing the amount of fluctuations.*

Most of the phenomena in our natural environment occur under far-from equilibrium conditions resulting in a rich behavior both in time and space. An important class of such processes takes place in so-called driven systems, which have attracted considerable interest recently. In many of these systems, particles are driven either by an external field (force) [1–4], or they are self-propelled [5–9], and their collective behavior manifests itself in new kinds of interesting transitions, including noise induced ordering [10,2–4] or an ordering in a continuous 2d velocity space [5].

Phase transitions are also common in equilibrium systems, and the related analogies have represented an important contribution to the understanding of non-equilibrium processes. In most cases, lattice models have been considered to demonstrate non-equilibrium transitions. For example, jamming transitions have been seen in discretized traffic models [11–14] and driven lattice gases [15–18]. However, off-lattice (i.e., continuum) symmetry is known to bring in qualitatively new behavior; in particular, this is definitely so in 2d, see, e.g., the XY-model [19] versus the Ising model (in equilibrium). A continuum model may lead to new effects due to the fact that the notions of order and disorder have extra facets in this case, and it can describe compressible systems in a more delicate way.

In this paper we will consider a simple continuum model exhibiting a paradoxical, new kind of transition that we call "freezing by heating", which is closely related to situations relevant from the practical point of view. The model consists of particles driven in opposite directions and interacting through a simple repulsive potential. The particles move off-lattice in a periodic channel and are subject to random forces as well. The most interesting feature of the transition we find for this system is, that a *crystallized state* with a higher total energy is achieved from a fluid state over a transient disordered state *by increasing the amount of fluctuations.* Pedestrians moving in a passage, a system of light (rising) and heavy (sinking) particles in a vertical column of fluid, or a system of oppositely charged colloidal particles in an electric field represent potential applications of our model.

We denote the location of particle i at time t by $\boldsymbol{x}_i(t)$ and its velocity $d\boldsymbol{x}_i(t)/dt$ by $\boldsymbol{v}_i(t)$. Furthermore, we assume the acceleration equation

$$m\frac{d\boldsymbol{v}_i(t)}{dt} = m\frac{v_0\boldsymbol{e}_i - \boldsymbol{v}_i(t) + \boldsymbol{\xi}_i(t)}{\tau} + \sum_{j(\neq i)} \boldsymbol{f}_{ij}\Big(\boldsymbol{x}_i(t), \boldsymbol{x}_j(t)\Big) + \boldsymbol{f}_{\mathrm{b}}\Big(\boldsymbol{x}_i(t)\Big) . \quad (1)$$

m is the mass of the particle, v_0 the velocity with which it tends to move in the absence of interactions, τ the corresponding relaxation time, and $\boldsymbol{e}_i \in \{(1,0),(-1,0)\}$ the direction into which particle i is driven. $\gamma = m/\tau$ may be interpreted as a friction coefficient. $\boldsymbol{f}_{ij}$ represents the repulsive interactions between particles i and j, $\boldsymbol{f}_{\mathrm{b}}$ the interactions with the boundaries, and $\boldsymbol{\xi}_i$ the fluctuations of the individual velocities. For the interactions between the particles, we have chosen the simple function

$$\boldsymbol{f}_{ij}(\boldsymbol{x}_i, \boldsymbol{x}_j) = \frac{m}{\tau}\boldsymbol{g}_{ij}(\boldsymbol{x}_i, \boldsymbol{x}_j) \ \text{ with } \ \boldsymbol{g}_{ij}(\boldsymbol{x}_i, \boldsymbol{x}_j) = -\boldsymbol{\nabla} A(d_{ij} - D)^{-B} , \quad (2)$$

depending on the parameters A and B, and the distance $d_{ij}(t) = \|\boldsymbol{x}_i(t) - \boldsymbol{x}_j(t)\| > D$ only. Thus, $\boldsymbol{f}_{ij}$ describes the (usually decelerating) effect of a soft repulsive potential of particle j with a hard core of diameter D, reflecting the space occupied by the particle. Our choice of these details of the model corresponds to a motion of finite sized particles tending to avoid collisions and maintaining, if possible, a given velocity v_0. We have also studied more complex potentials, e.g., velocity-dependent ones (see later). In addition, the interactions with the boundaries were assumed to be

$$\boldsymbol{f}_{\mathrm{b}}(\boldsymbol{x}_i) = \frac{m}{\tau}\boldsymbol{g}_{\mathrm{b}}(\boldsymbol{x}_i) \ \text{ with } \ \boldsymbol{g}_{\mathrm{b}}(\boldsymbol{x}_i) = -\boldsymbol{\nabla} A(d_{i\perp} - D/2)^{-B} , \quad (3)$$

where $d_{i\perp}$ denotes the shortest distance to the closest wall.

We would like to point out that, depending on the value of τ, (1) may describe different situations. We call the case of large τ *inertial*, while in the quasi-adiabatic limit of small τ, (1) describes *over-damped* motion and simply becomes

$$\boldsymbol{v}_i(t) = v_0\boldsymbol{e}_i + \sum_{j(\neq i)} \boldsymbol{g}_{ij}(\boldsymbol{x}_i, \boldsymbol{x}_j) + \boldsymbol{g}_{\mathrm{b}}(\boldsymbol{x}_i) + \boldsymbol{\xi}_i(t) . \quad (4)$$

In contrast to previous studies of similar models [8], we will now investigate the decisive role of the fluctuations $\boldsymbol{\xi}_i(t)$, which have been assumed to be uncorrelated and distributed according to a bivariate Gaussian with vanishing mean value and finite variance θ. However, fluctuations have been limited to a maximum of $\|\boldsymbol{\xi}_i\| = 3\sqrt{\theta}$.

We started our simulations with N (typically between 24 and 72) particles which were randomly distributed in the corridor without allowing overlaps. For half of the particles a driving into the $(-1,0)$ direction, and for the other half a driving into the (1,0) direction was assigned. Integrating equation (1), using periodic boundary conditions, has then produced the following results: For small noise amplitudes θ and moderate densities, our simulations lead to the formation of coherently moving linear structures (just as if the particles moved along traffic lanes) (Fig. 1a). For relatively large N, jamming occurs, depending on the respective initial condition. Nevertheless, lane formation, if at all, happens relatively quickly and is very dominant and robust in our model (in contrast with simpler lattice gas versions we have tested).

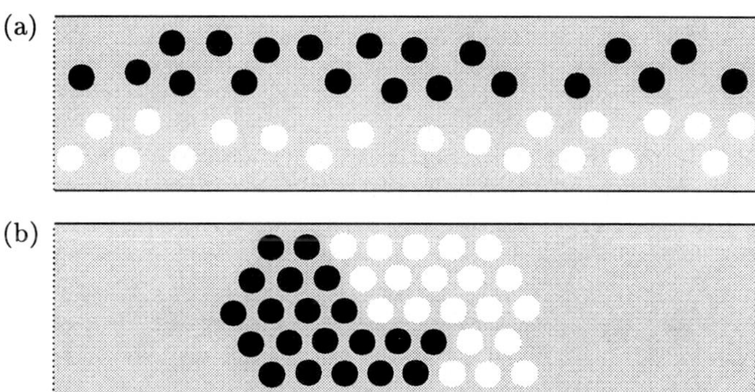

Fig. 1. Simulation of 20 particles moving over-damped from left to right (black) which interact with 20 particles moving over-damped from right to left (white) in a periodic channel of length 80 and width 20 at different noise intensities. The model parameters are $D = 3$, $v_0 = 5$, $A = 2$, $B = 5$, and $\lambda = 0.2$ (which make sense for pedestrians, if unit time is 1 s and unit length is chosen 0.25 m). (a) Lanes of uniform directions of motion forming for small noise intensity ($\theta = 10$). (b) Blocked, crystallized state resulting for large noise intensity ($\theta = 1000$).

The mechanism of lane formation can be understood as follows: Particles moving against the stream or in areas of mixed directions of motion will have frequent and strong interactions. In each interaction, the encountering particles move a little aside in order to pass each other. This side-wards movement tends to separate oppositely moving particles. Moreover, once the particles move in uniform lanes, they will have very rare and weak interactions. Hence, the tendency to break up existing lanes is negligible, when the fluctuations are small. Furthermore, the most stable configuration corresponds to a state with a minimal interaction rate [20].

Whereas lane formation was also observed in previous studies of related models *with deterministic dynamics only* [8], in the present, more realistic model we have discovered a surprising phenomenon when we increased the noise amplitude. If the fluctuations were large enough, they were able to prevent lane formation or even to destroy previously existing lanes, eventually causing a mutual blocking of the opposite directions of motion for given, sufficiently high densities (cf. Fig. 1b). In this blocked state, the particles were arranged in a hexagonal lattice structure, very much like in a crystal. In particular, there are densities, at which we find lanes at small noise amplitudes, but a crystallized state, if the noise amplitude is large (cf. Fig. 2).

Thus, one can drive the system from a "fluid" state with lanes of uniform directions of motion to a "frozen" state *just by increasing the fluctuations.* We call this hysteretic transition "freezing by heating". Although this transition exists for the simplest version of the model (defined by equations (1) through (3)), it is more pronounced if the gradients of the repulsive potentials are multiplied with factors

$$\{\lambda + (1 - \lambda)[1 + \cos(\varphi_i)]/2\}\,, \tag{5}$$

where φ_i denotes the angle between the direction of motion $\boldsymbol{e}_i$ of particle i and the direction of the object exerting the repulsive force. The parameter λ reflects the relative influence of forces "from behind", in order to mimic that the interactions may depend on the (relative) velocities of the particles, a situation quite common in realistic driven systems.

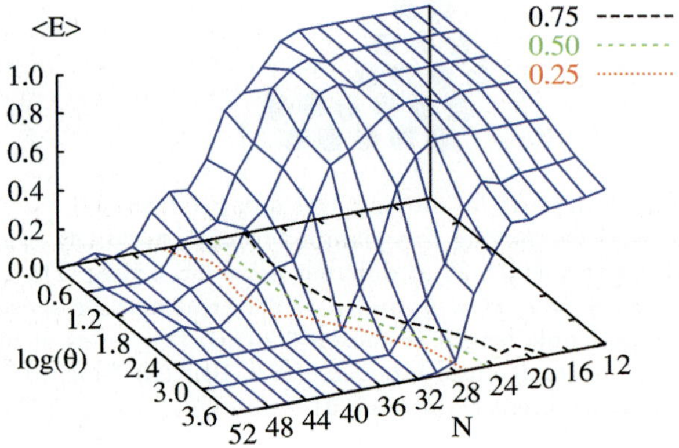

Fig. 2. The efficiency $\langle E \rangle$ of the system as a function of the particle number N and the noise intensity θ on a logarithmic scale, simulated for the over-damped case and the parameter values displayed in Figure 1. Shown are averages over 25 simulation runs with different random seeds. Broken lines are equipotential lines belonging to the labeled levels.

To characterize the state of the system, we calculated two different quantities. The expression

$$E = \lim_{T\to\infty} \frac{1}{T} \int_0^T dt\, \frac{1}{N} \sum_{i=1}^{N} \frac{\boldsymbol{v}_i(t) \cdot \boldsymbol{e}_i}{v_0}\,, \tag{6}$$

for which we expect the relation $0 \le E \le 1$, is a measure for the "efficiency" of motion, i.e., Ev_0 is the average speed at which the particles are able to move in their respective "target direction" $\boldsymbol{e}_i$. A representative simulation result as a function of the particle number N and the noise intensity θ is displayed in Figure 2.

Another important quantity is the sum of the potential and kinetic energies associated with a given state:

$$W = \lim_{T\to\infty} \frac{1}{T} \int_0^T dt \left[\sum_i \frac{m}{2} v_i^2 + \frac{1}{2} \sum_{i\neq j} \frac{mA}{\tau} (d_{ij} - D)^{-B} \right] . \quad (7)$$

The paradox here is that the above mentioned crystallized state is usually more unstable than the fluid state (in the sense that the total energy (7) of the system is higher in the crystallized state than in the fluid one). This becomes most obvious in the over-damped case of small τ, where the kinetic energy is negligible compared to the potential energy. The state with the highest total energy is then the most compressed state of the repulsively interacting particles, which corresponds to the crystallized state. This energetically less favorable state is maintained by the propulsion term $m(v_0 \boldsymbol{e}_i - \boldsymbol{v}_i)/\tau$ in (1). Note that the absolute value of this term becomes largest for $\boldsymbol{v}_i = \boldsymbol{0}$ (i.e., blocking), while it is small for the fluid state with $\boldsymbol{v}_i \approx v_0 \boldsymbol{e}_i$. The crystallized state can only be destroyed by fluctuations with noise amplitudes $\sqrt{\theta}$ which are significantly larger than v_0, giving rise to a disordered (gaseous) state.

We consider the transition to a stationary state with a higher total energy by increasing the noise intensity to be a signature of a new class of behavior in certain non-equilibrium systems, which may have interesting applications. Why is freezing by heating new? Some glasses may crystallize when slowly heated. However, this is a well understood phenomenon: the amorphous state is metastable for those temperatures, and crystallization means an approach to the more stable state with smaller total energy. In general, one can distinguish three cases when fluctuations (i.e., temperature or external perturbations) are increased: (i) Total energy increases and order is destroyed (e.g., melting). (ii) Total energy decreases, ordering takes place, and the system goes from a *disordered metastable to an ordered stable state* (e.g., in metallic glasses and some granular systems). (iii) Total energy increases and ordering takes place, while the system goes from a *partially ordered stable to a highly ordered metastable state*, which corresponds to the new situation presented here.

In our case, crystallization is achieved by spontaneously driving the system with the help of noise uphill towards higher total energy. The system would like to maximize its efficiency [20], but instead it ends up with minimal efficiency due to noise-induced crystallization. The role of "temperature" or noise here is to destroy the energetically more favorable fluid state, which inevitably leads to jamming. Crystallization occurs in this jammed phase, because it yields the densest packing of particles at the equilibrium distance of the particles. The corresponding transition seems to be related to the off-lattice nature of our model and is different from those reported for driven diffusive systems on a lattice [15–18,21]. It should be noted that the transition we find is not sharp, which is a consequence of the finite system sizes considered and of the disorder in the initial state.

We would like to point out that "freezing by heating" is likely to be relevant to situations involving pedestrians under extreme conditions (panics). Imagine a very smoky situation, caused by a fire, in which people do not know which is the right way to escape. When panicking, people will just try to get ahead, with a reduced tendency to follow a certain direction. Thus, fluctuations will be very large, which can lead to fatal blockings.

Our results demonstrate that, in self-driven systems, phenomena qualitatively different from those occurring in thermodynamical systems can be observed. While most non-equilibrium transitions have analogies to equilibrium ones, the noise-induced ordering observed in the effect of "freezing by heating" is just opposite to the transitions occurring in equilibrium systems. This suggests that future studies along the lines of the present approach are likely to lead to further unexpected findings.

Acknowledgement. D.H. wants to thank the DFG for financial support by a Heisenberg scholarship and D. Mukamel for helpful suggestions. This work was supported by OTKA F019299 and FKFP 0203/1997.

References

1. H.J. Herrmann, in: *3rd Granada Lectures in Computational Physics,* P.L. Garrido and J. Marro, (Eds.), (Springer, Heidelberg, 1995).
2. A.D. Rosato , K.J. Strandburg, F.Prinz, and R.H. Swendsen, *Why the Brazil Nuts are on Top: Size Segregation of Particulate Matter by Shaking*, Phys. Rev. Lett. **58,** 1038 (1987).
3. J.A.C. Gallas, H.J. Herrmann, and S. Sokolowski, *Convection Cells in Vibrating Granular Media,* Phys. Rev. Lett. **69,** 1371 (1992).
4. P.B. Umbanhowar, F. Melo, and H. L. Swinney, *Localized excitations in a vertically vibrated granular layer,* Nature **382**, 793 (1996).
5. T. Vicsek et al., Phys. Rev. Lett. **75**, 1226 (1995).
6. J. Toner and Y. Tu, Phys. Rev. Lett. **75,** 4326 (1995).
7. E. Ben-Jacob et al., Nature **368,** 46 (1994).
8. D. Helbing and P. Molnár, Phys. Rev. E **51,** 4282 (1995).
9. D. Helbing et al., Nature **388,** 47 (1997).
10. J.S. Walker and C.A. Vause, Sci. Am. **256**, 90 (1987).
11. K. Nagel and M. Schreckenberg, J. Phys. I France **2,** 2221 (1992).
12. J. Krug and P.A. Ferrari, J. Phys. A **29,** L465 (1996).
13. D. Helbing and B.A. Huberman, Nature **396,** 738 (1998).
14. D. Helbing and M. Schreckenberg, Phys. Rev. E **59,** R2505 (1999).
15. M.R. Evans et al., Phys. Rev. Lett. **74,** 208 (1995).
16. G. Szabó and A. Szolnoki, Phys. Rev. A **41,** R2235 (1990).
17. B. Schmittmann and R.K.P. Zia, in: *Phase Transitions and Critical Phenomena,* Vol. 17, C. Domb and J. Lebowitz, (Eds.), (Academic Press, New York, 1995).
18. B. Schmittmann and R.K.P. Zia, Phys. Rep. **301,** 45 (1998).
19. J.M. Kosterlitz and D.J. Thouless, J. Phys. C **6,** 1181 (1973).
20. D. Helbing and T. Vicsek, New J. Phys. **1,** (1999); (see `http://www.njp.org/`).
21. B. Schmittmann, K. Hwang, and R.K.P. Zia, Europhys. Lett. **19,** 19 (1992).

Traffic Data and Applications

Phase Transitions in Traffic Flow

B.S. Kerner

DaimlerChrysler AG, HPC: E224, 70546 Stuttgart, Germany

Abstract. A review of an experimental study of phase transitions in traffic flow is presented. A critical comparison of real features of phase transitions with recent numerical results is given. A qualitative theory of congested traffic flow recently developed is discussed.

1 Introduction

1.1 Traffic Flow Models

Traffic flow dynamics is a very old field. Up to recent time theoretical studies have mainly been based on applications and on further developments of such classical traffic models and theories as the Lighthill-Whitham-model [1], the Herman *et al.* and Gazis *et al.* car-following models [2,3], the Newell model [4], the Prigogine kinetic model [5], the Payne macroscopic model [6], the Wiedemann psycho-physical model [7], the Gipps model [8], different queuing models (e.g., [9,10]) and other theoretical approaches (see papers in [11,12]).

Last years a wide community of physicists has been involved in studying the traffic flow dynamics [13,14]. On the one hand, new traffic flow models and theoretical approaches, in particular the Nagel-Schreckenberg cellular automaton model [15] and a new approach to the optimal velocity (OV) models [4,16] made by Bando *et al.* [17] have been proposed and developed. On the other hand, last results of the non-linear physics of dissipative distributed systems, in particular a theory of autosolitons (localized dissipative patterns) [18], have been applied by Kerner and Konhäuser for the development of a non-linear theory of moving traffic jams. In the framework of this theory it has been predicted that [19]

(i) Wide moving jams possess some characteristic, i.e., unique, coherent, predictable and reproducible parameters.
(ii) Free traffic flow is in a metastable state with respect to jam emergence, if the flow rate in free flow is higher than the flow rate out from a wide moving jam, q_{out}.

These two results have indeed been found out by Kerner and Rehborn in their experimental studies of the jam's propagation [20]. Therefore the mentioned properties (i), (ii) of jams and free flows [19] can be considered as necessary requirements for each traffic flow model [21,22]. Later studies by Bando *et al.* [23], by Krauß *et al.* [24], by Barlovic *et al.* [25], by Helbing and Schreckenberg [26], by Treiber *et al.* [27], and by Mahnke and Kaupužs [28] showed that there can be a lot of traffic flow models and theoretical approaches which are able to reproduce the properties (i) and (ii) of moving jams and free flows which have first been found out from an investigation of a macroscopic traffic flow model [19]. Note that the existence of the characteristic parameters of moving jams has been used for a development of the method for tracing and prediction of moving traffic jams and time-dependent travel times [29,30]. This method uses essentially the fact that the velocities of the propagation of the moving jams' fronts can be calculated with the available measured traffic data. Besides, the method works without any validations of model parameters at different highway infrastructure and other road conditions. Some results of an application of the method are considered in [31].

1.2 Hypothesis about the Fundamental Diagram

All these and other later traffic flow models and theories (e.g., [1–17,19,22–28,32–48]) are based on *the hypothesis of the fundamental diagram*: The fundamental diagram is either a result of a traffic flow model or it is used in models explicitly. Recall that the fundamental diagram is a curve in the flow-density plane which gives a correspondence of the vehicle density to the flow rate in traffic flow. The form of the fundamental diagram should be related to the obvious results of observations: the more vehicles on the road, the lower the average speed with which they move. The product of the vehicle density and the average speed is the flow rate. Therefore, the fundamental diagram should start in the origin (when the density is zero, so is the flow rate), and should have at least one maximum (Fig. 1a) (e.g., [1,3,5–14,32–36]). It is often proposed that the fundamental diagram consists of two isolated curves which are separated by a gap (the problem of capacity drop): a curve with a positive slope for *free traffic flow* and a curve with a negative slope for *congested traffic flow* [49–54].

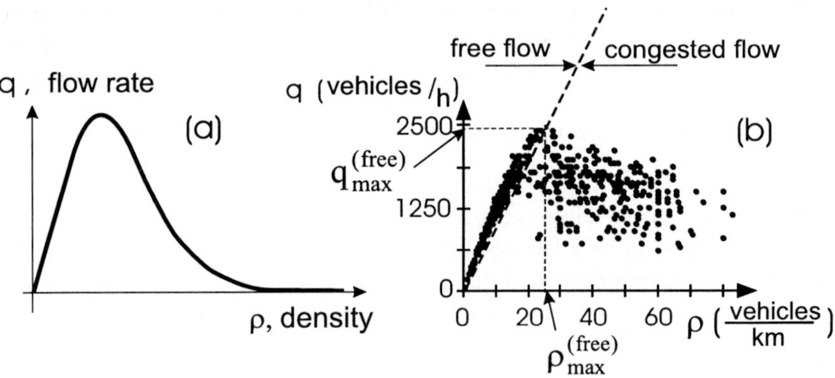

Fig. 1. Illustration of the fundamental diagram (a) and real patterns which have observed in different countries (e.g., [49,50]) (b).

1.3 Congested Flow

When the vehicle density is high enough, traffic is usually in a congested regime (e.g., [5,9–14,32–36]). In the paper, phenomena will be defined through their features rather than through the reasons of their occurrence. Often congested flow is defined as a traffic state which occurs only due to a highway bottleneck as the inevitable consequence of an upstream flow that exceeds the downstream capacity of the bottleneck. In this definition it is obvious to propose that congested flow is not self-organizing, but rigidly controlled by external constraints (e.g., [10]). However, observations have shown that although bottlenecks are the most frequent reason, they are not the only reason for the occurrence of congested traffic [55]. Besides, in [20,55–62] it has been shown that different self-organizing processes are responsible for features of *real* traffic congestion. For these reasons the mentioned definition of congested traffic (e.g., [10]) will not be used here.

Congested traffic states will be defined as counter states to states of free flow (e.g., [49]). It is well-known that experimental points which are related to free flow can be approximately presented by a curve with a positive slope on the flow-density plane. Congested traffic will be determined, therefore, as a state of traffic where the average vehicle speed is lower than the minimum possible speed which is related to the limit point $(q_{\max}^{\text{free}}, \rho_{\max}^{\text{free}})$ in free flow (Fig. 1b) (e.g., [49,50]). In this definition nothing is said about reasons of the occurrence of the congestion. It is linked to the mentioned experimental facts that a congested state can spontaneously occur also away from bottlenecks and that there are a lot of diverse self-organizing effects which are responsible for the congestion [20,56,57,62].

One of the well-known phenomena in congested traffic flow is the 'stop-and-go' phenomenon which has experimentally been studied by a lot of authors, in particular by Edie and Foote, Treiterer, Koshi *et al.* [63,49]. Koshi *et al.* [49] found out that the vehicle speed across different highway lanes within the 'stop-and-go' waves can be synchronized. Besides, for an explanation of the 'stop-and-go' waves Koshi *et al.* [49] introduced a-mirror-image-of-the-Greek-letter-λ-fundamental diagram which should describe both free and congested regimes of traffic flow. Different other explanations of the 'stop-and-go' phenomenon made in the frame of the hypothesis of the fundamental diagram have been proposed (e.g., [5,11,13,36]).

However, it is well-known that within congestion, the flow-density data do not form a neat relationship but show a broad and complex spreading of measurement points, i.e., the flow-density data cover a two-dimensional region on the flow-density plane (Fig. 1 (b)) (e.g., [49,50]). This complexity in traffic flow is usually interpreted either as fluctuations, or as an instability, or else as a traffic jam formation (e.g., [13,33,35,36,49,50,64]).

1.4 Synchronized Traffic Flow – A New Phase of Traffic Flow

It seems reasonable that away from bottlenecks as well within 'stop-and-go' waves [49] also in each other state of congested flow vehicles change to that lane for which the vehicle speed is higher. This increases the density on this lane and consequently decreases the vehicle speed, i.e., this process synchronized the vehicle speed with the speeds on the other lanes leading to synchronized traffic flow. The related tendency to equilibrate the speed across lanes in dense enough traffic flow is the well-known fact [49] which can show almost any multi-lane traffic flow model.

In the mid-1990s Kerner in collaboration with Rehborn [56] discovered that real synchronized traffic flow possesses qualitative different non-linear properties (in comparison both with free traffic flow and with wide moving traffic jams. As a result, synchronized traffic flow must be considered as a separate traffic phase. In other words, in congested traffic two different traffic phases must be distinguished: wide moving traffic jams and synchronized traffic flow. In particular, while states of free flow can really be represented by a curve on the flow-density plane, *even* the multitude of homogeneous states of synchronized flow (i.e., states which are homogeneous spatially and stationary in time; often such states are called 'steady speed' states) cover a two-dimensional region in the flow-density plane: A given vehicle speed (a steady speed) in a homogeneous and stationary state of synchronized flow may be related to an infinity multitude of vehicle densities, and a given density may be related to an infinity multitude of different steady speeds.

Thus, in contrast to traffic flow models and theories (e.g., [1,3–17,19,22–28,32–50,53,54]) there are *no* fundamental diagrams which may describe the whole multitude even of homogeneous (steady) states of traffic flow.

Because congested traffic consists of two traffic phases (synchronized flow and wide moving jams), there are three qualitatively different traffic phases [56]:

1. Free flow.
2. Wide moving jams.
3. Synchronized flow.

Note that a correlation analysis of single vehicle data which has recently been made by Neubert *et al.* [65] confirms macroscopic features of the traffic phases discovered by Kerner and Rehborn [56,57].

1.5 Phase Transitions

Due to the existence of three traffic phases, there are three different types of phase transitions in traffic [57]:

(1) 'Free flow ↔ Jam' which will be called as the 'F↔J'-transitions;
(2) 'Free flow ↔ Synchronized flow' which will be called as the 'F↔S'-transitions;
(3) 'Synchronized flow ↔ Jam(s)' which will be called as the 'S↔J'-transitions.

Observations show that all of these phase transitions are local first-order phase transitions [57]. This means that these transitions are accompanied by sometimes similar looking breakdown, hysteresis and nucleation effects. However, it must be distinguished between them because the non-linear features of these phase transitions are qualitatively different. This differentiation determines one of the main difficulties in the understanding of the observations and of the traffic flow theory. In particular, the well-known 'stop-and-go' phenomenon [49,63], as it has recently been found out by Kerner [55], can be explained based on the differentiation which has been made between the phase transitions (1), (2) and (3) in traffic flow.

Recently, Kerner found out that the existence of the limit point in free flow $(q^{\text{free}}_{\text{max}}, \rho^{\text{free}}_{\text{max}})$ (Fig. 1b) is linked to the F→S-transition rather than the F→J-transition [55,62]. As a consequent, it turns out that the cascade of two phase transitions 'Free flow → Synchronized flow → Jam(s)' is responsible for the jam emergence in free traffic flow [55]. This experimental result is in contrast to all numerical and analytical results of traffic flow models (e.g., [11,13–17,19,22–28,36–45,47,48]).

The experimental fact that there is no fundamental diagram which may describe the whole multitude even of hypothetical homogeneous states of synchronized traffic flow [56] could be one of the main reasons why mathematical models and theories (e.g., [1,5,7–17,19,22–28,32–48]) still cannot explain and predict the experimental features of the phase transitions in real traffic flow which have been found out in [21,55–57,59–62]. For this reason, results of observations of the phase transitions in real traffic flow, and a recent qualitative theory of congested traffic based on the hypotheses by Kerner [55,59–62] are covered in this review.

The article is organized as the following. First, definitions and some macroscopic features of traffic phases will be considered. Second, some peculiarities of phase transitions will be discussed. The main attention will be given to a consideration of the F→S-transition because this transition plays a very important role in traffic flow dynamics [62]. Third, features of the complex dynamics of moving jams will be considered. Finally, the mentioned hypotheses to the theory of congested traffic [55,59–62] will be discussed.

2 Features of Traffic Phases

Some features of these three traffic phases can be seen in Fig. 2 which shows the propagation of a sequence of two wide jams through the 13,2 km length section of the highway A5-North in Germany [20]. It can be seen that the sequence of two wide jams propagates with almost a constant velocity of the jam's downstream fronts both through free flow which exists in a part of the highway between the intersections I2 and I3 and through synchronized flow which exists between the intersections I1 and I2 and partially upstream of I2 (Fig. 2b-e). Based on the example shown in Fig. 2 some features of traffic phases will be illustrated in this section.

2.1 Free Traffic Flow

Free flow is the most investigated traffic phase (e.g., [33,35,49,50,67,68]). In particular, experimental points corresponding to free flow can be represented by a curve in the flow density plane (Fig. 3a-c). Because traffic in Germany

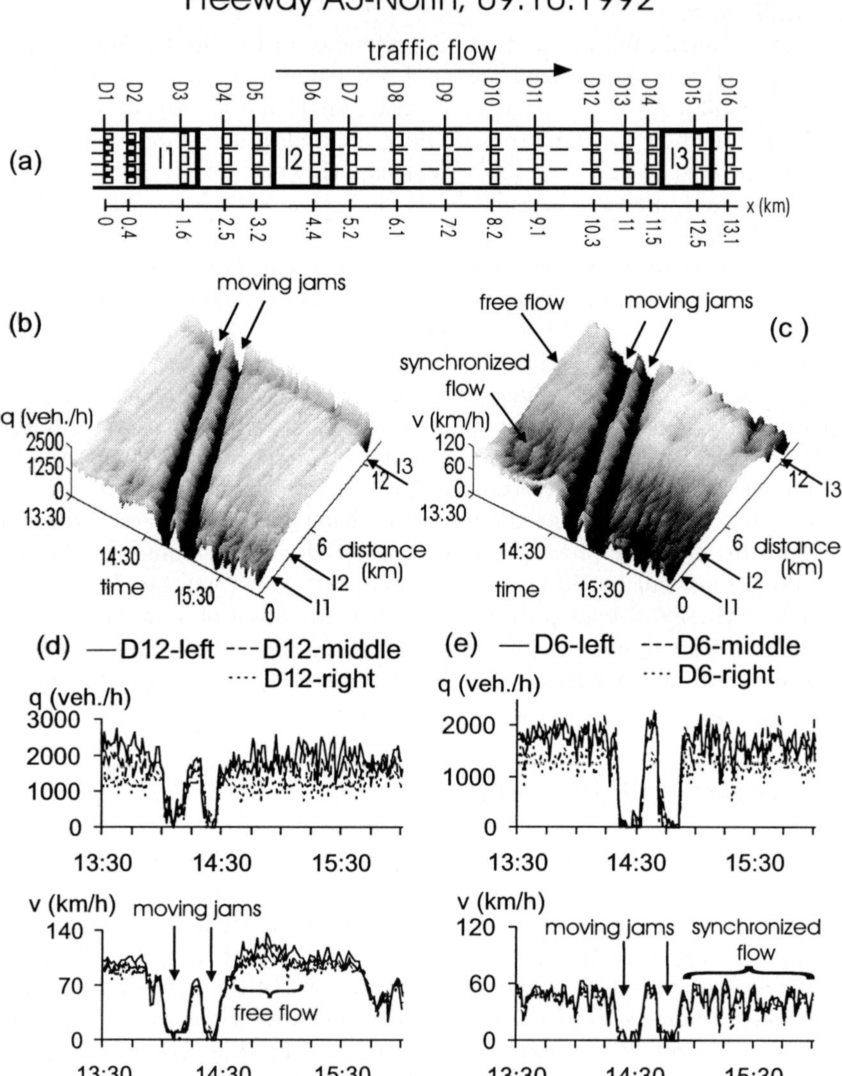

Fig. 2. Explanation of traffic phases: Free flow, moving traffic jams and synchronized flow (b-e) on a section of the highway A5-North in Germany (a) [20,66].

has a strong asymmetry between different lanes, the limit point in free flow $(q_{max}^{free}, \rho_{max}^{free})$ is different for the left (Fig. 3a), middle (Fig. 3b) and right lanes (Fig. 3c). The latter is linked to the different percentage of long vehicles on these lanes: on the left lane almost no long vehicles move, on the middle lane the percentage was about 10 %, on the left lane it was about 40 %. Note that Germany has a strong asymmetry between lanes, e.g., long vehicles must not move on the left lane of the highway section shown in Fig. 2a. Another well-known effect is also usual for free flow: With an increase in the flow rate the percentage of vehicles which move on the left lane becomes higher than the percentage of vehicles which move on the right lane beginning at some density of free flow (e.g., [68]) (Fig. 3d). These effects have been found out in cellular automata models (see references in [68,69]).

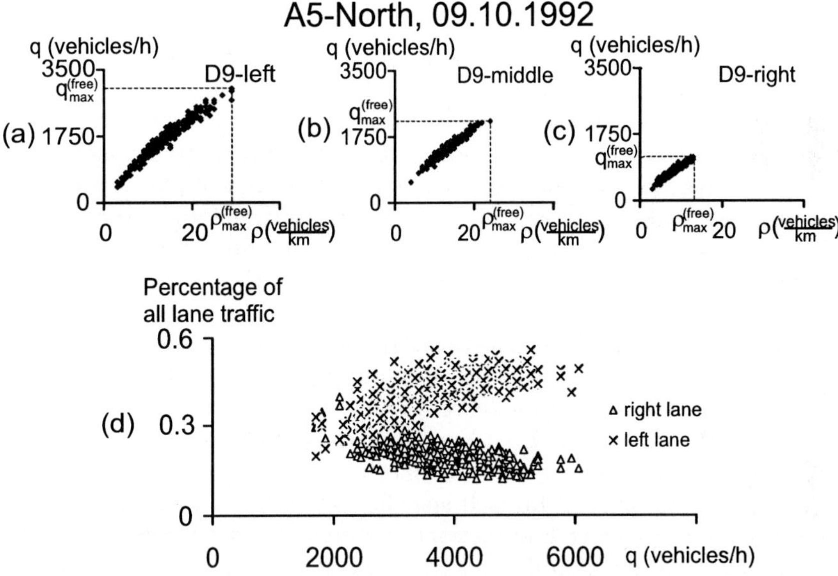

Fig. 3. Explanation of features of free traffic flow [33,35,49,50,67,68]: Free flow on the flow density plane for different lanes (a)-(c) and the percentage of vehicles which move on the left and right lane at the detectors D6 as a function of the total flow rate over the highway (d) during 7:00-12:00 on 09.10.1992 on the highway A5-North (see Fig. 2).

2.2 Synchronized Traffic Flow

In contrast to free flow, in synchronized flow the average vehicle speeds are nearly the same across different highway lanes. During the F→S-transition the initial

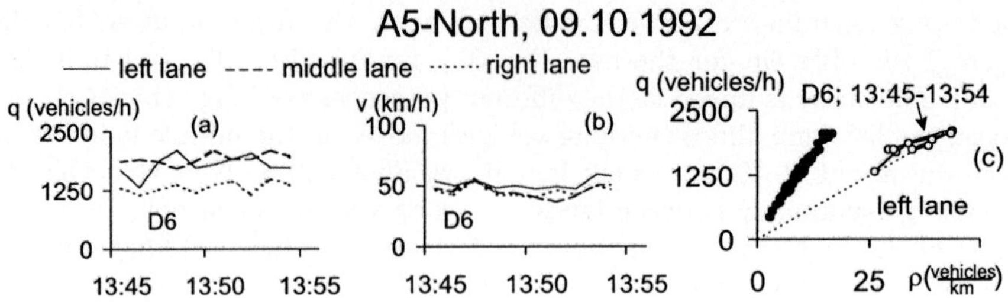

Fig. 4. Explanation of synchronized flow of the type (ii): The flow rate (a) and the vehicle speed (b); (c) Free flow (black points, during 7:00 – 8:00) and synchronized flow (circles, during 13:45 – 13:54) on the flow density plane for the left lane at the detectors D6 on 09.10.1992 on the highway A5-North (see Fig. 2)) [66].

different average speeds on different lanes decrease almost simultaneously up to nearly the same 'synchronized' speed (Figs. 4 and 5) [56,66].

Some other differences between free and synchronized flows are: While in free flow waves only with positive velocity may propagate, the dynamics of synchronized flow can be very complex: waves propagating both with positive and negative velocity and breakdown effects (e.g., the pinch effect [55]) are possible. As it has already been mentioned, in contrast to free flow, there are no curves (fundamental diagrams) on the flow-density plane which can describe the whole multitude even of homogeneous states of synchronized flow [56,21].

The whole multitude of hypothetical homogeneous and stationary states of traffic flow ('steady speed states') correspondingly to the related hypotheses by Kerner [21,55,60] (see Sect. 4 for more details) is shown in Fig. 6a for a multilane road and in Fig. 6 (b) for a one-lane road (hatched regions in Fig. 6). It must be noted that usually only a part of this multitude of the states of synchronized flow is realized in each specific case (see e.g., Fig. 3)). It can be supposed that homogeneous and stationary states of synchronized flow can not exists in a vicinity of the jam density $\rho_{\max}$, exactly if the vehicle speed is lower than some minimal possible vehicle speed $v^{\text{syn}}_{\text{min}}$ (Fig. 6) which may be about 5 – 10 km/h. Therefore, the related points on hatched regions in Fig. 6 should be excepted.

There are three types of states of synchronized flow [56,59]:

(i) Stationary and homogeneous states where both the average speed and the flow rate are stationary during a relatively long time interval.
(ii) States where only the average vehicle speed is stationary ('homogeneous-in-speed-states').
(iii) Non-stationary and non-homogeneous states.

Each of these three types of states of synchronized flow, either (i), or (ii) or else (iii), covers a two-dimensional region of the flow density plane. However, in each concrete situation usually only a part of this multitude is realized.

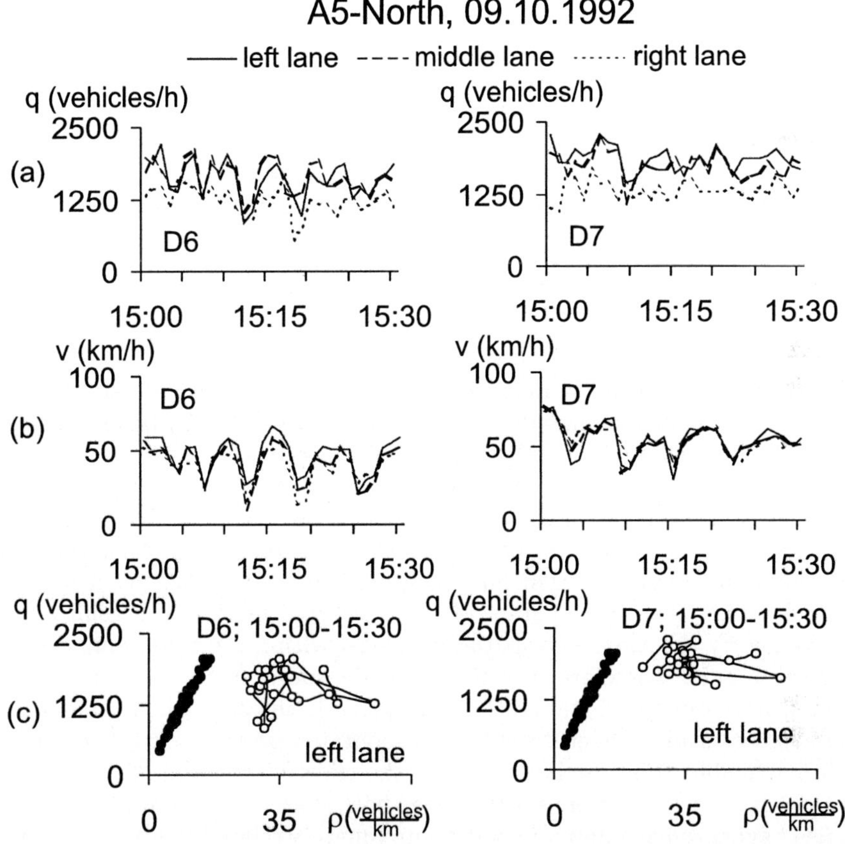

Fig. 5. Explanation of synchronized flow of the type (iii): The flow rate (a) and the vehicle speed (b); (c) Free flow (black points, during 7:00 – 8:00) and synchronized flow (circles, during 15:00 – 15:30) on the flow density plane for the left lane at the detectors D6 (left) and D7 (right) on 09.10.1992 on the highway A5-North (see Fig. 2)) [66].

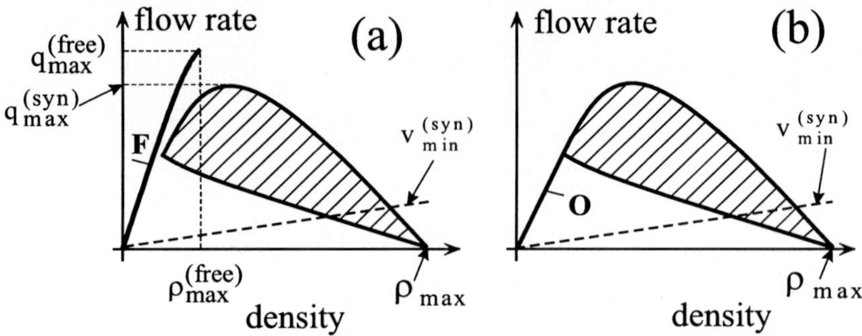

Fig. 6. A hypothesis about homogeneous states of traffic flow [59]: (a) Multitudes of homogeneous states of free (curve F) and of synchronized flow (hatched region) on a multi-lane road. (b) A multitude of homogeneous states of flow on a one-lane road (see explanations in Sect. 4).

In homogeneous-in-speed states (ii), the average speed of vehicles is nearly stationary during a relatively long time interval, but the flow rate and therefore the density, noticeably changes during this time interval. It may be assumed that waves of the flow rate (and of the density) propagate with positive velocity in such synchronized traffic flow. Often the values of the flow rate do not change synchronously in different lanes of a highway. It may be assumed additionally that different waves of the density propagate in different lanes. Note that if the vehicle speed in a synchronized flow measured at some location on a road is a stationary one, then in the state of the synchronized flow the vehicle speed should be nearly the same also spatially at least within some space upstream of the detection side. Therefore, the stationary in speed states (ii) can be called as 'homogeneous-in-speed states'. The exception is the case when the detection side is located inside a fixed front which separates synchronized flow and free flow. In particular, this case is realized at the effective location of a bottleneck, where vehicles accelerate from synchronized flow phase to free flow phase [70]. An example of synchronized flow which is approximately related to homogeneous-in-speed states is shown in Fig. 4. On the one hand, homogeneous-in-speed states of synchronized flow in the flow-density plane are related to a line whose slope equals the vehicle speed in the synchronized flow (dotted line in Fig. 4c). On the other hand, these states are related to an additional peak in the speed distribution function, i.e., they can be considered (as it has been proposed by Lubashevsky and Mahnke [71]) as a platoon of the vehicles in the speed-space.

In essentially non-stationary and non-homogeneous states (iii) both the average speed of vehicles and the flow rate abruptly change from one experimental point to the next one (Fig. 5). It may be assumed that waves with both negative and positive velocities may propagate in such synchronized traffic flow.

In states of synchronized flow (ii), changes in the flow rate and the density are correlated almost as good as in free flow (in the case shown in Fig. 4, the related coefficient of the correlation is $K \approx 0.9$). Also in the states (i) this correlation is usually relatively high. In contrast, in states (iii) there is almost no correlation between changes in the flow rate and the density: an increase in the flow rate can be accompanied both by an increase and by an decrease in the density [56] (in the case shown in Fig. 5c, left, $K \approx -0.11$). Note that this coefficient of the correlation between the flow rate and the density has recently been proposed by Neubert *et al.* [65] as a possible criterion for the identification of states of synchronized flow in comparison with free flow and jams. However, often a spatial-temporal sequence of all of these types of states of synchronized flow (i), (ii) and (iii) occurs. Therefore, it seems to be difficult to make the reliable identification of synchronized flow based on the cross-correlation coefficient.

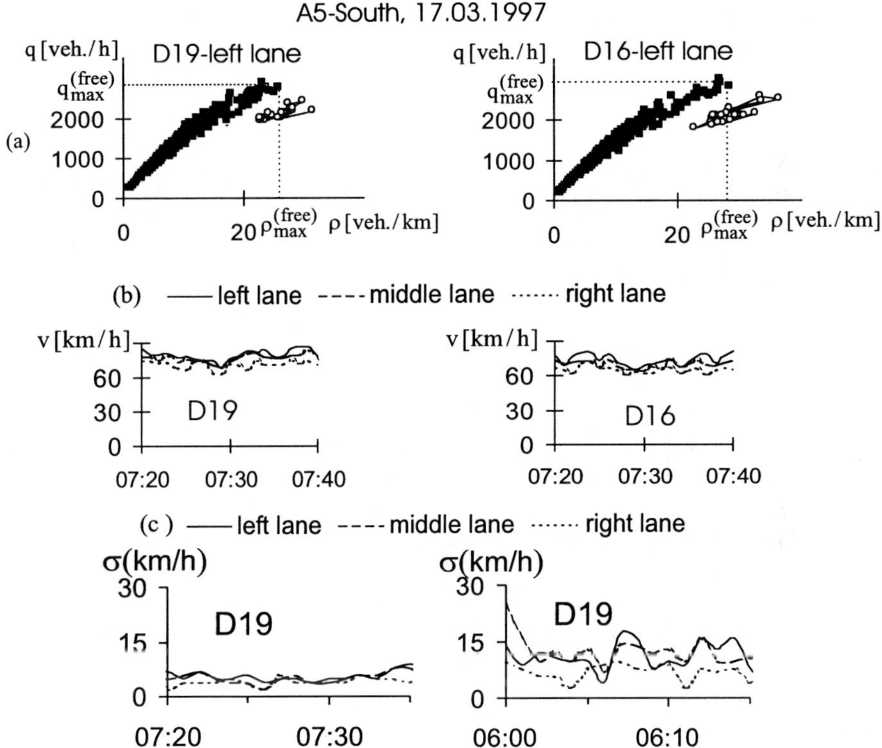

Fig. 7. Fluctuations in free and synchronized flow: An example of homogeneous-in-speed-states of synchronized flow (between D19 and D16 on A5-South, see Fig. 8)a,b [61]; (c) - the speed variations at D19 for synchronized flow (left) and free flow (right) [66].

Investigations of experimental data made by Kerner and Rehborn [57] showed that the speed variance which is linked to fluctuations in synchronized flow is

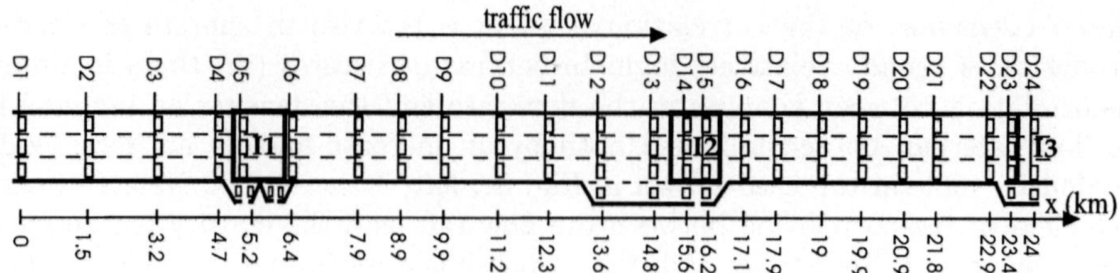

Fig. 8. Infrastructure of the A5-South.

considerably lower than in free flow (Fig. 7). Besides, in contrast to free flow, the speed variance on different lanes in synchronized flow is nearly the same. Both results may be linked to a bunching of vehicles in synchronized flow. Indeed, the bunching of vehicle, i.e., a building of vehicle platoons on the road restricts the speed variance between neighboring vehicles [57]. On the other hand, such a bunching of vehicles should lead to a long-space correlation of the vehicle speeds in synchronized flow. The latter has indeed been found out in single vehicle data by Neubert *et al.* (Fig. 9) [65]. Therefore this correlation can be used as one of the criteria for the identification of synchronized flow.

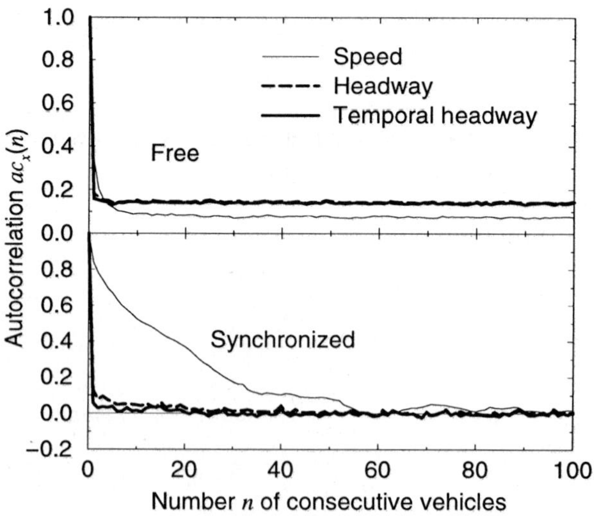

Fig. 9. A long-range signal obtained from the autocorrelation of spatial and temporal headways in free flow (top) and in synchronized flow (below) [65].

It is obviously that the average vehicle speed can be synchronized across different lanes in the congested regime usually only away from on-, off-ramps and other bottlenecks, and if there is the only one route for vehicles on a highway. However, *the tendency* to the synchronization of the average vehicle speeds across different lanes in the congested regime has an importance for the features of synchronized flow mentioned above. In general, each traffic flow in the congested regime which show this tendency will be called synchronized traffic flows even if the synchronization of the vehicle speeds across different lanes is locally destroyed, for example due to highway bottlenecks.

2.3 Wide Moving Traffic Jams

A wide moving jam is an upstream moving localized structure which is restricted by two fronts where the vehicle speed changes sharply (Fig. 2c, d) [20]. The vehicle speed and the flow rate inside a wide jam are either zero or negligible for the jam propagation: Exactly, there is no influence of the inflow into the jam on the jam's outflow. As in synchronized flow, the vehicle speed is usually synchronized across different highway lanes within the jam's fronts. On the flow-density plane measured points related to fronts of wide moving jams usually cannot be separated from points related to synchronized flow. As it has been found out in [21,59], in contrast to synchronized flow, after a wide jam has emerged, it *propagates* through either free flow or any states of synchronized flow and through any bottlenecks (e.g., at on- and off-ramps) *keeping* the velocity of the jam's downstream front. This behavior of wide moving jams can clearly be seen in Fig. 2c, where the sequence of two wide jams propagates through at least three bottlenecks (in the intersections I1, I2 and I3, Fig. 2a) and through different sometimes very complex states of synchronized flow (Fig. 2e, bottom). In contrast to the jam, after synchronized flow has occurred at an on-ramp, the downstream front of the synchronized flow is *fixed* at the on-ramp [57].

It has already been mentioned that wide moving jams possess characteristic, i.e., unique, coherent, predictable and reproducible parameters which do not depend on time and they are the same for different wide moving jams, if control parameters of traffic (weather, other road conditions) do not change [20]. These parameters can be represented on the flow density plane by the characteristic line for the downstream front called 'the line J' [20,56] (Fig. 10). The slope of the line J equals the velocity of the downstream front of the jam. Kerner and Rehborn found that the flow rate at the limit point for free flow $q_{\rm max}^{\rm free}$ is considerably higher than the flow rate out from a wide jam $q_{\rm out}$: $q_{\rm max}^{\rm free}/q_{\rm out} \approx 1.5$. This means that free flow in the range of the flow rate $[q_{\rm out}, q_{\rm max}^{\rm free}]$ (and in the related range of the density $[\rho_{\rm min}, \rho_{\rm max}^{\rm free}])$ (Fig. 10) is in a metastable state with respect to the jam emergence [20]. This macroscopic experimental proof of the theoretical result by Kerner and Konhäuser [19] has been recently found out also in a microscopic study by Neubert *et al.* (Fig. 11) [65]. In particular, the peak in the time headway distribution which emerges at ≈ 2 sec (Fig. 11) [65] corresponds to the flow rate out from the jam $q_{\rm out} \approx 1800$ vehicles/h in the macroscopic study shown on Fig. 10 [20].

As it has already been mentioned, the velocity of the downstream front of a wide jam v_g is one of the characteristic parameters and it does not depend on whether the jam propagates through free flow or synchronized flow [21]. On the contrary, the flow rate out from the jam q_{out}, the density $\rho_{\min}$ and the vehicle speed $v_{\max}$ in the outflow from the jam are the characteristic parameters only if free flow is formed in the outflow from the jam. In more detail the features of the line J have been considered in the review [21].

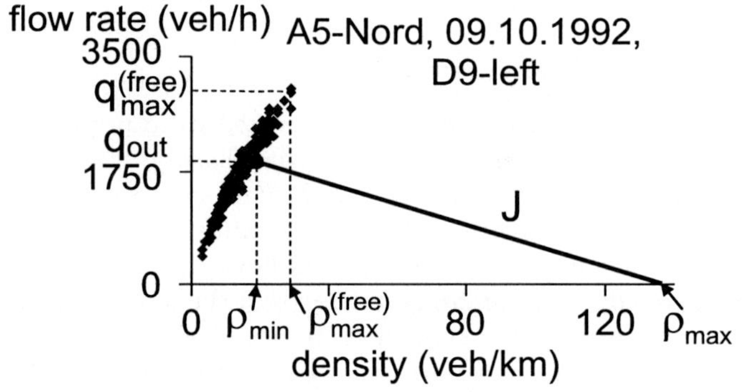

Fig. 10. The characteristic line for the downstream front of a wide jam (line J) and free flow (black points) on the flow-density plane. The line J is related to the second jam in Fig. 2(b-e) [20].

A qualitative difference between wide moving jams and synchronized flow is illustrated in Fig. 12. A jam which has emerged in the vicinity of the intersection I3 on the highway A5-South (Fig. 8) propagates through the whole highway section. Propagating through the bottleneck at the on-ramp (D16), the jam induces the F $\rightarrow$ S transition: The jam plays a role of a local perturbation with the highest possible amplitude (see Sect. 3). As well as in other cases (Fig. 13–15), the synchronized flow is further fixed at the on-ramp and it is self-maintained for a long time. The synchronized flow has no influence on the further propagation of the jam [21,58,59]. The flow rate may remain on an average nearly the same in free and in synchronized flows (Figs. 12, 13). In contrast, inside the jam both the speed and the flow rate decrease down to zero (Fig. 2 (b, c)).

3 Features of Phase Transitions

3.1 F$\rightarrow$J-Transition

All known models reach maximum free flow at the density where the critical amplitude of the F$\rightarrow$J-transition is zero (e.g., [13–17,19,22–28,32–45,47,48]). However, as it has been discovered by Kerner, traffic in most experiments can not get to this point since the F$\rightarrow$S-transition happens at a lower density [55,62].

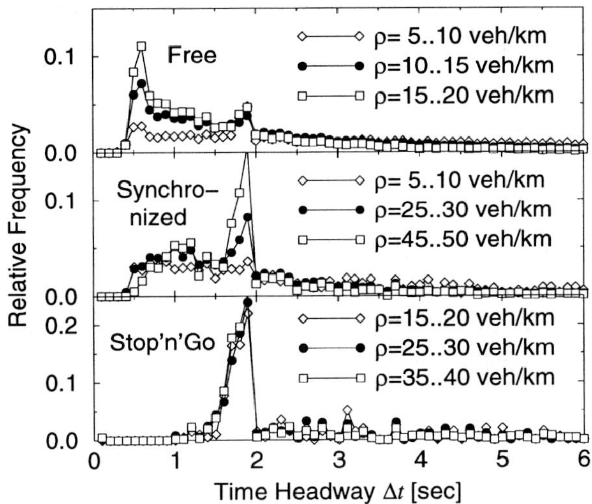

Fig. 11. Time headway distributions for different density regimes [65].

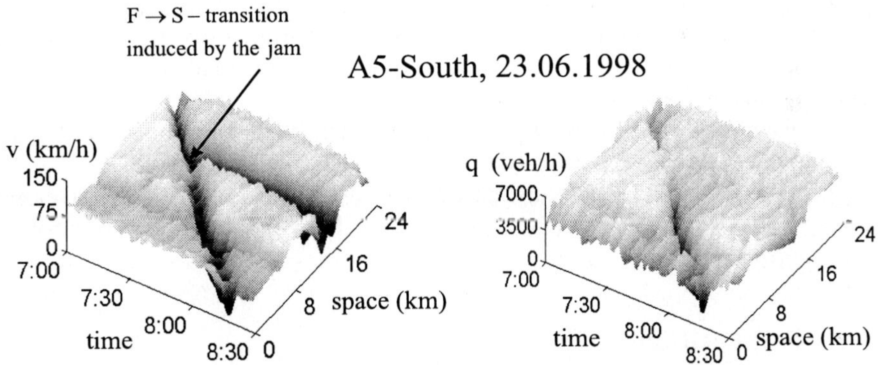

Fig. 12. An example of the F→S transition induced by a moving jam on A5-South (Fig. 8): Distributions of the average across the highway lanes vehicle speed (left) and the total flow rate on the highway (right) in space and time [72].

It must be noted that the vehicle density $\rho_{\rm min}$ which is one of the characteristic parameters and corresponds to the flow rate out of a wide jam, $q_{\rm out}$, determines the threshold density for jam formation: (i) traffic jams cannot form at vehicle densities below $\rho_{\rm min}$; (ii) and above $\rho_{\rm min}$ free flow is metastable with respect to jam formation [19,20] (see details in [21]). However, experimental observations of real traffic have shown that jams only emerge in free flow if the formation of the synchronized phase is somehow hindered [62,73]. In the most cases the phase transition from free flow to synchronized flow is observed instead. It can even be shown that the existence of the maximum point of free flow in the flow-density plane ($q^{\rm free}_{\rm max}$ and $\rho^{\rm free}_{\rm max}$ in Fig. 3), is linked to the F→S-transition rather than the F→J-transition [55,60–62]. Therefore there are two different phase transitions in free flow with different threshold densities and it is a challenge to discover the differences between them. Although this contradicts the results of all the mathematical models and seems to be highly counter-intuitive, it is an experimental result that is observed in real traffic flow.

3.2 F→S→J-Transitions

So how does a jam emerge in free flow? As it has been found out by Kerner [55], the first step is the F → S-transition. The second step is known as the 'pinch effect'. In this process the synchronized flow compresses itself into a very high density state in which spontaneous local perturbations are very likely to grow and lead to the formation of a traffic jam, i.e. the S → J-transition occurs leading to the jam emergence. In other words, the emergence of jams in initially free flow is related to a cascade of two phase transitions following one another: First from free flow to synchronized flow and then from synchronized flow (usually through the pinch effect) to jams. Note that these qualitative different non-linear phenomena can occur spontaneously on different locations on the road with a time delay between them. The related cascade of phase transitions is called the F→S→J-transition [55].

Examples of the F→S→J-transition which occur away from bottlenecks and at on-ramps are shown respectively in Fig. 5 and 6 in the review [62]. A qualitative theory of the pinch effect and of the S → J-transition have been proposed in [55,60] (see also Sect. 4). The consideration here will be restricted to the F→S-transition.

3.3 F→S-Transition

In [57], where the F→S-transition has been discovered, it has been stressed that this transition is a local first order phase transition. In particular, this phase transition occurs at an on-ramp. In this case synchronized flow is often self-maintained at the on-ramp during a long time. The downstream front which separates synchronized flow upstream and free flow downstream is fixed in the vicinity of the on-ramp [57,70]. While the F → S-transition is the local first order one, each local perturbation which amplitude exceeds some critical amplitude should cause this transition (the nucleation effect). Obviously, if these conclusions are correct, then the F→S-transition and the effect of self-maintaining of synchronized flow must occur and show qualitatively the same features in all following cases:

1) When a random local fluctuation whose amplitude exceeds the critical amplitude occurs in the vicinity of the on-ramp in an initial free flow (Fig. 15).
2) When an upstream moving jam which has initially occurred far enough downstream of the on-ramp reaches the on-ramp, and besides the F→S-transition can occur in the initial free flow (Fig. 12). Indeed, in this case the jam plays a role of the local perturbation of the highest possible amplitude.
3) When a localized region of synchronized flow propagating upstream which has initially occurred far enough downstream of the on-ramp reaches the on-ramp, and nearby the F → S-transition can occur in the initial free flow (Fig. 13 and 14). Also in this case the localized region of synchronized flow plays the role of a nuclei for the phase transition.

The case 1) is related to a random F→S-transition and the cases 2) and 3) are induced F→S-transitions. Fig. 12 – 15 show that all these cases really occur in real traffic flow. After the F→S-transition has occurred, the behavior of synchronized flow which is self-maintained at the on-ramp is qualitatively the same in all these cases (Fig. 12–15). This confirms the conclusions made in [57].

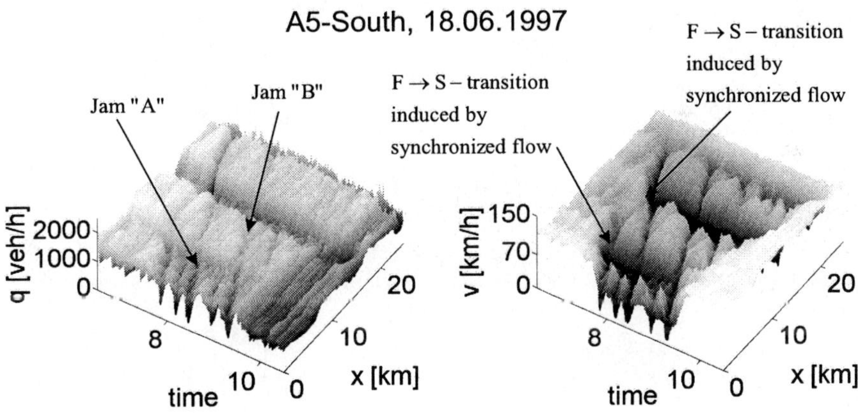

Fig. 13. Overview of two phenomena: (i) induced F→S-transitions at on-ramps caused by the synchronized flow and (ii) of the propagation of jams through the pinch region which is upstream of the intersection I1 on A5-South (Fig. 8): The average flow rate per lane (a) and the average across the highway lanes vehicle speed (b) in space and time [72].

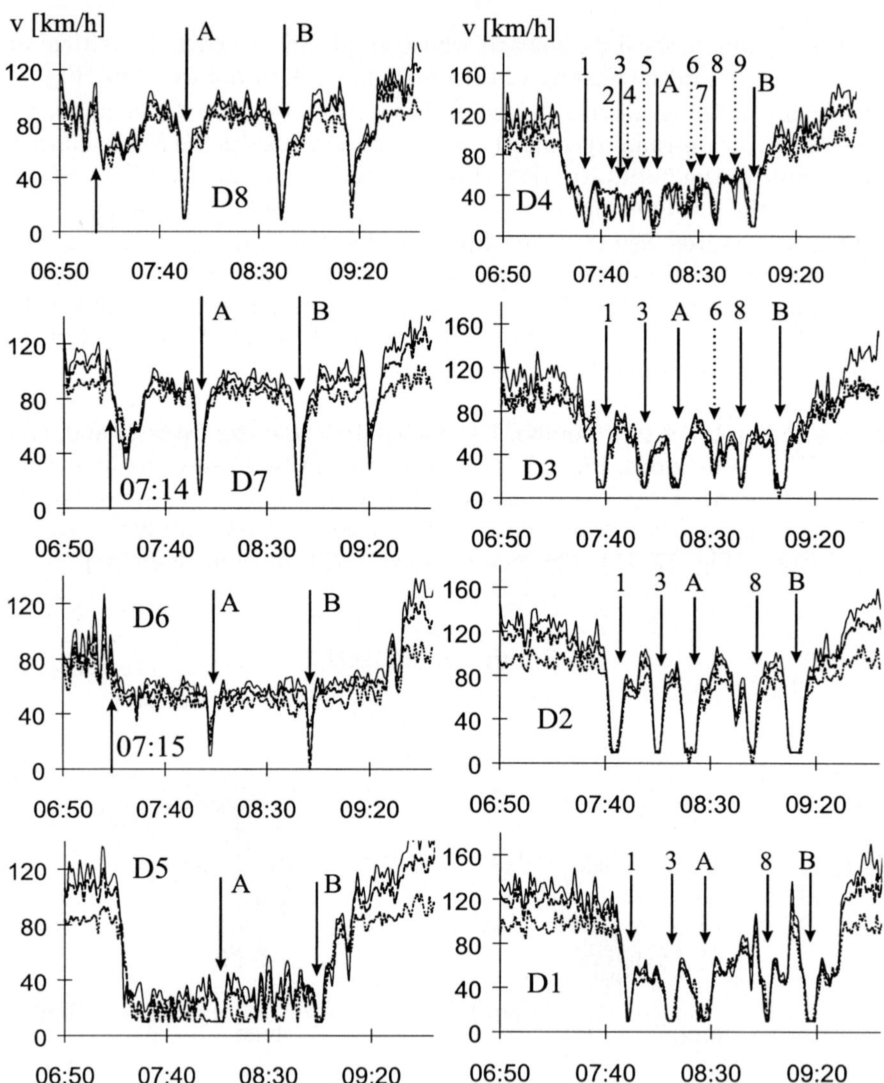

Fig. 14. The F→S-transitions at the on-ramp (D6) induced by the synchronized flow (up arrows) and the jam emergence and interaction in the pinch region (down arrows) (see also the overview in Fig. 13): The vehicle speed at different detectors [72].

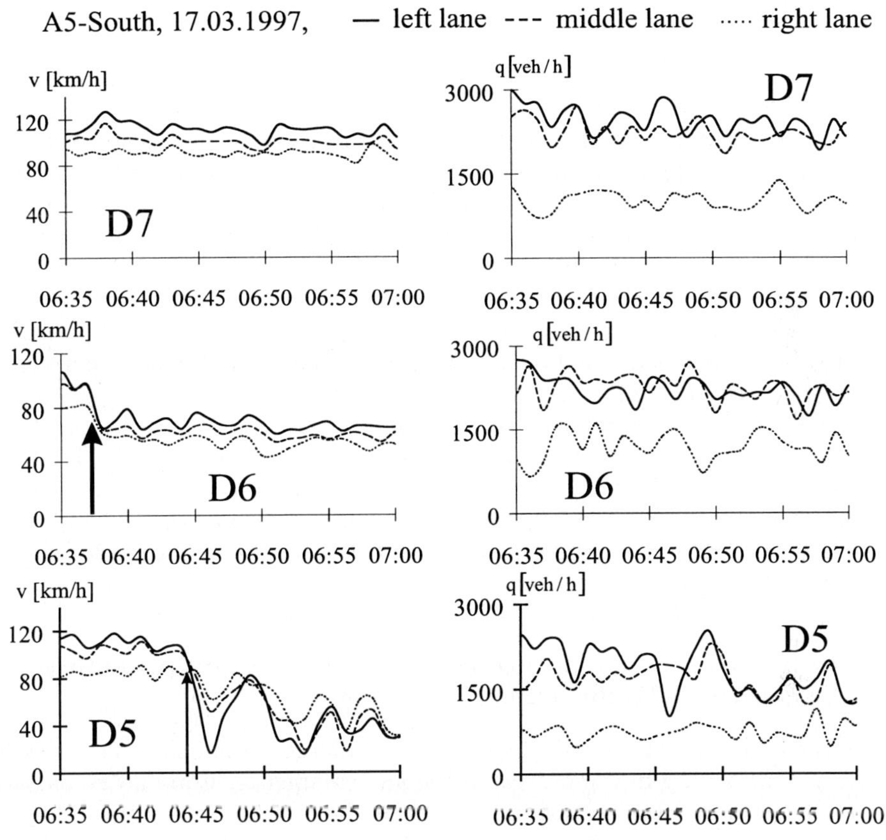

Fig. 15. The F→S-transition at the on-ramp (up arrow, D6) caused by a random perturbation [72].

3.4 A Comparison of Experimental Features of the F→S-Transition with Model Results

A common result of observations is that after the F → S-transition has occurred, the flow rate q_{out}^{bottl} in the outflow from the downstream boundary of synchronized flow (i.e. the boundary where synchronized flow is transferred into free flow downstream) can change in a very wide range. This result is true either when the F→S-transition occurs at the on-ramp (Fig. 16, left) or away from bottlenecks (Fig. 16, right) [55,73]. Besides, this flow rate q_{out}^{bottl} can considerable exceed the flow rate out from a wide jam, q_{out} (Fig. 16c, d). In particular, when for example $q_{\text{out}} \approx 1800$ vehicles/h, then q_{out}^{bottl} can approximately be changed in the range [1600 vehicles/h, 2600 vehicles/h] (Fig. 16c, d) [70]. The mentioned values are related to the left lane of the highway A5 (the percentage of long vehicles on the left lane is negligible because long vehicles must not move of the left lane on the section of the highway, Fig. 8). These results are in contradiction to all model results where a "theoretical phase diagram for traffic near on-ramps" has recently been derived by Helbing, Hennecke and Treiber [44,46] and Lee, Lee and Kim [45]. Indeed, it is important for this theory of the phase diagram for traffic near on-ramps that the flow rate q_{out}^{bottl} nearly equals q_{out}: at least, q_{out}^{bottl} cannot considerably exceed q_{out} [44–46]. However, this is in contrast to all experimental results (Fig. 16c, d) [70].

It turns out that either at on-ramps or away from bottlenecks the F→J-transition has *never* been observed: jam(s) emerge *only* due to the F→S→J-transition [55,62,73]. This experimental result is also in contradiction to all model results [15,17,19,22–28,36,38–48] derived from numerical and analytical investigations of the traffic flow models, which are based on the hypothesis about the fundamental diagram.

3.5 Propagation of Jams through the Pinch Region: Complex Dynamics of Traffic Jams Interaction

Recently, the nature of the emergence of "stop-and-go" traffic flow has been disclosed [55]. As well as in the process of the jam emergence, stop-and-go patterns occur due to the F→S→J-transition: First the F→ S-transition occurs. Then the pinch effect in synchronized flow, i.e., a self-compression of synchronized flow is realized. In the related pinch region states of flow (outside moving jams) are lying noticeably above the line J in the flow-density plane and a sequence of growing narrow moving jams spontaneously occurs. In other words, the pinch region is the source of narrow moving jams. Finally, a successive process of self-transformation of the narrow moving jams into wide jams can be realized. The latter non-linear process determines a scale in distance between stop-and-go patterns consisting of wide moving jams. Wide moving jams can move further upstream a lot of kilometers away from the pinch region where the narrow jams initially have occurred. Usually the F→S-transition occurs at a highway bottleneck, i.e., at off- or on-ramps [55]. In these cases the pinch region where narrow moving jams occur is formed upstream of the bottleneck.

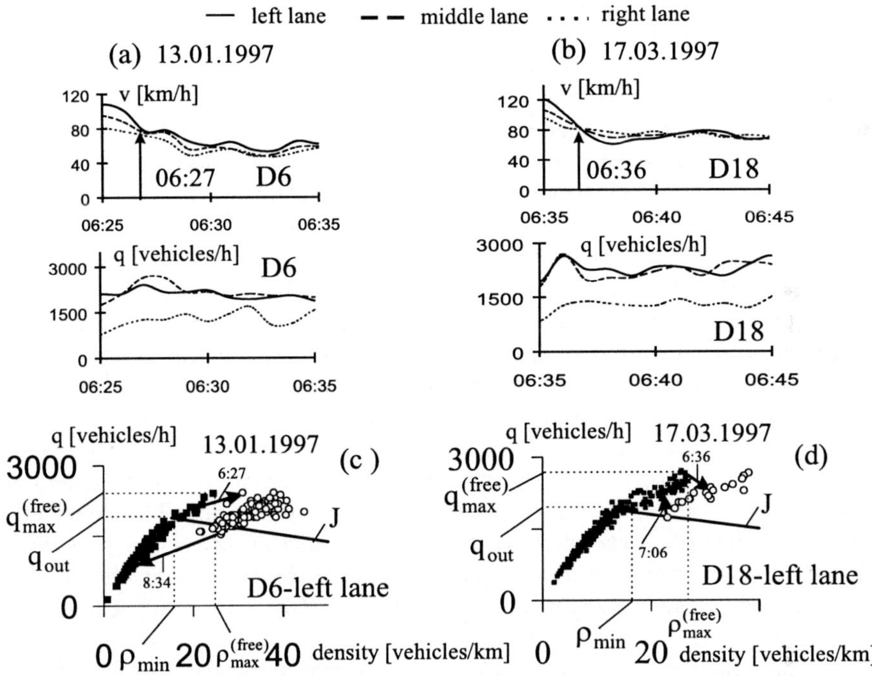

Fig. 16. Features of the F→S-transitions at the on-ramp (left) and away from bottlenecks (right) on the highway A5-South (Fig. 8): (a, b) - The vehicle speed and the flow rate during the F→S-transitions [55]; (c, d) - A part of the line J (line J), free flow (black points), synchronized flow (circles), the F→S-transitions (arrows related to 6:27 (c) and to 6:36 (d)) and the return S→F-transitions (arrows related to 8:34 (c) and to 7:06 (d)) on the flow-density plane [73].

Often two or more different separated pinch regions upstream of different highway bottlenecks appear almost simultaneously. For example, it can be seen in Fig. 13 (right) that a downstream pinch region located at about 22 km (upstream of the off-ramp at D23, Fig. 8), a downstream pinch region located at about 15 km (upstream of the on-ramp at D16), and an upstream pinch region located at about 5 km (upstream of the on-ramp at D6) appear on the highway section. In the spatially separated pinch regions related sequences of moving jams – a 'downstream' sequence(s) and an 'upstream' sequence(s) of moving jams – can occur. Such sequences of moving jams can really be seen upstream of the related pinch regions in Fig. 13 (right).

A complex dynamics of the jams interaction occurs when some jams from the downstream sequence of jams (the jams A and B in Figs. 13 and 14) due to their upstream propagation approaches the upstream pinch region where the upstream sequence of narrow moving jams is emerging [72]. It occurs that the downstream front of a jam propagating through synchronized flow in the pinch region changes parameters of synchronized flow on a way that this suppresses the emergence of new jams in the pinch region. As a result, instead of two isolated sequences of jams an *united* sequence of jams is usually built. The united sequence consists of (i) wide moving jams which have initially belonged to the downstream sequences of jams (the jams A and B in Fig. 13 and 14) and (ii) moving jams which are emerging in the upstream pinch region (down arrows 1–9 in Fig. 14). The latter jams can emerge in the upstream pinch region only far enough from the former jams in the downstream direction. This is linked to the fact mentioned above that a wide moving jam propagation through the pinch region prevents the emergence of a new jam in a some downstream vicinity of this wide jam. The latter effect has the same nature as the effect of the damping of the narrowing jams in the pinch region downstream of a wide jams [55]. Also the non-linear effects of the jam emergence in the upstream pinch region away from the jams A and B are qualitative the same as it has been found out in [55] (compare the evolution of narrow jams which are symbolically shown by down arrows 1–9 in Fig. 14 with the evolution of narrow jams shown by down arrows on Fig. 1 in [55]).

4 Hypotheses about Theory of Traffic Flow

Due to on- and off-ramps and other bottlenecks on a highway, real traffic flow is usually non-homogeneous. However, it must be noted that all mentioned types of phase transitions in traffic flow are also observed away from on- and off-ramps and other freeway bottlenecks [55]. Therefore, some intrinsic non-linear properties of homogeneous states of traffic flow, which do not depend on whether bottlenecks exists on a highway or not, are essentially responsible for the complex behavior of traffic flow. Due to this, for theoretical study of non-linear properties of real traffic flow the properties of homogeneous states (steady states) of traffic flow should first be understood. The hypotheses by Kerner [55,59–61] discussed below are based on the results of observations and are devoted to steady states of traffic flow and to features of phase transitions in traffic flow.

4.1 A Hypothesis about the Multitude of Homogeneous States (Steady States) of Traffic Flow [56,59]

In the flow-density plane, homogeneous states (steady states) of flow on a multi-lane-road are related to a curve (curve F) for free flow and to a two-dimensional region for synchronized flow (Fig. 6a). The multitudes of states of free flow on a multi-lane-road overlap homogeneous states (steady states) of synchronized flow in the density. They are separated by a gap in the flow rate at a given density.

The multitude of homogeneous states of synchronized flow on a multi-lane road (hatched region in Fig. 6a) is identical to the related multitude of homogeneous states of traffic flow on a one-lane road (hatched region in Fig. 6b).

4.2 A Hypothesis about the Behavior of Infinitesimal Perturbations in Initially Homogeneous States of Traffic Flow [59–61]

Independently of the vehicle density in an initial state of flow infinitesimal perturbations of traffic flow variables (the vehicle speed and/or the density) do not grow in any homogeneous states of either free or synchronized flow or else of homogeneous-in-speed states of synchronized flow: In the whole possible density range homogeneous states of flow can exist. In other words, in the whole possible density range (Fig. 6) there are no unstable homogeneous states of traffic flow with respect to infinitesimal perturbations of any traffic flow variables.

4.3 A Hypothesis about Continuous Spatial-Temporal Transitions between Different States of Synchronized Flow [59–61]

Local perturbations in synchronized flow can cause continuous spatial-temporal transitions between different states of both homogeneous and homogeneous-in-speed states of synchronized flow (hatched region in Fig. 6).

4.4 Explanation of the Hypotheses for Homogeneous (Steady) States of Traffic Flow

To explain these hypotheses, note that in synchronized flow spacing between vehicles is relatively low (i.e., the density is relatively high) in comparison with free flow at the same flow rate (Fig. 6). At low spacing a driver is able to recognize a change in the spacing to the vehicle in front of him, even if the difference in speed is negligible. In other words, the driver is able to maintain a time-independent spacing (without taking fluctuations into account) to the vehicle in front of him in an initially homogeneous state of synchronized flow. The ability of drivers to maintain a time-independent spacing should be valid for a finite range of spacing. Therefore, a given vehicle speed may be related to an infinite multitude of homogeneous states with different densities in a limited range (for example, the range $\rho_{\mathrm{min}}^{\mathrm{syn}} \leq \rho \leq \rho_{\mathrm{max}}^{\mathrm{syn}}$ is related to homogeneous-in-speed states of synchronized flow for a fixed vehicle speed, dotted line in Fig. 17a). For this reason, the multitude of homogeneous states of synchronized flow covers a two-dimensional region in the flow-density plane (hatched region in Figs. 6 and 17a, c). States of free flow (curve F in Fig. 17c) and states of synchronized flow (hatched region) overlap in densities, i.e., $\rho_{\max}^{\rm free} > \rho_S$, where ρ_S is the limit (minimum) density for homogeneous states of synchronized flow. However, in free flow on a multi-lane road due to the possibility for vehicles to overtake the average vehicle speed can be higher than the maximum speed in synchronized flow at the same density. Therefore, there is a gap in the flow rate between states of free flow and synchronized flow at a given density (Fig. 6a).

Small enough perturbations in spacing which are linked to an original fluctuation in the braking of a vehicle do not grow. Indeed, small enough changes in spacing are allowed, therefore drivers should not immediately react on it. For this reason even after a time delay, which is due to a finite reaction time of drivers τ_{reac}, the drivers upstream should not brake stronger than drivers in front of them to avoid an accident. As a result, a local perturbation of traffic variables (density or vehicle speed) of small enough amplitude does not grow. An occurrence of this perturbation may cause a spatial-temporal transition to another state of synchronized flow. In other words, local perturbations may cause continuous spatial-temporal transitions between different states of synchronized flow.

Independently of the vehicle density on a one-lane road vehicles cannot overtake. Therefore, homogeneous (steady) states of flow on the one-lane road at higher density are identical to homogeneous states of synchronized flow on a multi-lane road (hatched regions in Fig. 6). Lower density states of flow on the one-lane road can be approximately represented by a curve (curve O in Fig. 6b) which at low enough density has a slope equal to the same desired speed of vehicles as in free flow on a multi-lane road (Fig. 6a). In real traffic flow a spatial alternation of states of free and synchronized flow on a multi-lane-road in a middle density range can lead to the occurrence of 'mixture states' [59].

4.5 A Hypothesis about Two Different Kinds of Nucleation Effects in Traffic Flow [60–62]

There are two qualitatively different kinds of nucleation effects in traffic flow:

1. The nucleation effect which is responsible for the jam's formation, i.e., for the F→J-transition and the S→J-transition (Fig. 17a, b).
2. The nucleation effect which is responsible for the F→S-transition (Fig. 17c, d).

The nucleation effect which is responsible for the F→S-transition is linked to an avalanche self-decrease in the mean probability of overtaking in traffic flow, P. This self-decrease in the mean value of the probability of overtaking will be realized, if such a critical local perturbation of traffic variables (vehicle speed or/and density) occurs which causes an initial local decrease in the probability of overtaking below some critical value P_{cr} (dotted curve P_{cr} in Fig. 17d). A dependence of the critical amplitude of the density perturbation on the density

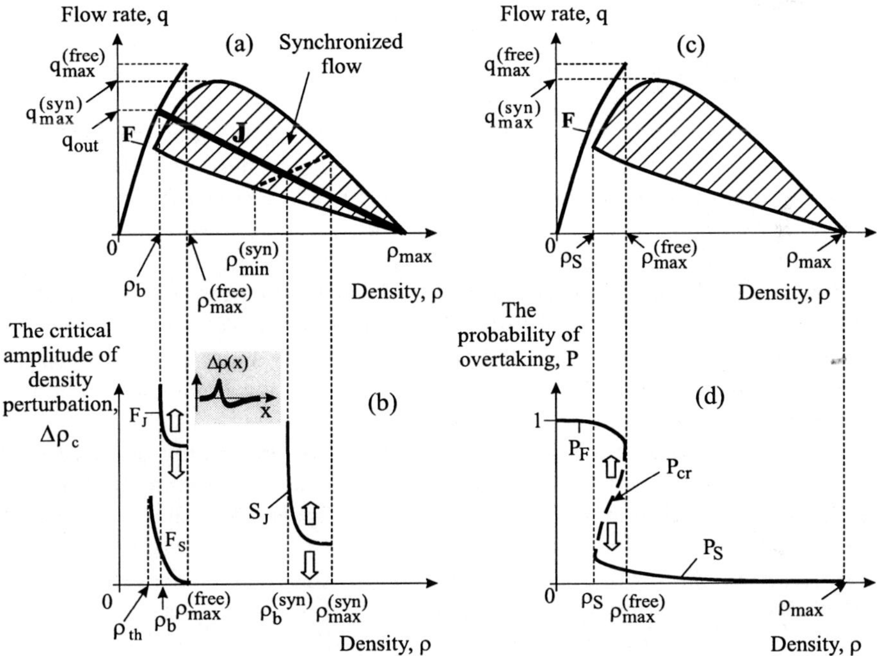

Fig. 17. Explanation of hypotheses to traffic flow theory [61]. States of free (curve F) and synchronized flow (hatched region) in Fig. 17a, c are the same as in Fig. 6 (a).

for the F→S-transition is shown by the curve F_S in Fig. 17b. In contrast, the nucleation effect which is responsible for the jam's formation is linked to an avalanche self-growth of the amplitude of a local critical perturbation of the traffic variables (density or/and vehicle speed), i.e., when the initial amplitude of the local perturbation exceeds some critical amplitude. Note that dependencies of the critical amplitude of the density perturbation on the density are shown by the curve F_J for the F→J-transition and by the curve S_J for the phase transition of the S→J-transition in Fig. 17b. The curve S_J is related to a fixed vehicle speed, dotted line in Fig. 17a.

4.6 A Hypothesis about the Critical Amplitude of Local Perturbations and the Probability of Phase Transitions in Free Flow [60–62]

At each given density in free flow the critical amplitude of a local perturbation of traffic variables (density or/and vehicle speed) which is needed for the realization of the F→S-transition (curve F_S in Fig. 17b) is considerably lower than the critical amplitude of a local perturbation which is needed for the realization of the F→J-transition (curve F_J in Fig. 17a). Therefore, at each given density in free flow the probability of an occurrence of the F→S-transition is considerably higher than of the F→J-transition. The threshold density ρ_{th} for the F→S-transition can differ both from the threshold density ρ_b (where $\rho_b = \rho_{\min}$) for the F→J-transition and from the limit density ρ_S for homogeneous states of synchronized flow.

Because the critical value of the probability of overtaking is an increasing function of the density (Fig. 17d, dotted curve P_{cr}), the higher the density in free flow the lower the amplitude of a local perturbation of the traffic variables (density (curve F_S in Fig. 17b) or/and vehicle speed) which causes an avalanche self-decrease in the probability of overtaking in the related local region. At the limit density for free flow $\rho = \rho^{\text{free}}_{\max}$ the critical amplitude of a local perturbation of traffic variables (density or/and vehicle speed) which is needed for the realization of the F→S-transition reaches zero, respectively the probability of the F→S-transition reaches one.

4.7 A Hypothesis about the Nucleation Effect which is Responsible for the Jam Emergence [55]

The line J (line J in Fig. 17a) determines the threshold of the jam's existence and excitation. In other words, all (**an infinite number !**) homogeneous states of traffic flow which are related to the line J in the flow-density plane are *threshold states* with respect to the jam's formation. The line J separates all homogeneous states of both free and synchronized flow into two qualitatively different classes:

1. In states which are related to points in the flow-density plane lying below (see axes in Fig. 17a) the line J *no* jams either can continue to exist or can be excited.

2. States which are related to points in the flow-density plane lying on and above the line J are *metastable states* with respect to the jam's formation where the related nucleation effect can be realized. Only the local perturbations of traffic variables whose amplitude exceeds some critical amplitude grow and can lead to the jam's formation (up arrows, curves F_J and S_J in Fig. 17b), otherwise jams do not occur (down arrows, curves F_J and S_J in Fig. 17b). The former critical local perturbations act as nucleation centers (nuclei) for the jam's formation.

The critical amplitude of the local perturbations is maximal at the line J and depends both on the density and on the flow rate above the line J. The critical amplitude of the local perturbations is considerably lower in synchronized flow than in free flow (compare the curve F_J with the curve S_J in Fig. 17b). The lower the vehicle speed in synchronized flow is the lower the critical amplitude of perturbations for states of synchronized flow being at the same distance above the line J in the flow-density plane.

All conclusions about spatio-temporal transitions between different states of synchronized flow, the metastable and the stable states, and the critical amplitude of local perturbations in synchronized flow (Sect. 4.4) on a multi-lane road are obviously valid also for the related states of flow on a one-lane road (i.e., for states in the hatched region in Fig. 6b).

4.8 Explanation of Hypotheses for Nucleation Effects in Traffic Flow [55,60–62]

To explain the hypotheses, note that in free flow of low enough densities a driver is not hindered to overtake, i.e., the related mean probability of overtaking is $P = 1$ (Fig. 17d). The higher the density in free flow is the lower the probability of overtaking. However, in free flow up to the limit point $\rho = \rho_{\rm max}^{\rm free}$ the probability of overtaking does not decrease drastically (solid curve P_F, Fig. 17d). Indeed, in free flow the difference between vehicle speeds on different lanes of a highway can be high enough. Therefore, the probability for a vehicle to be able to overtake is also high. In contrast to that in synchronized flow of high enough density drivers are not able to overtake at all, i.e., $P = 0$. The lower the density of synchronized flow is the higher the probability of overtaking. However in synchronized flow up to the limit point $\rho = \rho_S$ the probability of overtaking cannot increase drastically (solid curve P_S, Fig. 17d). Indeed, in contrast to free flow, in synchronized flow the difference between vehicles speeds on different lanes of a highway is low. This hinders the overtaking. Therefore, at each given vehicle density in the density range $[\rho_S, \rho_{\rm max}^{\rm free}]$, where states of free flow and synchronized flow overlap in the density, the probability of overtaking in free flow is considerably higher than in synchronized flow. As a result, the dependence $P(\rho)$ is **Z**-shaped, i.e., it has a hysteresis loop, and therefore it consists of three branches:

(i) The branch P_F for free flow.
(ii) The branch P_S for synchronized flow.

(iii) The branch P_{cr} which is related to the critical value of the mean probability of overtaking (Fig. 17d).

The sense of the branch P_{cr} is the following: If in a local region of free flow due to a local perturbation of traffic variables (speed or/and density) the probability of overtaking is decreased below the critical value P_{cr}, then an avalanche self-decrease in the probability of overtaking occurs leading to a self-formation of synchronized flow where $P = P_S$ (down arrow in Fig. 17d). If on the contrary in a local region of synchronized flow due to a local perturbation of traffic variables the probability of overtaking is increased above the critical value P_{cr}, then an avalanche self-increase in the probability of overtaking occurs leading to a self-formation of free flow where $P = P_F$ (up arrow in Fig. 17d).

Because the critical value P_{cr} of the mean probability of overtaking merges with the value P_F of the mean probability of overtaking for free flow at the limit density $\rho = \rho_{\text{max}}^{\text{free}}$ for free flow, already a local perturbation of traffic variables of a very small (but finite) amplitude causes an avalanche self-decrease in the mean probability of overtaking at the limit density $\rho = \rho_{\text{max}}^{\text{free}}$. Therefore, in free flow at the limit density $\rho = \rho_{\text{max}}^{\text{free}}$ the probability of the F→S-transition reaches one.

Above a threshold of any local first order phase transition, the higher the distance from the threshold the higher is the related probability of the occurrence of the phase transition. In other words, above the related threshold density $\rho = \rho_{th}$ (Fig. 17b) the probability of the F→S-transition should be an increasing function of the density in free flow reaching one at the limit density $\rho = \rho_{\text{max}}^{\text{free}}$ for free flow. Such behavior of the probability of the breakdown phenomenon in a freeway bottleneck has recently been found out by Persaud, *et al.* [74]. This experiment is an additional confirmation of the finding by Kerner and Rehborn in [57] that the latter phenomenon is caused by the first order local F→S-transition.

Therefore, the existence of the limit point $\rho = \rho_{\text{max}}^{\text{free}}$ for free flow is indeed linked to the occurrence of the F→S-transition. On the contrary, the critical amplitude of traffic variables (density or/and vehicle speed) which is needed for the F→J-transition (curve F_J in Fig. 17b) is a relatively high finite value even at the limit density $\rho = \rho_{\text{max}}^{\text{free}}$. Indeed, as it has been mentioned, at any density the probability of the F→J-transition is considerably lower than of the F→S-transition. To explain this [62], let us recall that synchronized flow occurs, if due to a local perturbation of traffic variables (density or/and vehicle speed) the probability of overtaking decreases below the critical value P_{cr} (Fig. 17d). The critical amplitude of a local perturbation which is able to cause the local avalanche self-decrease in the probability of overtaking can be considerably lower than the critical amplitude of a local perturbation which is needed for the F→J-transition. Indeed, for the occurrence of synchronized flow an initial perturbation itself does not necessarily have to increase avalanche-like: It must only cause an avalanche-like decrease in the mean value of the probability of overtaking. In contrast, for the jam formation the amplitude of the critical perturbation itself must begin to grow avalanche-like in an initial free flow.

An explanation of the hypothesis about the nucleation effect which is responsible for the jam's formation has been made in [55]. Other hypotheses about features of congested flow can be found in [59,60].

5 Conclusions

The physics of traffic flow is proving to be much richer than most physicists working in the field had expected. Recent experimental observations of real traffic patterns have discovered new phases of traffic flow, such as synchronized flow, and revealed a large number of features that are qualitatively similar to those observed in other non-linear systems. The next challenge for physicists in this field is to develop new mathematical and theoretical concepts to explain these features.

Acknowledgement. I would like to thank Hubert Rehborn, Mario Aleksic, Joachim Wahle and Sergey Klenov for their help, the public authorities of Hessen for the support in the preparation of data and the German Ministry of Education and Research for the support within the project "SANDY".

References

1. M.J. Lighthill and G.B. Whitham, Proc. R. Soc. A **229**, 317 (1955).
2. R. Herman, E.W. Montroll, R.B. Potts, and R.W. Rothery, Op. Res. **7**, 86–106 (1959).
3. D.C. Gazis, R.Herman, and R.W. Rothery, Op. Res. **9**, 545–567 (1961).
4. G.F. Newell, Op. Res. **9**, 209 (1961).
5. I. Prigogine and R. Herman, *Kinetic Theory of Vehicular Traffic,* (American Elsevier, New York, 1971).
6. H.J. Payne, in: *Mathematical Models of Public Systems*, G.A. Bekey (Ed.), Vol. 1, (Simulation Council, La Jolla, 1971).
7. R. Wiedemann, *Simulation des Verkehrsflusses,* (University of Karlsruhe, Karlsruhe, 1974).
8. P.G. Gipps, Trans. Res. B. **15**, 105–111 (1981).
9. G.F. Newell, *Applications of queueing theory,* (Chapman Hall, London, 1982).
10. C.F. Daganzo, *Fundamentals of Transportation and Traffic Operations,* (Elsevier Science Inc., New York, 1997).
11. J.-B. Lesort, (Ed.), *Transportation and Traffic Theory,* (Elsevier Science Ltd, Oxford, 1996).
12. A. Ceder, (Ed.), *Transportation and Traffic Theory,* (Elsevier Science Ltd, Oxford, 1999).
13. D.E. Wolf, M. Schreckenberg, and A. Bachem, (Eds.), *Traffic and Granular Flow,* (World Scientific, Singapore, 1995).
14. M. Schreckenberg and D.E. Wolf, (Eds.), *Traffic and Granular Flow '97,* (Springer, Singapore, 1998).
15. K. Nagel and M. Schreckenberg, J Phys. I France **2**, 2221 (1992).
16. G.B. Whitham, Proc. R. Soc. London A **428**, 49 (1990).

17. M. Bando, K. Hasebe, A. Nakayama, A. Shibata, and Y. Sugiyama, Phys. Rev. E **51**, 1035–1042 (1995).
18. B.S. Kerner and V.V. Osipov, *Autosolitons: A new approach to problems of self-organization and turbulence,* (Kluwer, Dordrecht, Boston, London, 1994).
19. B.S. Kerner and P. Konhäuser, Phys. Rev. E **50**, 54–83 (1994).
20. B.S. Kerner and H. Rehborn, Phys. Rev. E **53**, R1297–R1300 (1996).
21. B.S. Kerner, in [14], p. 239–267.
22. M. Herrmann and B.S. Kerner, Physica A **255**, 163–188 (1998).
23. M. Bando, K. Hasebe, A. Nakayama, A. Shibata, and Y. Sugiyama, J. Phys. I France **5**, 1389 (1995).
24. S. Krauß, P. Wagner, and C. Gawron, Phys.Rev. E **53**, 5597 (1997).
25. R. Barlovic, L. Santen, A. Schadschneider, and M. Schreckenberg, Eur. Phys. J. B. **5**, 793 (1998).
26. D. Helbing and M. Schreckenberg, Phys. Rev. E **59**, R2505 (1999).
27. M. Treiber, A. Hennecke, and D. Helbing, Phys. Rev. E **59**, 239.
28. R. Mahnke and J. Kaupuzs, Phys. Rev. E **59**, 117 (1999).
29. B.S. Kerner, H. Rehborn, and H. Kirschfink, German patent DE 196 47 127, (www.depanet.de); US-patent US 5861820, (dips-2.dips.org), 1998.
30. B.S. Kerner, H. Rehborn, and H. Kirschfink, *Straßenverkehrstechnik* **9**, 430–438 (1997).
31. B.S. Kerner, M. Aleksic, and H. Rehborn, in this book.
32. G.B. Whitham, *Linear and Nonlinear Waves,* (Wiley, New York, 1974).
33. A.D. May, *Traffic Flow Fundamental,* (Prentice Hall, Inc., New Jersey, 1990).
34. M. Cremer, *Der Verkehrsfluß auf Schnellstraßen,* (Springer, Berlin, 1979).
35. W. Leutzbach, *Introduction to the Theory of Traffic Flow,* (Springer, Berlin, 1988).
36. D. Helbing, *Verkehrsdynamik,* (Springer, Berlin, Heidelberg, 1997).
37. R. Kühne, in: *Highway Capacity and Level of Service*, U. Branolte (Ed.), p. 211 (A.A. Balkema, Rotterdam, 1991).
38. M. Rickert, K. Nagel, M. Schreckenberg, and A. Latour, Physica A **231**, 534–550 (1996).
39. K. Nagel, D.E. Wolf, P. Wagner, and P. Simon, Phys. Rev. E **58**, 1425–1437-(1999).
40. B.S. Kerner, S. L. Klenov, and P. Konhäuser, Phys. Rev. E. **56**, (1997).
41. B.S. Kerner, P. Konhäuser, and M. Schilke, Phys. Rev. E **51**, 6243–6246 (1995).
42. H.Y. Lee, H.-W. Lee, and D. Kim, Phys. Rev. Lett. **81**, 3042–3045 (1998).
43. D. Helbing and M. Treiber, Phys. Rev. Lett. **81**, 3042–3045 (1998).
44. D. Helbing, A. Hennecke, and M. Treiber, Phys. Rev. Lett. **82**, 4360 (1999).
45. H.Y. Lee, H.-W. Lee, and D. Kim, Phys. Rev. E **59**, 5101 (1999).
46. M. Treiber, A. Hennecke, and D. Helbing, cond-mat/0002177; submitted to Phys. Rev. E.
47. D. Chowdbury, L. Santen, and A. Schadschneider, Reports on Progress in Physics (in press).
48. T. Nagatani, Phys. Rev. E **48**, 3290 (1993); **51**, 922 (1995); B.S. Kerner, P. Konhäuser, and M. Schilke, Phys. Lett. A **215**, 45–56 (1996); in [11] pp. 119–145; in: *Modeling Transport Systems*, D. Hensher, J. King, and T. Oum, (Eds.), Vol.2, pp. 167–182 (Elsevier Science Ltd, Oxford, 1996); B.S. Kerner, P. Konhäuser, and M. Rödiger, in: *Proceedings of the Second World Congress on the Intelligent Transport Systems*, Vol.IV, pp. 1911–1914 (VERTIS, Tokyo, 1995); D. Helbing, Phys. Rev. E **51**, 3164 (1995); **53**, 2366 (1996); P. Nelson, Transp. Theory Stat. Phys. **24**, 383 (1995); Transp. Res. B **29**, 297 (1995); R. Wegener and A. Klar, Transp. Theory Stat. Phys. **25**, 785 (1996); H. Lehmann, Phys. Rev. E **54**, 6058 (1996); R.

Mahnke and N. Pieret, Phys. Rev. E **56**, 2666–2671 (1997); P. Wagner, K. Nagel, and D.E. Wolf, Physica A **234**, 687 (1996); V. Shvetsov and D. Helbing, Phys. Rev. E **59**, 6328 (1999); D. Helbing, Phys. Rev. E **55**, 5498 (1997); D. Helbing, Physica A **242**, 175 (1997).
49. M. Koshi, M. Iwasaki, and I. Ohkura.: in: *Proceedings of 8th International Symposium on Transportation and Traffic Theory*, V.F. Hurdle, E. Hauer, and G.N. Stewart, p. 403 (University of Toronto Press, Toronto, Ontario, 1983).
50. F.L. Hall, B.L. Allen, and M.A. Gunter, Trans. Res. A **20**, 197 (1986).
51. J.H. Banks, Transp. Res. Record **1225**, 53–60 (1989).
52. F.L. Hall, V.F. Hurdle, and J.H. Banks, Transp. Res. Record, 12–18 **1365** (1992).
53. F.L. Hall and F.O. Montgomery, Traffic Eng. and Control, 420–425 (1993).
54. W. Brilon and M. Ponzlet, in [13], pp. 23–40.
55. B.S. Kerner, Phys. Rev. Lett. **81**, 3797 (1998).
56. B.S. Kerner and H. Rehborn, Phys. Rev. E **53**, R4275–R4278 (1996).
57. B.S. Kerner and H. Rehborn, Phys. Rev. Lett. **79**, 4030 (1997).
58. B.S. Kerner and H. Rehborn, Int. Verkehrswesen **50**, 196–203 (1998).
59. B.S. Kerner, in: *Proc. of the* 3rd *Symp. on Highway Capacity and Level of Service*, R. Rysgaard, (Ed.), Vol. 2, pp. 621–642 (Road Directorate Denmark, 1998).
60. B.S. Kerner, in: [12], p. 147.
61. B.S. Kerner, Transportation Res. Record **1678**, 160 (1999).
62. B.S. Kerner, Physics World **12**, No. 8, 25–30 (1999).
63. L.C. Edie and R.S. Foote, in: *Proc. of Symp. on the Theory of Traffic Flow,* R. Herman (Ed.), pp. 175–192 (Elsevier, New York, 1959); J. Treiterer, *Investigation of traffic dynamics by aerial photogrammetry techniques,* Ohio State University Technical Rep. PB 246 094, (Columbus, Ohio, 1975).
64. P.H.L. Bovy, (ed.) *Motorway Analysis: New Methodologies and Recent Empirical Findings*, (Delft University Press, Delft, 1998).
65. L. Neubert, L. Santen, A. Schadschneider, and M. Schreckenberg, Phys. Rev. E **60**, 6480 (1999).
66. B.S. Kerner, M. Aleksic, and H. Rehborn, in preparation.
67. *Highway Capacity Manuel*, Special Report 209, Third Edition, (TRB, Washington, D.C., 1998).
68. W. Brilon and N. Wu, in: *Traffic and Mobility,* W. Brilon, F. Huber, M. Schreckenberg, and H. Wallentowitz, (Eds.), p.163 (Springer, Berlin, 1999).
69. P. Wagner, in: [13], p. 199.
70. B.S. Kerner, in: *Preprints of the 79 th TRB Annual Meeting*, Washington D.C., TRB Paper No. 00-1573 (TRB, Washington, D.C., 2000).
71. I.A. Lubashevsky and R. Mahnke, in this book.
72. B.S. Kerner, in preparation.
73. B.S. Kerner, submitted.
74. B. Persaud, S. Yagar, and R. Brownlee, Transportation Res. Record **1634**, 64 (1998).

An Integrated Model of Transport and Urban Evolution (ITEM) – Traffic and City Development in Emergent Nations

G. Haag

Steinbeis-Transferzentrum Angewandte Systemanalyse, Schönbergstraße 22, 70599 Stuttgart, Germany

Abstract. By the year 2000, more than half of the world's population will live in cities. This means that not only an extensive exchange of population, goods and information in a globalised world can be expected but also that the transport of goods, population and information must be effectively managed. A city which wants to represent an important node in this network must provide beside appropriate economic, social and cultural conditions and political stability, qualified labour, and an urban as well as an internationally operating interurban system of transport. Moreover, all cities face a common problem: they must possess the capacity to sustain unprecedented numbers of citizens within limited budgets and severe environmental constraints.

1 Introduction

"The Third World faces a big challenge with respect to the development of the transportation of persons and goods, especially in the very fast growing conurbation areas. Currently there exists a good chance for the emergent nations to create well-designed transportation and urban/regional structures concerning their technology, economy and ecology". This statement of Prof. Rolf Scharwächter may explain the initiation of the sponsoring project "Traffic and City Development in Emergent Nations" by the DaimlerChrysler AG in 1996. The project was performed under the guidance of Prof. Wolfgang Weidlich, University of Stuttgart and Prof. Günter Haag, Steinbeis Transfer Centre Applied Systems Analysis, Stuttgart. The results of the project have been summarised in an "Integrated Model of Transport and Urban Evolution (ITEM)" with an application to a metropole of an emergent nation [29].

One of those emergent nations facing a rapid urbanisation process is China with a substantial increase in population, to an estimated 1.3 billion by the year 2000, and to 1.5-1.6 billion by the year 2050 (State Planning Commission, 1994). For all Chinese commentators these statistics dominate every consideration of global development. Nanjing, the former capital of Southern China is the regional capital of the Jiangsu Province. It has 2.61 million inhabitants and is therefore one of the big cities in China. The city of Nanjing (the old capital of six dynasties of ancient southern China) is located about 300 km to the north-west of Shanghai on the Yangtse-River (31° north latitude, 118° east meridian).

Nanjing City is an important traffic node, since it has the biggest river port in China that links Nanjing with the ocean. Furthermore, in Nanjing there is one of the few bridges crossing the Yangtse River. Hence, it connects Xuzhou (and farther away Beijing) in the north and Zhengzhou in the north-west, Hefei (the capital of Anhui Province) in the west, and Wuxi, Suzhou and Shanghai in the east. Its spatial location underlines its strategic position within the global transport network of China.

In Greater Nanjing nine industrial development zones exist. These nine development zones are located outside the urban area. Hence, they have to be accounted for by the six main traffic directions. For the traffic situation in the urban area those industrial development zones are most important that are located nearby the exit points of the various directions.

The volume of passenger transport generated by the location of work heavily depends on where the employees live. Furthermore, the amount of freight traffic generated depends on the transport intensity of the production process and on the composition of the firms located in an industrial development zone.

The primary aim of the project is to understand the essential characteristics of the interrelated dynamics between the traffic system and the regional settlement structure of metropoles in emergent nations in order to be able to make scenario-based forecasts supporting the political decision processes. The area of Greater Nanjing have been chosen because of its importance within the traffic network of China.

In order to understand the complex processes and interactions between traffic and urban development, both systems will be analysed in a first step by stochastic decision models. For that purpose the considered area will be divided into homogenous regions (cells or locations). The development of these areas is described by "macro-variables". Examples for these variables are above all those which describe the traffic infrastructure of the cells but also the momentary distribution of the population, the distribution of lodgings, factories, shopping centres, leisure centres, to mention a few. The modelling concept establishes a connection between the dynamics of these macro-variables and the behaviour of the individuals of the considered population ("micro-level").

The interactions between micro- and macro-level are captured as follows: On the one hand the dynamics on the macrolevel - i.e. the development of the traffic and of the regional settlement structure - is determined by the behaviour of the individuals on the micro-level. On the other hand the "attractivity" differences between the cells which depend on the macro-variables influence the decisions of the individuals as well. Apart from rational motives of the actors several elements of uncertainty, e.g. irrational behaviour as a result of insufficient information, have to be taken into account. Hence, the description of decision processes is based on a stochastical and dynamical decision model within the master equation approach. The traffic system as well as the settlement structure and the population of Greater Nanjing form a complex intertwined system. Its dynamics take place on different time scales but are modelled by making use of the same principles:

1. The daily flows of traffic in Nanjing are the result of very quick decision processes of the actors to realize a trip between two traffic cells (origin -

destination) with a special purpose. Decision processes for a certain destination, the moment of the setting out, the mode of transportation, the choice of the route etc. take place on a very short time scale.
2. The development of the regional configuration (e.g. spatial population distribution) is a process on a long-term time scale. The population distribution changes because of migration, i.e. moves between the cells. The equations of motion which describe the migratory behaviour contain transition rates, i.e. migration flows between the cells. These flows depend on distance effects and "attractivity" differences as a result of different regional advantages (e.g. rents, available living space, traffic links etc.).

Empirical data such as traffic flows between the regions, population numbers or migration flows are based on the traffic system as well as on the development of the population distribution. The system parameters as "attractivities" and "distance" (or resistance) parameters are then determined with the help of a non-linear estimation procedure. In a further step the attractivities can be connected with the macro-variables or key-variables making use of a multiple regression. At the same time the relevance of the single key variables is determined. The two last steps consist of the numerical simulation of the results and the recommendations arising from these simulations.

2 The Integrated Transport and Evolution Model (ITEM)

Traffic demand is the result of social and individual activities as well as the evolution of land use patterns. Accordingly, individual decision processes of the agents form the basis (individuals, companies etc.) of all transport phenomena. In this way, the traffic volume is influenced by the total number of activities, by the places and starting time of activities, by the means of transport and route selection, by the co-ordination of activities with other agents (e.g. formation of car pools) as well as by feedback effects.

In total therefore, traffic demand is the result of a great number of individual decision processes. These complex processes can take place on a short-term or long-term basis. In order to represent the demand behaviour of the agents in a model the following has to be taken into account: "The important point here is that travel demands on any particular route or network are not pre-ordained. They arise out of tens of thousands of individual choices being made every day, with each person in the population deciding where, when and how they wish to travel. Quite reasonably, they make these decisions in their own interest, rather than that of the community as a whole, and on the basis of the best information available at the time, which is often rather poor. The purpose of building a mathematical model of the travel demand that results from all these choices has to be able to predict how that demand might change, as the circumstances determining those choices change." [22–24].

2.1 The Micro-Level

The micro level of the society is determined by the individual decision behaviour of the different agents of the economic system. So-called "attitudes" are in the foreground of every individual concerning specific "aspects" of the society, such as the present earnings, the kind of work, the present living conditions, the family background, leisure facilities, politics, the traffic situation, to mention a few performance-influencing factors. Let

$$p_{ij}^{\alpha r(l)}(\overrightarrow{E}, \overrightarrow{\kappa}_l) \tag{1}$$

be the probability per unit of time that an individual $l, (l = 1, .., I)$ carries out a trip from area i to area j, given a certain population distribution $\overrightarrow{E}$ in the considered region. The individuals might be allocated to trip purpose specific subpopulations α. The different attitudes of each individual are summarised in the "attitude vector" $\overrightarrow{\kappa}_l$. The corresponding individual transition rates $p_{ij}^{\alpha r(l)}(\overrightarrow{E}, \overrightarrow{\kappa}_l)$ can be determined e.g. via panel data.

Dynamic micro models, based on panel data, have the advantage to be able to represent the modification of policy relevant variables in the course of time. However, modelling of traffic flows requires the preparation of a comprehensive (representative) data base, combined with corresponding costs [3].

2.2 The Macro-Level

The number of the system variables necessary for the description of the society is reduced drastically if one passes to the macro level. It is our aim to derive the trip frequency via a behavioural decision model describing individuals and households on the basis of sub-populations. These groups are characterised by comparable socio-economic features so that the traffic behaviour of the individual decision makers within the same group is approximately uniform, however quite different between individuals of different groups. Correspondingly, the entire population of the region under investigation is subdivided into groups [17,18]. The activity patterns and the corresponding trip distribution can therefore be taken as group-related. The region under investigation is subdivided into L non-overlapping traffic cells $i, i = 1, ..., L$. The total population might be further disaggregated (persons, budgets and so forth) into trip purposes (e.g. home - work, home - education, home - shopping, ...) $\alpha, \alpha = 1, .., P$. A subdivision according to traffic modes (e.g. on foot, bicycle, bus, car...) $r, r = 1, .., R$, is also introduced, since modifications of the modal split have to be considered and evaluated as well. The number of trips for the trip purpose α from traffic cell i to traffic cell j at time t (hour of the day) using traffic mode r is denoted by $F_{ij}^{\alpha r}(t)$.

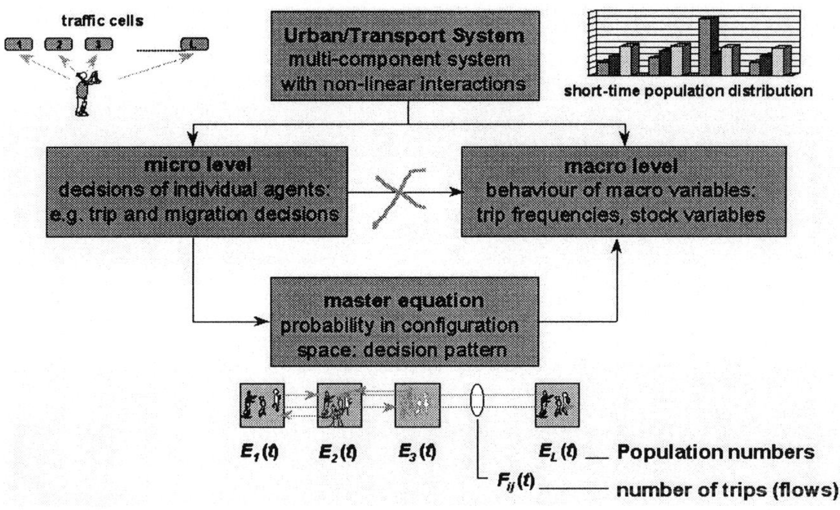

Fig. 1. The relation between the micro-level and the macro-level.

Then

$$F_{ij}^{\alpha}(t) = \sum_{r=1}^{R} F_{ij}^{\alpha r}(t) \tag{2}$$

is the number of trips between i and j for trip purpose α and

$$F_{ij}^{r}(t) = \sum_{\alpha=1}^{P} F_{ij}^{\alpha r}(t) \tag{3}$$

is the number of trips between i and j using mode r, and

$$F_{ij}(t) = \sum_{r=1}^{R} \sum_{\alpha=1}^{P} F_{ij}^{\alpha r}(t) \tag{4}$$

is the total number of trips between i and j. Therefore, accordingly the total number of trips in the system is given by

$$F(t) = \sum_{i=1}^{L} \sum_{j=1}^{L} F_{ij}(t) \,. \tag{5}$$

The traffic volume $O_i^{\alpha}(t)$ (source volume) of the traffic cell i is gained by summation from the traffic flows:

$$O_i^{\alpha}(t) = \sum_{j=1}^{L} F_{ij}^{\alpha}(t) \,. \tag{6}$$

The number of trips $D_j^\alpha(t)$ for the trip purpose α into the traffic cell j at time t reads correspondingly:

$$D_j^\alpha(t) = \sum_{i=1}^{L} F_{ij}^\alpha(t)\,. \tag{7}$$

The population distribution is denoted by

$$\vec{E} = \{E_1, ..., E_i, ..., E_L\}\,, \tag{8}$$

where E_i is the number of agents of traffic cell i. E_i will be modified by the decisions of the individual agents to carry out a trip between the cell i and anyone of the other cells. Therefore, the population distribution $\vec{E}$ is connected via traffic related activities of the agents with a great number of individual decision processes and finally with the attractivities of the specific traffic cells.

The micro level (individuals, households) appears at the first look to determine the dynamics of the macro level completely, while there is no feedback of the macro level to the micro level. However, this is not the case. Rather, a mutual influence of the two levels occurs:

Among other things, the actions (activities) of the individual agents of the economic system express themselves in the dynamics of the traffic flows and therefore in the time-dependent population redistribution of the system. Therefore, micro behaviour and macro dynamics in itself are strongly coupled. Mathematically the dependence of the individual trip decisions on the macro state appears in the fact that the attractivities of the traffic cells are among other things depending on the respective macro state, characterised e.g. by the time-dependent population distribution.

2.3 The Stochastic Framework

In addition to rational motives, uncertainties have to be considered in the case of decision processes also. Therefore, a stochastic consideration of these processes is reasonable. For this reason the configurational probability $P\left(\vec{E},t\right)$ to find a certain spatial population distribution $\vec{E}$ at time t is introduced. Of course, $P\left(\vec{E},t\right)$ must satisfy at all times the probability normalisation condition

$$\sum_{\vec{E}} P\left(\vec{E},t\right) = 1, \tag{9}$$

where the sum extends over all possible population-configurations $\vec{E}$. If the configurational transition rates $F_t(\vec{E}+\vec{k},\vec{E})$ from any $\vec{E}$ to all neighbouring $\vec{E}+\vec{k}$ are given, an equation of motion for $P(\vec{E},t)$ can be derived [27]. It reads:

$$\frac{d}{dt}P\left(\vec{E},t\right) =$$

$$\sum_{\vec{k}} F_t\left(\vec{E},\vec{E}+\vec{k}\right)P\left(\vec{E}+\vec{k},t\right) - \sum_{\vec{k}} F_t\left(\vec{E}+\vec{k},\vec{E}\right)P\left(\vec{E},t\right), \quad (10)$$

where the sum on the right hand side extends over all $\vec{k}$ with non vanishing configurational transition rates $F_t(\vec{E}+\vec{k},\vec{E})$ and $F_t(\vec{E},\vec{E}+\vec{k})$, respectively. This Equation is denoted as a master equation and has a very direct and intuitively appealing interpretation. The change in time of P, $\frac{dP(\vec{E},t)}{dt}$, is due to two effects of opposite direction, firstly to the probability flux from all neighbouring configurations $\vec{E}+\vec{k}$ into the considered configuration $\vec{E}$ namely $\sum_{\vec{k}} F_t\left(\vec{E},\vec{E}+\vec{k}\right)P\left(\vec{E}+\vec{k},t\right)$ and secondly to the probability flux out of the configuration $\vec{E}$ into all neighbouring configurations $\vec{E}+\vec{k}$, namely $\sum_{\vec{k}} F_t\left(\vec{E}+\vec{k},\vec{E}\right)P\left(\vec{E},t\right)$. Consequently the master equation represents a balance equation for probability fluxes.

The transition rates occurring in the master equation (transition probabilities per unit of time) are associated directly with the short-term evolution of the conditional probability. For further explanations see [27]. The solution of the master equation, namely the time-dependent distribution contains in detailed manner all information about the population distribution process, about the evolution of the urban/regional level and therefore indirectly also about the trip frequencies. The configuration assigned to the time-dependent maximum of the distribution represents the most probable population distribution and/or the most probable distribution of the population and of the stocks in the urban/regional system at the time t, for given economic constraints.

The probability distribution contains such a tremendous amount of information compared with the empirical information (data base) available concerning transport events and/or urban/regional processes that a less exhaustive description in terms of mean values seems to be adequate in most applications. Therefore, it is highly desirable to derive self-contained equations of motion for the mean decision behaviour of agents under given boundary conditions of the system. The mean population in traffic cell i at time t is defined as:

$$E_i(t) = \sum_{\vec{E}} E_i P\left(\vec{E},t\right), \quad (11)$$

where again the summation extends over all possible population-configurations.

It is now possible to derive equations of motion for the mean values directly from the master equation. For this purpose the master equation is multiplied by $\vec{E}$ from the left and subsequently summed up via all states $\vec{E}$. However, the equations received consequently are not yet self-contained since the determination of the right side demands the knowledge of the probability distribution $P\left(\vec{E},t\right)$. If it can be assumed that the probability distribution is a well behaved, sharply peaked unimodal distribution quasi-closed approximate mean value equations can be derived. In this line first the respective process-specific transition rates have to be represented and explained. This is carried out in the following for the traffic level.

The configurational transition rate $F_t(\vec{E}+\vec{k},\vec{E})$ from population distribution $\vec{E}$ to the neighbouring distribution $\vec{E}+\vec{k}$ is the sum of all the contributions $F_{ij}^{\alpha r}(\vec{E}+\vec{k},\vec{E})$:

$$F_t(\vec{E}+\vec{k},\vec{E}) = \sum_{r=1}^{R}\sum_{\alpha=1}^{P}\sum_{i,j=1}^{L} F_{ij}^{\alpha r}(\vec{E}+\vec{k},\vec{E}), \tag{12}$$

where $F_{ij}^{\alpha r}(\vec{E}+\vec{k},\vec{E})$ indicates the number of trips between the traffic cells $i \to j$ for trip purpose α using traffic mode r. In the total transition rate $F_t\left(\vec{E}+\vec{k},\vec{E}\right)$ we have neglected contributions referring to simultaneous changes of e.g. trip purpose, traffic mode and location. The explicit dependence of the individual terms on $\vec{E}$ indicate that merely those contributions related to a change of the population distribution $\vec{E} \to \vec{E}+\vec{k}$ are summed up. In this way, a summation via all such terms yields the total transition rate. The index t indicates the possibility of an explicit time dependence. The transition rates have to be specified depending on the kind of the interactions/processes. The further procedure is based on the STASA traffic model which has to be extended and generalised for ITEM [3]. After insertion of the transition rates into the master equation and carrying out of above-mentioned approximations, one receives equations of motion for the short-run population redistribution:

$$E_i(t+1) = E_i(t)\sum_{\alpha}\sum_{r}\sum_{j}\left(F_{ji}^{\alpha r} - F_{ij}^{\alpha r}\right), \tag{13}$$

where $E_i(t)$ is the population number of traffic cell i, and t specifies the daily hour group. Because empirical data are available only for certain hour groups t, a transition to a difference equation in place of a differential equation is adequate.

The term $F_{ij}^{\alpha r}\left(\vec{E}+\vec{k},\vec{E}\right)$ describes purpose specific trips from $i \to j$ using traffic mode r. Therefore, this transition rate corresponds directly to the empirically countable traffic flows. Now the question about the driving forces or motives which cause an individual to carry out a trip must be raised.

If, however, panel data are available on the trip-decision behaviour of the different agents of the system (micro-level), the configurational transition rates can directly be computed via:

$$F_{ij}^{\alpha r}\left(\vec{E}+\vec{k},\vec{E}\right) = \sum_{l\in\Gamma_i} p_{ij}^{\alpha r(l)}\left(\vec{E},\vec{\kappa}_l\right), \tag{14}$$

where one has to sum up over all individual contributions of all agents belonging to subpopulation α and travelling from traffic cell i to traffic cell j using mode r.

This procedure is however very extensive because of the required immense data base [4]. Therefore, a less exhaustive procedure which is more amenable to the analysis and the simulation work is indicated.

Since $E_i(t)$ agents are at time t in traffic cell i, the probability for a trip to another traffic cell will be proportional $E_i(t)$. Let $p_{ij}^{\alpha r}\left(\overrightarrow{E},\overrightarrow{x}\right)$ be the transition rate from i to j for trip purpose α. Of course, this transition rate depends among others on the explicit distribution of the agents $\overrightarrow{E}$ and cell specific characteristics $\overrightarrow{x}$ of the infrastructure, for example, job supply, the housing market, services available for companies and households as well as leisure facilities, to mention a few [13]. In this way, the number of the trips between i and j is given by:

$$F_{ij}^{\alpha r}(t) = E_i(t)\, p_{ij}^{\alpha r}\left(\overrightarrow{E},\overrightarrow{x},t\right), \tag{15}$$

where $\overrightarrow{k} = \left\{0,...,1_j^{\alpha r},...,0,...,(-1)_i^{\alpha r},...,0,...\right\}$ represent trip probabilities from one traffic cell i into j, and $F_{ij}^{\alpha r}\left(\overrightarrow{E}+\overrightarrow{k},\overrightarrow{E}\right) = 0$ for all other $\overrightarrow{k}$. The dynamics of the (mean) population redistribution on the macro-level can be derived directly by an averaging procedure from the master equation:

$$\frac{dE_i(t)}{dt} = \sum_{\alpha,r}\sum_{j=1}^{LMAX} E_j(t)\, p_{ji}^{\alpha r}\left(\overrightarrow{E},\overrightarrow{x}\right) - \sum_{\alpha,r}\sum_{j=1}^{LMAX} E_i(t)\, p_{ij}^{\alpha r}\left(\overrightarrow{E},\overrightarrow{x}\right). \tag{16}$$

The change of the population number $E_i(t)$ of a traffic cell i of one hour to the next hour can be calculated on the basis of the traffic flows $F_{ij}^{\alpha r}$ and $F_{ji}^{\alpha r}$ between the different traffic cells i and j. The flows themselves in turn depend on the respective attractivities $u_i^{\alpha}\left(\overrightarrow{E},\overrightarrow{x}\right)$ of the traffic cells, the "resistance" functions $g^{\alpha r}\left(w_{ij}^{\alpha r}\right)$ and the scaling parameter (mobility parameter) $\varepsilon^{\alpha r}(t)$. In this way, the population distribution is represented in a time-dependent manner. This is required in order to count traffic demand and traffic supply in an appropriate manner. In particular the determination and estimation of daily traffic peaks can only be carried out using a time-dependent transport model.

The individual transition rates $p_{ij}^{\alpha r}\left(\overrightarrow{E},\overrightarrow{\kappa}\right)$ represent group related trip probabilities from one traffic cell to another one, where three factor sets essentially are important:

- attractivities $u_i^{\alpha}\left(\overrightarrow{E},\overrightarrow{x}\right)$ of the particular traffic cells which depend across-the-board on e.g. the population distribution, the distribution of the work places and dwellings. The significance of the different socio-economic variables $\overrightarrow{x}$ is determined by means of a multivariate regression procedure. Obviously the composition of the particular set of key (significant) variables depends strongly on the trip purpose α .
- resistances $g^{\alpha r}\left(w_{ij}^{\alpha r}\right)$ depending on trip purpose α, and the transport mode r, where $w_{ij}^{\alpha r}$ represents a generalised resistance or cost parameter $w_{ij}^{\alpha r} = t_{ij}^{\alpha r} + b_1^{\alpha} c_{ij}^{r} + b_2^{\alpha} K_{ij}^{\alpha r}$ for a trip from i to j, where t_{ij} is the traffic density dependent travel time t_{ij}, c_{ij}^{r}, the travel costs, $K_{ij}^{\alpha r}$ appropriate comfort

parameters, and b_1^α, b_2^α are the corresponding weight factors. The following resistance function seems to be adequate in order to describe modal choice behaviour (second term) as well as decisions concerning the time of start of the trip (third term):

$$g^{\alpha r}\left(w_{ij}^{\alpha r}(t)\right) = \\ c_0^{\alpha r}\left(w_{ij}^{\alpha r}(t)\right)^{c_1^{\alpha r}} \frac{\exp\left(-c_2^{\alpha r} w_{ij}^{\alpha r}(t)\right)}{\sum_{s=1}^{R} \exp\left(-c_2^{\alpha s} w_{ij}^{\alpha s}(t)\right)} \frac{\exp\left(-c_3^{\alpha r} w_{ij}^{\alpha r}(t)\right)}{\sum_{t'=t-1}^{t+1} \exp\left(-c_3^{\alpha r} w_{ij}^{\alpha r}(t')\right)},$$

where the parameters $c_0^{\alpha r}, c_1^{\alpha r}, c_2^{\alpha r}$ and $c_3^{\alpha r}$ depend on trip purpose α, the travel mode r and the specific trip hour and day t, respectively. In a first step only travel times t_{ij}^r are considered as proxy in the resistance parameter $w_{ij}^{\alpha r}$. In accordance with [16] and [26], the following simplified resistance function of the gamma type seems to be adequate:

$$g^{\alpha r}\left(t_{ij}^r\right) = t_{ij}^{c_1^\alpha} \exp\left(-c_2^{\alpha r} t_{ij}^r\right) \tag{17}$$

- and a time-dependent scaling parameter $\varepsilon^{\alpha r}(t)$ that correlates with the mean mobility behaviour of the population (e.g. mean number of trips per day).

For the modelling of the traffic flows, the following functional dependence proved to be effective:

$$F_{ij}^{\alpha r}(t) = E_i(t)\, p_{ij}^{\alpha r}\left(\vec{E}, \vec{x}\right) = \\ = E_i(t)\, \varepsilon^{\alpha r}(t)\, f^{\alpha r}, b_t g^{\alpha r}\left(w_{ij}^{\alpha rr}\right) \exp\left[\gamma^\alpha u_j^\alpha\left(\vec{E}, \vec{x}\right) - u_i^\alpha\left(\vec{E}, \vec{x}\right)\right], \tag{18}$$

where the flexibility of the agents to undertake a trip is considered by the scaling parameter $\varepsilon^{\alpha r}(t)$, the width of the hour group by b_t and the parameter $f^{\alpha r}$ indicates the accessibility of traffic mode r.

The described modelling of the traffic flows is based on the general experience that agents compare the "attractivities" of the traffic cells with respect to the trip purpose α and that the probability for a trip $i \to j$ increases with increasing differences $\left[\gamma^\alpha u_j^\alpha\left(\vec{E}, \vec{x}\right) - u_i^\alpha\left(\vec{E}, \vec{x}\right)\right] > 0$ of "attractivities" per unit of time and simultaneously the trip probability into the opposite direction $j \to i$ decreases. In this sense the "attractivity" differences are the "driving forces" of the transport/traffic system. The different weighting of the source cell and the target cell, and therefore the differently strong influence of the site factors on the generation and distribution of traffic is considered by the parameter γ^α.

The macroscopic traffic flows are usually different even for identical "attractivities" of the traffic cells, since they have to be multiplied by the population numbers E_i and E_j, respectively. On the other hand, a simultaneous increase of the "attractivities" of all traffic cells by a common factor may not change the trip distribution. This is guaranteed by the normalisation of the "attractivities"

$$\sum_{i=1}^{L} u_i^{\alpha}\left(\overrightarrow{E}, \overrightarrow{x}\right) = 1. \tag{19}$$

The transportation system is in its (macroscopic) balance, if simultaneously for all traffic cells the sum of inflows into a traffic cell equals the sum of outflows.

The internal traffic within a traffic cell is given by:

$$F_{ii}^{\alpha r} = E_i p_{ii}^{\alpha r}\left(\overrightarrow{E}, \overrightarrow{x}\right) = E_i \varepsilon^{\alpha r}(t) f^{\alpha r}, b_t g^{\alpha r}\left(w_{ii}^{\alpha r}\right) \exp\left[\left(\gamma^{\alpha} - 1\right) u_i^{\alpha}\left(\overrightarrow{E}, \overrightarrow{x}\right)\right]. \tag{20}$$

As expected, the internal traffic of a cell i depends directly on the corresponding socio-economic data of that particular cell and via the normalisation of the "attractivities" on the data of all other cells as well.

The dependence of the attractivities $u_i^{\alpha}\left(\overrightarrow{E}, \overrightarrow{x}\right)$ of the traffic cells on the population distribution $\overrightarrow{E}$ and further socio-economic variables $\overrightarrow{x}$ has the consequence that the evolution equation for the population redistribution by means of traffic now becomes a non-linear differential equation or difference equation. In this case, non-linearities reflect the complexity of the involved decision processes. Depending on the initial conditions i.e. the agents distribution at a given time and the further system parameters of the urban/regional areas, the non-linear dynamics can lead to a complex variety of self-organising traffic flows [19].

Consequently, the dynamics of the population distribution is described by traffic flows and feedback effects resulting from the land-use patterns [10].

This transport model contains trip-chains in aggregated form. Here a decisive difference appears to traditional traffic models, since the hourly (day time-controlled) redistribution of the agents can be considered and simulated. More precisely, the so called four-stage-algorithm is not able to describe trip-chains as well as time-of-day dependent effects. Because the decisions concerning the transport system depend on the current distribution of the agents, a dynamic redistribution of the population is not only plausible it improves simultaneously the model quality. It is a further advantage of the this sub-model of ITEM that traffic generation and traffic redistribution occurs at one model step.

2.4 Parameter Estimation of the Traffic Model

The system parameters, such as mobilities, attractivities and the parameters of the resistance function can be directly linked to the trip decision processes by different optimisation procedures. For this aim the transport model has to be matched to the empirical traffic flow matrices $F_{ij}^{\alpha re}(t)$ (index e), the population numbers $E_i^e(t)$ and travel time matrices t_{ij}^r, which in turn are depending on the traffic volume of the hour groups, respectively. The minimisation of the functional

$$H_t\left[\varepsilon^{\alpha r}, c_1^{\alpha r}, c_2^{\alpha r}, \overrightarrow{u}^{\alpha}\right] = \\ = \sum_{i,j=1}^{L} \left[F_{ij}^{\alpha re}(t) - \varepsilon^{\alpha r}(t)\, b_t E_i^e(t)\, g^{\alpha r}\left(w_{ij}^{\alpha r}(t)\right) \exp\left(u_j^{\alpha} - u_i^{\alpha}\right)\right]^2 = \text{Min.} \tag{21}$$

enables to calculate the optimal "attractivities" $u_i^{\alpha}\left(\overrightarrow{E}, \overrightarrow{x}\right)$ and all further system parameters for each trip purpose α and trip mode r.

Depending on the time of day t, the trip purpose distribution changes within the (macroscopic) traffic flows. For example, trips between home - work and work - home may predominate in the morning hours and the late afternoon, while leisure time trips or shopping trips show another temporal distribution over the day. Therefore, the estimated parameters are depending on the trip purpose and time of day.

The estimated attractivities obtained in the first step, may now depend on cell specific key-factors $x_i^n, n = 1, \ldots$ gained from two different fields: on the one hand from the class of the so-called synergy variables, describing general group effects, such as pigeons effects and on the other hand from a sequence of potential attractivity indicators, e.g. number of jobs available, the number of vacant dwellings, regional income per capita, shop distribution and other local infrastructure depending factors. The set of key-factors (selected socio-economic variables) is in a second step determined via a multiple regression analysis:

$$u_i^{\alpha}\left(\overrightarrow{E}, \overrightarrow{x}\right) = \sum_n b_n^{\alpha} x_i^n . \tag{22}$$

The elasticity's b_n^{α} assigned to the socio-economic variables x_i^n are dimensionless numbers and indicate the influence of the independent variables on the dependent variable. The selection of relevant indicators is performed using appropriate statistical characteristics (T-values, other significance tests).

However, up to now the portrayed formalism describes only short-term effects like redistribution of the population due to traffic events. Long-term effects, e.g. migration of the population between cells as well as changes in the land-use pattern have to be considered too.

2.5 Migration Flows and Population Distribution

Due to transport processes there is a redistribution of the population over the day (short-time process), $E_i(t)$, described by the transport sub-model. In the long-time, by migration events and because of the demographic development, the "housing population", denoted by E_{i0}, is changing as well. This slowly changing "housing population" E_{i0} provides the starting conditions for the short-time redistribution:

$$E_i(t = 6 \text{ am}) = E_{i0} . \tag{23}$$

The analysis and forecasting of the long-time population redistribution, via the master equation approach has become a common tool in regional dynamics over the years [14]; [28], [20,21,25,12]. Following [15] the migration flow between i and j is given by

$$\begin{aligned} W_{ij}^{\gamma}(\tau) &= E_{i0}^{\gamma}(\tau)\, p_{ij}^{\gamma}\left(\vec{E}_0, \vec{M}\right) \\ &= E_{i0}^{\gamma}(\tau)\, v^{\gamma}(\tau)\, f^{r}(d_{ij}) \exp\left[V_j^{\gamma}\left(\vec{E}_0, \vec{M}\right) - V_i^{\gamma}\left(\vec{E}_0, \vec{M}\right)\right], \end{aligned} \tag{24}$$

where γ separates different age groups of the population, $v^{\gamma}(\tau)$ is a general mobility indicator of the population, and the distance-dependence of the migration flow is described by the deterrence function $f^{\gamma}(d_{ij})$, and the "migration attractivity" of the cell (region) i is introduced by $V_i^{\gamma}\left(\vec{E}_0, \vec{M}\right)$. The log-time intervals (e.g. one year) are indicated by τ. In order not to underestimate long-distance moves, the deterrence function

$$f^{\gamma} = a_{\tau} \exp\left(\frac{-\beta_{\tau} d_{ij}}{1 + \eta_{\tau} d_{ij}}\right) \tag{25}$$

has to be proved very efficient, where d_{ij} measures the "distance" of the cells, and $\alpha_{\tau}, \beta_{\tau}, \eta_{\tau}$ are scaling parameters.

The migration attractivities $V_i^{\gamma}\left(\vec{E}_0, \vec{M}\right)$ have to be distinguished of the attractivities $u_i^{\alpha}\left(\vec{E}, \vec{x}\right)$ involved in the transport model, even if some "driving forces" (key-factors) seem to be relevant for both sub-models. In general the key-factors $\vec{M}$ involved in the migration attractivities may be related to synergy effects (population of cell/region i), housing market (e.g. number of vacant dwellings), labour market (e.g. number of jobs available), services, different accessibility indicators (coupling with the traffic sub-model), and indicators related to infrastructure and environment, to mention a few.

The long-time population redistribution reads

$$\begin{aligned} E_{i0}^{\gamma}(\tau + 1) = E_{i0}^{\gamma}(\tau) + \delta E_{\text{natural},i}^{\gamma}(\tau) + \sum_{j=1}^{L} W_{ji}^{\gamma}(\tau) - \sum_{j=1}^{L} W_{ij}^{\gamma}(\tau) \\ + \text{IMMIG}_i^{\gamma}(\tau) - \text{EMIG}_i^{\gamma}(\tau), \end{aligned} \tag{26}$$

where the age dependent birth- and death-processes are described by the function $\delta E_{\text{natural},i}^{\gamma}(\tau)$. Immigration and emigration events are exogenously introduced via the terms $\text{IMMIG}_i^{\gamma}(\tau)$ and $\text{EMIG}_i^{\gamma}(\tau)$, respectively.

The estimation procedure for the parameters of the "urban" sub-model depends on the data base. If empirical migration flows are available, the estimation can be performed in analogy to the transport model. In case that only population numbers are observed, a multinomial logit model (MNL) seems to be appropriate.

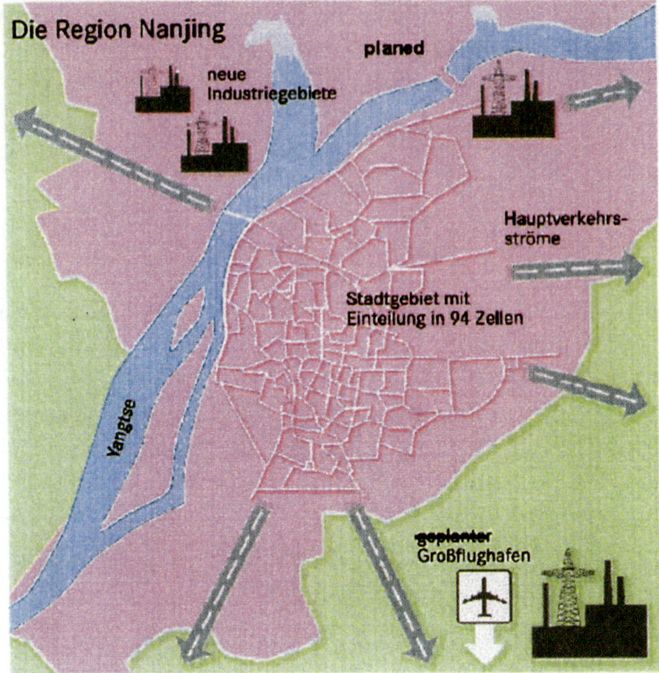

Fig. 2. The Region of Nanjing.

For the MNL the population number E_{i0} of cell i (housing population distribution) is given by

$$E_{i0}(\tau) = E_{\text{ges}} \frac{\exp(V_i(\tau))}{\sum_i \exp(V_i(\tau))} \tag{27}$$

with the total population number E_{ges} and the "population attractivities" $V_i(\tau)$ of the cells. The population attractivities can be connected with characteristics (appropriate indicators) of the spatial areas by a multiple regression. Possible explanatory indicators (key indicators) are the same as for the "attractivities" in the migration model, e.g. the number of work places, the number of dwellings, but also accessibility indicators of the traffic cells. Therefore a link to the traffic system can be identified.

2.6 The Stepwise Procedure of ITEM

In the figures 2.3 the framework of the ITEM model is illustrated for the region of Nanjing (Fig. 2).

In the following eight steps we summarise the procedure in schematic terms:

Step 1 - Traffic assignment

In the first step the traffic flows $F_{ij}^{\alpha r}$ (1987) are assigned on the corresponding transport network (1987). This determines the travel times t_{ij}^{r}(1987) between the single cells as well as accessibility measures of the cells for the considered modes. In the same way the travel time matrices t_{ij}^{r} (iteration 1) and the corresponding accessibility measures are determined for the traffic flows $F_{ij}^{\alpha r}$ (1987) and the network for the year 2000 for the planning scenarios.

Step 2 - Estimation of the parameters

Given the traffic flows $F_{ij}^{\alpha r}$ (1987), the corresponding travel times t_{ij}^{r} (1987) and the population numbers E_i (1987) the parameters of the ITEM transport model are estimated by a non-linear estimation procedure:

- attractivities $u^{\alpha r}$ (1987) of the cells i,
- parameters $\gamma^{\alpha r}$,
- scaling parameters (mobility parameters) ε^{r} for the modes r, and
- the parameters of the resistance function b^{r} and c^{r}.

Statistical tests (correlation, Fisher's F-value) are used to evaluate the quality of the parameter estimation.

Step 3 - Attractivity parameters (multiple regression)

By means of a multiple regression the dependence of the attractivities of the cells on the characteristics of the cells (for example population and employee numbers, accessibilities) is determined. That means the corresponding parameters b_i^{n} and β_i^{n} (influence of the independent variables, elasticities) as well as statistical test measures are calculated.

Step 4 - Analysis of the population distribution

To analyse the population distribution and its dependence on the accessibilities of the cells the "population attractivities" (MNL model) are calculated from the population distribution 1990. The influence of the changed accessibility measures of the cells for the considered planning scenarios on the "population attractivities" can be determined by a multiple regression using accessibility measures for a basic scenario.

In the fifth step an iteration procedure (step 5 to 7) to forecast the evolution of the transport system and the population distribution for the planning scenarios is employed.

Step 5 - Simulation of the population development

If it is known which accessibility measures are significant for the population attractivities, in the next step these attractivities and therefore the changed population distribution E_i (iteration 1) can be calculated for the planning scenarios till the year 2010. Furthermore changed migration flows into or out of the considered region can be taken into account.

Step 6 - Simulation of the traffic flows

In this step the (transport) attractivities $u^{\alpha r}$ (iteration 1) of the cells, that can also depend on the population distribution and on accessibilities, can be calculated for the different considered planning scenarios as well.

With these attractivities u_i (iteration 1), with the travel times t_{ij}^{r} (iteration 1) and the population numbers E_i (iteration 1) of step 5, the traffic flows $F_{ij}^{\alpha r}$

(iteration 1) are determined for both modes r. The mobility parameter and the parameters of the resistance function were assumed to be constant.

Step 7 - Assignment of the traffic flows

The traffic flows $F_{ij}^{\alpha r}$ (iteration 1) are now assigned for each planning scenario on the corresponding network to get the new travel times t_{ij}^{r} (iteration 2) and accessibility measures. With these travel times and accessibility measures for each mode the population configuration (dependent on the accessibilities) and the traffic flows change again, so that an iteration procedure has to be applied.

To accelerate this iteration procedure the average between the new travel times t_{ij}^{r} (iteration 2) and the travel times t_{ij}^{r} (iteration 1) of the step before were taken to calculate the next step of the procedure. Step 5 to 7 are repeated till the change of the flows and travel times can be ignored, in other words when the procedure converges. As result one gets the traffic flows $F_{ij}^{\alpha r}$ for the single modes r and trip purposes α for the different planning scenarios until 2010.

Step 8 (Results)

The comparison of the traffic flows $F_{ij}^{\alpha r}$ for the planning scenarios with the basic scenario shows the effects of the changed modal split and the interaction with the population development. Even direct and indirect induced traffic shares are considered by the model.

From the traffic flows $F_{ij}^{\alpha r}$ and their assignment on the corresponding network the traffic performances and so on can be determined for the single schemes.

The general framework for the scenarios in addition to the methodology so far is presented in Fig. 3.

3 Summary and Recommendations

In Greater Nanjing nine industrial development zones exist. These nine development zones are located outside the urban area. Hence, they have to be accounted for by the six mail traffic directions. For the traffic situation in the urban area those industrial development zones are most important that are located nearby the exit points of the various directions.

According to these criteria, the following industrial zones have been considered in detail:

1. The Headquarter of the Nanjing High and New Technology Industry Development Zone.
2. The Nanjing Economic-Technical Development Zone including the Nanjing Xingang High and New Technology Industry Park.
3. The Nanjing Jiangning Economic and Technological Development Zone including the Nanjing Jiangning High and New Technology Industry Park and the Nanjing Non-State-Run Science and Technology District.

The volume of passenger transport generated by the location of work evidently heavily depends on where the employees live. Furthermore, the amount

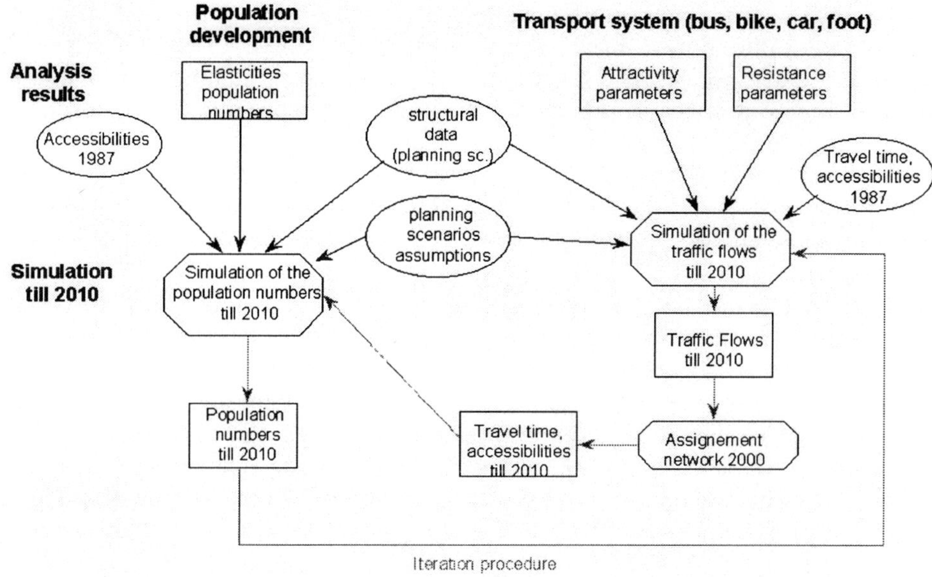

Fig. 3. The General Framework.

of freight traffic generated depends on the transport intensity of the production process and on the composition of the firms located in an industrial development zone.

In the light of the many uncertainties concerning future transport systems as well as population development, it has proved to be valuable to employ scenarios as a vehicle for the communication with policy-makers. Scenarios are richer in scope than conventional forecasting tools and offer more innovative investigation to policy-makers. The scenario approach employed includes the description of four scenarios designed to compare and examine alternative futures with respect to traffic development and population redistribution.

External factors influencing the transport sector of Nanjing are e.g. global institutional/political developments inside (or outside) China, or the introduction of new technologies.

In the post-war period we observe a steady increase of mobility, not only for daily home-to-work trips, but also for business trips or leisure trips. In the age of globalisation it seems likely that a continued pressure will exist towards higher mobility levels.

In this context, it should be emphasised that mobility as such is not bad, because it creates many economic benefits to both consumers and producers. The problem is however, that there are many social costs involved in the form of pollution, accidents, traffic congestion etc. which are not (fully) charged to the sources of the externalities. Thus, the main question for many governments nowadays is whether it will be possible to ensure that mobility rates are compatible with sustainability criteria. This would mean that the so-called de-coupling hypothesis would have to be implemented, which means a de-coupling of economic growth and environmental charges caused by the transport sector. This would of course require drastic changes in our modes of production, consumption, transportation and technological innovation. Clearly, there are many uncertainties involved in developing such new (policy) strategies, and hence it seems wise to rely on appropriate scenarios which might depict some of the future developments and allowing us to identify possible bottlenecks.

Of course, a possible de-coupling of economic growth and transport infrastructure requires at least a minimal transportation system and an effective information network. Only under those conditions a city or region can maintain its importance in the global economy.

In the Chinese context, it is clear that decisive external factors for scenarios depend on the demography, the economic policy (e.g., the open door policy), as well as on the settlement and industrialisation policy. Thus, these factors of uncertainty will influence the transportation system in China, on the urban level as well as on the national level.

At this point we summarise the main results of scenarios E, A, B, C:

Scenario E, the Empirically Given Basic Scenario, describes the current situation of the transport system and of the population distribution purely based on empirical data. In the subsequent scenarios A, B and C the same considerable increase from 1990 until 2010 of private car ownership and the same moderate increase of the total population numbers have been assumed.

Scenario A, the Mixed Land use Scenario, describes the straight forward extrapolation of scenario E which amounts to a continuation of the given spatial mixture of housing and business land use. The result is that no dramatic changes occur concerning the transportation system. In this respect the mixed land use proves to be advantageous .

Scenario B, the Separated Land use Scenario, describes a separation of housing and business areas according to plans for the year 2010 of the Nanjing municipal planning administration. The result is that the traffic load in the centre and North of Nanjing increases dramatically. This means: Increase of private car ownership in combination with separation of landuse leads to a partial overcharge of the transportation network.

Scenario C, the Separated Land use Scenario with Periphery Shifted Traffic, describes, a periphery shifted traffic, e.g. engendered by an intense development of the industrial zones around the urban area of Nanjing. The result is that the shifted traffic, after standard assignment, leads to congestions on specific primary roads, in particular on the main outlets.

Recommendations:

A comparison of the different scenarios clearly demonstrates that the development of a strong CBD with many jobs located in the centre of Nanjing and with a transfer of housing units to the periphery of the City generates in the long run a tremendous amount of traffic, mainly in the morning and evening hours. Therefore planners should avoid a strict separation of the spatial location of workplaces and housing/living opportunities. On the other hand synergy aspects have to be considered as well. This means a concentration of services and light industries in the CBD bears many synergies because of a high accessibility of different agents/firms. This high accessibility levels are related to the short distances.

What should the planners do in this conflicting situation?

- Nanjing should develop towards a multi-centre municipality instead of a single-centred central place.
- Each of the different small centres of the metropolitan area should provide all facilities needed for daily living: Local administrative offices, small shops, housing units, market places, cinemas, theatres and so on. This means within those small centres a good mixture of different services should be provided.
- The accessibility of the different centres of the metropolitan area of Nanjing must be considerably improved. Therefore, the capacities of the corresponding traffic network to the CBD must be increased in order to be able to manage the growing transport demand, especially during the rush hours.
- The expansion of the traffic network should have the same priority as the development of the CBD in order to avoid intermediate phases of chaotic and costly traffic jams.
- The expected increase of the volume of transit traffic from the south of China (Hongkong) to north (Beijing) and from the east (Shanghai) to west (Hefei) requires the development of a very effective ring road around the City of Nanjing. In order to avoid long time traffic jams on this ring road we recommend the reduction of the number of possible exits (junctions) considerably. Only with very few and well designed exits it is possible to keep a main part of the intra urban traffic outside of the ring road.
- As our simulations demonstrate, the second bridge over the river Yangtse which is already under construction is certainly an important step toward an effective management of the transit traffic and of the inter urban traffic.
- The urban traffic crossing the river Yangtse should be minimized. This should be an objective of urban/regional policy. An adequate mixture of workplaces, services and housing units at both river banks is therefore recommended.
- The transport system of Nanjing bears some bottlenecks with respect to the transport modes. Especially the public transport system (mainly busses) must be considerably improved. The development of the public transport system of Curitiba, the capital of the state of Paraná in the south of Brazil, could be partially used as a guideline (cf. Rabinovich and Leitman 1996). The essentials of the urban traffic system of Curitiba are:
  - Busses cover the main part of the public transport because subways are too expensive.
  - Separate bus lanes on the main roads
  - Efficient bus entering and bus leaving system realized by bus stop stations with automatic ticket sales
  - The bus companies are private and paid for the covered distance per day and not for the number of sold tickets

- A consequent separation of the modes by foot, bike, busses and cars for the main roads should be organised. The typical existing main roads of Nanjing have already a structure which could easily be transformed into a more effective road system. In any case it is not necessary to cut down the beautiful old trees which separate already different lanes (tracks) and which represent one of the nicest and best known sights of Nanjing.
- The management of car parking within Nanjing bears some further innovative intervention possibilities. Of course there are multi-storey car park possibilities needed in the CBD with high accessibility. On the other side Park-and-Ride opportunities on the periphery of the City could help to reduce the traffic volume in the centre. However, Park-and-Ride works effectively only, if the transaction costs from private car to public transport system are less than the private travelling costs. The choice of an appropriate tariff structure is needed. New telematic tools should be used to give at least estimates of expected waiting time and number of vacant parking places. The public authorities should develop a uniform Park-and-Ride management structure for Nanjing in order to support its acceptance. Moreover, intermodality aspects should not only consider the mode-change from car to bus but also from car to bike.

The above mentioned development aspects of the urban and the transport system of Nanjing have to be seen in the light of the rapid changes in the field of transport and communications techniques. Several trends combine to diminish the impacts of transport infrastructure on regional development and are therefore helpful in de-coupling economic growth and transport infrastructure:

- Telecommunication may reduce the need for some goods transport and person trips. However, they may also increase transport by their ability to create new markets.
- With the shift from heavy-industry manufacturing to high-tech industries and services other location factors have become relevant and replace partially the traditional ones. These new location factors include factors concerning leisure, culture, image and environment, i.e. quality of life, and factors concerning access to information, high-level services, qualified labour and to municipal institutions.

The city of Nanjing is in a fortunate situation in so far as it is in a transitional and fast developing phase. This situation allows of realising many innovative ideas concerning an integrated city and traffic development. The recommendations drawn from the results of the ITEM project indicate that good chances exist for Nanjing to arrive at a sustainable city evolution guaranteeing a sufficiently high level of mobility.

Acknowledgement. We thank the DaimlerChrysler AG for sponsoring our project by a generous financial grant and by providing all means for implementing our scientific project in full independence. Our special thanks go to Dr. Uli Kostenbader.

We gratefully acknowledge the intense co-operation with our partners Prof. Wang Wei and Prof. Deng Wei from the Southeast University of Nanjing.

Finally, we are grateful to our Senior-Advisors Prof. Dr. Dr. h.c. mult. Ilya Progogine and Prof. Dr. Dr. h.c. mult. Hermann Haken for their continuous interest and supervision of the project.

References

1. Der Bundesminister für Verkehr (Ed.), *Gesamtwirtschaftliche Bewertung von Verkehrswegeinvestitionen*, Schriftenreihe Heft **72**, (1993).
2. Der Bundesminister für Verkehr (Ed.), *Bewertung von Verkehrswegeinvestitionen im ÖPNV*, 1987.
3. Der Bundesminister für Verkehr (Ed.), *Qualifizierung, Quantifizierung und Evaluierung wegebauinduzierter Beförderungsprozesse*, FE-Nr. 90436/95.
4. D. Courgeau, *Migrants and Migrations, Population,* Selected Papers, Editions de I.N.E.D., Paris, 1985.
5. Daimler-Benz AG, *Straßenverkehr in China*, Forschungsprojekt 1990.
6. R.H. Fazio and M.P. Zanna, *Direct Experience and Attitude-Behaviour Consistency*, Berkovitz, L. (Ed.), Advances in Experimental Social Psychology **51**, 161-202 (1981).
7. G.T. Fechner, *In Sachen der Psychophysik*, Leipzig, 1877.
8. M. Fishbein and I. Ajzen, *Belief, Attitude, Intention and Behaviour,* An Introduction to Theory and Research, Reading, 1975.
9. M. Gaudry, B. Mandel, and W. Rothengatter, *Introducing Spatial Competition through an Autoregressive Contiguous Distribution (AR-C-D)*, Process in Intercity Generation-Distribution Models within a Quasi-Direct Format (QDF), Paper presented at the Transport Econometrics Conference of the AEA in Calais, 1994.
10. P.B. Goodwin, *Traffic Growth and the Dynamics of Sustainable Transport Policies*, ESRC Transport Studies unit, University of Oxford, 1994.
11. P.B. Goodwin, *Empirical Evidence on Induced Traffic A Review and Synthesis*, ESRC Transport Studies unit, University of Oxford, 1995.
12. G. Haag, *Dynamic Decision Theory Applications to Urban and Regional Topics*, (Dordrecht, 1989).
13. G. Haag, *Transport: A Master Equation Approach*, in: *Urban Dynamics, Designing an Integrated Model*, C.S. Bertuglia, G. Leonardi, and A.G Wilson, (Eds.), (London und New York, 1990).
14. G. Haag and W. Weidlich, *A Stochastic Theory of Interregional Migration*, Geographical Analysis **16**, 331-357 (1984).
15. G. Haag, M. Munz, D. Pumain, L. Sanders, and Saint-Julien, *Therese Interurban Migration and the Dynamics of a System of Cities*, Environment and Planning A **24**, 181-198 (1992).
16. H. Hautzinger, *Reiseweite- und Reisezeiteffekte von Geschwindigkeitszuwächsen im Personenverkehr*, Int. Verkehrswesen **34**, 227-249 (1982).
17. E. Kutter, *Demographische Determinanten städtischen Personenverkehrs*, Braunschweig, 1972.
18. E. Meier, *Neuverkehr infolge Ausbau und Veränderung des Verkehrssystems*, Schriftenreihe des Instituts für Verkehrsplanung, Transporttechnik Straßen- und Eisenbahnbau der ETH Zürich **81**, (1990).

19. P. Nijkamp and A. Reggiani, *Interaction, Evolution and Chaos in Space*, (Berlin, Heidelberg, New York, 1992).
20. D. Pumain, *La Dynamique des Villes*, (Economica, Paris, 1982).
21. D. Pumain and G. Haag, *Urban and Regional Dynamics - Towards an Integrated Approach*, Environment and Planning A **23**, 1301-1313 (1991).
22. *The Standing Advisory Committee on Trunk Road Assessment Trunk Roads and the Generation of Traffic*, (the Department of Traffic, London, 1994).
23. *The Standing Advisory Committee on Trunk Road Assessment Trunk Roads and the Generation of Traffic*, Response by the Department of Traffic to the Standing Advisory Committee on Trunk Road Assessment, (the Department of Traffic, London, 1994).
24. *The Standing Advisory Committee on Trunk Road Assessment Transport Investment, Transport Intensity and Economic Growth,* Response by the Department of Traffic to the Standing Advisory Committee on Trunk Road Assessment, (the Department of Traffic, London, 1998).
25. L. Sanders, *Systeme de Villes et Synergetique*, (Anthropos, Paris, 1992).
26. Steierwald, Schönharting und Partner GmbH, *Nachbarschaftsverband Stuttgart*, (Personen-/Kfz-Verkehr 1990, Stuttgart 1993).
27. W. Weidlich and G. Haag, *Concepts and Models of a Quantitative Sociology*, (Berlin, Heidelberg, New York, 1983).
28. W. Weidlich and G. Haag, *Interregional Migration - Dynamic Theory and Comparative Analysis*, (Berlin, Heidelberg, New York, 1988).
29. W. Weidlich and G. Haag, *An Integrated Model of Transport and Urban Evolution - With an Application to a Metropole of an Emergent Nation*, (Berlin, Heidelberg, New York, 1999).

Statistical Analysis of Freeway Traffic

L. Neubert[1], L. Santen[2], A. Schadschneider[2], and M. Schreckenberg[1]

[1] Physik von Transport und Verkehr, Gerhard-Mercator-Universität, Duisburg, Germany
[2] Institut für Theoretische Physik, Universität zu Köln, Germany

Abstract. Single-vehicle data of freeway traffic as well as selected *Floating-Car* (*FC*) data are analyzed in detail. Traffic states are distinguished by means of aggregated data. We propose a method for a quantitative classification of these states. The data of individual vehicles allow for insights into the interaction of vehicles. The time-headway distribution reveals a characteristic structure dominated by peaks and controlled by the underlying traffic states. The tendency to reach a pleasant level of "driving comfort" gives rise to new $v(\Delta x)$-diagrams, known as *Optimal-Velocity* (*OV*) curves. The insights found at locally fixed detectors can be confirmed by *FC* data. In fundamental diagrams derived from local measurements new high-flow states can be observed.

1 Introduction

Experimental and theoretical investigations of traffic flow have been the focus of extensive research interest during the past decades. Various theoretical concepts have been developed and numerous empirical observations have been reported (see e.g., [1–15] and references therein). However, most of them are still under debate. Recent experimental observations suggest the existence of three qualitatively different phases [5]: 1) *Free-flow states* with a large mean velocity, 2) *synchronized states*, where the mean velocity is considerably reduced compared to the free-flow states, but all cars are moving, and 3) *stop-and-go states* with broad jams. The last two states are summarized as *congested states*. Following [5] three different types of synchronized traffic exist. Long time intervals of constant density, flow, and velocity characterize the first type (i). In the second type (ii) variations of density and flow are observable, but the velocity of the cars is almost constant. Finally, there is no functional dependence between density and flow in the third type (iii) of synchronized traffic.

This work focuses basically on two points. First, we present a *direct analysis of single-vehicle* data which leads to a more detailed characterization of the different microscopic states of traffic flow, and second, we use standard techniques of time-series analysis, i.e., utilization of autocorrelation and cross-correlations, in order to establish *objective* criteria for an identification of the different states [15]. These investigations have been carried out using extended data sets from locally fixed detectors on a German highway. Concluding, the insights we got are compared with results drawn from car-following experiments.

2 Data from Locally Fixed Detectors

The data set is provided by 12 counting loops all located at the German highway A1 near Köln (Fig. 1) [14,15]. At this section of the highway a speed limit of 100 km/h is valid – at least theoretically. A detector consists of three individual detection devices, one for each lane. By combining three devices covering the three lanes belonging to one direction (except D2) one gets the cross-sections. D1 and D4 are installed nearby the highway intersection "Köln-Nord", while the others are located close to a junction. These locations are approximately 9 km apart. Within two weeks in June 1996 more than two million single-vehicle measurements were recorded. Nearly 16 % of the detected vehicles were trucks. During this period the traffic data set was not biased due to road constructions or bad weather conditions. The traffic computer recorded the distance-headway Δx as well as the velocity v of the vehicles passing a detector. The time elapsed between two consecutive vehicles has to be calculated via $\Delta t_n = \Delta x_n / v_n$ for the vehicle n and its predecessor $n-1$[1].

For a sensible discussion it is plausible to split up the data set according to the different traffic states. Time-series of the speeds allow for a manual separation into free flow and congested states. The congested states, in turn, are composed of stop-and-go traffic as well as of synchronized states (here only type (iii)). A distinction is possible by utilizing the cross-correlation as described later. Beside

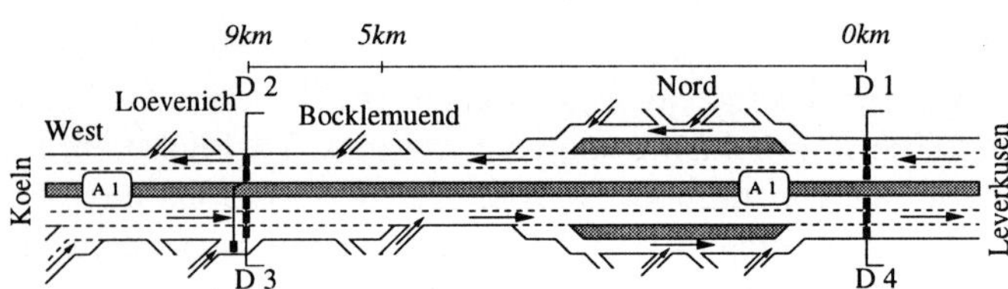

Fig. 1. The considered section of the highway A1 near Köln. The detector installations D1/D4 are nearly one kilometer apart from a highway intersection.

the separation with regard to the traffic states the data can be arranged in density intervals. The density ρ is a global quantity like flow J or mean velocity $\langle v \rangle$. In the present case the density can only be determined by the hydrodynamic relation $\rho = J / \langle v \rangle$. Due to the event-driven measurements one makes an error in the high-density regime. Standing cars lead to an overestimation of the velocity of passing cars and an underestimation of the density. Therefore, the data points in the fundamental diagram $J(\rho)$ (Fig. 2) are shifted to the left in stop-and-go

[1] In fact the velocity of a car passing the detector and the time elapsed since the leading vehicle has passed are the direct measurements (see [15] for an elaborate discussion).

traffic. In the same way other methods must fail, since the prime reason is the event-driven measurement method [17].

Correlation functions can detect and quantify coupling effects between variables. With ψ the variable of interest the autocorrelation

$$ac_{\psi}(\tau) = \frac{1}{\sigma(\psi)} \left[\langle \psi(t)\psi(t+\tau) \rangle - \langle \psi(t) \rangle^2 \right] \tag{1}$$

provides information about the temporal evolution of ψ ($\sigma(\psi)$ is the variance of ψ). In Fig. 2 the typical behavior for 1-min-averages is depicted. In free-flow traffic there are only slight changes in both flow and density. These correlations break down during the crossover to synchronized flow – the corresponding data points in the fundamental diagram are scattered over an extended plane. In stop-and-go traffic an oscillating structure can be found, a hint on the alternating occurrence of stall and movement [18].

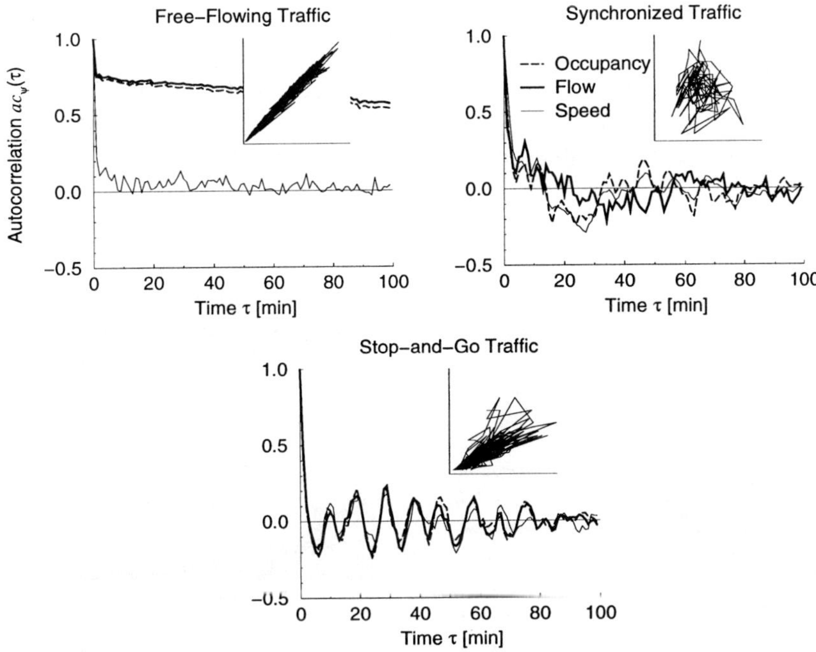

Fig. 2. Typical autocorrelation functions of aggregated variables depending on the traffic state. The correlation collapses in the case of synchronized flow, in stop-and-go traffic one observes oscillations [18].

Now, τ is substituted with η in (1) as well as t with n, where η is the number of consecutive vehicles. The results are plotted in Fig. 3. In the case of free-flowing traffic streams the vehicles move quite independent of each other, which

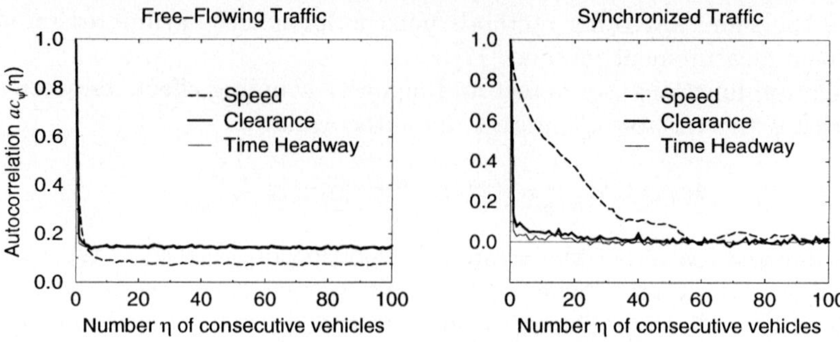

Fig. 3. If single-vehicle data are applied to the autocorrelation function, one can obtain a synchronization effect along the road.

supports the assumption of *random headway states* for the analytical description of time-headways. If synchronized traffic emerges an adaption of vehicular speeds between lanes [5] as well as among consecutive cars is to notice.

On the other hand, the cross-correlation detects, whether two variables ξ and ψ or coupled or not. It is defined as

$$cc_{\xi\psi}(\tau) = \frac{1}{\sqrt{\sigma(\xi)\sigma(\psi)}} \left[\langle \xi(t)\psi(t+\tau)\rangle - \langle \xi(t)\rangle\langle \psi(t+\tau)\rangle\right] . \tag{2}$$

By calculating the cross-correlation $cc_{J\rho}$ between flow and density an objective parameter is given for discriminating between the several states of traffic. From investigations of the present data base one may conclude the following:

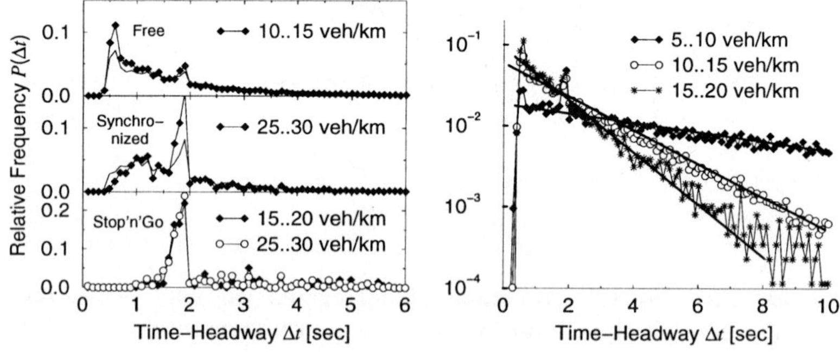

Fig. 4. At least one peak emerges in all traffic states, which position is independent of the density of traffic. In free-flow traffic the background signal can be described through a Poisson process with exponential decay [3].

Traffic state	$cc_{J\rho}$	Interpretation and remarks
Free flow	≈ 1	Density controls flow
Synchronized	≈ 0	Data points cover a plane in the fundamental diagram
Stop-and-Go	$\approx \pm 1$	Data points close to a line with positive or negative slope (with regard to the applied ρ-method)

Further insights into the car-car-interaction are provided by an analysis of the histograms of clearances, speed differences and time headways. Time headways of unbounded moving cars (Fig. 4) can be described through so-called *random headway states* which yields an exponentially decaying background signal of the probability distribution $P(\Delta t)$ [3] in the form

$$P(\Delta t) = \frac{1}{\tau'} e^{-\Delta t/\tau}, \tag{3}$$

where τ is directly linked to the present traffic flow J and $\tau' \approx \tau/\nu$ with ν the number of bins in an interval of 1 sec (here typically $\nu = 10$). Additionally, two peaks show up, the weaker one near $\Delta t = 2$ sec and the higher one below 1 sec and rising with ascending density. The latter corresponds to high-current states in the fundamental diagram, since small time-headways are necessary to generate them. With the transition to congested traffic states this background looses its evidence. In stop-and-go states there is only the peak at $\Delta t \approx 2$ sec – it reflects the fact that only cars leaving a jam are to detect with the typical time headway for this procedure.

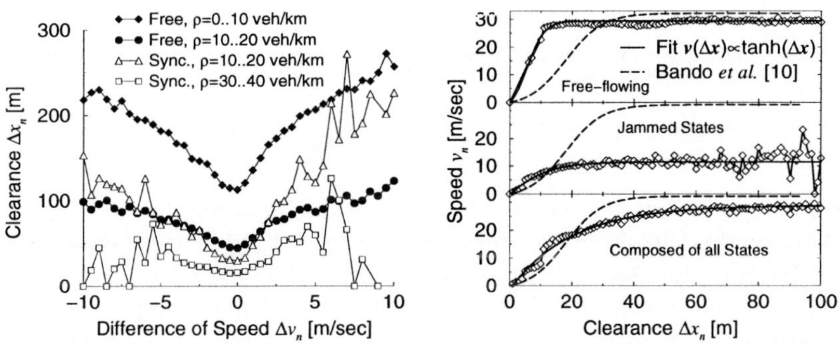

Fig. 5. Left: "Driving comfort" is characterized by a minimization of the difference of speeds between two consecutive cars with a minimal distance. Consistently, only data points with $\Delta v_n \approx 0$ m/sec should contribute to the OV curve [10].

The data are also suitable to generate OV diagrams. Before this, it is worth to have a look at the speed differences Δv_n. By means of Fig. 5 one realizes that the drivers tend to minimize driving actions and simultaneously to improve their "driving comfort": With vanishing Δv_n they follow their predecessors closer than suggested with a mean gap $\propto \rho^{-1}$, and this is the most probable situation. For deriving an *optimal* velocity only data points with $\Delta v_n \approx 0\,\mathrm{m/sec}$ are taken into account (right diagram in Fig. 5). Additionally, the data sets again were distinguished according to the underlying global traffic state. This leads to different *OV* relations, which can be fitted nicely by a `tanh` ansatz.

3 *Floating-Car* Data

The *FC* data were collected during field studies in the framework of the Research-Cooperative "NRW-FVU" of North-Rhine Westfalia [6]. The data sets recorded during such studies contain the own velocity v, the difference of velocities Δv to the predecessor and the clearance Δx to him.

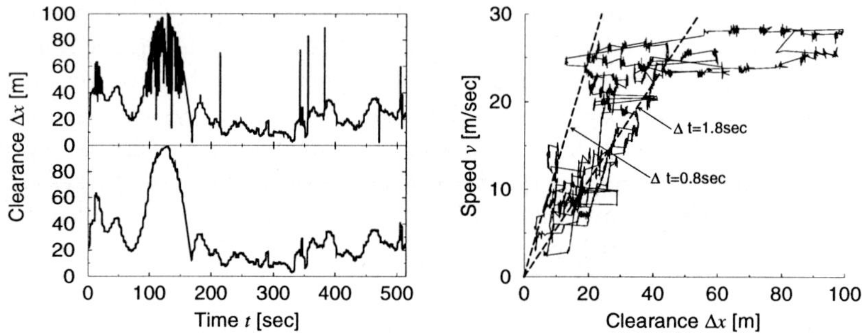

Fig. 6. After a clean-up all errors in the time-series were erased. With this data base the speed-distance relation (right) is plotted.

The distance to the leading vehicle and its velocity is measured using a laser beam. This method leads to very precise results as long as the beam does not miss the leading car. But nevertheless a part of the data set is not correct, e.g., data recorded during lane-changing maneuvers. These faulty items can be easily identified and have to be omitted before analyzing the data set (see Fig. 6). Afterwards the data set is diluted in a way that only one measurement per second remains. The $v(\Delta x)$-diagram (here without discrimination due to traffic states) now is comparable to the previously presented one. In free flow the data points are arranged on a line parallel to the ordinate, very small clearances (down to $\Delta x \approx 15\,\mathrm{m}$ at high speeds ($100\,\mathrm{km/h}$)) can be obtained. From this microscopic level of view also macroscopic quantities can be computed. Since $J \propto \Delta t^{-1}$ and $\rho \propto \Delta x^{-1}$ a fundamental diagram can be plotted (Fig. 7). Two important facts have to be mentioned: There are stable high-flow states persisting for minutes and caused by two consecutive vehicles with less or no fluctuations in the driver's behavior. These results support the view that already in the free-flow regime small platoons can be observed which move at small distances. The transition into congested states takes place only at reduced traffic volumes and not at or nearby the maximum.

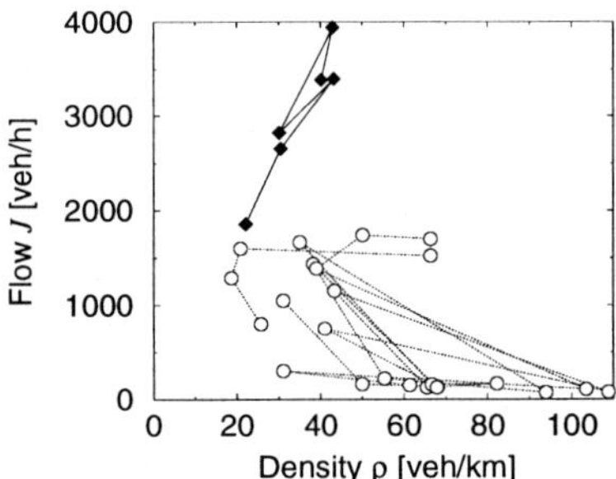

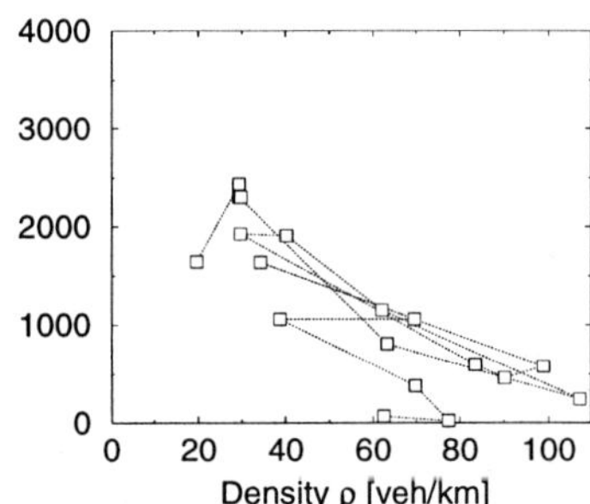

Fig. 7. Fundamental diagrams drawn from local measurements. Of special interest is the long-lasting high-flow state (left), which corresponds to a stable follow-the-leader formation with short time-headways for minutes.

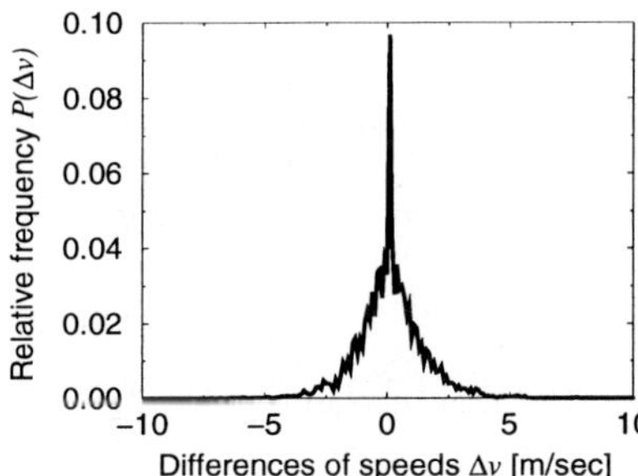

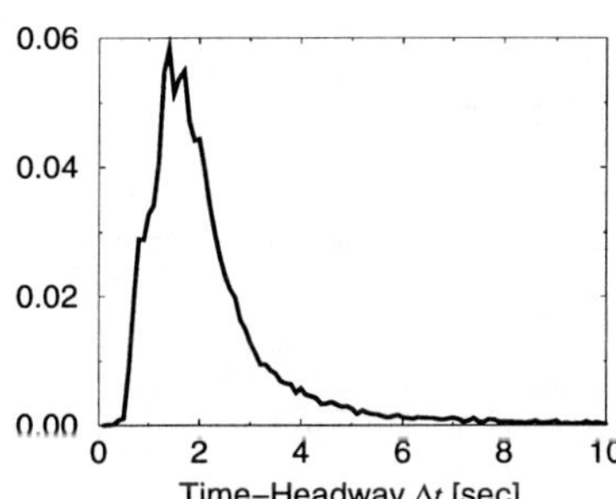

Fig. 8. By *FC* data previous results can be verified. The driver tends to make Δv disappear, his preferred temporal headway is around 2 sec.

The histograms in Fig. 8 confirm the findings of the previous section. Especially, one recognizes that the state $\Delta v \approx 0\,\mathrm{m/sec}$ is the most attractive one, and the most probable time headway is located near $\Delta t = 2$ sec. Crucial contributions to the distribution come also from time headway shorter than 1 sec.

Acknowledgement. It is our pleasure to thank B.S. Kerner and D. Chowdhury for fruitful discussions. The authors are grateful to the "Landschaftsverband Rheinland" (Köln) for data support, to the "Systemberatung Povse" (Herzogenrath) for technical assistance, to the Ministry of Economic Affairs, Technology and Transport of North-Rhine Westfalia as well as to the Federal Ministry of Education and Research of Germany for the financial support (the latter within the BMBF project "SANDY"). Parts of this work were done in the framework of the research-cooperative "NRW-FVU" of North-Rhine Westfalia.

References

1. D.E. Wolf, M. Schreckenberg, and A. Bachem (Eds.), *Traffic and Granular Flow* (World Scientific, Singapore, 1996); M. Schreckenberg and D.E. Wolf (Eds.), *Traffic and Granular Flow '97* (Springer, Singapore, 1998).
2. H. Leutzbach, *Introduction to the theory of traffic flow* (Springer, Berlin, 1988); D. Helbing, *Verkehrsdynamik* (Springer, Berlin, 1997); A. Ceder (Ed.), *Proceedings of the 14th International Symposium on Transportation and Traffic Theory* (Pergamon, Amsterdam, 1999).
3. A.D. May, *Traffic Flow Fundamentals* (Prentice Hall, Englewood Cliffs, 1990).
4. P.H.L. Bovy (Ed.), *Motorway analysis: New methodologies and recent empirical findings* (University Press Delft, Delft, 1998).
5. B.S. Kerner, Phys. Rev. Lett. **81**, 3979 (1998); B.S. Kerner and H. Rehborn, Phys. Rev. E **53**, R4275 (1996); Phys. Rev. Lett. **79**, 4030 (1998); Phys. Rev. E **53**, R1297 (1996).
6. W. Brilon, F. Huber, M. Schreckenberg, and H. Wallentowitz (Eds.), *Traffic and Mobility: Simulation – Economics – Environment* (Springer, Heidelberg, 1999).
7. D. Chowdhury, L. Santen, A. Schadschneider, S. Sinha, and A. Pasupathy, J. Phys. A **32**, 3229 (1999).
8. A. Schadschneider, Eur. Phys. J. B **10**, 573 (1999).
9. M. Koshi, M. Iwasaki, and I. Okuhra, *Some findings and an overview on vehicular flow characteristics*, in: V.F. Hurdle, E. Hauer, and G.N. Steward (Eds.), *Proceedings of the 8th International Symposium on Transportation and Traffic Theory* (University of Toronto Press, Toronto, 1981).
10. M. Bando, K. Hasebe, A. Nakayama, A. Shibata, and Y. Sugiyama, Jap. J. Indust. Appl. Math. **11**, 203 (1994).
11. R. Barlovic, L. Santen, A. Schadschneider, and M. Schreckenberg, Eur. Phys. J. B **5**, 793 (1998).
12. B.S. Kerner and P. Konhäuser, Phys. Rev. E **48**, R2335 (1993).
13. M. Treiber and D. Helbing, cond-mat/9901239 (1999).
14. P. Wagner and J. Pleinke, Z. Naturforsch. **52a**, 600 (1997).
15. L. Neubert, L. Santen, A. Schadschneider, and M. Schreckenberg, Phys. Rev. E **60**, 6480 (1999).
16. D. Chowdhury, A. Pasupathy, and S. Sinha, Eur. Phys. J. B **5**, 781 (1998).
17. H. Rehborn, private communications within the research project "SANDY".
18. R.D. Kühne, *Freeway speed distribution and acceleration noise – Calculations from a stochastic continuum theory and comparison with measurements*, in: N.H. Gartner and N.H.M. Wilson (Eds.), *Proceedings of the 10th International Symposium on Transportation and Traffic Theory* (Elsevier, New York, 1987).

Nonlinear Control of Stop-and-Go Traffic

R. Sollacher and H. Lenz

Siemens AG, Corporate Technology, Information and Communications, 81730 Mnchen, Germany

Abstract. In highway traffic one observes metastabiltiy in a certain range of the traffic density. Perturbations of a large enough amplitude develop into jams or stop-and-go waves. Observations and simulations indicate that the latter behave like solitary waves. Current traffic control systems based on varying speed limits cannot always prevent the creation of stop-and-go waves nor can they damp them. We have designed a nonlinear controller which is able to extinguish stop-and-go waves; its performance is demonstrated within simulations using a realistic traffic model. The construction is based on a transformation to a coordinate system moving with the propagation speed of the stop-and-go waves.

1 Highway Traffic and Control

One of the most interesting features of highway traffic is its metastability: once the traffic density exceeds a certain critical value, perturbations of a large enough amplitude lead to the formation of jams or stop-and-go waves [4]. Both of these seem to be remarkably stable structures persisting for several hours until the traffic flux feeding them becomes too low. Having a constant propagation velocity and constant shape fully established stop-and-go waves can be considered as solitary waves [1]. Although very appealing from a physicists point of view these phenomena of self-organisation in traffic reduce the efficiency of transport on highways: as the flow out of a jam is considerably smaller than the maximum possible flow of homogeneous traffic they effectively decrease the highway capacity. In addition, stop-and-go waves constitute a potential source of accidents due to their sharp transitions from free flow to stopping propagating even upstream.

Concerning the control of traffic flow, there are two major research directions. In North-America the complete control of a car by the AHS (Automated Highway System) is favoured (e.g., [7]), while in Europe control actions have to be realized by mandatory limiting speed signs displayed by gantries above the highway [8–14]. Early investigations by Zackor showed that an increase of the maximum flow can be achieved by speed limitations [13]. However, none of these approaches is based on the characteristic features of stop-and-go traffic.

In the following we present an approach for the design of a nonlinear control law creating a homogeneous traffic flow by mandatory speed signs. The approach is based on a macroscopic model showing the characteristic properties of traffic flow [4]. In particular, it is able to describe stop-and-go waves. It is given by the following partial differential equations for the local density $\rho(x,t)$ and the local average speed $v(x,t)$:

$$\dot{\rho} + q' = 0$$

$$\dot{v} + vv' = \frac{1}{\tau}\left(V(\rho, u) - v\right) - \frac{c_0^2}{\rho}\rho' + \frac{\eta_0}{\rho}v'' \ . \tag{1}$$

Here, $q = \rho v$ denotes the traffic flux and $\dot{}$ and $'$ denote the partial derivatives with respect to time t and space x. The parameters τ, c_0 and η_0[1] have to be chosen such that the model describes real world traffic phenomena as appropriately as possible. In a homogeneous state drivers tend to travel with a desired velocity $V(\rho, u)$ depending on the density ρ respectively distance between the cars and on a control action u which represents the effect of, e.g., a speed limit.

In a first attempt a standard linear feedback control rule was investigated. It stabilises traffic flow but only with unrealistically large control actions [2]. Therefore, two well known approaches from nonlinear control theory have been applied to this problem: input-state-linearisation [18–20] and the sliding-mode approach [18,21,22]. The resulting control laws are almost identical. Therefore only the sliding-mode approach is presented.

2 Controller Design

Due to their robust properties, sliding-mode controllers are increasingly used in order to control nonlinear systems of the form given by

$$\frac{d^n y}{dt^n} = f(\boldsymbol{y}) + b(\boldsymbol{y})u \ , \tag{2}$$

where $\boldsymbol{y} = (y, \dot{y}, \ldots, y^{(n-1)})$ is the state vector depending on some "time"-variable t, n is the order of the system given, and $u = u(t)$ is the control signal [22]. However, the traffic model defined in (1) consists of a set of partial differential equations, while (2) describes a set of ordinary differential equations.

In order to make use of established results from nonlinear control theory one has to cast the dynamical traffic model (1) into a form equivalent to (2). Now, as already mentioned in the beginning, due to the metastability of traffic stop-and-go waves can develop under certain conditions. These stop-and-go waves have the character of solitary waves, i.e. they have a constant shape and propagate upstream with a constant velocity. This implies that a hypothetical observer moving with this constant velocity of the stop-and-go wave sees a stationary structure. Therefore, the description of these asymptotic structures is essentially one-dimensional and can be cast into the form of (2).

Let us consider this derivation in more detail. The assumption of a solitary wave implies

$$\rho(x,t) = \tilde{\rho}(x - v_s t), \tag{3}$$

$$v(x,t) = \tilde{v}(x - v_s t). \tag{4}$$

Defining

$$z = x - v_s t,$$

[1] We took the values $\tau = 6s$, $c_0 = 47.91km/h$ and $\eta_0 = 59.33m/s$.

$$V(\rho, u) = (1+u)V(\rho), \tag{5}$$

with the obvious definition $V(\rho) = V(\rho, u=0)$, the equations (1) can be cast into the following form:

$$\tilde{\rho} = \frac{q_0}{\tilde{v} - v_s}, \tag{6}$$

$$\frac{d^2\tilde{v}}{dz^2} = -\frac{q_0}{\eta_0}\left(\frac{c_0^2}{(\tilde{v}-v_s)^2} - 1\right)\frac{d\tilde{v}}{dz} - \frac{q_0}{\eta_0}\frac{V(\tilde{\rho}) - \tilde{v}}{\tau\,(\tilde{v}-v_s)} - \frac{q_0}{\eta_0}\frac{V(\tilde{\rho})}{\tau\,(\tilde{v}-v_s)}u. \tag{7}$$

Using the equivalent control method to design the nonlinear sliding-mode controller [22] leads to the following two components of the control action:

$$u = u_e + u_n, \tag{8}$$

$$u_e = \frac{\tilde{v} - V(\tilde{\rho}) + \frac{\tau}{(\tilde{v}-v_s)}\left((\tilde{v}-v_s)^2\left(1+\lambda\frac{\eta_0}{q_0}\right) - c_0^2\right)\frac{d\tilde{v}}{dz}}{V(\tilde{\rho})}, \tag{9}$$

$$u_n = K\frac{\tau\eta_0(\tilde{v}-v_s)}{q_0 V(\tilde{\rho})}\mathrm{sgn}(s). \tag{10}$$

Here, u_e is the "equivalent" control, u_n is the non-continuous part and $s = \lambda v + dv/dz$ is a control variable with $s=0$ defining the sliding surface; λ and K are parameters. In general, K is chosen sufficiently large such that u_n accounts for system uncertainties. However, such an imprecise knowledge of the system could also be given by the approximation

$$\hat{u}_e = \frac{\tilde{v} - V(\tilde{\rho})}{V(\tilde{\rho})}. \tag{11}$$

This approximation is motivated by the fact that the acceleration dv/dz in practice is not measured and therefore would have to be estimated.

So far, the design of the controller has been performed in a specific coordinate frame and for a situation in which we have reached a specific asymptotic state characterised by stop-and-go waves. As the controller was constructed such that this asymptotic state becomes a homogeneous one, one might expect that the controlled system no longer exhibits stop-and-go waves. In order to show this we have to re-transform the controlled system to the original coordinate system. Identifying

$$\frac{dv}{dt} = \dot{v} + vv' \approx (\tilde{v} - v_s)\frac{d\tilde{v}}{dz}, \tag{12}$$

replacing $\tilde{v}$ by v, $\tilde{\rho}$ by ρ and using (11) we arrive at the following equation for the acceleration:

$$\dot{v} + vv' = -\frac{c_0^2}{\rho}\rho' + \frac{\eta_0}{\rho}v''. \tag{13}$$

This system turns out to be stable with respect to small perturbations [2]. However, the system is also stable with respect to large perturbations as shown in Fig. 1b.

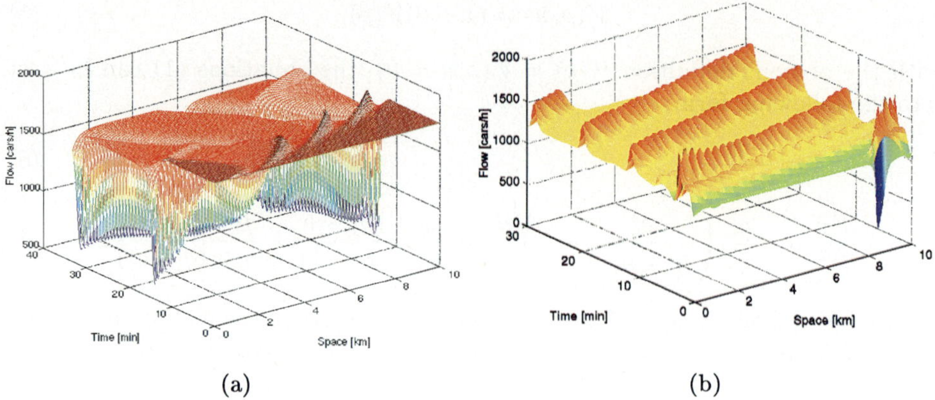

(a) (b)

Fig. 1. (a) Stop-and-go waves develop from a small initial perturbation in the uncontrolled system. (b) An initial stop-and-go wave is damped in the controlled system.

3 Towards Practical Realization

In order to meet constraints set by existing traffic controlling systems, further approximations are necessary. Up to now, mandatory speed limits have discrete values and cannot change continuously in space and time. A closer look at the control rule described above reveals that stop-and-go waves are damped by increasing the flow above some density and by decreasing the flow below this density. This effect can be mimicked by a suitable change of speed limits.

Extending an analysis of Zackor [13] using a large set of data from highway A9 North of Munich one finds different averaged flow-density relations for each speed limit. These fitted curves can be seen in Fig. 2a. Switching to another speed limit at certain threshold densities yields an effective flow-density relation (bold curve in Fig. 2a). The threshold densities are chosen such that stop-and-go waves are damped but also that the system is linearly stable.

Finally, speed limits are piecewise constant in space because the displays are located some distance apart. In a simulation we have investigated the effect of this distance on the quality of the controller measured in difference between maximum and minimum density on the road which is finally reached. The threshold density is determined either locally, i.e., at the position of the speed limit display, or non-locally, i.e., as an average of the local density and the one at the next gantry. As shown in Fig. 2b the nonlocal version shows a dramatically better performance up to a distance between gantries of about 800 m. This distance is probably related to the typical wavelength of stop-and-go waves. In order to determine this value further data analysis is needed.

Acknowledgement. This work was partially supported by the German Ministry of Education and Research (BMBF-Project "SANDY", grant 13N7092/2). One of the authors (H. Lenz) is grateful for receiving a PhD-scholarship by

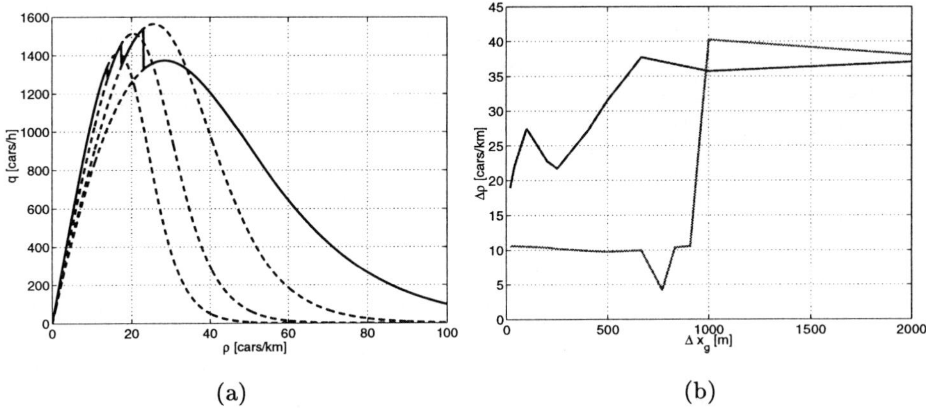

(a) (b)

Fig. 2. (a) Fitted flow-density relations for speed limits 120 km/h, 100 km/h, 80 km/h and 60 km/h (from left to right). The bold line is the effective flow-density relation resulting from switching speed limits. (b) The maximum difference in density resulting from an application of speed limit control using local (blue/dark) and nonlocal (red/bright) threshold densities.

Siemens AG, Corporate Technology, Department Information and Communications, Munich, Germany.

References

1. J.S. Russell, *Report of the fourteenth meeting of the British Association for the Advancement of Science*, York, Sept. 1844, London 1845, 311-390, Plates XLVII-LVII.
2. H. Lenz, *Entwicklung nichtlinearer, diskreter Regler zum Abbau von Verkehrsflussinhomogenitäten mit Hilfe makroskopischer Verkehrsmodelle,* PhD Thesis, (Shaker, Aachen, 1999).
3. H. Lenz, R. Sollacher, and M. Lang, *Nonlinear Speed-Control for a Continuum Theory of Traffic Flow,* pp. 67–72 (IFAC'99, 14th World Congress, Beijing, China, Vol. Q, 1999).
4. B.S. Kerner and P. Konhäuser, *Structure and parameters of clusters in traffic flow,* Phys. Rev. E **50**, 54–83 (1994).
5. D. Helbing, *Fundamentals of Traffic Flow,* Phys. Rev. E **55**, 3731–3738 (1997).
6. C. Wagner, C. Hoffmann, R. Sollacher, J. Wagenhuber, and B. Schürmann, *Second-Order Continuum Traffic Flow Model,* Phys. Rev. E **54**, 5073–5085 (1996).
7. R. Horowitz, *Automated Highway Systems: The Smart Way to Go,* pp. 452-463 (8th IFAC Symp. on Transportation Systems, Chania, Greece, 1997).
8. A. Stotsky, *Adaptive/Variable Structure Control of Traffic Flow,* pp. 307–312 (IFAC, 13th World Congress, San Francisco, 1996).
9. M. Cremer and S. Fleischmann, *Traffic Responsive Control of Freeway Networks by a State Feedback Approach,* pp. 357–376 (Proc. 10th Int. Symp. Transpn. and Traffic Theory, 1987).

10. M. Cremer, *Traffic Flow on Highways: Models, Observation and Control,* (Springer, Berlin, 1979).
11. M. Papageorgiou, *Applications of Automatic Control Concepts to Traffic Flow Modeling and Control,* (Lecture Notes in Control and Information Sciences, Vol. 50, Springer, Berlin, 1983).
12. S. Smulders, *Control by Variable Speed Signs - The Dutch Experiment,* pp. 99–103 (6th IEE Conf. on Traffic Monitoring and Control, 1992).
13. H. Zackor, *Judgement of traffic-dependent speed limits on highways,* (Forschung, Straßenbau und Verkehrstechnik, Heft 128, 1972).
14. H. Zackor, R. Kühne, and W. Balz, *Investigation of traffic flow when the maximum capacity is reached and when the flow is instable,* (Forschung, Straßenbau und Straßenverkehrstechnik, Heft 524, 1988).
15. B.S. Kerner and H. Rehborn, *Experimental features and characteristics of traffic jams,* Phys. Rev. E **53**, R1297–R1300 (1996).
16. B.S Kerner and H. Rehborn, *Experimental Properties of Complexity in Traffic Flow,* Phys. Rev. E **53**, R4275–R4278 (1996).
17. B.S. Kerner, *Experimental Characteristics of Traffic Flow for Evaluation of Traffic Modelling,* pp. 793–798 (IFAC, Transportation Systems, Chania, Greece, 1997).
18. J.-J.E. Slotine and W. Li, *Applied Nonlinear Control,* (Prentice Hall, Englewood Cliffs, New Jersey, 1991).
19. H. Lenz and D. Obradovic, *Global Control of Lorenz Chaos,* pp. 1486–1487 (Proc. 36th IEEE CDC, San Diego, Vol. 2, 1997).
20. H. Lenz and D. Obradovic, *Robust Control of the Chaotic Lorenz System,* Int. J.of Bifurcation and Chaos **7**, 1847–2854 (1997).
21. H. Lenz, R. Berstecher, and M. Lang, *Adaptive Sliding-Mode Control of the Absolute Gain,* pp. 667–672 (Proc. of the IFAC Nonlinear Control Systems Design Symp., Enschede, The Netherlands, Vol. 3, 1998).
22. J.H. Hung, W. Gao, and J.C. Hung, IEEE Transactions on Industrial Electronics **40**, 2–22 (1993).

Online Simulation and State Estimation for a Traffic Flow Model

J. Meier

University of Stuttgart, Institute for System Dynamics and Control Engineering, Pfaffenwaldring 9, 70550 Stuttgart, Germany

Abstract. For advanced traffic information systems and traffic control systems, the detailed and accurate knowledge of the current state of a road network is of great importance for their efficiency. In real traffic, measurements are mostly collected at local points, from sensors like loops, infrared sensors, or video cameras. To get the overall state of a traffic system, online traffic simulations or traffic state estimations are used. This paper focuses on the main differences between these two approaches. An example with real traffic data shows, that the state estimation can cope better with inaccurate initial conditions and measurement errors.

1 Problem Formulation

Considered is a section i of a motorway between two local sensors, e.g., inductive loops. This section is influenced by the neighbour sections $i-1$ and $i+1$. State variables are the traffic density c and the spatial mean speed v. The traffic flow is $q = c \cdot v$.

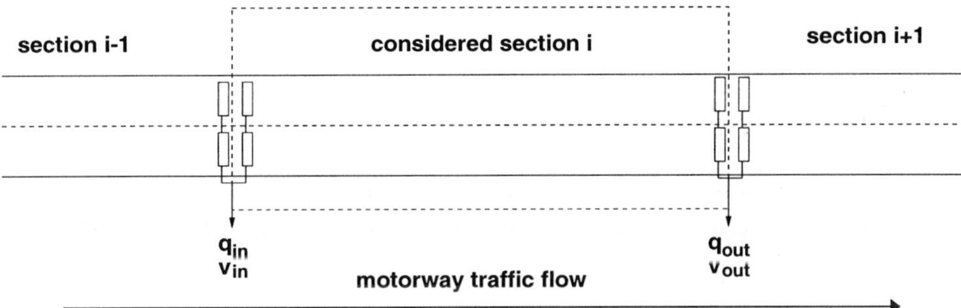

Fig. 1. Considered section of a motorway.

Measurement signals at the boundaries are:

- traffic flow q_{in} and local mean speed v_{in} at upstream loop,
- traffic flow q_{out} and local mean speed v_{out} at downstream loop.

It is intended to compute the variables c, v and q along the section i from the measurement signals by means of online simulation and of state estimation.

2 Macroscopic Traffic Flow Model

A macroscopic traffic flow model, proposed by Payne [4] and later refined and validated by Cremer [1], is used. The model is discrete in time ($t_k = kT$, $k = 0, 1, \ldots$) and in space ($j = 1, \ldots, n$). The considered section i is divided into n segments of length Δ_j. This leads to traffic density c_j and mean speed v_j in segment j and to the traffic flow q_j from segment j to segment $j+1$.

The time discrete model equations for the states are:

$$c_j(k+1) = c_j(k) + \frac{T}{\Delta_j}\left[q_{j-1}(k) - q_j(k)\right] , \qquad j = 1, \ldots, n \tag{1}$$

$$\begin{aligned} v_j(k+1) = v_j(k) &+ \frac{T}{\tau}\left[V(c_j(k)) - v_j(k)\right] \\ &+ \frac{T}{\Delta_j}\left[v_j(v_{j-1(k)} - v_j(k))\right] + \frac{\nu}{\Delta_j}\frac{T}{\tau}\left[\frac{c_j(k) - c_{j+1}(k)}{c_j(k) + \kappa}\right] . \end{aligned} \tag{2}$$

The equation for the densities is derived by a simple balance for the vehicles. The difference equation for the mean speed consists of a relaxation term, a convection term, and an anticipation term. $V(c)$ in (2) is the speed-density-characteristic. The traffic flow is computed as a weighted mean:

$$q_j(k) = \alpha c_j(k) v_j(k) + (1-\alpha) c_{j+1}(k) v_{j+1}(k) , \qquad j = 1, \ldots, n-1 . \tag{3}$$

The whole time discrete model is comprising $3n - 1$ equations. These model equations are completed by the measured variables $q_{\text{in}}, v_{\text{in}}, q_{\text{out}}, v_{\text{out}}$.

3 Online Simulation

For the online simulation, the measurements are used as boundary conditions to complete the model equations for the computation of the $3n + 3$ variables:

$$q_0 = q_{\text{in}},\ q_n = q_{\text{out}},\ v_0 = \frac{1}{\alpha}(v_{\text{in}} - (1-\alpha)v_1),\ c_{n+1} = \frac{q_{\text{out}}}{v_{\text{out}}} . \tag{4}$$

In order to initialize the online simulation, suitable initial conditions of c_j, v_j for $k = 0$ have to be chosen. In contrast to space continuous traffic flow models, all measurements are used as boundary conditions.

The problems of the online simulation are:

- errors of initial conditions are persistent in the simulation,
- measurement errors can accumulate.

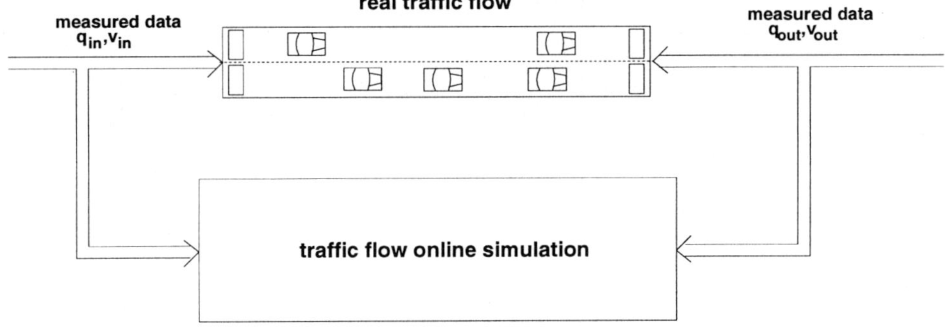

Fig. 2. Scheme of online simulation.

4 State Estimation with Kalman Filter

Another approach to solve the formulated problem concerns the estimation of the state variables of the traffic flow model with a Kalman filter [2,3].

The input variable u and output variable y (measurements) are introduced as:

$$u = q_{\text{in}} = q_0 \ , \qquad y = \begin{pmatrix} v_{in} \\ q_{out} \\ v_{out} \end{pmatrix} \ .$$

The output variables of the state estimation are obtained by means of extrapolation from the estimated states $\hat{c}$ and $\hat{v}$:

$$\hat{y} = \begin{pmatrix} \hat{v}_{in} \\ \hat{q}_{out} \\ \hat{v}_{out} \end{pmatrix} = \begin{pmatrix} \hat{v}_1 - \alpha\varepsilon(\hat{v}_2 - \hat{v}_1) \\ \hat{c}_n\hat{v}_n + (1-\alpha)\varepsilon(\hat{c}_n\hat{v}_n - \hat{c}_{n-1}\hat{v}_{n-1}) \\ \hat{v}_n + (1-\alpha)\varepsilon(\hat{v}_n - \hat{v}_{n-1}) \end{pmatrix} \ .$$

In a similar way, the boundary values are determined from the estimates:

$$\hat{v}_0 = \hat{v}_1 + \varepsilon(\hat{v}_1 - \hat{v}_2), \hat{c}_{n+1} = \hat{c}_n + \varepsilon(\hat{c}_n - \hat{c}_{n-1}), \hat{q}_n = \hat{q}_{out} \ .$$

For the application of the state estimation

- the traffic flow model must be observable by the measurements and
- the Kalman filter must be designed such that the difference between the output of real traffic flow and traffic flow simulation is minimized.

The basic scheme of a state estimation is shown in Fig. 3. Using the traffic flow simulation a prediction $\hat{y}$ is made for the actual measurements. This prediction is compared with the real measurements y. The resulting error signal $\hat{y}-y$ is used by a correction rule for the estimated states $\hat{c}_j, \hat{v}_j$ $(j = 1, \ldots, n)$ to improve the prediction.

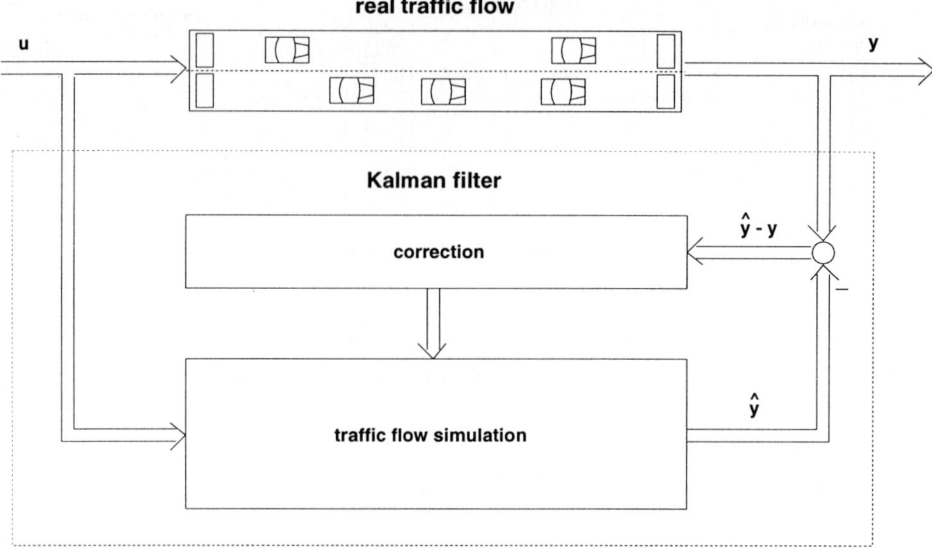

Fig. 3. Scheme of state estimation with Kalman filter.

The advantages of the state estimation are:

- the influence of initial errors is minimized
- measurement errors do not accumulate.

The accuracy of the estimation depends on the number of state variables that have to be estimated. So the length of the section and the accuracy of the spatial discretization are limited.

5 Application Example with Real Traffic Data

The two algorithms were tested with real traffic data from a 3.83 km long section at Motorway A9. In order to investigate the influence of initial errors, inaccurate initial conditions with higher densities have been used. Figure 4 shows the comparison of measured real traffic data, online simulation and state estimation at two cross sections 1.27 km and 2.66 km behind the first inductive loop.

Due to the higher densities, vehicles accumulate in the online-simulation in front of the downstream loop (Fig. 5a). This results in a decreased speed at the second cross section (Fig. 4a). The state estimation leads to realistic density and mean speed profiles (Fig. 5c,d). The local mean speeds at both cross sections are estimated sufficiently well (Fig. 4).

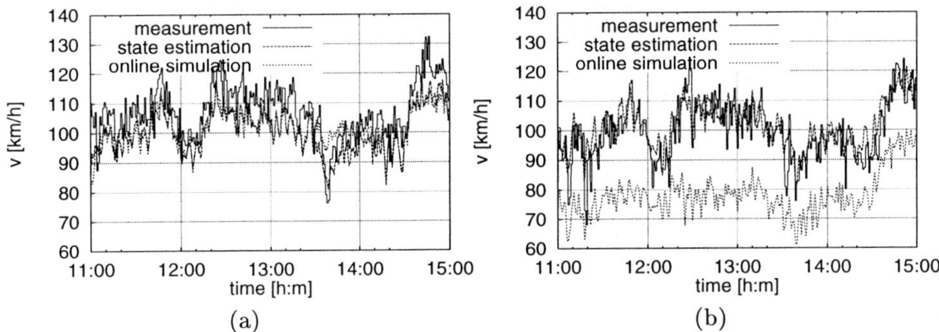

Fig. 4. Comparison between measurement, online simulation, and state estimation.

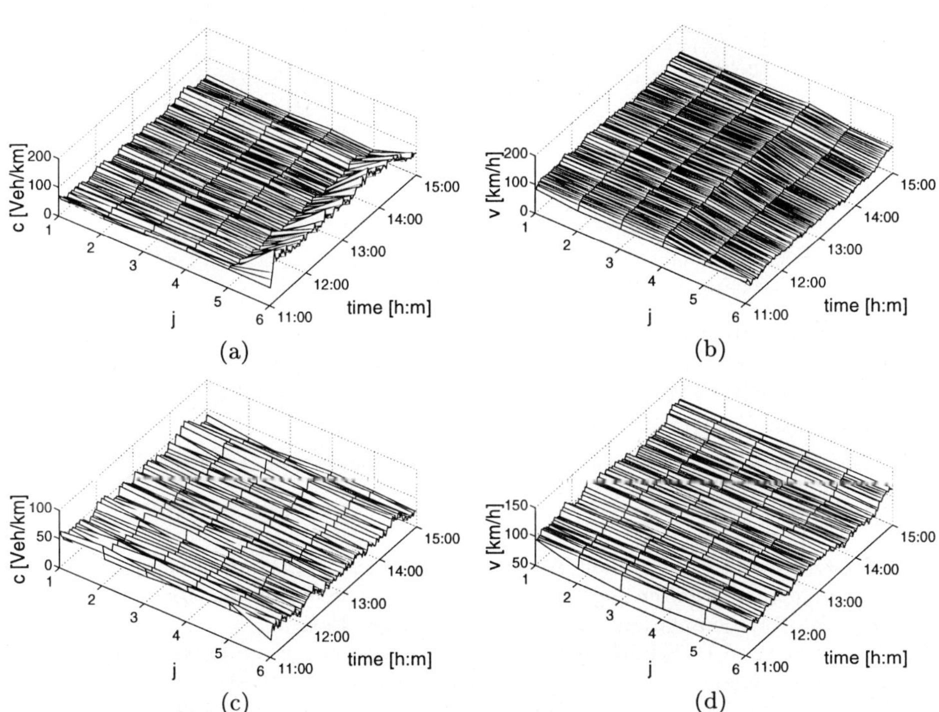

Fig. 5. Online Simulation: traffic density (a) and mean speed (b), state estimation: traffic density (c) and mean speed (d).

6 Application

The state estimation, in an augmented configuration, is used for automatic incident detection and estimation of the congestion tail for the COMPANION warning system in a project with the car manufacturer BMW AG. A field test on the Motorway A92 near Munich will be carried out. Figure 6 shows a demonstration tool for this application.

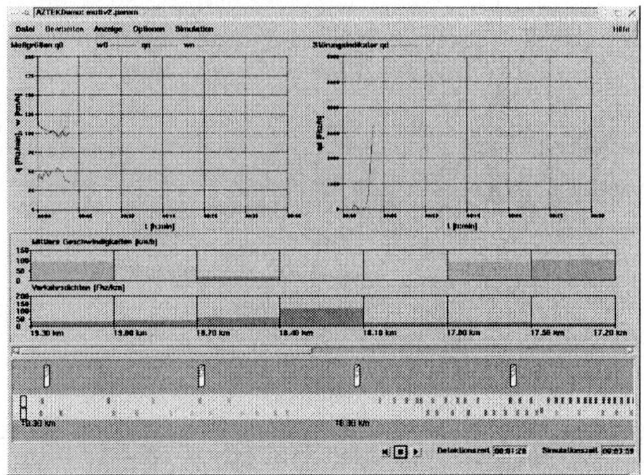

Fig. 6. Demonstration tool for automatic incident detection.

Acknowledgement. The work on this topic was started with Prof. Cremer at the Automatic Control Group, Technical University Hamburg-Harburg and is carried on with Prof. Wehlan and Prof. Zeitz at the Institute for System Dynamics and Control Engineering, University of Stuttgart.

References

1. M. Cremer and F. Meissner, *Traffic prediction and optimization using an efficient macroscopic simulation tool,* in: *Modelling and Simulation,* pp. 515–519 (Proc. of the 1993 Eur. Simulation Multiconference, Lyon, 1993).
2. M. Cremer and H. Schütt, *A comprehensive concept for simultaneous state observation, parameter estimation and incident detection,* pp. 95–111 (Proc. of the 11th Int. Symp. on Transportation and Traffic Theory, Yokohama, 1990).
3. D.C. Gazis and C.H. Knapp, *On-line estimation of traffic densities from time-series of flow and speed data,* Transp. Sci. **5**, 283-301 (1971).
4. H.J. Payne, *Models of freeway traffic control,* Simulation Council Proc. **1**, 51–61 (1971).

Traffic News by Dynamic Fuzzy Classification

C. Schnörr

DDG Gesellschaft für Verkehrsdaten mbH, Niederkasseler Lohweg 20,
40547 Düsseldorf, Germany

Abstract. A necessary prerequisite for numerous services in traffic telematics is a good knowledge of the current traffic situation by location and time. An interdisciplinary solution[1] is presented, which provides a location-time-dynamical description of the traffic situation by revealing traffic domains with uniform traffic states. For this pattern recognition task in a dynamical process with even chaotic traffic state transitions, the underlying measurements are classified into different traffic states. Some specific problems arising in this context are solved: 1) Data from different sources and of different units of measurement, e.g., velocities, flows and densities, which are only sparsely available with respect to location and time are combined ("data fusion"). 2) "Floating Car Data (FCD)" is integrated using morphological filters and recognizing its location-time relation compared to other data. 3) The stochastic data are filtered without suppressing significant state transitions. 4) The subjective feeling of road-users, that it is difficult to distinguish between different traffic states, is taken into account by fuzzy-classification. 5) Using a region growing method the segmentation problem of traffic domains is solved without being restricted to a predefined coarse road segmentation. 6) Good stability is obtained despite contradictory demands for a high resolution, a short reaction time and the differentiation of more than two traffic states. The chosen approach is confirmed by results with actual traffic data. The author knows of no other procedure with comparable capabilities.

1 Introduction and Problem Description

Due to the complexness of traffic dynamics states can change rapidly causing obstructions and even road accidents. Therefore there is a high demand for timely and location-exact information about traffic states. Especially important is the recognition of traffic states which typically precede traffic jams, for example dense and stop-and-go traffic.

Traffic domains, for which a particular uniform traffic state is predominant, spread, grow, shrink, move, divide themselves, until they finally disappear and free traffic prevails again. **Therefore the challenge is to find, localize, classify and keep track of these traffic domains**. This information must then be prepared in a form suitable for presentation to road-users. This paper presents a new and robust procedure for the automatic generation of traffic news in which **interdisciplinary methods from diverse fields like pattern recognition, segmentation, data fusion and morphology are combined**.

[1] It's patent pending in some countries.

2 Properties of Measurements

At present, essentially only three types of traffic measurement systems are being used on a large scale: 1) induction loops embedded in the roadway (VIZ), 2) stationary installed infrared or radar systems (SES), and 3) mobile systems, which provide so-called "Floating Car Data (FCD)". Induction loops provide average traffic volumes (in cars/time) and velocities from fixed locations and constant averaging periods. SES-detectors provide the same, but for economic reasons their data transmission is event triggered. On the other hand floating-cars provide only velocities from non-predefined locations. Due to the limited number of detectors and their different characteristics and due to temporary failures, traffic measurements are only sparsely and irregularly available with respect to location and time.

3 Other Traffic State Detection Methods

In some traffic information centers (VIZ) a traffic news generating procedure is used which classifies single velocity values by a binary threshold: congested or free. Consequently it produces relatively instable traffic news.

There are also methods based on road-segments, which balance the traffic volume at the beginning and the end of a segment [1,2]. Such methods can detect a congestion before it is seen at a measuring point. Substantial drawbacks however are that 1) these methods pose higher demands on the properties of the available measurements. 2) Asynchronous data streams (SES) or asynchronous data from variable locations (FCD) are difficult to handle. 3) Measuring errors cause stability problems for flow balance. 4) They are restricted to predefined road-segments.

4 Difficulties Describing Dynamic Traffic States

The special challenge when describing dynamic traffic states is to differentiate between several traffic states quickly, stably and consistently using incomplete and stochastic traffic data originating from different physical measurements. To compound matters, according to the subjective feeling of road-users does not make clear distinctions between different traffic states.

Since traffic domains are highly variable with respect to location and time, we are here not only faced with a dynamic segmentation problem, but also with stably tracking these domains. To do this, a description is necessary which allows a similarity comparison between traffic domains with respect to their traffic states. In addition, a restriction to a predefined coarse road segmentation has to be avoided. Otherwise the true dynamic of the traffic cannot be adequately captured and described.

The art is to achieve a high stability for traffic announcements despite the contradictory demands for:

1. a high local resolution of the traffic description,
2. a short reaction time to changing traffic conditions and
3. the differentiation of more than two traffic states.

5 The Procedure for the Generation of Traffic News

The procedure allows spatial resolution information and reaction times to be parameterized over a wide range while preserving exact information about the length and the position of a traffic event within the traffic announcement.

Preprocessing of Measurements and Feature Extraction: In order to determine the traffic situation the spatial and temporal properties of traffic measurements have to be considered without reducing the reaction time. The target here is the extraction of features from the traffic measurements or patterns, which characterize the traffic situation in a unique way.

First, incorrect measurements are eliminated. Then, all measurement vectors are written in sliding history windows spanning the locations and times of interest, which are updated with the aggregation period of the induction loops. Values older than 20 minutes are deleted. These windows are evenly divided in cells, each of them is 200m by 1 minute. This gives sliding time functions of the measurements by maintaining the location/time relation, which is essential for the incorporation of FCD (Fig. 1 left/middle).

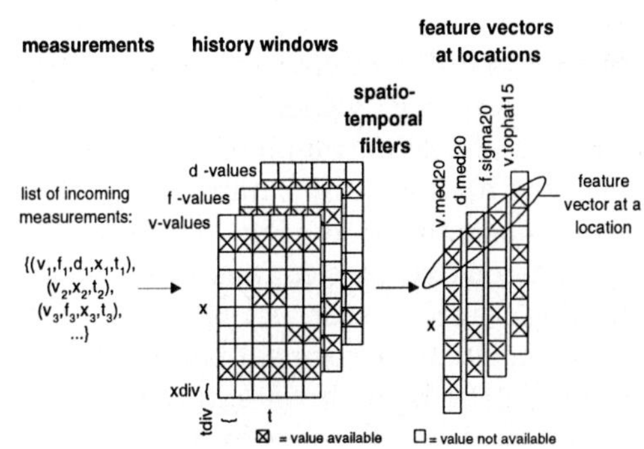

Fig. 1: Sketch of the processing line.

Due to the stochastic properties of the data, spatial and temporal filters are necessary to reduce the strong fluctuations in the velocity and density values without suppressing significant state transitions. The analysis of time functions shows that a median filter is more suitable than the common averaging filter.

The standard deviation of the traffic flow and the dilated TopHat function of the velocity quantify temporal changes of the traffic and are used as features to identify stop-and-go traffic: $(TopHat(v) = v - Opening(v))$, [3]. Temporal gaps can be easily taken into account by these filters.

This preprocessing provides a feature vector (Fig. 1 right): $(v.med20, d.med20, f.sigma20, v.tophat15)^T(x)$. This description based on features is the necessary prerequisite for stably classifying traffic states with good spatial and temporal resolution.

Fuzzy Classification: Now the traffic state is classified in one of four classes: congested, stop-and-go, dense and free traffic.

Taking into account the subjective feeling of road-users, that it is difficult to distinguish between different traffic states, the feature space is partitioned by discriminant fuzzy-functions (Fig. 2) [4]. Each sub-function evaluates a feature and quantifies the degree of existence of a traffic state.

The corresponding borders in the feature space smoothly turn from one state to the other. The continuous description is necessary to get stable classification results. Otherwise gradual changes in the input could lead to repeating state transitions, and the gradual spatial-temporal information of the traffic states would be lost. This is why binary decision techniques are completely inadequate.

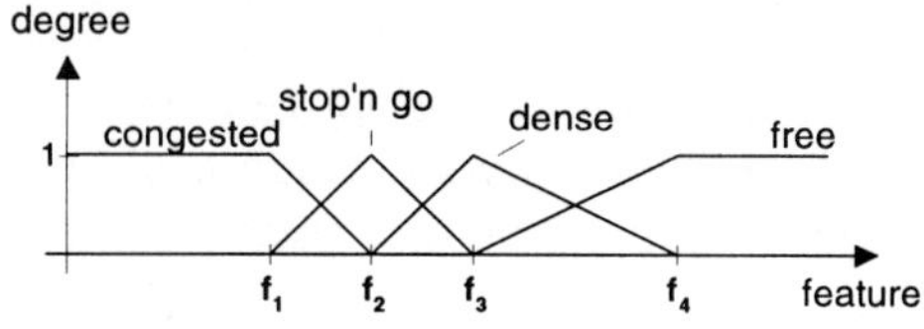

Fig. 2: Example of a fuzzy-division of a feature.

In addition, by using gradual state vectors, classification results for features of different physical units of measurement, e.g., velocities, flows and densities, can be combined in a consistent way, e.g., by summation and normalization (Fig. 3). All the different features contribute to the final state vector. At this point the results from segment based balancing methods can be integrated to provide additional information between measurement locations.

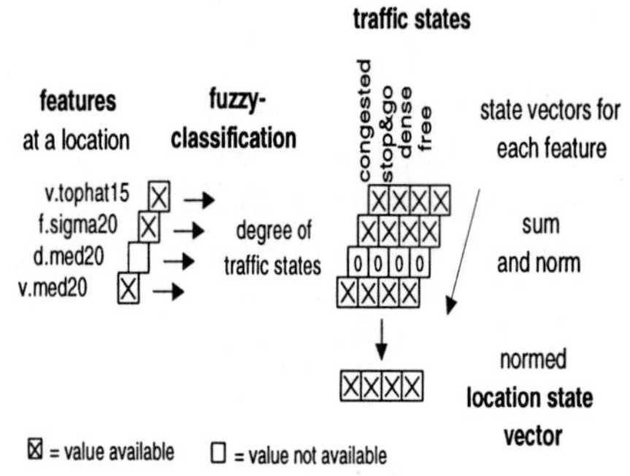

Fig. 3: Normalized local state vectors from state vectors for each feature.

The result of the fuzzy-classification is a state description at locations, where measurements were available and features could be calculated.

Extrapolation of Local Traffic State Vectors: In order to bridge the gaps in the traffic state description of a road, the state vectors available at measurement locations are extrapolated along the road by a Gaussian filter, which allows the locally reliable fuzzy-descriptions to be expanded to other locations, but with a decreasing amplitude reflecting the increasing distance from a real measurement location. The spatial resolution of the traffic news can be adjusted by the parameters of the Gaussian filter. In addition, a temporal autoregressive smoothing of the state vectors is carried out [5].

Dynamic Segmentation of Traffic Domains: The aim is to capture the actual spatial and temporal dynamics of the traffic without being limited to predefined fixed divisions of the road. This dynamic segmentation is achieved by a region growing method, which starts at a location and successively combines locations with similar traffic state descriptions to a traffic domain. The result is a complete partition of a road in traffic domains.

Up to now all calculations on the traffic state descriptions by the filter and the domain-segmentation have been based on fuzzy concepts. No binary decisions have taken place. Therefore, the domain state vectors still contain the complete information which is particularly important in the case of gradual state transitions and is a requirement for the stability of the following management system for traffic news.

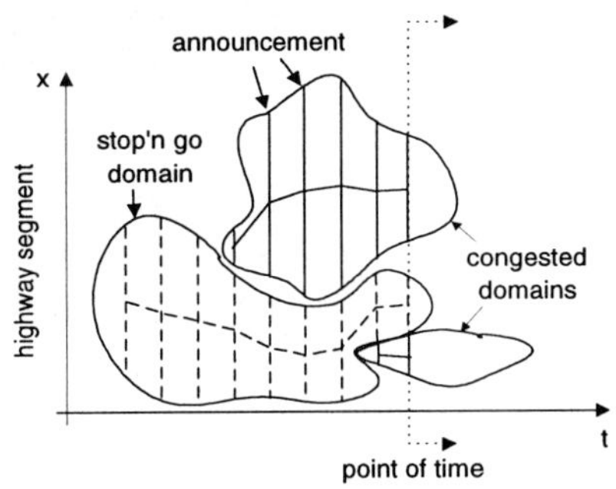

Fig. 4: Visualization of the results.

Domain Tracking and Management of Traffic News: The traffic domains are updated every 3 minutes and stored in a list. Using a continuous similarity measure a matching of the previous domains to the actual domains which considers the domain state vectors, the locations and the lengths of the domains is undertaken. Then the new domains, those which no longer exist and those which have undergone substantial changes are identified. Accordingly, delete-, new- and update-announcements are generated.

At this point the final decision about the traffic state based on the maximum component of the domain state vector is made. The result is a dynamic tracking of domains with similar traffic states and a list of announcements which is permanently kept up to date (Fig. 4).

Visualization: For visual monitoring during operation of the traffic news generating system the domains that belong to the traffic announcements are displayed in the foreground along with the interpolated traffic measurements in the background. The time increases from left to right. Domain regions are depicted by vertical lines, and horizontal lines connecting them show their temporal assignment during the tracking in successive updates (Fig. 4). In Fig. 5 an actual congestion situation is shown. In the background, interpolated velocities are displayed as grey values (dark = low v-values).

This presentation provides a visual impression of the quality of the results of the traffic dynamics analysis and an overview of the traffic conditions in the network of roads.

6 Results from Current Operation

The DDG already has considerable experience with the implementation and operation of this classification procedure. Typical reaction times are 5-10 minutes starting from the moment when a congestion situation is "seen" by a detector.

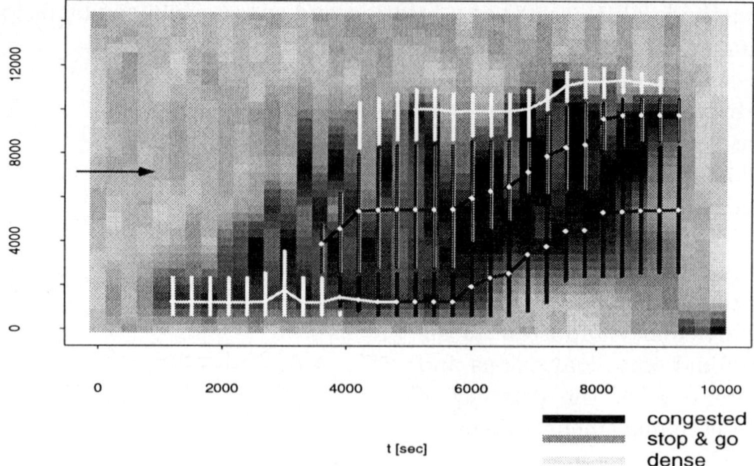

Fig. 5: Traffic messages produced for a real world congestion situation.

In most cases however, it's possible to warn of an overload congestion earlier by announcing dense or stop-and-go traffic, which usually precedes the congestion. First experience with FCD measurements shows that some congestion situations can be detected earlier than by other detector systems.

7 Outlook

An increase in the traffic information through additional measurements, e.g., by FCD, will allow more precise traffic news spatially as well as temporally without any change in the presented procedure. Further improvements can be achieved by the use of traffic flow models which will provide calculated measurements as additional input.

References

1. B. Steinauer, W.Krux, and F. Offermann, *Fuzzy-Logik kontra Schwellwertanalyse bei der Störfallergennung*, Straßenverkehrstechnik **7**, 323–330 (1997).
2. B. Krause and M. Pozybill, *Fuzzy-Logik in der Verkehrstechnik*, Straßenverkehrstechnik **8**, 356–362 (1996).
3. J. Serra, *Image Analysis and Mathematical Morphology*, (Academic Press, 1982).
4. R.O. Duda and P.E. Hart, *Pattern Classification and Scene Analysis*, (Wiley & Sons, New York, 1972).
5. A. Papoulis, *Probability, Random Variables, and Stochastic Processes*, (McGraw-Hill Series In Systems Science, McGraw-Hill, 3rd ed., 1991).

Evaluation of Single Vehicle Data in Dependence of the Vehicle-Type, Lane, and Site

B. Tilch[1] and D. Helbing[1,2]

[1] II. Institute of Theoretical Physics, University of Stuttgart, Pfaffenwaldring 57/III, 70550 Stuttgart, Germany
[2] Collegium Budapest – Institute for Advanced Study, 1014 Budapest, Hungary

Abstract. In this paper we study dependencies of fundamental diagrams, time gap distributions, and velocity-distance relations on vehicle types, lanes and/or measurement sites. We also propose measurement and aggregation methods that have more favourable statistical properties than conventional methods.

In the recent two years, traffic flow modelling has been more and more stimulated by empirical studies. In particular, single-vehicle data may shed some more light on the mechanisms of the transition from free to congested traffic flow. In contrast to early studies of, for example, the distribution of time gaps between successive cars (for an overview see [1]), Bovy *et al.* [2] have carried out separate analyses for different sample periods (morning/noon/evening) and different vehicle types (passenger-cars, articulated and non-articulated trucks). In addition, they have determined the relations between vehicle speeds and the distance gaps to the respective car in front separately for free and congested traffic, and for different cross sections of a Dutch freeway. Neubert *et al.* [3] mainly restrict to the analysis of one cross section of a German freeway. However, they go one step beyond the Dutch study by distinguishing not only free and congested traffic, but also different density regimes, which leads them to interesting conclusions. In the following, we try to combine both efforts by distinguishing free and congested traffic, different cross sections, and different vehicle types as well. At several cross sections of this freeway (Fig. 1), double induction loops record the time of passing, the lane, the velocity, and the length of each passing vehicle. In the following, vehicles longer than 6 m are denoted as trucks, shorter vehicles as cars. On average, there were about 20% of trucks in the right lane, and less than 1% trucks in the left lane, but the proportion was strongly varying in time [4].

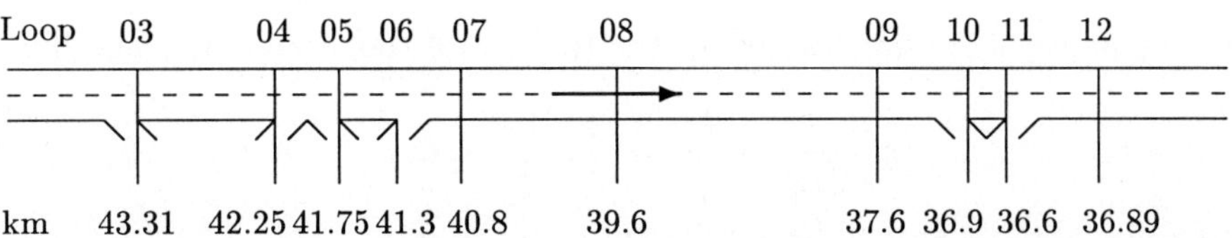

Fig. 1. Investigated section of the Dutch freeway A9 from Haarlem to Amsterdam, where a speed limit of 120 km/h applies, and locations of the measurement sites.

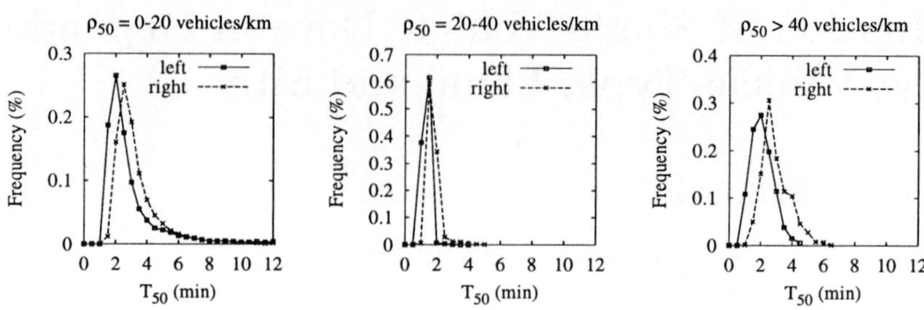

Fig. 2. Distribution of the measurement intervals T_{50} at different vehicle densities ρ_{50}. (The data were taken from all loops.)

Although this is not significant for the results of this study, we have not, as usual, determined macroscopic quantities like the traffic flow, the average velocity, or the vehicle density by averaging over fixed time periods (like 1 minute). In order to have comparable sample sizes, we have instead averaged over a fixed number N of cars, as suggested in [5]. Otherwise the *statistical* error at small traffic flows (i.e. at small and large densities) would be quite large. This is compensated for by a flexible measurement interval T_N (Fig. 2). It is favourable that T_N becomes particularly small in the (medium) density range of unstable traffic, so that the method yields a good representation of traffic dynamics. However, choosing small values of N does not make sense, since the temporal variation of the aggregate values will mainly reflect statistical variations, then. In order to have a time resolution of about 2 minutes on each lane, one should select $N = 50$, while $N = 100$ can be chosen when averaging over both lanes. Aggregate values over both lanes for $N = 50$ are comparable with 1-minute averages, but show a smaller statistical scattering at low densities (compare results in [5] and [6]). With increasing N, the maxima move to higher values, and the distributions of T_N become broader. The distribution for the left lane has its maxima at lower values of T_N, probably because of the smaller number of trucks. Throughout this paper, we use $N = 50$, but we have checked that our results don't change significantly for $N = 30$ or $N = 100$. Based on the passing times t_i of successive vehicles i in the same lane, we are able to calculate the time gaps $\Delta t_i = (t_i - t_{i-1}) > 0$. The (measurement) time interval

$$T_N = \sum_{i=i_0+1}^{i_0+N} (t_i - t_{i-1}) \equiv \sum_{i=i_0+1}^{i_0+N} \Delta t_i \quad (1)$$

for the passing of N vehicles defines the (inverse of the) traffic flow Q_N by:

$$\frac{1}{Q_N} = \frac{T_N}{N} = \frac{1}{N} \sum_{i=i_0+1}^{i_0+N} \Delta t_i \,. \quad (2)$$

Note that the traffic flow is very much dependent on the measurement site (Fig. 3). Consequently, the following graphs are different for other sites as well, but only in the quantitative details, not in a qualitative (fundamental) way.

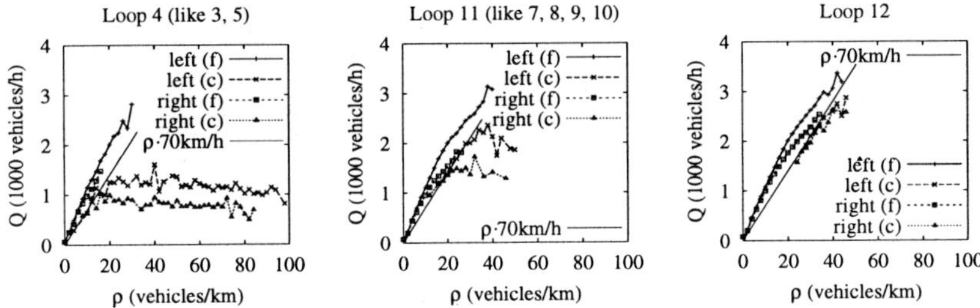

Fig. 3. Free (f) and congested (c) traffic flow as a function of the density for the left and the right lane at different cross sections of the road: upstream (left), downstream (middle), and far enough away from an on-ramp or bottleneck (right). From left to right we observe a relaxation from congested to free traffic.

Now, we approximate the (brutto) distance gap Δx_i by $\Delta x_i = v_i \Delta t_i$, where v_i is the actual velocity of vehicle i. This assumes that the vehicle velocities do not considerably change during the time interval Δt_i and implies

$$\frac{1}{Q_N} = \langle \Delta t_i \rangle_N = \left\langle \frac{\Delta x_i}{v_i} \right\rangle_N = \langle \Delta x_i \rangle_N \left\langle \frac{1}{v_i} \right\rangle_N + C_N \, , \tag{3}$$

where C_N is the covariance between the distance gaps Δx_i and the inverse velocities $1/v_i$. We expect that this covariance is particularly relevant at large vehicle densities (Fig. 4, left). The density ρ_N and the average velocity V_N are defined by

$$\frac{1}{\rho_N} = \langle \Delta x_i \rangle_N = \frac{1}{N} \sum_{i=i_0+1}^{i_0+N} \Delta x_i \qquad \text{and} \qquad \frac{1}{V_N} = \left\langle \frac{1}{v_i} \right\rangle_N = \frac{1}{N} \sum_{i=i_0+1}^{i_0+N} \frac{1}{v_i} \, . \tag{4}$$

Then, we obtain the fluid-dynamic flow relation $Q_N = \rho_N V_N$ by the conventional assumption $C_N = 0$ which, however, tends to overestimate the density (Fig. 4, middle). The fact that V_N is defined as the harmonic mean value of the vehicle velocities v_i automatically corrects for the fact that the *spatial* velocity distribution differs from the *locally measured* one (see [7,6] for details). The common method of determining the density via $Q_N / \langle v_i \rangle_N$ underestimates the density (Fig. 4, right).

In our investigation of *time gap distributions*, we have not only distinguished different density regimes, but also free traffic (f) and congested traffic (c) (Fig. 5). Measurement intervals with $V_N < 70\,\text{km/h}$ were classified as congested, otherwise traffic was considered to be free.

In all cases we found practically continuous, unimodal time gap distributions. The distribution for the right lane is usually broader than for the left lane, and its maximum lies at higher time gaps Δt. This is a consequence of the higher percentage of trucks. The maximum for the left lane varies from $\Delta t_{\text{max}}^{\text{free}} \approx 0.9\,\text{s}$ for $\rho_{50} = 0 - 20$ vehicles/km up to $\Delta t_{\text{max}}^{\text{congested}} \approx 2.3\,\text{s}$ for $\rho_{50} = 30 - 160$ vehicles/km.

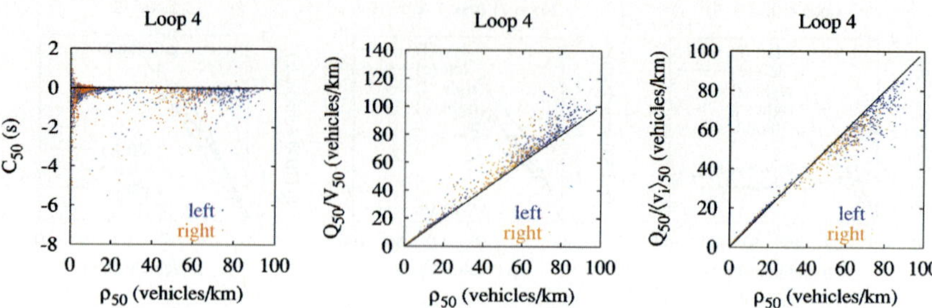

Fig. 4. Correlation C_{50}, density Q_{50}/V_{50} according to the fluid-dynamic formula, and conventionally determined density $Q_{50}/\langle v_i \rangle_{50}$ as a function of the density ρ_{50} according to the proposed definition (4), in comparison with the usually assumed relations (—).

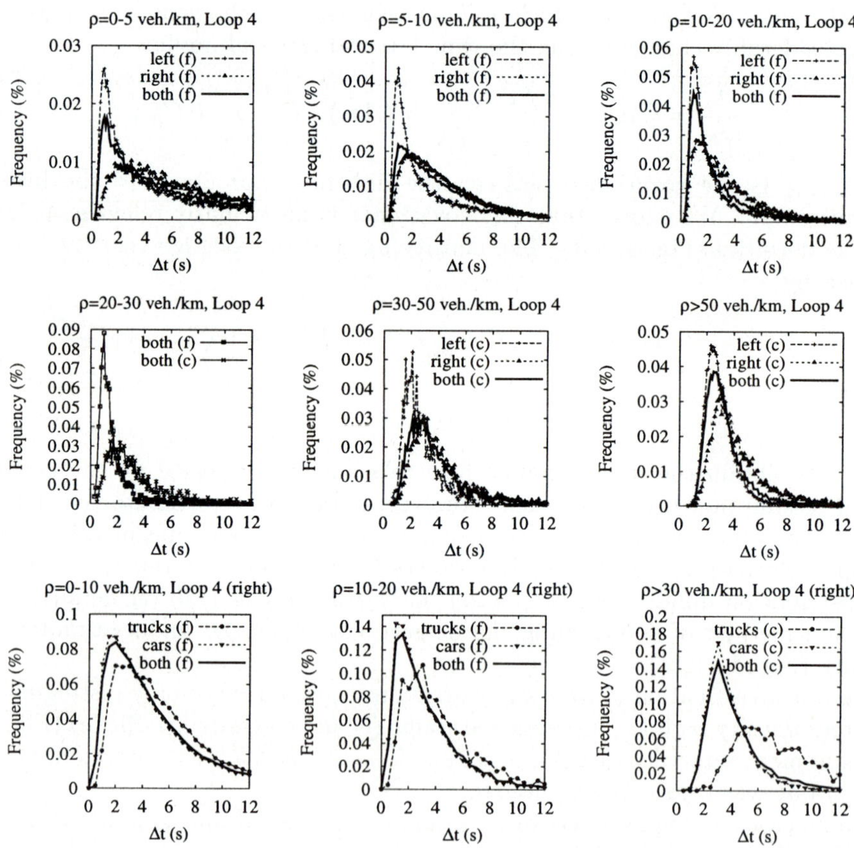

Fig. 5. Time gap distributions in different density regimes for different lanes and vehicle types.

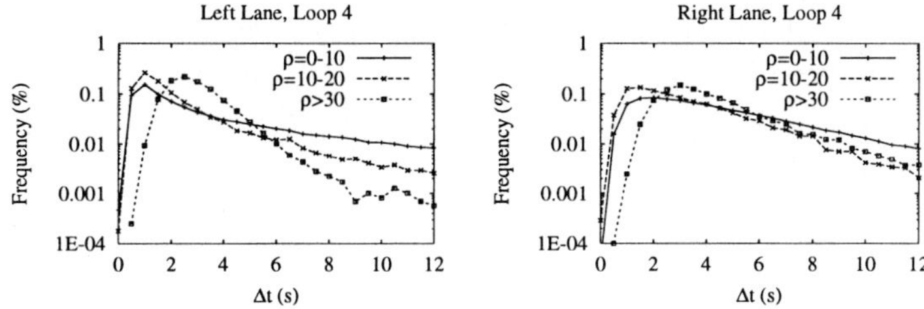

Fig. 6. Half-logarithmically plotted time gap distributions for different density regimes, separately for the left and the right lane.

Due to the existence of very large gaps, the time gap distribution is broad at small densities. It is sharpest immediately before the transition to congested traffic, and becomes much broader afterwards. While, in the right lane, the time gap distributions decay exponentially for $\Delta t > 3\,\mathrm{s}$, this seems to be different in the left lane (Fig. 6). Various suggestions for fitting the time gap distributions can be found in [1,2]. We have also plotted the average speeds v of vehicles

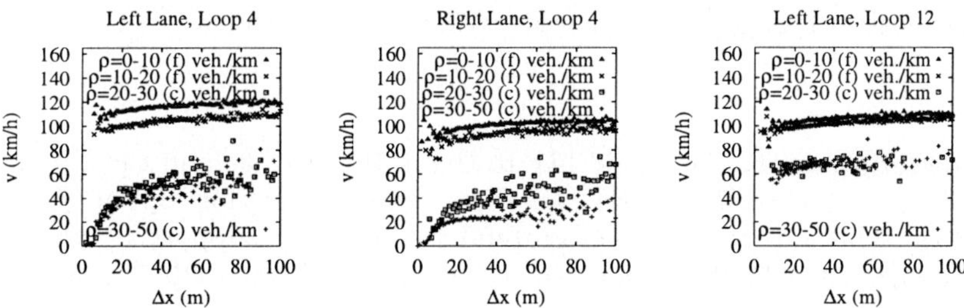

Fig. 7. Average vehicle velocities as a function of the distance gap for various density regimes, separately for the left and the right lane.

over their distance headways Δx to the respective car in front (Fig. 7). The plots show the tendency of a density-dependent saturation of vehicle speeds with large headways, which may be interpreted as frustration effect [3]. Like in [2], we prefer to look at the data inversely (Fig. 8, upper graphs), since the circumstance that we don't find typical headway-dependent velocities could just reflect large individual differences in the "optimal" velocity-distance relations $v_i^{\mathrm{d}}(\Delta x)$, or better: distance-velocity relations, which drivers prefer [8]. This would be consistent with the broad distance gap distributions (Fig. 8, lower graphs). Note that some variation of the distance gaps Δx comes already from the fact that gaps with regard to faster vehicles tend to be larger, because of the increasing distance. Gaps with regard to slower vehicles tend to be larger as well, since faster cars require some additional safety distance to decelerate. (Fig. 5 of [3].)

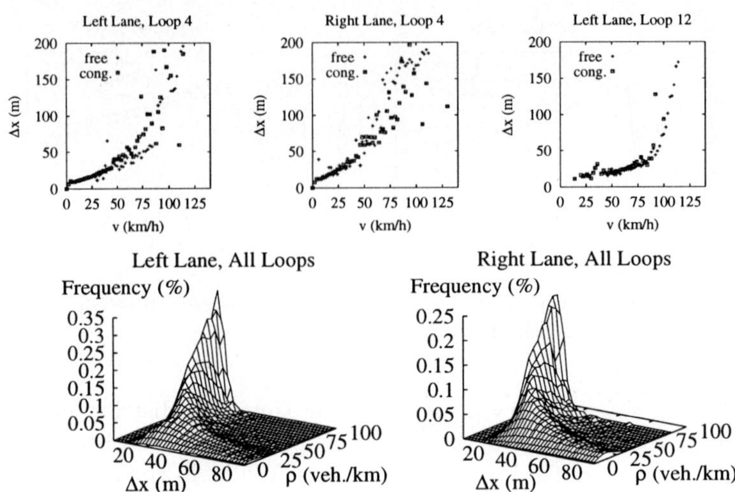

Fig. 8. Upper graphs: Average distance gaps as a function of the velocity at different measurement sites. Differences between free and congested traffic are found upstream of a bottleneck (left, middle), but not away from it (right). Below: The broad distance gap distributions change rather smoothly with increasing density.

Acknowledgement. We would like to thank Henk Taale and the Dutch *Ministry of Transport, Public Works and Water Management* for supplying the freeway data. The authors are also grateful for financial support by the BMBF (research project SANDY, grant No. 13N7092) and by the DFG (grant No. He 2789).

References

1. A.D. May, *Traffic Flow Fundamentals,* (Prentice Hall, Englewood Cliffs, NJ, 1990).
2. P.H.L. Bovy, *Motorway Traffic Flow Analysis,* (Delft University Press, Delft, 1998).
3. L. Neubert, L. Santen, A. Schadschneider, and M. Schreckenberg, Phys. Rev. E **60**, 6480 (1999).
4. M. Treiber and D. Helbing, J. Phys. A **32**, L17-L23 (1999).
5. D. Helbing, *Traffic data and their implications for consistent traffic flow modeling,* in: *Transportation Systems,* M. Papageorgiou and A. Pouliezos, (Eds.), Vol. II, pp. 809-814 (International Federation of Automatic Control, Chania, Greece, 1997).
6. D. Helbing, *Verkehrsdynamik,* (Springer, Berlin, 1997).
7. W. Leutzbach, *Introduction to the Theory of Traffic Flow,* (Springer, Berlin, 1988).
8. T. Bleile, *Modellierung des Fahrzeugfolgeverhaltens,* (PhD Thesis, University of Stuttgart, 1999).

Forecasting of Traffic Congestion

B.S. Kerner, H. Rehborn, and M. Aleksic

DaimlerChryslerAG, HPC: E224, 70546 Stuttgart, Germany

Abstract. Results of investigations of a recent method for the automatic tracing of moving traffic jams and of the prediction of time-dependent vehicle trip times are presented using different levels of data inputs. The method is based on the previous findings that moving jams possess some characteristic parameters, i.e., the parameters are unique, coherent, predictable and reproducible. Based on available data it is found that the method, which performs without any validation of the parameters of a model under different infrastructures of a highway, weather, etc., can be applied for a reliable forecasting of traffic congestions on a highway.

1 Introduction

Traffic forecasting based on an application of different dynamic traffic flow models whose parameters should be validated corresponding to real traffic conditions is the well-known approach (e.g., [1-4]). However, due to very complex dynamics of real traffic flow which strongly depend on the infrastructure of a highway, weather, etc., it is difficult to choose a set of model parameters in a way that the model shows results compatible with real traffic flow. Recently, it has experimentally been found out that moving traffic jams possess some characteristic parameters [5]. "Characteristic" means that the parameters do not depend on initial conditions, they are the same for different moving jams and do not depend on time. In other words, the characteristic parameters are unique, coherent, predictable and reproducible. Obviously, these properties of moving jams can be used for a development of models and methods of traffic forecasting performing without any validation of model parameters. Based on available data such a method has been proposed which allows to predict the propagation of moving traffic jams and time-dependent travel times [6,7]. The method allows a prediction of traffic jams and a forecasting of time-dependent vehicle trip times without parameter calibration under different infrastructures and environmental conditions.

In this paper the results of experimental investigations of the method are presented using different amounts of traffic data as inputs. However, before the results are described, recent experimental discoveries of characteristic parameters of traffic jams [5] and of phases in the congested regime [8] will be considered briefly (see also a recent review [9]).

1.1 Characteristic Parameters of Moving Traffic Jams

In [5] it has been found that the downstream front of a moving jam possesses some characteristic parameters which do not depend on initial conditions and on time and are also the same for different moving jams. These characteristic parameters of moving traffic jams are: (i) the velocity of the downstream front of the moving jam v_{gr}, (ii) the vehicle flow rate q_{out} and other parameters of the flow in the outflow of the moving jam, (iii) the vehicle density inside the moving jam ρ_{max}.

It is important to note that the flow rate q_{out} and other parameters of the flow in the outflow of the moving jam keep the property of being characteristic parameters only if free traffic flow is formed in the outflow from the jam. However, the velocity of the downstream front of the moving jam is always the characteristic parameter independent of the traffic parameters downstream from the moving jam [9-12]. The latter property allows to make a prediction of the propagation of the moving jam without any validating of model parameters from available traffic data [6].

1.2 Three Phases of Traffic

Recently, it has been found [8] that in congested traffic flow two different phases have to be distinguished: wide moving traffic jams and synchronized traffic flow. Therefore, there are three phases of traffic: (i) free traffic flow, (ii) synchronized traffic flow and (iii) moving jams. In contrast to free traffic flow where vehicles are able to overtake one another freely the probability of overtaking in synchronized traffic flow is very low [9].

Experimental investigations have shown that moving traffic jams propagating through synchronized traffic flow maintain the average velocity of the downstream front which is characteristic for moving jams propagating through free flow. This property of moving traffic jams allows to predict the jam propagation on the highway independently of the state in which the traffic actually is.

2 Method for the Automatic Tracing of Traffic Jams

Because a wide moving jam propagates through either free or any states of synchronized flow and bottlenecks keeping the velocity of the downstream front [10-12], this velocity can be found from available data (measured through stationary detection sites or floating car data or else other alternative measurement techniques) after a jam has been detected.

The velocity can be used for the further prognosis of the propagation of the traffic jam. This basic idea has been used in the method for prediction of traffic congestions and travel times where also some other important properties of traffic flow have been taken into account. The detailed description of the model can be found in [6]. Here only a qualitative discussion of the method is given for the case when data from stationary detection sites on a highway are used. The development of the method for tracing and prediction of moving traffic jams uses essentially the fact that the velocities of the propagation of the moving jams' fronts can be calculated with the available measured traffic data. The method

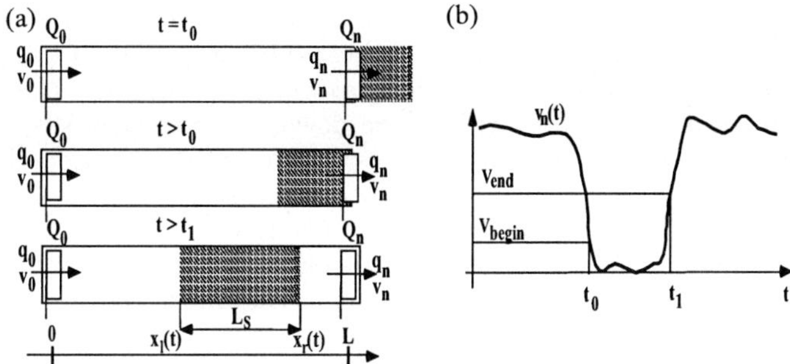

Fig. 1. Illustration of the method: (a) schematic movement of a wide traffic jam on a highway. (b) registration of the times t_0 and t_1 of the upstream and downstream fronts of the jam, respectively.

is able to handle lane reductions, on- and off-ramps and other kinds of possible influences of the infrastructure [6].

The schematic illustration for the jam tracing method in Fig. 1a shows as measurements two detection sites Q_0, Q_n on a road section. After the registration of a moving traffic jam in the measured data at Q_n at time t_0, the method starts to calculate continuously the positions of the downstream $(x_r(t))$ and upstream $(x_l(t))$ fronts, the length L_s of the moving jam and vehicle trip times even if none of the detection sites is within or close to the moving jam. For this calculation the measured traffic data and/or a time-series prediction based on historical traffic data of the vehicle flow rate $q_0(t)$ and $q_n(t)$ and the vehicle speeds $v_0(t)$ and $v_n(t)$ at the detection sites Q_0 and Q_n are needed.

One sub problem of the method of tracing and prediction of moving traffic jams is the determination of the time t_0, when a jam occurs. In the example shown, a sharp decrease in the vehicle speed $v_n(t)$ at the detectors Q_n below some given value, v_{begin}, can be chosen as the criterion (Fig. 1b). The value t_0 could also be determined via other well-known incident detection methods. The same is true for the time t_1 which determines the appearance of the downstream front of a jam at the detector Q_n: t_1 is determined by the time when the vehicle speed increases above some given value v_{end}.

The aim of the method is to trace and predict a moving traffic jam at all times between detection sites using the actual measured traffic data or a time-series prediction of the data. The advantages of the method are shown in Table 1: (i) the propagation of the traffic jam can be traced and predicted at any time – even if the moving traffic jam is in between detectors on a road section, (ii) the prediction of vehicle trip times can be done according to the position and length of the moving traffic jam and (iii) the jam disappearance can be predicted if it is within the horizon of the prognosis (see Table 1).

As can be seen from Table 1, many useful applications of the method are possible: the output information of the method can be used for driver information

Input of the method	Advantages of the method
* permanent measurements or time-series prediction of traffic speeds and traffic volumes of detection sites or other sensors (floating car data, infra-red, cameras, etc.) information about the infrastructure (on-, off-ramps, lane reductions, intersections, lanes, etc.) percentages of trucks	* short-time prediction of the jam and calculation of the speeds of upstream and downstream front of the jam * calculation of the position of up- and and downstream front and jam length * prediction of the vehicle trip time * calculation of the loss-time * prediction of the time of the jam disappearance

Table 1. Advantages of the method [7].

systems or for traffic control systems. It is possible to predict the traffic jam dissolution: if the jam's downstream front moves faster than the upstream front, the length of the moving traffic jam decreases in time and the dissolution time can be calculated.

3 Description of Results

The method has been tested on a lot of experimental data and it shows very good results. For the verification of the method's results experimental data of an infrastructure with many detection sites (Fig. 2) has been chosen. First, the data of all detectors has been used for the jam tracing method. Then, the data of some detectors has been omitted and the process of tracing has been repeated. The data which has not been used is taken for comparison to the method's results. This procedure has been performed continuously with less and less detectors up to the point until the method is no more accurate enough for acceptable practical results. The results of these tests of the method are summarized in Table 2 and one example is given in Fig. 3.

Suitability of the infrastructure	Distance of the detection sites	Measurements within intersections and on- and off-ramps
Very good	< 6 km	Fully detected
Good	> 6 km	Fully detected
Useful	< 3 km	No detection
Possibly useful	3 - 10 km	No detection
Bad	> 10 km	No detection

Table 2. Suitability of the infrastructure

Accurate results can be achieved with distances between detection sites of up to 6 km on highway stretches without intersections. Even when there are

somewhat larger distances between detection sites the accuracy of the results of the method are still useful as long as on- and off-ramps are fully detected.
Below we discuss results which are related to data from the highway A5-South in Germany whose infrastructure is schematically shown in Fig. 2.

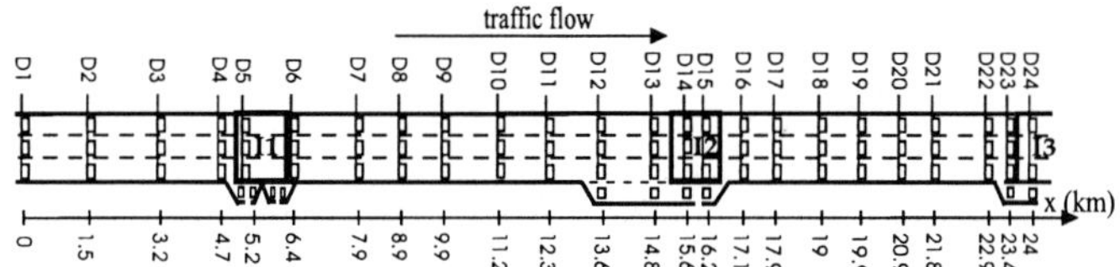

Fig. 2. Infrastructure of the A5-South.

In these examples for a determination of the times t_0 and t_1 very simple criteria are used: traffic at a detection site is considered to be in the phase "jam" (not to be in the phase "jam") if the flux is below (above) 1000 veh/h and the average speed is below (above) 30 km/h during a one-minute measurement interval, i.e., $v_{\text{begin}} = v_{\text{end}} = 30$ km/h.
Fig. 3 visualizes the application of the method for the automatic tracing of traffic jams on real traffic data from the highway A5-South.
The x-axis in both graphs denotes the time, the y-axis denotes the location, measured in kilometers from the beginning of the highway stretch under consideration as seen in Fig. 2. Solid black circles represent those measurement intervals (x-coordinate) during which the incident detection method reports a jam, at the detection site at a specific location (y-coordinate). Thin solid lines represent the upstream and downstream fronts of the traffic jam traced by the method. Note that these fronts can be anywhere within the considered stretch of the highway while circles can only be at the exact locations of the detection sites. These two examples show that in general the predicted positions of the jam fronts are in good accordance with the incidents detected at the detection sites. The predicted position of the upstream front of the jam in Fig. 3a, for example, reaches position 12,3 km at 9:14 and detector D11 at position 12,3 km (see Fig. 2) first reports a jam at 9:14, too (see corresponding solid circle in Fig. 3a). In Fig. 3b, the moving jam is traced using only detection site data from the positions D18, D14, D11 and D3.

The jam grows approximately during the time interval 8:49 - 8:58, remains about constant in size during 8:59 - 9:31 and shrinks afterwards until its extinction at approximately 9:43. This is because the flow rate into the jam changes over time. In comparison, the positions of the jam's fronts in Fig. 3b are in good agreement with Fig. 3a: for a useful application of the method the number of used measurements of the highway can be reduced according Table 2.
Taking only 4 of 18 detectors on the 19 km stretch into account, the method

is still able to trace and predict the jam's fronts within the accuracy of the one-minute-interval (0.3 km).

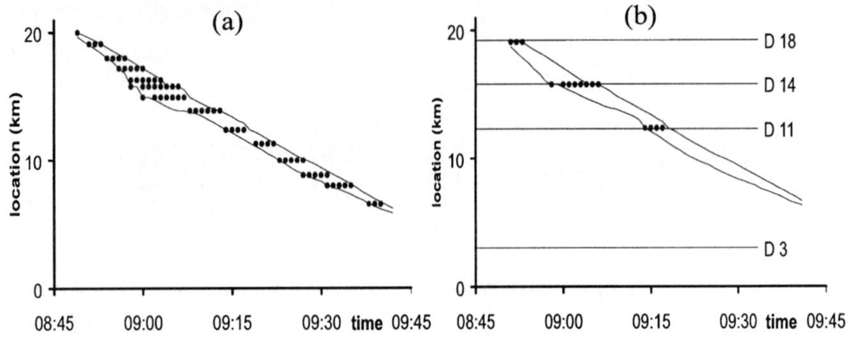

Fig. 3. Results of the method: (a) using all available traffic data: propagation of a wide traffic jam, (b) using reduced input data (4 of 18 detectors on 19 km)

References

1. M. Cremer, *Der Verkehrsfluss auf Schnellstrassen*, (Springer, Berlin, 1979).
2. J. Esser, L. Neubert, J. Wahle, and M. Schreckenberg, in: *Transportation and Traffic Theory*, A. Ceder, (Ed.), pp. 535–554 (Elsevier Science Ltd., Oxford, 1999).
3. M. Ben-Akiva, E. Cascetta, H. Gunn, D. Inaudi, and J. Whittaker, in: *Proceedings 7th World Conference on Transport Research*, (Sydney, 1995).
4. W. Kronjäger and P. Konhäuser, in: *Preprints 8th IFAC/IFIP/IFORS Symposium on Transportation Systems*, M. Papageorgiou and A. Pouliezos, (Eds.), Vol. 2, pp. 805-809 (1997).
5. B.S. Kerner and H. Rehborn, *Phys. Rev. E* **53**, 1297–1300 (1996).
6. B.S. Kerner, H. Rehborn, and H. Kirschfink, German patent DE 196 47 127, (www.depanet.de), US-patent US 5861820 (dips-2.dips.org), (1998).
7. B.S. Kerner, H. Kirschfink, and H. Rehborn, *Strassenverkehrstechnik* **9**, 430–438 (1997).
8. B.S. Kerner and H. Rehborn, *Phys. Rev. E* **53**, R4257–R4278 (1996).
9. B.S. Kerner, *Physics World* **12**, No. 8, 25–30 (1999).
10. B.S. Kerner and H. Rehborn, *Int. Verkehrswesen* **50**, No. 5, 196–203 (1998).
11. B.S. Kerner, in: *Proc. of the 3rd Symp. on Highway Capacity and Level of Service*, R. Rysgaard, (Ed.), Vol. 2, pp. 621–642, (Road Directorate, Denmark, 1998).
12. B.S. Kerner, in: *Traffic and Granular flow '97*, M. Schreckenberg and D.E. Wolf, (Eds.), pp. 239-283 (Springer, Singapore, 1998).

Empirical Phase Diagram of Traffic Flow on Highways with On-Ramps

H.Y. Lee[1], H.-W. Lee[2], and D. Kim[1]

[1] Department of Physics,
Seoul National University, Seoul 151-742, Korea
[2] School of Physics, Korea Institute for Advanced Study,
207-43 Cheongryangri-dong, Dongdaemun-gu, Seoul 130-012, Korea

Abstract. Multiple congested traffic states are found from empirical data. A preliminary empirical phase diagram of congested traffic flow is constructed. This paper also discusses connections between observations and theoretical analysis.

1 Introduction

Empirical and theoretical studies of traffic flow on highways have attracted considerable interest. Especially, investigations based on physical concepts such as dynamic phase, phase transitions, and hysteresis have revealed interesting properties of traffic flow [1]. In recent theoretical studies, several different congested traffic phases are predicted and a phase diagram of congested traffic flow is constructed [2,3].

This paper presents empirical congested traffic phases and a preliminary phase diagram of congested traffic flow [4]. We compare the empirical results with theoretical ones. In the next section, we present the characteristics of the observed congested traffic states. Section 3 discusses the empirical phase diagram, and Section 4 concludes the paper.

2 Congested Traffic States

We analyze the traffic flow on one section of the Olympic Highway, in Seoul, Korea (Fig. 1). Although there are many ramps, there is a 5 km long lane-divider which almost isolates the two inner-lanes from the outer lanes. Therefore the two inner-lanes in this section can be regarded as a ramp-free highway. In the section, 15 image detectors record the traffic data for the flux q, the arithmetic mean velocity v, and the harmonic mean velocity u in every 30 sec interval. From the two inner-lane traffic data for 14 different days in June and July, 1998, we search for congested traffic states that are long lived ($\sim$ 1 hour). As a result, we identify four congested traffic states which appear practically everyday, and they are compared with theoretically predicted states [2,3]. We note that an exact continuity equation can be set up using the measured quantities. Here, q/u plays the role of the density. Therefore, we identify q/u with the density in Ref. [2,3] and plot the spatio-temporal structure of q/u instead of q/v. However, our results are not sensitive on the choices.

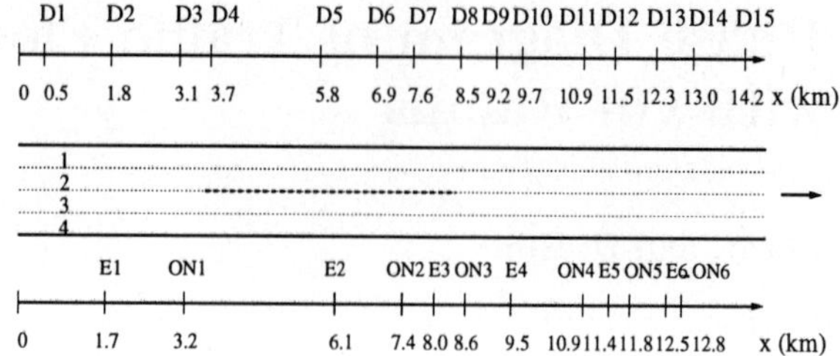

Fig. 1. The Olympic Highway section under study. Dn, ONn, and En mark the locations of detectors, on-ramps, and off-ramps, respectively. The dashed line in the middle denotes the lane divider.

The first type of the congested traffic states on Olympic Highway, CT1, is shown in Fig. 2. The congested region is localized between 2 km and 10 km but the boundaries of the region fluctuate. As we approach the upstream boundary of the congested region, the velocity oscillation is amplified and its period increases, which is shown in Fig. 2(b). We find much similarities between the CT1 state and the recurring hump (RH) state in Ref. [3] Both in CT1 and RH states, we observe spontaneous development of systematic oscillation in the localized congested region. The amplifications of velocity oscillation during the propagation are common to both states, but, the period enhancement of the oscillation in the CT1 state is not observed in the RH state.

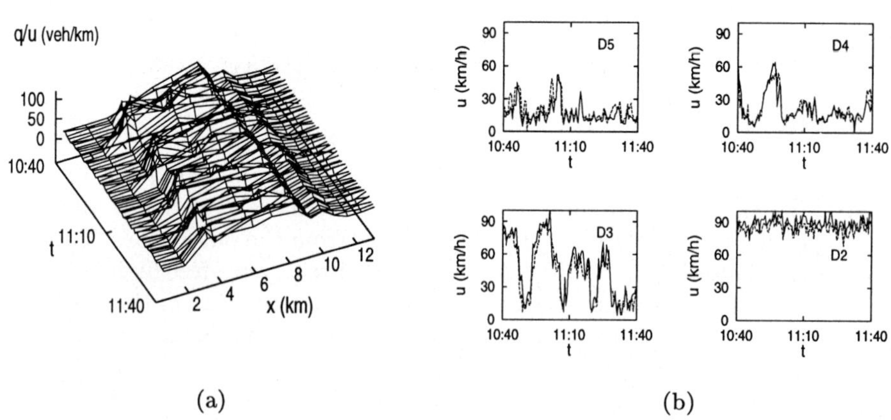

Fig. 2. (a) The spatio-temporal evolution of q/u for the CT1 state. (b) Velocity vs. time plot at different locations (solid line for the lane 1 and dashed line for the lane 2). The stop-and-go pattern appears at 3.1 km but the free flow appears at 1.8 km.

Figure 3(a) shows the second kind of the congested traffic states, CT2. In contrast to the CT1 state, we observe motionless boundaries of the congested region and we do not find the development of systematic oscillation in the congested region. The CT2 state can be related to the pinned localized cluster (PLC) state [2,3]. In both states, the extension of the congested region does not vary with time and systematic oscillations do not develop. Figure 3(b) shows the spatial variation from 5.8 km to 12.3 km of the 1 hour averaged q/u-q relations of the CT2 state. This pattern is essentially identical to the one found for the PLC state (Fig. 2(b) in Ref. [3]).

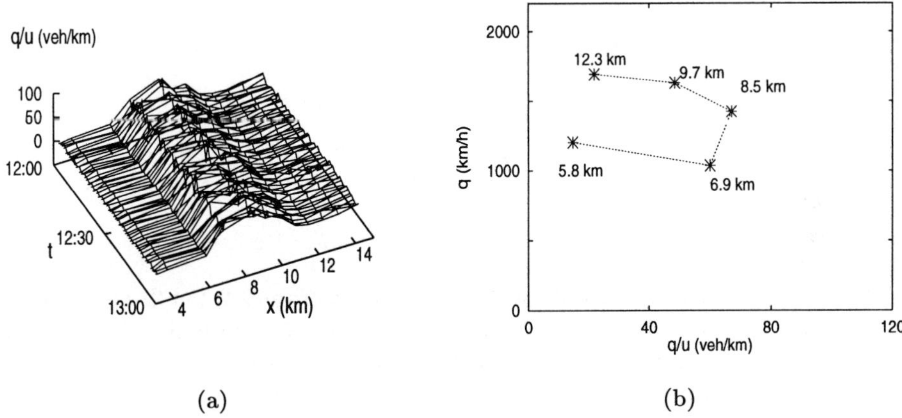

Fig. 3. (a) The spatio-temporal evolution of q/u for the CT2 state. (b) The spatial variation from 5.8 km to 12.3 km of the averaged q/u-q relations of the CT2 state. Two-dimensional covering occurs in the congested region of the CT2 state when 1 min data for q/u and q are used instead.

The third congested traffic state, CT3, is shown in Fig. 4(a). The boundaries of the congested region are motionless, but the congested region is much shorter compared to those in the CT1 and CT2 states. Figure 4(b) shows an interesting feature of the CT3 state. The velocity remains almost constant even under significant fluctuations of the flux. As a result, the q/u-q relation forms a straight line even in the congested region. This property is also found in the analysis of German highway data [5]. The motionless boundaries of the congested region of the CT3 state agree with those of the PLC state. However, the local q/u-q relations of the CT3 state which all almost straight lines in the congested region are not realized in the PLC state.

The fourth congested traffic state, CT4, is depicted in Fig. 5(a). The CT4 state appears usually during morning rush hours. In contrast to the CT1, CT2, and CT3 states, the upstream boundary of the congested region propagates

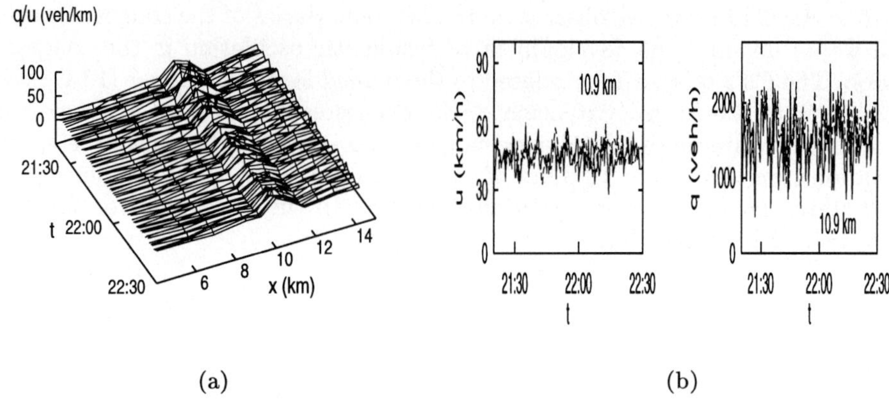

Fig. 4. (a) The spatio-temporal evolution of the CT3 state. (b) Velocity vs. time plot and flux vs. time plot at 10.9 km. The velocity remains almost constant but the flux fluctuates significantly.

backwards. The average propagation rate ranges from 2.2 km/h to 8.8 km/h. We observe that the higher the upstream flux level is, the higher the propagation rate is. The expansion of the congested region is also predicted in the two theoretically predicted states, the oscillating congested traffic (OCT) state and the homogeneous congested traffic (HCT) state. The velocity evolutions of the CT4 state appears in Fig. 5(b). Since we do not detect the spontaneous formation of large amplitude oscillation in the CT4 state, the HCT state seems to be the closest theoretical counterpart of the CT4 state. The output flux from the congested region of the CT4 state is almost constant ($\approx$ 1900 veh/h). This property is also predicted for the HCT state [2].

3 Empirical Phase Diagram

Theoretical studies [2,3] predict that the upstream flux level $f_{\rm up}$ at the far upstream of the congested region, where the free flow is maintained, and the on-ramp flux level $f_{\rm rmp}$ are the two important control parameters for different congested traffic states. Although there exist metastable regions, where different initial conditions may lead to different congested traffic states, the occupying region of each congested traffic state is clearly distinguished from each other in the $f_{\rm rmp}$ and $f_{\rm up}$ plane. Motivated by the model results, we examine whether the values of $f_{\rm rmp}$ and $f_{\rm up}$ can serve as important control parameters for the empirical traffic states. For the empirical data, $f_{\rm up}$ and $f_{\rm rmp}$ fluctuates with time unlike the ideal situation with fixed values of $f_{\rm up}$ and $f_{\rm rmp}$ in Refs. [2,3]. However, the fluctuations of $f_{\rm up}$ and $f_{\rm rmp}$ are greatly suppressed in long time scale (ten minutes or longer). Hence, we use $f_{\rm up}$ and $f_{\rm rmp}$ by averaging over the time

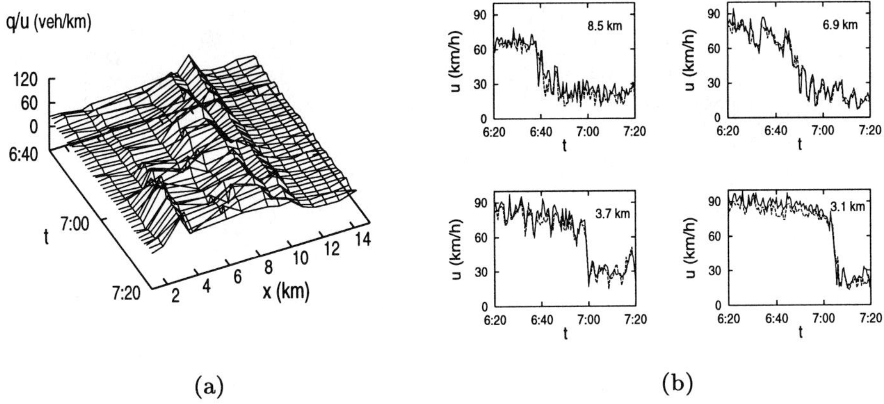

(a) (b)

Fig. 5. (a) Spatio-temporal evolution of the CT4 state. While the downstream boundary of the congested region is fixed, the upstream boundary of the congested region propagate backwards. (b) Velocity vs. time plots at different locations. The large amplitude oscillation does not develop in the congested region of the CT4 state.

during which a particular state is maintained. In this way, we obtain a point $(f_{\text{rmp}}, f_{\text{up}})$ from each time interval with a particular congested traffic state. Figure 6 shows the phase diagram of the congested traffic flow. Different symbols represent different congested traffic states. Below the dashed line, the free flow is observed. Each congested traffic state occupies a distinguishable region in the f_{rmp}-f_{up} plane although there are some overlaps.

In addition to the existence of the phase diagram of empirical data, the empirical phase diagram show much similarities with theoretical phase diagram studied in Ref. [2,3]. Both in the theoretical and empirical phase diagrams, metastability between the free flow and the congested traffic states with localized structure is observed. And the data locations of the CT2 state are to the left of those of the CT1 state, which agrees with the relationship between the PLC and the RH states. Also both CT4 and HCT states appear when f_{rmp} is sufficiently large.

4 Conclusion

We identify four distinct congested traffic states by considering both temporal traffic patterns at fixed locations and spatial structure of the congested region. The empirical phase diagram is obtained in the upstream and on-ramp flux plane. We find qualitatively good agreements between observed empirical states and theoretically predicted states in Ref. [2,3]. Also the empirical and theoretical phase diagrams agree with each other qualitatively. However, we need more studies to improve the empirical phase diagram. Since different geometries may

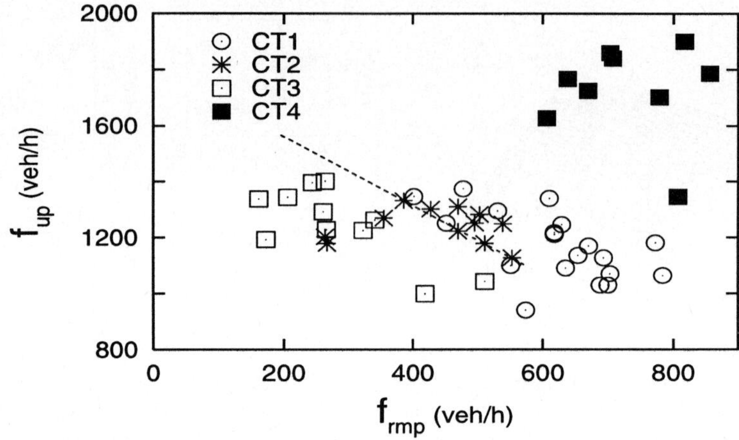

Fig. 6. The phase diagram of the four congested traffic states, cf. Ref. [3].

affect the evolution of the congested traffic states, we should construct the phase diagram by focusing on a fixed on-ramp region. More extensive empirical studies are also required to determine the phase boundaries more accurately and to examine meta-stability among the congested traffic states.

References

1. B.S. Kerner, Physics World **12**, 25–30 (1999).
2. D. Helbing, A. Hennecke, and M. Treiber, Phys. Rev. Lett. **82**, 4360-4363 (1999).
3. H.Y. Lee, H.-W. Lee, and D. Kim, Phys. Rev. E **59**, 5101-5111 (1999).
4. H.Y. Lee H.-W. Lee, and D.Kim, cond-mat/9905292.
5. B.S. Kerner and H. Rehborn, Phys. Rev. E **53**, R4275-4278 (1996).

On-Line Simulation of the Freeway Network of North Rhine-Westphalia

O. Kaumann, K. Froese, R. Chrobok, J. Wahle, L. Neubert, and M. Schreckenberg

Physik von Transport und Verkehr, Gerhard-Mercator-Universität, Duisburg, Germany

Abstract. Recently, an on-line simulation for traffic flow in urban areas has been presented [1,2]. In this contribution, a framework for on-line simulations of freeways is proposed. It is based on cellular automaton models which are supplemented by real world traffic data stemming from about 2,500 inductive loops distributed over the freeway network of North Rhine-Westphalia (NRW). We propose and analyse different methods to tune the simulation with regard to the real world measurements. The possibility of a short-term traffic forecast is discussed.

1 Introduction

Oversaturated freeways and congested main roads in cities reflect the fact that the existing road networks are not able to cope with the demand for mobility which will further increase in future. Especially, in densely populated regions, like the state of North Rhine-Westphalia, it is on the one hand socially untenable to expand the existing infrastructure further in order to relax the situation. On the other hand mobility is a vital good for the economic development of this region.

Therefore, the existing road network has to be used more efficiently using Advanced Traveller Information Systems (ATIS) which inform the road user about traffic conditions or provide route guidance. The basic requirements for these advices are precise spatially and temporally resoluted data about the *actual traffic state*, e.g., link travel times or traffic densities. Usually, the traffic state is measured locally using various detection technologies, mostly inductive loops. In order to provide network-wide information it is convenient to combine the measured data with a suitable traffic flow model, i.e., perform on-line simulations. The simulated data can be processed by route guidance systems which allow the road users to organise their trips with regard to individual preferences [3].

The outline of the paper is as follows: In the next section we show the requirements for the simulation of freeway traffic. The difference between urban and freeway traffic and its implications for modelling is discussed. In the third section we describe the network and the underlying data base. In Sect. 4 we focus on the methods to tune the simulation with real world data and discuss three different approaches. We conclude with a discussion and an outlook on the possibility of traffic forecast.

2 Modelling of Freeway Traffic

In general, traffic flow simulators are used to simulate traffic on various spatial and temporal scales[1]. However, for the on-line simulation of freeway traffic a few basic requirements have to be fulfilled. If one is interested in information about single vehicles, e.g., link travel times or velocities, a microscopic model is necessary.

In order to provide real-time information to the road user the computational performance of the simulations is crucial. In several previous works it has been pointed out that cellular automata are a promising concept for high-speed micro-simulations [2,4–6]. In the present work we use the standard cellular automaton model proposed by Nagel and Schreckenberg [7]. For modelling multilane traffic we employed the lane changing rules of [8].

In general, the traffic dynamics on urban roads are governed by the intersections, mainly traffic lights, whereas on freeways dynamic phases, e.g., synchronised flow or stop-and-go traffic emerge (for an overview see [9]). The analysis of single-vehicle data [10] yields that for highway traffic a more detailed description of the dynamics seems to be necessary [11].

3 Network and Real World Data

The simulations are based on the freeway network of North Rhine-Westphalia , an area of about 34,000 km^2. The roads of the network have a length of 6,000 km. There are 67 highway intersections and 830 on- and off-ramps. The digital version of the network consists of 3,560 edges and 1.4 million sites of a length of 7.5 m each. Similar to urban areas [1] the topology of the network was constructed using basic elements, namely main tracks and transfer tracks [12]. To provide precise travel times the length of every piece of topology, especially transfer tracks, was determined using a Geo-Information System (GIS)[2].

Currently, data from about 2,500 inductive loops is accessible. Every minute the aggregated amount of cars and trucks as well as their velocities are sent via permanent lines from the traffic control centres in Recklinghausen and Leverkusen to the controller of the simulation. These data are incorporated in the simulation in order to calculate dynamic turning probabilities and to tune the simulation, i.e., to adapt the simulated results to the real world data. This is done at the so-called check-points which are described in the following section.

In the simulation the vehicles are guided in the network randomly, according to the turning percentages calculated on the basis of the measured data. This method is used, since origin-destination information with a sufficient resolution in time and space is not available.

[1] For an overview of about simulators see: `http://www.its.leeds.ac.uk/smartest`.

[2] The basis of the GIS is the NW-SIB provided by the state of NRW.

4 Tuning Strategies

For realistic results it is crucial to incorporate the real world data in the simulation without perturbing the dynamics which are present in the network. In difference to urban networks, the sinks and sources of a freeway network are well defined, namely the on- and off-ramps. To adapt the simulated results and the real world data we additionally introduce check-points. They are located at those places where a complete cross-section is available, i.e., all lanes are covered by an inductive loop. Here, it is convenient to perform adjustments. In principle, the last minute's results of the simulation have to be compared with the measured data. This can be done using different methods: sink- and source-strategy, flow-tuning or the tuning of the mean gap, which are described in the following.

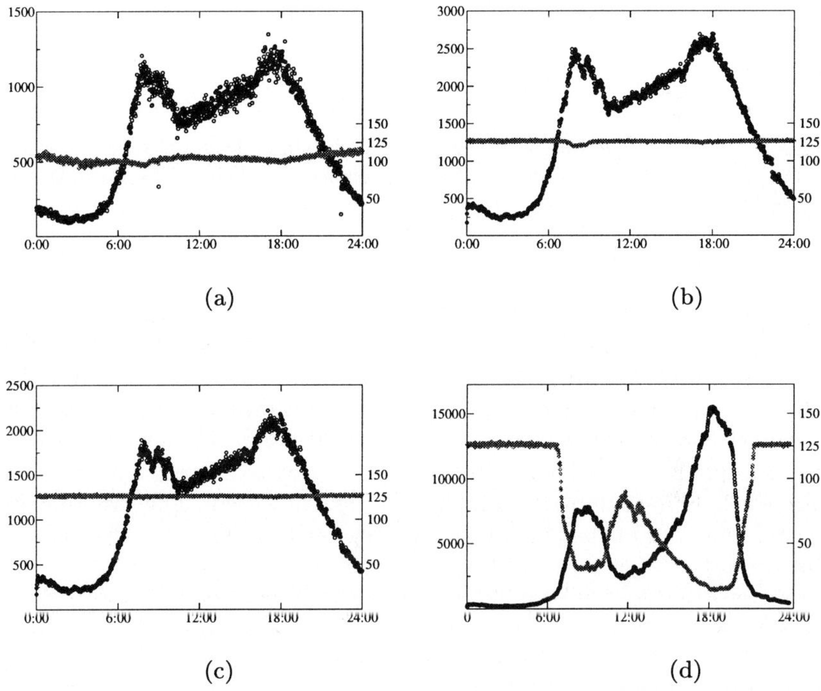

Fig. 1. Number of cars and their velocity vs. time: (a) data collected at inductive loops (for a part of the highway network A1 and A43 from Münster to the Ruhr-area); reproduced by applying (b) the sink- and source-strategy; (c) the tuning of the mean-gap; (d) the flow-tuning. The sink- and source-strategy as well as the mean-gap strategy reproduce the characteristics of the measured data quite well. The flow-tuning method tends to introduce artificial jams in the system.

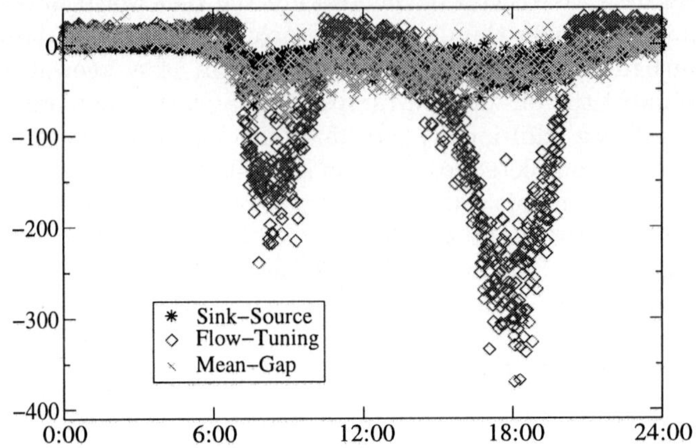

Fig. 2. Differences of vehicles at the check-points vs. time. It can be seen that all strategies do not fill in the complete number of measured vehicles. The flow-strategy performs worst.

In Fig. 1 empirical data detected at the check-points and the simulation results at a typical working day for a part of the freeway network are presented. The time series shows typical features like rush-hours which are also found in urban traffic [2]. The velocity seems to be very stable throughout the whole day. Note that we use a maximum velocity of 5 cells/time step $\approx$ 125 km/h. Since the simulation extrapolates the data it is clear that the number of cars measured in the network is higher than the number of cars detected. The difference of measured cars and cars in the simulation are depicted in Fig. 2 for the different methods of tuning.

4.1 Sink- and Source-Strategy

The sink- and source-strategy is the simplest approach to the problem of adapting the simulation to real world data. Every check-point consists of a sink at the beginning and a source at the end. The incoming vehicles are deleted at the sink and the source adds the measured number of vehicles. Although the method is very crude the results are quite good (Fig. 1b). But it is very disadvantageous to completely destroy the dynamics at this place. In addition, information about the travel time is lost since no car can pass the check-point.

4.2 Flow-Tuning

Instead of removing all incoming vehicles and add the measured ones, tuning strategies try to adjust only the difference between the simulation and real world data. According to this difference cars are added or removed. There are two feasible ways for tuning, either by adapting densities or flows. Density-tuning is more difficult and demands inductive loops which measure the velocity. Using flow-tuning the cars are alway added with maximum velocity. It gives reliable results only for low densities. But if the number of measured cars increases rapidly, like in the morning rush-hour, larger number of cars have to be added at the check-point, at the same time a lot of vehicles approach the check-point. In high density regimes this introduces artificial jams behind the check-points, the average velocity breaks down and the number of cars increases rapidly (Fig. 1d). Additionally, the check-point is blocked and no cars can be added (Fig. 2).

4.3 Tuning of the Mean-Gap

Both strategies presented above are disadvantageous because they strongly intervene in the dynamics of the system. Therefore, we propose the so-called tuning of the mean-gap which follows the idea to add the cars to the network "adiabatically", i.e., without perturbing the system.

Similar to the flow-tuning the input to the system is the difference between simulated and real world data. But different to the flow-tuning the vehicles are introduced in an area around the check-point. In this area the mean-gap $\langle g \rangle$ of the cars is calculated. From the real world data a velocity v_{in} in cells/time step is determined. Now, the cars are added into the system with regard to the mean gap $\langle g \rangle$ and their velocity v_{in}. This is done without disturbing the cars which are already on the track; in other words in a way that no car has to brake because of the added vehicles. If it is not possible to add the necessary number some cars are left out. Altough this is not correct, it is done in order to keep the dynamics of the system. It turns out that this strategy reproduces the traffic state quite well (Fig. 1d) and reduces the differences of vehicles at the check-points (Fig. 2).

5 Discussion and Conclusion

A framework for performing on-line simulation of a freeway network was presented and applied to the network of North Rhine-Westphalia. For reasons of efficiency a cellular automaton approach is used. In order to achieve realistic results, the real world data has to be incorporated in the simulation very carefully. We proposed three methods for tuning: the sink- and source-strategy, the flow-tuning strategy, and the tuning of the mean-gap. Simulations for a part of the network showed that the tuning of the mean-gap performs best. This performance is due to the fact that cars are added "adiabatically", i.e., without perturbing the dynamics of the network.

Since more sophisticated models seem to be necessary for a realistic description of freeway traffic [11] we will study the influence of the model in such a huge network in the future. The efficiency of the underlying model allows to perform simulations in multiple real-time. This is a basic requirement for traffic forecasts. Also the influence of ramp-metering in parts of the network will be studied. The framework can also be a powerful tool for traffic flow control.

Acknowledgement. We thank N. Eissfeldt and C. Vogt for helping us digitising the network of NRW. The authors are also grateful to the "Landschaftsverband Rheinland" and "Landschaftsverband Westfalen Lippe" for data support and to the Ministry of Economic Affairs, Technology and Transport of North Rhine-Westphalia for financial support.

References

1. J. Esser and M. Schreckenberg, *Microscopic simulation of urban traffic based on cellular automata*, Int. J. of Mod. Phys. C **8**, 1025 (1997).
2. J. Esser, L. Neubert, J. Wahle, and M. Schreckenberg, *Microscopic online simulation of urban traffic*, in: *Proc. of the 14th ISTTT Int. Symp. on Transportation and Traffic Theory 99,* A. Ceder, (Ed.), (Pergamon, Amsterdam, 1999).
3. J. Wahle, O. Annen, C. Schuster, L. Neubert, and M. Schreckenberg, *A dynamic route guidance system based on real traffic data*, accepted, Eur. J. Op. Res.
4. K. Nagel, *Individual adaption in a path-based simulation of the freeway traffic of Northrhine-Westfalia*, Int. J. of Mod. Phys. C **7**, 883–892 (1996).
5. M. Rickert and P. Wagner, *Parallel real-time implementation of large-scale, route-plan-driven traffic simulation*, Int. J. of Mod. Phys. C **7**, 133 (1996).
6. K. Nagel, J. Esser, and M. Rickert, *Large-scale traffic simulations for transport planning*, in: *Ann. Rev. of Comp. Phys. VII,* D. Stauffer, (Ed.), (World Scientific Pubsh. Company, 2000).
7. K. Nagel and M. Schreckenberg, *A cellular automaton model for freeway traffic*, J. Phys. I France **2**, 2221 (1992).
8. K. Nagel, D.E. Wolf, P. Wagner, and P. Simon, *Two-lane traffic rules for cellular automata: A systematic, approach*, Phys. Rev. E **58**, 1425 (1998).
9. B.S. Kerner, *Phase transitions in traffic flow*, in: *these proceedings.*
10. L. Neubert, L. Santen, A. Schadschneider, and M. Schreckenberg, *Single-vehicle Data of highway traffic: A Statistical Analysis*, Phys. Rev. E **60**, 6480 (1999).
11. W. Knospe, L. Santen, A. Schadschneider, and M. Schreckenberg, *CA models for traffic flow: Comparison with empirical single-vehicle data*, in: *these proceedings.*
12. K. Froese, *Simulation von Autobahnverkehr auf der Basis aktueller Zähldaten*, Diploma thesis (Universität Duisburg, 1998).

Traffic Data Collection Using Image Processing Technology

P. Molnár[1] and T.R. Collins[2]

[1] Clark Atlanta University, Atlanta/Georgia, USA
[2] Georgia Tech Research Institute, Atlanta/Georgia, USA

Abstract. The primary purpose of this project[1] is to collect data required to adjust the parameters of a traffic model, allowing realistic predictions of the traffic in Atlanta. This project will provide information with a completeness and accuracy which has not been collected elsewhere. The acquired data will serve the traffic simulation as well as providing an empirical basis for future development in traffic modeling. The verified and calibrated traffic flow model will be implemented in a computer simulation program, which then can be used to focus on the development of strategies for improvement of the traffic flow, optimal positioning for driver-information displays, and evaluation of the significance of the drivers' response rate for the success of control measures.

1 Background and Motivation

In recent years many models and computer simulations have been developed that helped to understand the behavior of traffic flow. As models become more sophisticated, their demand on empirical data for verification and calibration grows drastically.

The ideal way to verify traffic flow models would be to track a large number of vehicles over several miles, for example, by filming them from an airplane or helicopter, then reconstruct trajectories for each vehicle. These data would satisfy virtually any traffic model. Studies using aerial photogrammetric techniques have been conducted by Treiterer *et al.* [2], but the method is expensive and labor intensive. In other studies test vehicles flow with the traffic, and keep track of their position and velocity via GPS[2]. However, essential information like the traffic density cannot be obtained by this method. Also, the drivers of the test vehicles might not behave in an unbiased manner.

The most practical and also most common way is to collect data of single vehicles as they pass a sequence of cross-sections along the road. Each time a vehicle passes a detector at that cross-section, the velocity and the length of the vehicle will be recorded.

In order to compare empirical data with the outcome of simulation models, quantities such as traffic flow and average velocity have to be aggregated for a specific time interval ΔT. A comprehensive empirical study should also include the variance of the velocity distribution. The right choice of the time interval

[1] This work is funded by the Georgia Department of Transportation.

[2] Global Positioning System, or the more accurate Differential GPS

ΔT is very important in order to see temporal changes of the traffic condition. However, if the time interval is too short, statistical fluctuations will affect the outcome.

The density function can be obtained from empirical data in two ways, here distinguished by two different symbols ρ and ρ'. The first definition is based on the flow-relation $Q(x,t) = \rho(x,t)V(x,t)'$, while the second definition, $\rho'(x,t) = \frac{O(x,t)}{L(x,t)}$ uses the relative occupancy time O and the average length of the vehicles L. Variables x and t denote the cross section in the road and the time, respectively. Usually the first definition behaves more realistically in the data analysis, although the second might be technically more reasonable. On multilane roads, detectors should be installed for each lane separately. Multilane traffic models, such as [1], require additional information about the rate of lane changes.

2 Site Atlanta

The Atlanta metropolitan area is a good candidate for such a traffic study, since highway traffic is heavy due to enormous growth in the region. Highways I-75 and I-85 merge in the downtown area and intersect with I-20. The "Perimeter" I-285, takes most of the long distance and freight traffic off the inner city roads (Fig. 1). Traffic on I-20, I-75, and I-85 is caused mostly by commuters.

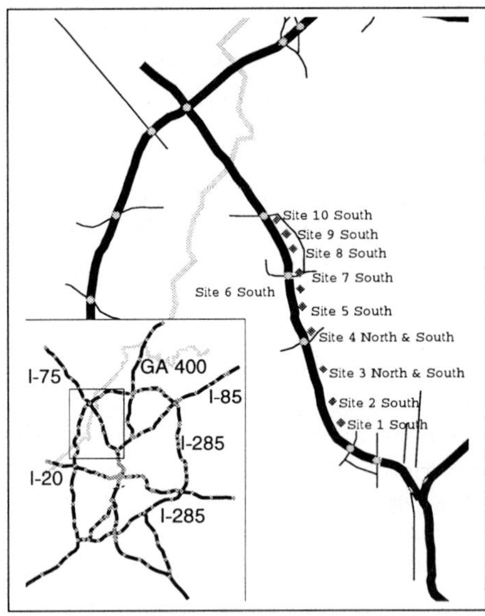

Fig. 1. The freeway network in Atlanta. The location of the ten test sites are shown on the right. Site 3 and Site 4 have two cameras each – one pointing south and one pointing north. All other sites have one camera pointing south towards oncoming traffic.

We selected a suitable stretch of about 4.3 miles on I-75 North, starting from Howell Mill Road to Mt. Paran Road. Data will be collected from the northbound lanes, which experience their major traffic load during the afternoon rush hour (see Fig. 1). The site includes two exits, both with metered entrance ramps. This particular location was chosen because of several factors: 1) the road segment should be fairly straight, and without large slopes, 2) the highway has only five lanes in this section, a limit given by the video tracking system, 3) availability of electrical power, and 4) frequent traffic congestion.

3 Experimental Setup

The equipment used to detect and track vehicles is the VideoTrak-900 from Peek Traffic Systems. This system uses patented image processing technology from Sarnoff Corporation. All of the image processing resides inside the VideoTrak unit, which is directly connected to a black and white CCD camera. The system is self-contained and can gather aggregated statistical data, such as volume (number of vehicles), lane occupancy (time lane is occupied), average speed, average density (defined as volume/speed), average headway in seconds, and average vehicle length.

The VideoTrak unit can also be operated in an *Academia Mode*, which produces *per vehicle records* (PVRs). The on-board memory, however, is not large enough to store the data, and a computer is required to retrieve the records continuously and save them on its hard disk.

We use twelve VideoTrak systems in this study. Each of them produces PVRs on a separate computer. In order to synchronize the data, time-stamps from a GPS receiver[3] are periodically included in the data stream.

After the cameras are installed, each unit has to be carefully configured: *Calibration Points* are used to define the perspective dimensions of the real-world coordinate system, and *Tracking Strips* distinguish between the various lanes of the road.

It is essential for any image processing algorithm to define a transformation matrix between the real-world coordinate system and the pixel coordinates from the video image. The VideoTrak unit creates the transformation matrix based on four *calibration points* that are placed in the video image (Fig. 3). The distances between these points in the real world, the height of the camera, and its distance to the first calibration point have to be measured. The accuracy of these measurements is critical for the proper reporting of distances in the image, as well as any quantities derived from distance (e.g., velocities).

The designation of calibration points was hampered by several factors: inability to access the highway median (two of the four points are essentially required to be there), limited field of view interacting with the most desirable area for vehicle tracking strips (i.e., the best view of vehicles may contain few distinguishing features like poles), and a software limit of less than 255 feet for any of the calibration distances.

[3] An alternative would be to use radio signals like WWV in the US or DCF77 in Europe.

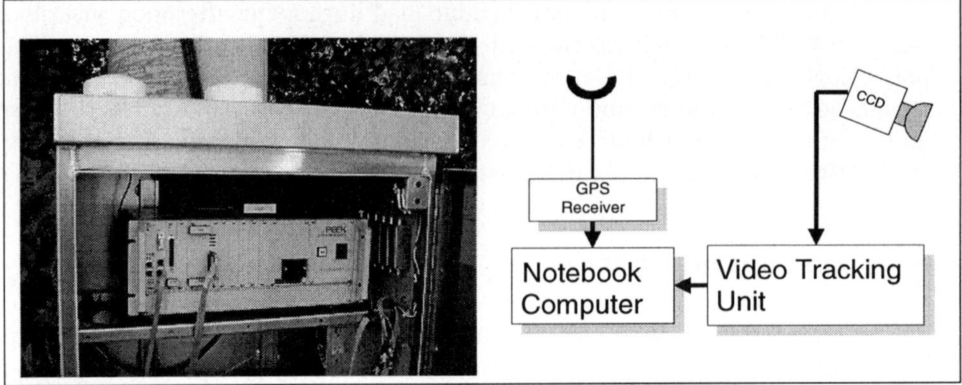

Fig. 2. The VideoTrak unit and the notebook PC are housed in a cabinet that is mounted to the camera pole. The two white plastic domes on top of the cabinet cover the antennas of the GPS receivers. The figure on the right-hand side shows a diagram of the setup.

We started by selecting a point near the bottom of the field of view along the shoulder, usually point A (but sometimes point D, if pointing north, in the direction of receding traffic). The second point can be selected further down the shoulder, but not so far that it cannot be aligned with visible lane markers (e.g., the leading or trailing edge of a painted line segment in the nearest lane). For the calibration points on the other side of the highway we had to choose significant features in or near the HOV lane or median which can be reliably lined up with shoulder markings (i.e., so that distance parallel to traffic flow can be estimated within a foot).

After the points have been identified, the various distances had to be measured or estimated. Camera-to-A distance and camera height were usually measured with a laser range-finder. A-B distance (along shoulder) could be easily measured with a laser range-finder, tape, or wheel. B-C and C-D were measured either with a range-finder or with knowledge of lane widths (or both, for confirmation). We aligned the calibration points in a rectangle, so diagonal distances can be computed by trigonometry.

Tracking strips serve to distinguish between the lanes of the road. The system produces PVRs for all vehicles that are inside a tracking strip. Tracking strips are treated independently, which can yield to multiple counts of the same vehicle in different lanes. This happens in particular for larger vehicles like trucks. The concept of tracking strips is a restriction for our purposes because it does not allow us to differentiate two vehicles next to each other, or one vehicle swerving into the other lane. The setup of tracking strips turned out to be difficult – small changes have significant impact on the volume count.

Every 300 ms a complete set of PVRs of all vehicles within the tracking strips will be stored on the hard-disk. Vehicles appear several times in the database

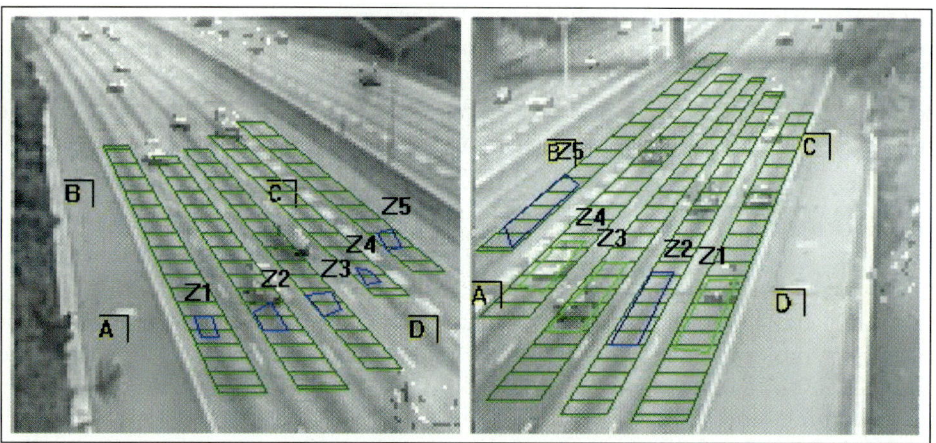

Fig. 3. The figures show the digitized video signal from the camera at Site 2 (left) and the North camera at Site 3 (right). The image is produced by the VT-900 system, and can be down-loaded to a PC allowing the definition of calibration points (A-D), and the tracking strips. Detection zones Z1-5 can be defined to simulate loop detectors; they are only used for testing purposes.

while they pass the site. In addition to generating the information that loop detectors would produce, we can also infer lane changes, at least within the field of view of the camera.

4 Image Processing Algorithm

The VideoTrak-900 uses Sarnoff pyramid vision algorithms to perform real-time video monitoring and tracking. Although the specifics of the algorithms used within this equipment are proprietary, the general approach has been described not only for image stabilization and tracking [3], but also for extended applications such as image mosaic construction [4], [5]. In the traffic application, the key problem is the detection of moving objects while rejecting slowly varying motion such as that which may result from camera sway or normal daily shadow variations. This is complicated by the need to provide robust operation under varying lighting conditions, including complete darkness. Typically, the Sarnoff approach begins by aligning a series of consecutive frames at sub-pixel resolution, essentially performing electronic image stabilization similar to what is done in consumer video camcorders. Differences are then computed for successive frames, providing an indication of motion energy relative to the background, and the regions of motion are segmented and tracked from frame to frame.

Pyramid techniques are computationally efficient because they allow for multi-resolutional representations of the image and can selectively apply processing at a resolution no finer than what is required for a specified level of detail. Often, these algorithms proceed in a coarse-to-fine manner, with computationally inexpensive coarse resolutions being processed first to determine where to concentrate finer resolution. Image alignment, for example, can converge very quickly to a solution when the images are represented coarsely, and the result can be used as a starting point for the next finer stage of the process.

5 Conclusion and Outlook

There is a price to pay for using proprietary hardware – we cannot entirely verify the performance of the tracking system. The system we chose includes the most recent technology, but it was designed for a different purpose. The use of the VideoTrak units in academia mode stretches its capabilities in terms of accuracy and reliability to the limits.

We believe that we can provide the community with a unprecedented detailed data set of single vehicle records in a major traffic situation once this project is successfully concluded. We hope that there will be similar field studies in other areas.

References

1. D. Helbing and A. Greiner, *Modeling and simulation of multi-lane traffic flow,* Phys. Rev. E **55**, 5498-5507 (1997).
2. J. Treiterer and J. I. Taylor, *Traffic Flow Investigations by Photogrammetric Techniques,* in: *Highway Research Record* **142***: Photogrammetry and Aerial Surveys*, pp. 1–12 (Highway Research Board, 1966).
3. R. Mandelbaum, M. Hansen, P. Burt, and S. Baten, *Vision for autonomous mobility: image processing on the VFE-200,* pp. 671–676 (Proc. of the 1998 IEEE Int. Symp. on Intelligent Control, 1998).
4. M. Hansen, P. Anandan, K. Dana, G. van der Wal, and P. Burt, *Real-time scene stabilization and mosaic construction,* pp. 457–465 (Proc. of 23rd Image Understanding Workshop **1**, 1994).
5. H. S. Sawhney, S. Ayer, and M. Gorkani, *Model-based 2D&3D dominant motion estimation for mosaicing and video representation,* pp. 583–590 (Proc. of the Fifth IEEE Int. Conference on Computer Vision, 1995).

Modelling of Traffic Flow

Microscopic Simulation of Congested Traffic

M. Treiber[1], A. Hennecke[1], and D. Helbing[1,2]

[1] II. Institute of Theoretical Physics, University of Stuttgart, Pfaffenwaldring 57/III, 70550 Stuttgart, Germany
[2] Collegium Budapest – Institute for Advanced Study, 1014 Budapest, Hungary

Abstract. We present simulations of congested traffic in open systems with a new car-following model. The model parameters are all intuitive and can be easily calibrated. Microsimulations with identical vehicles on a single lane produce the same traffic states as recent macrosimulations of open systems with on-ramps, which also qualitatively agree with real traffic data. The phase diagram in the phase space spanned by the traffic flow and the bottleneck strength is nearly equivalent to the macroscopic phase diagram. In agreement with macroscopic models, we found hysteresis, coexistent states, and a small region of tristability. We simulated the process of obtaining time-averaged traffic data by "virtual detectors". While for identical vehicles, the resulting flow-density data do not look very realistic, microsimulations of heterogeneous (multi-species) traffic offer a natural explanation of the observed wide scattering of congested traffic data.

1 Introduction

For about fifty years, now, researchers model freeway traffic by means of continuous-in-time microscopic models (car-following models) [1]. Since then, a multitude of car-following models have been proposed, both for single-lane and multi-lane traffic including lane changes.

In the simplest case, the acceleration of an individual vehicle depends only on the distance to the vehicle in front. Well-known models of this type include the model of Newell [2], or the "optimal-velocity model" by Bando *et al.* [3]. To achieve a better anticipative driver behavior and to avoid collisions, the acceleration in other models depends also on the velocity and on the approaching rate to the front vehicle [4–6]. Besides these simple models intended for basic investigations, there are also highly complex "high-fidelity models" with plenty of parameters like the Wiedemann model [7] or MITSIM [8], which try to reproduce traffic as realistically as possible.

From a physics point of view, a microscopic traffic model should be as simple as possible. The parameters should be intuitive, easy to calibrate, and the corresponding values should be realistic. The collective dynamics should reproduce all observed localized and extended traffic states [9], including synchronized traffic and the wide scattering of congested traffic data [10]. Furthermore, the observed hysteresis effects [11,12], complex states [10,13], and the existence of self-organized quantities like the constant propagation velocity of stop-and-go waves or the outflow from a traffic jam [14] should be reproduced. To be consistent with macroscopic models, a deterministic instability mechanism is favorable.

Finally, the dynamics must not lead to vehicle collisions and the model should allow for a fast numerical simulation.

In this paper, we check some of these criteria for the recently proposed intelligent-driver model (IDM) [15]. In particular, we give microscopic implementations of bottleneck inhomogeneities and present the phase diagram of congested traffic states for open, inhomogeneous systems [16]. We show that, for identical vehicles, the collective dynamics is qualitatively the same as that resulting from macroscopic traffic models like the non-local, gas-kinetic-based traffic model (GKT model) [17], or the Kerner-Konhäuser-Lee model (KKL model) [14,18]. Finally, we investigate heterogeneous traffic (composed of cars and trucks) and propose a natural explanation of the observed wide scattering of congested traffic data.

2 The Microscopic Intelligent-Driver Model (IDM)

The acceleration assumed in the IDM is a continuous function of the velocity v_α, the (netto) gap s_α, and the velocity difference (approaching rate) Δv_α of vehicle α to the leading vehicle

$$\dot{v}_\alpha = a^{(\alpha)} \left[1 - \left(\frac{v_\alpha}{v_0^{(\alpha)}} \right)^\delta - \left(\frac{s_\alpha^*(v_\alpha, \Delta v_\alpha)}{s_\alpha} \right)^2 \right] . \tag{1}$$

This expression is a superposition of the acceleration $a^{(\alpha)}[1 - (v_\alpha / v_0^{(\alpha)})^\delta]$ on a free road, and a braking deceleration $-a^{(\alpha)}[s_\alpha^*(v_\alpha, \Delta v_\alpha)/s_\alpha]^2$, describing the interactions with other vehicles. The deceleration term depends on the ratio between the "desired gap" s_α^* and the actual gap s_α, where the desired gap

$$s_\alpha^*(v, \Delta v) = s_0^{(\alpha)} + s_1^{(\alpha)} \sqrt{\frac{v}{v_0^{(\alpha)}}} + T_\alpha v + \frac{v \Delta v}{2\sqrt{a^{(\alpha)} b^{(\alpha)}}} , \tag{2}$$

is dynamically varying with the velocity and the approaching rate, reflecting an intelligent driver behavior. The IDM parameters are the desired velocity v_0, safe time headway T, maximum acceleration a, comfortable deceleration b, acceleration exponent δ, and the jam distances s_0 and s_1. Furthermore, the vehicles have a finite length l, which, however, has no dynamical influence. In Sections 2 to 4, we will assume identical "cars", while in Sect. 5 we assume two different types, "cars" and "trucks", see Table 1. For better readability, we will drop the vehicle index α in the following discussion of the model.

2.1 Equilibrium Traffic of Identical Vehicles

In equilibrium traffic ($\dot{v}_\alpha = 0$, $\Delta v_\alpha = 0$), drivers tend to keep a velocity-dependent equilibrium gap $s_\mathrm{e}(v_\alpha)$ to the front vehicle given by

$$s_\mathrm{e}(v) = s^*(v, 0) \left[1 - \left(\frac{v}{v_0} \right)^\delta \right]^{-\frac{1}{2}} . \tag{3}$$

Table 1. Model parameters of the IDM model used throughout this paper.

Type	v_0	T	a	b	s_0	s_1	l
Cars	120 km/h	1.2 s	0.8 m/s^2	1.25 m/s^2	1 m	10 m	5 m
Trucks	80 km/h	1.7 s	0.4 m/s^2	0.8 m/s^2	1 m	10 m	8 m

In particular, the equilibrium gap of homogeneous *congested* traffic ($v \ll v_0$) is essentially equal to the desired gap, $s_e(v) \approx s^*(v, 0) = s_0 + s_1\sqrt{v/v_0} + vT$, i.e., it is composed of a (small) high-density contribution, and a contribution vT corresponding to a time headway T.

Solving (3) for the equilibrium velocity $v = v_e$ leads to simple expressions only for $s_1 = 0$ and $\delta = 1$, $\delta = 2$, or $\delta \to \infty$. In particular, the equilibrium velocity for the special case $\delta = 1$ and $s_0 = s_1 = 0$ is

$$v_e(s) = \frac{s^2}{2v_0T^2}\left(-1 + \sqrt{1 + \frac{4T^2v_0^2}{s^2}}\right). \tag{4}$$

Macroscopically, homogeneous traffic consisting of identical vehicles can be characterized by the equilibrium traffic flow $Q_e(\rho) = \rho V_e(\rho)$ (vehicles per hour and lane) as a function of the traffic density ρ (vehicles per km and lane). For $\delta = 1$ and $s_0 = s_1 = 0$, this "fundamental diagram" follows from (4) together with the micro-macro relation between gap and density,

$$s = 1/\rho - l = 1/\rho - 1/\rho_{\max}. \tag{5}$$

The result is identical with the equilibrium velocity of the GKT model, if the GKT parameter ΔA is set to zero, (cf. (23) in [17]), which is a necessary condition for a micro-macro correspondence. Figure 1a shows that the acceleration coefficient δ influences the transition region between the free and congested regimes. For $\delta \to \infty$ and $s_1 = 0$, the fundamental diagram $Q_e(\rho) = \min(v_0\rho, [1 - \rho(l + s_0)]/T)$ becomes triangular-shaped. For decreasing δ, it becomes smoother and smoother.

2.2 Dynamic Single-Vehicle Properties

Figures 1b and 1c show how the parameters a and b determine the acceleration and braking behavior of single vehicles. At $t = 0$, one vehicle starts with zero velocity and 2.5 km of free road ahead. Initially, it accelerates with a and approaches smoothly the desired velocity v_0. At $x = 2.5$ km, we assume a standing obstacle, e.g., the end of a traffic jam. When approaching the obstacle, the model mimics "intelligent" drivers who anticipate necessary braking decelerations and brake so as not to exceed the comfortable deceleration b in normal situations [15].

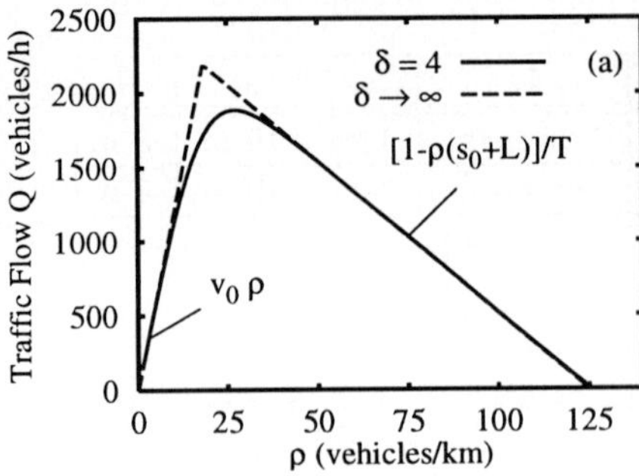

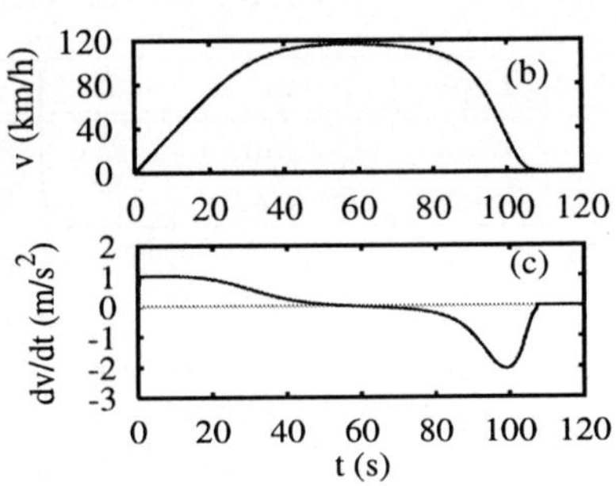

Fig. 1. (a) Equilibrium flow-density diagram of identical IDM vehicles with variable acceleration exponent δ and the "car" parameters of Table 1 otherwise. (b), (c) Temporal evolution of the velocity and acceleration of a single vehicle approaching a standing obstacle, which is reached after $t = 106$ s (cf. main text). The IDM parameters are $a = 1\ \mathrm{m/s^2}$, $b = 2\ \mathrm{m/s^2}$, and the "car" parameters of Table 1 otherwise.

2.3 Collective Behavior and Stability Diagram

Although we are interested in realistic *open* systems, it turned out that many features can be explained in terms of the stability behavior in a *closed* system. Figure 2a shows the stability diagram of homogeneous traffic on a circular road. The control parameter is the homogeneous density $\overline{\rho}$. We applied both a very small and a large localized perturbation to check for linear and nonlinear stability, and plotted the resulting minimum (ρ_{out}) and maximum (ρ_{jam}) densities after a stationary situation was reached. The resulting diagram is very similar to that of the macroscopic KKL and GKT models [14,17]. In particular, it displays the following realistic features: (i) Traffic is stable for very low and high densities, but unstable for intermediate densities. (ii) There is a density range $\rho_{\mathrm{c1}} \le \overline{\rho} \le \rho_{\mathrm{c2}}$ of metastability, i.e., only perturbations of sufficiently large amplitudes grow, while smaller perturbations disappear. Note that, for most IDM parameter sets, there is no second metastable range at higher densities, in contrast to the GKT and KKL models. (iii) The density inside of traffic jams and the associated flow $Q_{\mathrm{jam}} = Q_{\mathrm{e}}(\rho_{\mathrm{jam}})$, cf. Fig. 2b, do not depend on $\overline{\rho}$. As further "traffic constants", at least in the density range 20 veh./km $\le \overline{\rho} \le$ 40 veh./km, we observe a constant outflow $Q_{\mathrm{out}} = Q_{\mathrm{e}}(\rho_{\mathrm{out}})$ and propagation velocity $v_{\mathrm{g}} = (Q_{\mathrm{out}} - Q_{\mathrm{jam}})/(\rho_{\mathrm{out}} - \rho_{\mathrm{jam}}) \approx -15$ km/h of jams. Figure 2b shows the stability diagram for the flows. In particular, we have $Q_{\mathrm{c1}} < Q_{\mathrm{out}} < Q_{\mathrm{c2}}$, where $Q_{\mathrm{c}i} = Q_{\mathrm{e}}(\rho_{\mathrm{c}i})$, i.e., the outflow from congested traffic is metastable, while in the GKT, $Q_{\mathrm{out}} \approx Q_{\mathrm{c2}}$ is only marginally stable.

In *open* systems, a third type of stability becomes relevant. Traffic is *convectively* stable, if, after a sufficiently long time, all perturbations are convected out of the system. Both in the macroscopic models and in the IDM, there is a considerable density region $\rho_{\mathrm{cv}} \le \overline{\rho} \le \rho_{\mathrm{c3}}$, where traffic is linearly unstable but convectively stable.

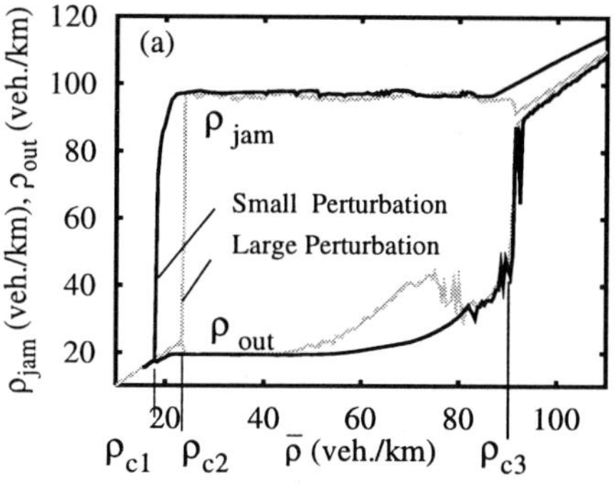

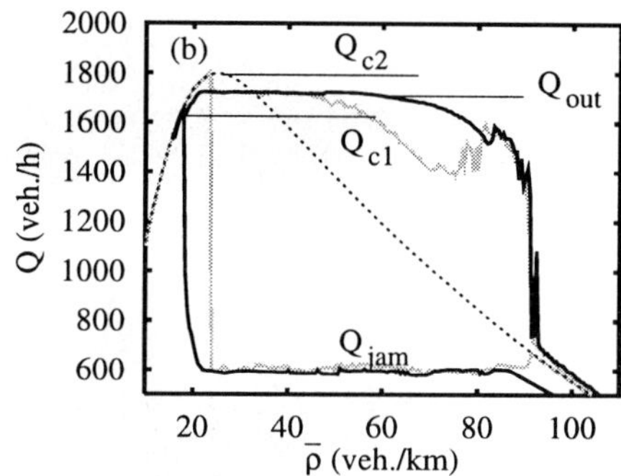

Fig. 2. Stability diagram of homogeneous traffic (for "car" parameters) in a closed system as a function of the homogeneous density $\overline{\rho}$ for small (grey) and large (black) initial perturbations of the density. In plot (a), the upper two lines display the density inside of density clusters after a stationary state has been reached. The lower two lines represent the density between the clusters. Plot (b) shows the corresponding flows and the equilibrium flow-density relation (dotted). The critical densities ρ_{ci} and flows Q_{ci} are discussed in the main text.

2.4 Calibration

The *fundamental relations* of homogeneous traffic are calibrated with v_0 (low density), δ (transition region), T (high density), and s_0 and s_1 (jammed traffic). The *stability behavior* of traffic in the IDM model is determined mainly by the model parameters a, b, and T. The density in and the outflow from traffic jams are also influenced by s_0 and s_1. Since the accelerations a and b do not influence the fundamental diagram, the model can be calibrated essentially independently with respect to traffic flows and stability. As in the GKT model, traffic becomes more unstable for decreasing a (which corresponds to an increased acceleration time $\tau = v_0/a$), and for decreasing T (corresponding to reduced safe time headways). Furthermore, the instability increases with growing b. This is also plausible, because an increased desired deceleration b corresponds to a less anticipative or less defensive braking behavior.

3 Microscopic Implementation of Bottlenecks

In *macroscopic* simulations, a natural implementation of road inhomogeneities is given by on- and off- ramps, which appear as a source term in the continuity equation for the density. An explicit *microscopic* modeling of ramps, however, would require a multi-lane model with explicit simulation of lane changes. In order to avoid the associated complications, one can either apply the micro-macro link and simulate the ramp section macroscopically [19], or introduce *flow-conserving* inhomogeneities by making one or more model parameters dependent on the location x of the road. Suitable parameters for the IDM are v_0 or T [15]. Local parameter variations act as a bottleneck, if the outflow Q'_{out} from congested traffic in the downstream section is reduced with respect to the outflow Q_{out}

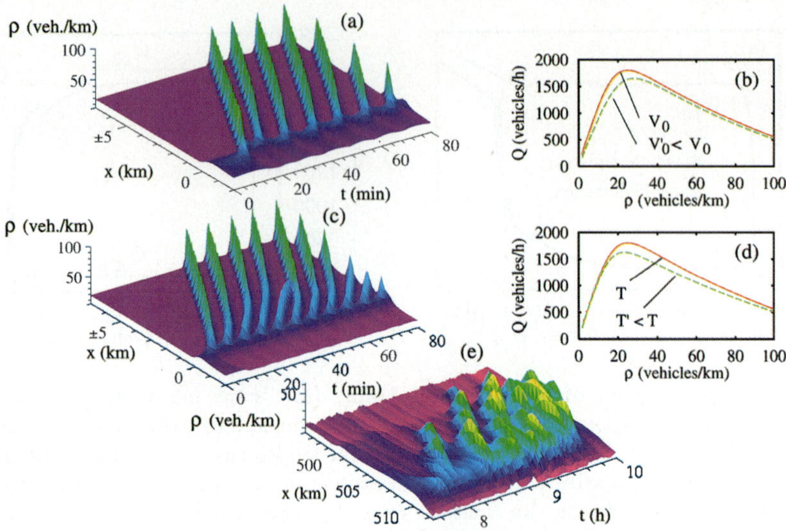

Fig. 3. Traffic breakdowns at differently implemented bottlenecks: (a) IDM simulation of the spatio-temporal density (inflow $Q_{\text{in}} = 1670$ vehicles/h) for a bottleneck corresponding to a decrease of v_0 in the downstream region. The density is derived from the microscopic distance s via relation (5) and a subsequent linear interpolation. (b) Related equilibrium flows upstream and downstream. (c), (d) Bottleneck corresponding to an increase of the safe time headway T. (e) Density obtained from 1-min detector data of the German freeway A9-South on Oct. 29, 1998. The traffic breakdown takes place upstream of the intersection "Neufahrn" at $x = 512$ km.

in the upstream section. This requires a reduced desired velocity $v_0' < v_0$ or increased time headway $T' > T$, or both. Figure 3a shows a traffic breakdown induced by a linear decrease of the desired velocity from $v_0 = 120$ km/h for $x \leq -L/2$ to $v_0' = 95$ km/h for $x \geq L/2$, while Fig. 3c shows the same effect for an increase of the safe time headway from $T = 1.2$ s to $T' = 1.45$ s ($L =$ 200 m). Both flow-conserving bottlenecks result in a similar traffic dynamics which, however, depends strongly on the amplitude of the parameter variation. Qualitatively the same dynamics is observed in real traffic data [Fig. 3(e)] and in macroscopic models including on-ramps with a ramp flow of $Q_{\text{rmp}} = Q_{\text{out}} - Q'_{\text{out}}$. This suggests to define a general "bottleneck strength" δQ by

$$\delta Q := Q_{\text{rmp}} + Q_{\text{out}} - Q'_{\text{out}}. \tag{6}$$

In particular, we have $\delta Q = Q_{\text{rmp}}$ for on-ramp bottlenecks, and $\delta Q = Q_{\text{out}} - Q'_{\text{out}}$ for flow-conserving bottlenecks. In the following, we will vary v_0. The regions with locally decreased desired velocity can be interpreted as sections with uphill gradients (which reduce the maximum velocities of vehicles).

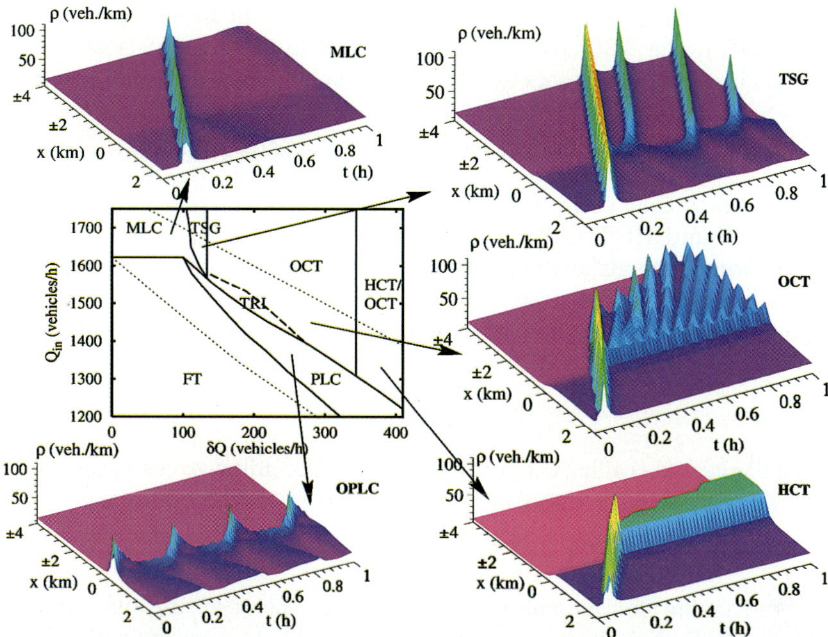

Fig. 4. Phase diagram of congested traffic states and corresponding spatio-temporal density plots as simulated with the IDM. The control parameters are the inflow Q_{in} and the bottleneck strength δQ, which increases with $(v_0 - v_0')$. The traffic states FT, HCT, OCT, TSG, MLC, and (O)PLC are explained in the main text. "TRI" indicates a tristable region.

4 Phase Diagram of Congested Traffic

In contrast to *closed* systems, in which the long-term behavior and stability is essentially determined by the average traffic density, the dynamics of *open* systems is controlled by the inflow Q_{in}. Furthermore, traffic congestions depend on road inhomogeneities and, because of hysteresis effects, on the history of previous perturbations. For a given history, the traffic states can be summarized by a phase diagram spanned by Q_{in} and δQ. Figure 4 shows the IDM phase diagram for traffic states that develop after a single density cluster crosses the inhomogeneity. Depending on Q_{in} and δQ, the initial perturbation (i) dissipates, resulting in free traffic (FT), (ii) travels through the inhomogeneity as a moving localized cluster (MLC) and neither dissipates nor triggers new breakdowns, (iii) triggers a traffic breakdown to a pinned localized cluster (PLC), which remains localized near the inhomogeneity for all times and either is stationary, cf. Fig. 5a for $t < 0.2$ h, or oscillatory (OPLC). (iv) Finally, the initial perturbation can induce extended congested traffic (CT), whose downstream boundaries are fixed at the inhomogeneity, while the upstream front propagates further upstream in the course of time. This kind of congested traffic can be homogeneous (HCT), oscillatory (OCT), or consist of triggered stop-and-go waves (TSG). In contrast to

OCT, where there is permanently congested traffic at the inhomogeneity ("pinch region" [13,15]), the TSG state is characterized by a series of isolated density clusters, each of which triggers a new cluster as it passes the inhomogeneity.

4.1 Boundaries between and Coexistence of Traffic States

Simulations show that the outflow $\tilde{Q}_{\text{out}}$ from the nearly stationary downstream fronts of OCT and HCT satisfies $\tilde{Q}_{\text{out}} \leq Q'_{\text{out}}$, where Q'_{out} is the outflow from clusters in homogeneous systems for the downstream model parameters. If the bottleneck is not too strong, we have $\tilde{Q}_{\text{out}} \approx Q'_{\text{out}}$. Then, for all types of bottlenecks, the congested traffic flow is given by $Q_{\text{cong}} = \tilde{Q}_{\text{out}} - Q_{\text{rmp}} \approx Q'_{\text{out}} - Q_{\text{rmp}}$, or

$$Q_{\text{cong}} \approx Q_{\text{out}} - \delta Q. \tag{7}$$

Extended congested traffic (CT) only persists, if the inflow exceeds the congested traffic flow. Otherwise, it dissolves to PLC. This gives the boundary

$$\text{CT} \to \text{PLC} : \delta Q \approx Q_{\text{out}} - Q_{\text{in}}. \tag{8}$$

If congested traffic flow is *convectively* unstable, the resulting oscillations lead to TSG or OCT. If it is *linearly* stable, $Q_{\text{cong}} < Q_{\text{c3}}$, we have HCT. If it is convectively stable, but linearly unstable, $Q_{\text{cong}} \in [Q_{\text{c3}}, Q_{\text{cv}}]$, one has a spatial *coexistence* of states with HCT near the bottleneck and OCT further upstream [15] (Fig. 5), which is frequently found in empirical data of congested traffic. In the IDM, this frequent occurrence is reflected by the wide range of flows falling into this regime. For the "car" parameters, we have $Q_{\text{c3}} = 600$ vehicles/h and $Q_{\text{cv}} = 1340$ veh./h.

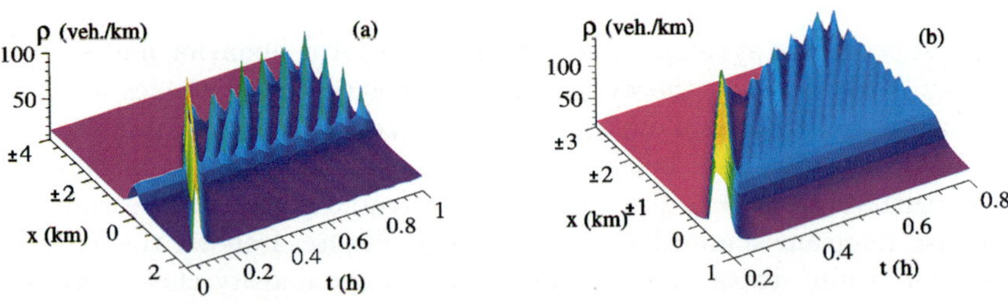

Fig. 5. (a) Transition from PLC to OCT in the tristable region ($Q_{\text{in}} = 1480$ vehicles/h, $v'_0 = 24$ m/s corresponding to $\delta Q = 220$ vehicles/h), triggered by a large perturbation. (b) Spatial coexistence of HCT and OCT for $Q_{\text{in}} = 1350$ veh./h and $v'_0 = 16$ m/s corresponding to $\delta Q = 400$ veh./h.

4.2 Multistability

In general, the local phase transitions between free traffic, pinned localized states, and extended congested states are hysteretic. In the regions between the two dotted lines of the phase diagram in Fig. 4, both, free and congested traffic is possible, depending on the previous history. In particular, for all five indicated phase points (but not for the simulations of Fig. 3), free traffic would persist without the downstream perturbation. In contrast, the transitions PLC-OPLC, and HCT-OCT-TSG seem to be non-hysteretic, i.e., the type of pinned localized cluster or of extended congested traffic, is uniquely determined by $Q_{\rm in}$ and δQ.

In a small subset of the meta stable region, labeled "TRI" in Fig. 4, we even found *tristability* between FT, PLC, and OCT. We obtained qualitatively the same also for the GKT model, and it has been found for the KKL model with OPLC instead of PLC for the pinned localized state [18]. Figure 5a shows that a single moving localized cluster passing the inhomogeneity triggers a transition from PLC to OCT. Starting with free traffic, the same perturbation would trigger OCT as well, while we never found reverse transitions OCT $\rightarrow$ PLC or OCT $\rightarrow$ FT (without a reduction of the inflow). That is, FT and PLC are meta stable in the tristable region, while OCT is stable.

4.3 Pinch Effect and Merging of Clusters

Careful investigations of traffic data related to OCT states [13] showed two phenomena: (i) In a narrow region near the inhomogeneity, there is nearly stationary congested traffic ("pinch region"), whereas further upstream, oscillations lead to temporarily lower traffic densities. (ii) While propagating upstream, the oscillations grow and merge to a few large-amplitude density clusters with free traffic in between. Detector data of other freeways, however, show this pinch effect without mergers, cf. Fig. 3(e). Simulating the IDM with the "car" parameters leads to very few mergers, cf. Figs. 3c and 4. For other parameters, however, the IDM reproduces mergers ending up with stop-and-go traffic [15]. A possible explanation is the *"starvation effect"*: For the parameters chosen in this article, the outflow $Q_{\rm out}$ from density clusters is in the middle of the meta stable region (Fig. 2b), so medium-sized and large density clusters persist. For the parameters in [15], however, the outflow from density clusters satisfies $Q_{\rm out} \approx Q_{\rm c1}$, so only large-amplitude clusters survive, while all others dissipate.

5 Multi-Species Single-Lane Traffic

In this section, we assume heterogeneous single-lane traffic consisting of 70% "cars" and 30% "trucks", where the latter are characterized by a lower desired velocity, lower accelerations, and a larger safe time headway compared to cars (Table 1). Again, we simulate a traffic breakdown to HCT at a flow-conserving inhomogeneity, where the desired velocity of cars is reduced from 120 km/h to 65 km/h, and that of trucks from 80 km/h to 36 km/h. (Similar results are found for less drastic increases of T.) To compare the result with real traffic data, we implement "virtual" detectors at several fixed locations. The detectors record passage times and velocities of each vehicle to determine the macroscopic flow $Q = n_\tau/\tau$ (n_τ is the number of passing vehicles in the averaging interval

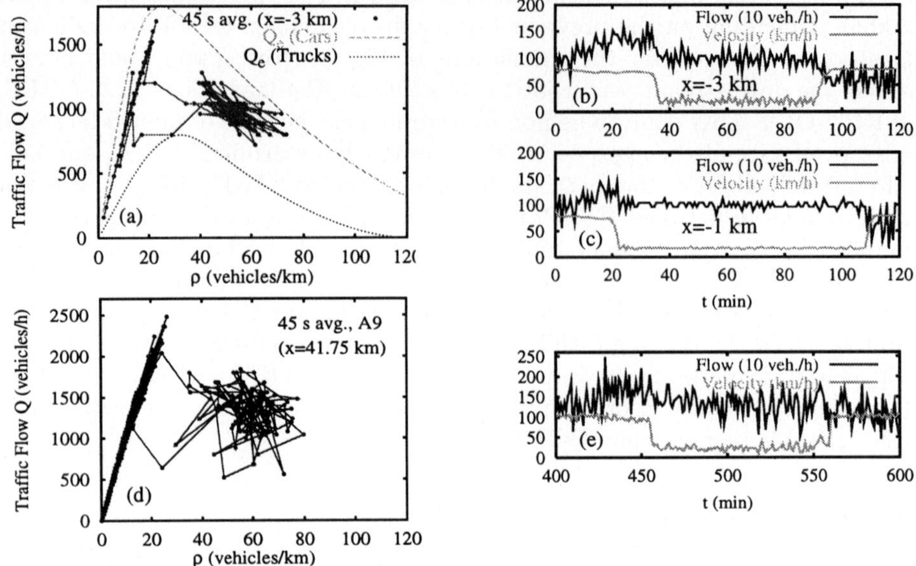

Fig. 6. (a) Flow-density diagram and (b)-(c) time series of single-lane heterogeneous traffic. (d)-(e) Empirical data of extended congested traffic on the Dutch freeway A9.

$\tau = 45$ s), the arithmetic velocity average V, and the density $\rho = Q/V$. Figure 6 shows the resulting fundamental diagram 3 km upstream of the bottleneck, and time series of Q and V at three upstream locations. For comparison, Fig. 6d and e show 45 s averages of *real* single-vehicle data of the Dutch freeway A9 from Haarlem to Amsterdam on October 14, 1994. The detector is about 0.7 km upstream of the on-ramp causing the traffic breakdown. The simulated and real traffic data agree qualitatively, in particular in the following respects: (i) There is a wide scattering of flow-density data in the congested regime (looking like an anisotropic two-dimensional random walk), while the data occupy a nearly one-dimensional region in the free regime. (ii) The distribution of flow-density data shows the typical inverse-λ form with a distinct gap between free and congested traffic data. (iii) During the breakdown, the velocity drops to 10-20 km/h, while the flow is reduced by only about 20 %. (iv) In all regions, the relative fluctuations of the velocity are smaller than those of the flow. Near the bottleneck, the fluctuations of congested traffic flow are much smaller than those of free traffic, while further upstream, the fluctuations grow.

Notice that the fluctuations of the simulated data come essentially from the different vehicle types and not from deterministic instabilities. Macroscopically, the situation of Fig. 6 corresponds to a spatial coexistence of HCT and OCT. (For a pure OCT or TSG state, there would be no gap between the flow-density data of free and congested traffic, both in real traffic data and in simulations.)

6 Discussion

Although there are also other proposals for an explanation of the various transitions between free traffic, congested traffic, and stop-and-go waves [13], we see the following advantages of our approach: The transitions result *naturally* from a model for traffic flow on *homogeneous* roads, just by adding speed limits (like above), source terms (in case of on-ramp flows) [16], or other kinds of inhomogeneities, which are *known* to exist. The implementation of these inhomogeneities is *straightforward,* without the requirement of additional refined model ingredients. Moreover, analogous to the other parameters of the model, the inhomogeneities are easily measurable quantities. There is no model ingredient, which could not be relatively easily be verified or falsified by empirical studies. Furthermore, *empirical results* confirming the existence of a phase diagram are already available [9]. Finally, we think that the simulated traffic states related to inhomogeneities of the road arise *so* naturally, that any explanation of empirical data must take these states into account. Our simulations show that, simply by assuming a mixture of different vehicle types, one can reproduce the observed scattering of flow-density data by a *deterministic* model having a *unique* equilibrium relation. The same has also been found for a macroscopic model [20]. Generalizations to multi-lane traffic result in an even better agreement with empirical findings, e.g., a larger scattering of "virtual" detector data and flow-stabilizing effects of speed limits [21,22]. To us, the most interesting open question is which kind of observable phenomena are produced by the heterogeneity of driver-vehicle units *in addition* to the scattering of traffic data [23].

Acknowledgement. The authors want to thank for financial support by the BMBF (research project SANDY, grant No. 13N7092) and by the DFG (grant No. He 2789). We are also grateful to the *Autobahndirektion Südbayern* and the *Hessisches Landesamt für Straßen und Verkehrswesen* for providing freeway data.

References

1. A. Reuschel, *Fahrzeugbewegungen in der Kolonne,* Österr. Ingenieur-Archiv **4**, 193–215 (1950).
2. G.F. Newell, *Nonlinear effects in the dynamics of car following,* Op. Res. **9**, 209 (1961).
3. M. Bando, K. Hasebe, A. Nakayama, A. Shibata, and Y. Sugiyama, *Dynamical model of traffic congestion and numerical simulation,* Phys. Rev. E **51**, 1035–1042 (1995).
4. P.G.A. Gipps, *Behavioural car-following model for computer simulation,* Trans. Res. B **15**, 105–111 (1981).
5. S. Krauß, *Microscopic Modelling of Traffic Flow,* FB 98-08, (DLR, Cologne, 1998).
6. D. Helbing and B. Tilch, *Generalized force model of traffic dynamics,* Phys. Rev. E **58**, 133–138 (1998).

7. R. Wiedemann, *Simulation des Straßenverkehrsflusses,* (Institut für Verkehrswesen, Universität Karlsruhe, 1974).
8. See internet page `http://hippo.mit.edu/products/mitsim/main.html`.
9. H.Y. Lee, H.W. Lee, and D. Kim, *Empirical phase diagram of congested traffic flow,* preprint, cond-mat/9905292.
10. B.S. Kerner and H. Rehborn, *Experimental properties of complexity in traffic flow,* Phys. Rev. E **53**, R4275–R4278 (1996).
11. J. Treiterer and J.A. Myers, *The hysteresis phenomenon in traffic flow,* in: *Proc. 6th Int. Symp. on Transportation and Traffic Theory,* D.J. Buckley, (ed.), pp. 13 (Elsevier, New York, 1974).
12. C.F. Daganzo, M.J .Cassidy, and R.L. Bertini, *Some traffic features at freeway bottlenecks,* Trans. Res. B **33**, 25–42 (1999).
13. B.S. Kerner, *The physics of traffic,* Physics World, Aug., 25–30 (1999).
14. B.S. Kerner and P. Konhäuser, *Structure and parameters of clusters in traffic flow,* Phys. Rev. E **50**, 54–83 (1994).
15. M. Treiber and D. Helbing, *Explanation of observed features of self-organization in traffic flow,* preprint, cond-mat/9901239.
16. D. Helbing, A. Hennecke, and M. Treiber, *Phase diagram of traffic states in the presence of inhomogeneities,* Phys. Rev. Lett. **82**, 4360–4363 (1999).
17. M. Treiber, A. Hennecke, and D. Helbing, *Derivation, properties, and simulation of a gas-kinetic-based, non-local traffic model,* Phys. Rev. E **59**, 239–253 (1999).
18. H.Y. Lee, H.W. Lee, and D. Kim, *Origin of synchronized traffic flow on highways and its dynamic phase transition,* Phys. Rev. Lett. **81**, 1130–1133 (1998).
19. A. Hennecke, M. Treiber, and D. Helbing, *Macroscopic simulation of open systems and micro-macro link,* in this volume.
20. M. Treiber and D. Helbing, *Macroscopic simulation of widely scattered synchronized traffic states,* J. Phys. A **32**, L17–L23 (1999).
21. Interactive simulations of the multi-lane IDM are available at `http://www.uni-stuttgart.de/treiber/MicroApplet/`.
22. R. Sollacher and H. Lenz, *Nonlinear control of stop-and-go traffic,* in this volume.
23. D. Helbing and B.A. Huberman, *Coherent moving states in highway traffic,* Nature **396**, 738–740 (1998).

Order Parameter as an Additional State Variable of Unstable Traffic Flow

I.A. Lubashevsky[1] and R. Mahnke[2]

[1] Theory Department, General Physics Institute, Russian Academy of Sciences, Vavilov str., 38, Moscow 117942, Russia
[2] Fachbereich Physik, Universität Rostock, D–18051 Rostock, Germany

Abstract. We discuss a phenomenological approach to the description of unstable vehicle motion on multilane highways that could explain in a simple way such observed self-organizing phenomena as the sequence of the phase transitions "free flow → synchronized motion → jam" and the hysteresis in them.
We introduce a new variable called order parameter that accounts for possible correlations in the vehicle motion at different lanes. So, it is principally due to "many-body" effects in the car interaction in contrast to such variables as the mean car density and velocity being actually the zeroth and first moments of the "one-particle" distribution function. Therefore, we regard the order parameter as an additional independent state variable of traffic flow and formulate the corresponding evolution equation governing the lane changing rate.
In this context we analyze the instability of homogeneous traffic flow manifesting itself in both of these phase transitions and endowing them with the hysteresis. Besides, the jam state is characterized by the vehicle flows at different lanes being independent of one another.

1 Introduction

The existence of a new basic phase in vehicle flow on multilane highways called the synchronized motion was recently discovered by Kerner and Rehborn [1], impacting significantly the physics of traffics as a whole. In particular, it turns out that the spontaneous formation of moving jams on highways proceeds mainly through a sequence of two transitions: "free flow → synchronized motion → stop-and-go pattern" [2]. Besides, all these transitions exhibit the hysteresis [2–4]. As follows from the experimental data [1,3,4] the synchronized mode is essentially a multilane effect. Recently Kerner [5,6] assumed that the transition "free flow → synchronized mode" is caused by "Z"-like form of the overtaking probability depending on the car density.

There have been proposed several macroscopic models dealing with multilane traffic flow [7–14]. Both these models specify the traffic dynamics completely in terms of the car density ρ, mean velocity v, and, may be, the velocity variance θ or ascribe these quantities to the vehicle flow at each lane individually. Nevertheless, a quantitative description of the synchronized mode is far from being developed well because of its complex structure [5,6]. In particular, it can form the totally homogeneous (*i*) and homogeneous-in-speed (*ii*) flows [1]. Especially in the latter case there is no explicit relationship between the mean car velocity

v and density ρ, with the value of v being actually constant and less then that of free flow. The other important feature is the key role of some cars bunched together and traveling much faster than the typical ones, which enables to regard them as a special car group [1]. Therefore, in the synchronized mode the function of car distribution in the velocity space should have two maxima and we will call such fast car groups platoons in speed. These features of the synchronized mode have been substantiated also in [15] using single-car-data. In particular, it has been demonstrated that the synchronized mode exhibits small correlations between fluctuations in the car flow, velocity and density. There is only a strong correlation between the velocities at different lanes taken at the same time and decreasing sufficiently fast as the time difference increases. By contrast, there are strong long-time correlations between the flow and density in the free flow state as well as the stop-and-go mode.

Keeping in mind a certain analogy with aggregation processes in physical systems Mahnke et al. [16,17] proposed a kinetic model for the formation of the synchronized mode treated as the motion of a large car cluster. In the present paper following practically the spirit of the Landau theory of phase transitions we develop a phenomenological approach to the description of this process. We ascribe to the vehicle flow an additional *internal* parameter will be called below the order parameter $h \in (0,1)$ characterizing the possible correlations in the vehicle motion at different lanes and write for it a governing equation. For the car motion where drivers do not change lane at all we set $h = 0$, in the opposite limit $h = 1$.

2 Order Parameter and the Individual Driver Behavior

For fixed values of ρ and v the order parameter h is assumed to be uniquely determined, thus, for a uniform vehicle flow we write:

$$\tau \frac{dh}{dt} = -\Phi(h, \rho, v), \tag{1}$$

where τ is the delay time and the function $\Phi(h, \rho, v)$ fulfills the inequality:

$$\frac{\partial \Phi}{\partial h} > 0. \tag{2}$$

We note that the time τ characterizes the delay in the driver decision of changing lanes but not in the control over the headway, so, this delay can be prolonged. The particular value $h(v, \rho)$ of the order parameter results from the compromise between the danger of an accident during changing lanes and the will of driver to move as fast as possible. Obviously, the lower is the mean vehicle velocity v for a fixed value of ρ, the weaker is the lane-changing danger and the stronger is the will to move faster. Besides, the higher is the vehicle density ρ for a fixed value of v, the stronger is this danger (here the will has no effect). Thus, the dependence $h(v, \rho)$ is an decreasing function of v and ρ, so, due to (2):

$$\frac{\partial \Phi}{\partial v} > 0\,, \quad \frac{\partial \Phi}{\partial \rho} > 0, \tag{3}$$

with the latter inequality being caused by the danger effect only. Equation (1) describes actually the behavior of the drivers that prefer to move faster than the statistically mean vehicle and whose readiness for risk is greatest. Exactly this group of drivers (platoons in speed) govern the value of h.

There is, however, another characteristics of the driver behavior, it is the mean velocity $v = \vartheta(h, \rho)$ chosen by the *statistically mean* driver taking into account also the danger resulting from the frequent lane changes by the "fast" drivers. Following typical assumptions the velocity $\vartheta(h, \rho)$ as a function of ρ is considered to be decreasing:

$$\frac{\partial \vartheta}{\partial \rho} < 0 \quad \text{and} \quad \rho\vartheta(\rho) \to 0 \quad \text{as} \quad \rho \to \rho_0, \tag{4}$$

where ρ_0 is the upper limit vehicle density on road. In general, the dependence of $\vartheta(h, \rho)$ on h should be increasing for small values of the vehicle density, $\rho \ll \rho_0$, because in this case the lane-changing makes no substantial danger to traffic and practically all the drives can pass by vehicles moving at lower speed without risk. By contrast, when the vehicle density is sufficiently high, $\rho \sim \rho_0$, the lane-changing is due to the car motion of the most "impatient" drivers whose behavior makes an additional danger to the main part of other drivers and the velocity $\vartheta(h, \rho)$ has to decrease as the order parameter h increases. For certain intermediate values of the vehicle density, $\rho \approx \rho_c$, this dependence is to be weak as well as near the boundary points, so:

$$\frac{\partial \vartheta}{\partial h} > 0 \text{ for } \rho < \rho_c\,, \quad \frac{\partial \vartheta}{\partial h} < 0 \text{ for } \rho > \rho_c\,, \quad \frac{\partial \vartheta}{\partial h} = 0 \text{ at } h = 0, 1. \tag{5}$$

Then the governing equation (1) takes the form:

$$\tau \frac{dh}{dt} = -\phi(h, \rho)\,, \text{ where } \phi(h, \rho) \stackrel{\text{def}}{=} \Phi[h, \rho, \vartheta(h, \rho)], \tag{6}$$

and the condition $\phi(h, \rho) = 0$ specifies the steady state dependence $h(\rho)$ of the order parameter on the vehicle density.

Let us, now, study properties and stability of this steady state solution. From (6) we get

$$\frac{\partial \phi}{\partial h} = \frac{\partial \Phi}{\partial h} + \frac{\partial \Phi}{\partial v}\frac{\partial \vartheta}{\partial h}\,, \qquad \frac{\partial \phi}{\partial \rho} = \frac{\partial \Phi}{\partial \rho} + \frac{\partial \Phi}{\partial v}\frac{\partial \vartheta}{\partial \rho}. \tag{7}$$

As mentioned above, the value of $\partial\Phi/\partial\rho$ is solely due to the danger during changing lanes, so this term can be ignored until the vehicle density ρ becomes sufficiently high. Thus, in a certain region $\rho < \rho_h < \rho_0$ the derivative $\partial\phi/\partial\rho \sim (\partial\Phi/\partial v)(\partial\vartheta/\partial\rho) < 0$ by virtue of (3) and (4) and the function $h(\rho)$ is increasing or decreasing for $\partial\phi/\partial h > 0$ or $\partial\phi/\partial h < 0$, respectively. This statement follows directly from the relation $dh/d\rho = -\,(\partial\phi/\partial\rho)\,(\partial\phi/\partial h)^{-1}$.

For long-wave perturbations of the vehicle distribution on a highway the density ρ can be treated as a constant. So, according to the governing equation (6), the steady-state traffic flow is unstable if $\partial\phi/\partial h < 0$. Due to (2) and (5) the

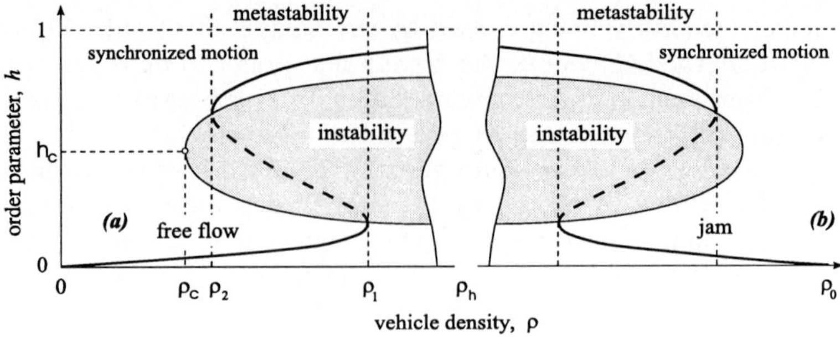

Fig. 1. The region of the traffic flow instability in the $h\rho$-plane and the form of the curve $h(\rho)$ displaying the dependence of the order parameter on the vehicle density.

first term in the expression for $\partial\phi/\partial h$ in (7) is dominant in the vicinity of the lines $h = 0$ and $h = 1$, thus, in these regions the curve $h(\rho)$ is increasing and the steady state traffic flow is stable. For $\rho < \rho_c$ the value $\partial\vartheta/\partial h > 0$, inequality (5), and, thereby, the region $\{0 < h < 1,\ 0 < \rho < \rho_c\}$ corresponds to the stable vehicle motion. However, for $\rho > \rho_c$ there can be an interval of the order parameter h where the derivative $\partial\phi/\partial h$ changes the sign and the vehicle motion becomes unstable. Therefore, as the car density ρ grows causing the increase of the order parameter h it can go into the instability region wherein $dh/d\rho < 0$. Under these conditions the curve $h(\rho)$ is to look like "S" (Fig. 1a) and its decreasing branch corresponds to the unstable vehicle flow. The lower increasing branch matches the free-flow state, whereas the upper one should be related to the synchronized phase because it is characterized by the order parameter coming to unity.

3 Phase Transitions and the Fundamental Diagram

The obtained dependence $h(\rho)$ actually describes the first order phase transition in the vehicle motion. Indeed, when increasing the car density exceeds the value ρ_1 the free flow becomes absolutely unstable and the synchronized mode forms through a sharp jump in the order parameter. If, however, after that the car density decreases the synchronized mode will persist until the car density attains the value $\rho_2 < \rho_1$. It is a typical hysteresis and the region (ρ_2, ρ_1) corresponds to the metastable phases of traffic flow. It should be noted that the stated approach to the description of the phase transition "free flow $\rightarrow$ synchronized mode" is rather similar to the hypothesis by Kerner [5,6] about "Z"-like dependence of the overtaking probability on the car density that can cause this phase transition.

Let us, now, discuss a possible form of the fundamental diagram showing $j = \rho\vartheta[\rho]$ where, by definition, $\vartheta[\rho] = \vartheta[h(\rho), \rho]$. Fig. 2a displays the dependence $\vartheta(h, \rho)$ of the mean vehicle velocity on the density ρ for the fixed limit values of the order parameter $h = 0$ or 1. For small values of ρ these curves practically coincide with each other. As the vehicle density ρ grows and until it comes close to the critical value ρ_c where the lane change danger becomes substantial, the

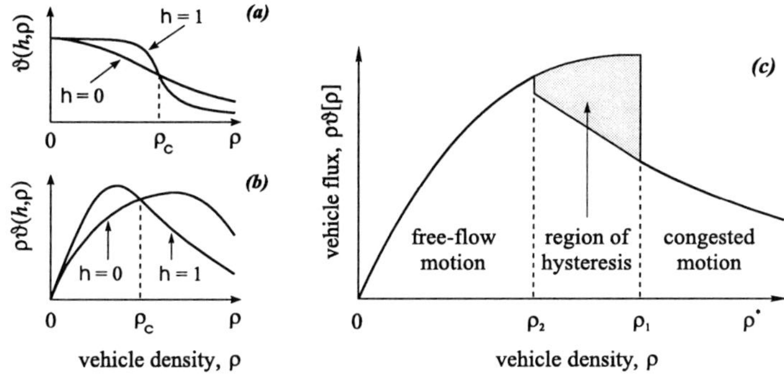

Fig. 2. The mean vehicle velocity (***a***) and the vehicle flux (***b***) vs. the vehicle density for the limit values of the order parameter $h = 0, 1$ as well as the resulting fundamental diagram (***c***).

velocity $\vartheta(1, \rho)$ practically does not depend on ρ. So at the point ρ_c at which the curves $\vartheta(1, \rho)$ and $\vartheta(0, \rho)$ meet each other the former curve, $\vartheta(1, \rho)$, is to exhibit sufficiently sharp decrease in comparison with the latter one. Therefore, on one hand, the function $j_1(\rho) = \rho\vartheta(1, \rho)$ has to be decreasing for $\rho > \rho_c$. On the other hand, at the point ρ_c for $h \ll 1$ the effect of the lane change danger is not extremely strong, it only makes the lane change ineffective, $\partial\vartheta/\partial h \approx 0$ (compare (5)). So it is reasonable to assume the function $j_0(\rho) = \rho\vartheta(0, \rho)$ increasing neat the point ρ_c. Under the adopted assumptions the relative arrangement of the curves $j_0(\rho)$, $j_1(\rho)$ is demonstrated in Fig. 2b, and Fig. 2c shows the fundamental diagram of traffic flow resulting from Fig. 1 and Fig. 2b.

The developed model predicts also the same type phase transition for large values of the order parameter. In fact, in an extremely dense traffic flow changing lanes is sufficiently dangerous and the function $\Phi(h, v, \rho)$ describing the driver behavior is to depend strongly on the vehicle density as $\rho \to \rho_0$. In addition, the vehicle motion becomes slow. Under such conditions the former term in the expression for $\partial\Phi/\partial\rho$ in (7) should be dominant and, so, $\partial\phi/\partial\rho > 0$ and the stable vehicle motion corresponding to $\partial\phi/\partial h > 0$ is characterized by the decreasing dependence of the order parameter $h(\rho)$ on the vehicle density ρ for $\rho > \rho_h$. Therefore, as the vehicle density ρ increases the curve $h(\rho)$ can again go into the instability region (in the $h\rho$-plane), which has to give rise to a jump from the synchronized mode to a jam. The latter matches small values of the order parameter h (Fig. 1b), so, it should comprise the vehicle flows along different lane where lane changing is depressed, making them practically independent of one another.

4 Conclusion

We have introduced an additional state variable of the traffic flow, the order parameter h, that accounts for internal correlations in the vehicle motion caused by the lane changing. Since such correlations are due to the "many-body" effects in the car interaction the order parameter is regarded as an independent state variable. Keeping in mind general properties of the driver behavior we have written the governing equation for this variable.

It turns out that in this way such characteristic properties of the traffic flow instability as the sequence of the phase transitions "free flow $\rightarrow$ synchronized motion $\rightarrow$ jam" can be described without additional assumptions. Moreover, in this model both the phase transitions are of the first order and exhibits hysteresis. Besides, the synchronized mode corresponds to highly correlated vehicle flows along different lanes, $h \approx 1$, whereas in the free flow and the jam these correlations are depressed, $h \ll 1$. So, the jam phase actually comprises mutually independent car flows along different lanes.

References

1. B.S. Kerner and H. Rehborn, Phys. Rev. E **53**, R4275 (1996).
2. B.S. Kerner, Phys. Rev. Lett. **81**, 3797 (1998).
3. B.S. Kerner and H. Rehborn, Phys. Rev. E **53**, R1297 (1996).
4. B.S. Kerner and H. Rehborn, Phys. Rev. Lett. **79**, 4030 (1997).
5. B.S. Kerner, Physics World **12**, 25 (1999).
6. B.S. Kerner, in: *Transportation and Traffic Theory,* A. Ceder (Ed.), p. 147 (Pergamon, Amsterdam, 1999).
7. D. Helbing, *Verkehrsdynamik,* (Springer, Berlin, 1997).
8. D. Helbing and M. Treiber, Phys. Rev. Lett. **81**, 3042 (1998).
9. M. Treiber, A. Hennecke, and D. Helbing, Phys. Rev. E **59**, 239 (1999).
10. V. Shvetsov and D. Helbing, Phys. Rev. E **59**, 6328 (1999).
11. A. Klar and R. Wegener, SIAM J. Appl. Math. **59**, 983, 1002 (1999).
12. H.Y. Lee, D. Kim, and M.Y. Choi, in: *Traffic and Granular Flow'97,* M. Schreckenberg and D.E. Wolf, (Eds.), p. 433 (Springer, Singapore, 1998).
13. T. Nagatani, Physica A **264**, 581 (1999).
14. T. Nagatani, Phys. Rev. E **60**, 1535 (1999).
15. L. Neubert, L. Santen, A. Schadschneider, and M. Schreckenberg, Phys. Rev. E **60**, 6480 (1999).
16. R. Mahnke and N. Pieret, Phys. Rev. E **56**, 2666 (1997).
17. R. Mahnke and J. Kaupužs, Phys. Rev. **59**, 117 (1999).

Macroscopic Simulation of Open Systems and Micro-Macro Link

A. Hennecke[1], M. Treiber[1], and D. Helbing[1,2]

[1] II. Institute of Theoretical Physics, University of Stuttgart, Pfaffenwaldring 57/III, D-70550 Stuttgart, Germany
[2] Collegium Budapest – Institute for Advanced Study, 1014 Budapest, Hungary

Abstract. We discuss the numerical treatment of boundary conditions for a non-local macroscopic model of uni-directional traffic. Furthermore, we propose a general scheme to derive non-local macroscopic models from given microscopic car-following models. Assuming identical (macroscopic) initial and boundary conditions, we show that there are microscopic models for which the corresponding macroscopic version displays a very similar dynamics. This enables us to combine micro- and macrosimulations of road sections by simple algorithms and even to simulate them simultaneously.

1 Treatment of Boundaries in Open Systems

As an example of a non-local macroscopic traffic model with well-known properties, we discuss the "gas-kinetic-based traffic model" (GKT model) [1]. It simulates freeway traffic by a set of two equations, the continuity equation

$$\frac{\partial \rho}{\partial t} + \frac{\partial (\rho V)}{\partial x} = 0 \tag{1}$$

for the traffic density $\rho(x,t)$, and a non-local equation for the macroscopic (locally averaged) velocity $V(x,t)$:

$$\left(\frac{\partial}{\partial t} + V\frac{\partial}{\partial x}\right) V + \frac{1}{\rho}\frac{\partial(\rho\theta)}{\partial x} = \frac{V_0 - V}{\tau} - \frac{C(\rho_\mathrm{a})\rho_\mathrm{a}(\theta+\theta_\mathrm{a})}{2} B(\delta_V)\,. \tag{2}$$

We approximate the velocity variance by a relation of the form $\theta = A(\rho)V^2$. $C(\rho)$ is a dimensionless effective cross-section reflecting the finite-space requirements of vehicles. The Boltzmann interaction term $B(\delta_V)$ results from the gas-kinetic approach and describes deceleration processes. It is an increasing function of the scaled velocity-difference $\delta_V = (V - V_\mathrm{a})/\sqrt{\theta + \theta_\mathrm{a}}$ with respect to the advanced "interaction point" $x_\mathrm{a} = x + \gamma(1/\rho_{\mathrm{max}} + TV)$. The functions $f_\mathrm{a}(x,t) = f(x_\mathrm{a},t)$ with $f \in \{\rho, V, \theta\}$ reflect the anticipation behavior of drivers similar to microscopic models.

The left-hand sides of the continuity and velocity equations constitute a hyperbolic set of partial differential equations. The direction of information flow due to convective and dispersive processes is given by the characteristic lines [2]. Specifically, the velocities of information flow are given by

$$\lambda_{1,2} = V - \frac{1}{2\rho}\frac{\partial P}{\partial V} \pm \sqrt{\frac{1}{4\rho^2}\left(\frac{\partial P}{\partial V}\right)^2 + \frac{\partial P}{\partial \rho}} \quad \text{with} \quad P = \rho\theta\,. \tag{3}$$

For all physically reasonable parameter sets they are always positive. Information flow *against* the direction of motion of the simulated vehicles is caused by the non-local interaction term.

The consequences are: (i) There is always a transport of information in both directions, upstream and downstream. (ii) The most suitable differencing scheme for the numerical integration is the upwind scheme [3]. By choosing the upwind scheme, the downstream boundary conditions are only relevant for the non-local interaction term. This is in accordance with the boundary conditions for microscopic traffic simulations, where the first car at the downstream boundary always needs a "phantom" leading car for its update process. In particular, the microscopic downstream boundary conditions determine only the *acceleration* of this first car, but do not determine the *flow* of exiting vehicles. At the same time, there always has to be a defined inflow of new vehicles at the upstream boundary. The same is true for the simulation of non-local macroscopic models with the upwind scheme. Therefore, it is no problem to formulate downstream boundary conditions which are always satisfied.

At the upstream boundary, unphysical situations can emerge when a backwards moving jam approaches the boundary, but the assumed boundary (in)flow is higher than the flow inside the traffic jam. Then the equations are no longer well-posed in a mathematical sense. Such situations may occur, whenever there is a slight offset in the arrival times of simulated and real jams at the boundary. This means the boundary conditions have to be implemented in a way that, on the one hand, copes with unexpected or offset density waves and, on the other hand, returns to the exact solution as soon as possible. This can be done by an algorithm that implements Dirichlet boundary conditions whenever the local group velocity

$$v_{\rm g}(0,t) = \frac{\rho(\delta,t)V(\delta,t) - \rho(0,t)V(0,t)}{\rho(\delta,t) - \rho(0,t)} \tag{4}$$

(with some small $\delta > 0$) is non-negative, and homogeneous von Neumann boundary conditions otherwise.

This algorithm has been tested as follows: First, we simulated a circular road of 20 kilometers length. The simulated scenario contains a local dipole-like perturbation in the initial conditions which develops to a backwards moving cluster (Fig. 1). Then we used the time-dependent results at the locations $x = 5\,\mathrm{km}$ and $x = 15\,\mathrm{km}$ as Dirichlet boundary conditions for the simulation of a new, 10 kilometer long *open* system. Figure 2 shows a comparison of the density profiles of the two systems at two times t_1 and t_2. At $t_1 = 8\,\mathrm{min}$, the cluster is not yet fully developed, and at $t_2 = 80\,\mathrm{min}$, it has propagated around the circle once. The profiles belonging to the same time points agree so well, that they would lay upon each other. For this reason we have depicted them with a slight artificial offset.

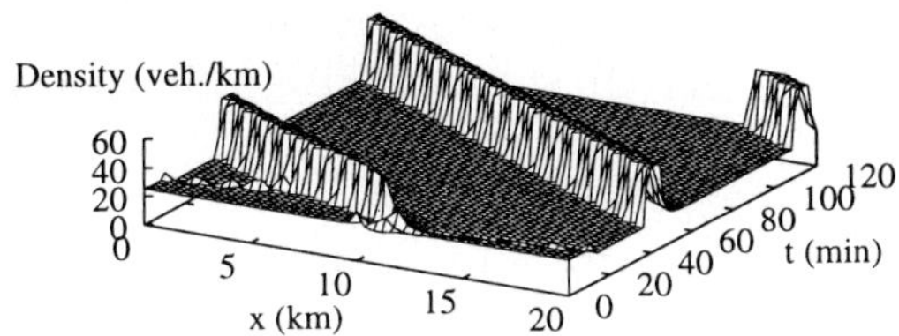

Fig. 1. Circular reference system simulated in order to obtain boundary conditions for the simulation of a comparable open system.

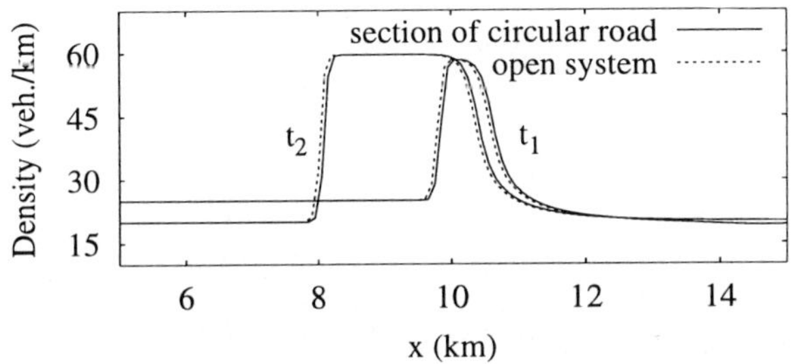

Fig. 2. Comparison of density profiles at two times $t_1 = 8\,\text{min}$ and $t_2 = 80\,\text{min}$, obtained in the circular reference system (solid) and in an open system which was simulated with boundary conditions taken from the reference scenario (dashed). In order to make the curves distinguishable, one of the profiles was shifted by 100 m.

2 Micro-Macro Link

One of the biggest problems in comparing microscopic and macroscopic models via coarse graining is that a separation between the microscopic scale (of single vehicles) and the macroscopic scale (allowing to determine densities and average velocities) is hardly possible. Either the averaging intervals are too small and aggregate quantities like the density cannot be defined consistently, or the averaging intervals are too large so that the dynamics of the model is smoothed out.

Generalizing an idea sketched in Ref. [4], we will now propose a way of obtaining macroscopic from microscopic traffic models, which is different from the gas-kinetic approach and (at least) applicable to the case of identical driver-vehicle units. For this, we define an average velocity by linear interpolation between the velocities of the single vehicles α:

$$V(x,t) = \frac{v_{\alpha+1}(t)[x_\alpha(t) - x] + v_\alpha(t)[x - x_{\alpha+1}(t)]}{x_\alpha(t) - x_{\alpha+1}(t)}\,, \tag{5}$$

where $x_\alpha \geq x \geq x_{\alpha+1}$, and the index α of driver-vehicle units increases against the driving direction. While the derivative with respect to x gives us $\partial V/\partial x = [v_\alpha(t) - v_{\alpha+1}(t)]/[x_\alpha(t) - x_{\alpha+1}(t)]$, the derivative with respect to t gives us the *exact* equation

$$\left(\frac{\partial}{\partial t} + V\frac{\partial}{\partial x}\right) V = A(x,t)\,, \tag{6}$$

where

$$A(x,t) = \frac{a_{\alpha+1}(t)[x_\alpha(t) - x] + a_\alpha(t)[x - x_{\alpha+1}(t)]}{x_\alpha(t) - x_{\alpha+1}(t)} \tag{7}$$

is the linear interpolation of the single-vehicle accelerations a_α characterizing the microscopic model. For most car-following models, the acceleration function can be written in the form $a_\alpha = a_{\text{mic}}(v_\alpha, \Delta v_\alpha, s_\alpha)$, where $\Delta v_\alpha = (v_\alpha - v_{\alpha-1})$ is the approaching rate, and $s_\alpha = (x_{\alpha-1} - x_\alpha)$ the distance to the vehicle in front. In order to obtain a macroscopic system of partial differential equations for the average velocity and density, the arguments v_α, Δv_α and s_α of the single-vehicle acceleration have to be expressed in terms of macroscopic fields as well. Specifically, we make the following approximations: $A(x,t) \approx a_{\text{mic}}(V(x,t), \Delta V(x,t), S(x,t))$, with $\Delta V(x,t) = [V_{\text{a}}(x,t) - V(x,t)]$, $S(x,t) = \frac{1}{2}[\rho(x,t) + \rho_{\text{a}}(x,t)]^{-1}$, and the non-locality given by $g_{\text{a}}(x,t) = g(x + 1/\rho(x,t), t)$ with $g \in \{\rho, V\}$. In this way, we obtain a non-local macroscopic velocity equation, which supplements the continuity equation (1) for the vehicle density and defines, for a given microscopic traffic model, a complementary macroscopic model. Note that it does not contain a pressure term $(1/\rho)\partial P/\partial x$ with $P = \rho\theta$, in contrast to the GKT model described above.

As Fig. 3 shows for the "intelligent driver model" (IDM) [5], there are microscopic models for which the above procedure leads to a remarkable agreement. First, we have simulated an open freeway stretch with the macroscopic counterpart of the IDM with given macroscopic initial and boundary conditions. Then, we have simulated the first three kilometers of the stretch microscopically with the IDM (Fig. 3a, dark grey), using the same upstream boundary condition. As downstream boundary condition, we used the macroscopic simulation result at the interface ($x = 3\,\text{km}$) to the remaining macroscopic simulation section (Fig. 3a, light grey). The density profile at the interface is illustrated in Figure 3b. Figure 3c shows the density profile 1 km upstream of the interface obtained by the microscopic simulation, compared with that obtained from the separate macroscopic simulation of the whole system. The density in the microscopic simulation section was determined via the formula $\rho(x,t) = (\{s_{\alpha+1}(t)[x_\alpha(t) - x] + s_\alpha(t)[x - x_{\alpha+1}(t)]\}/[x_\alpha(t) - x_{\alpha+1}(t)])^{-1}$.

The micro-macro link of traffic models is only of practical relevance, if micro- and macrosimulations can be carried out *simultaneously*. For this, we have derived rules how to translate macroscopic initial and boundary conditions into microscopic ones, with almost identical simulation results. In addition, we have developed methods to determine and apply the interface conditions (i.e. the re-

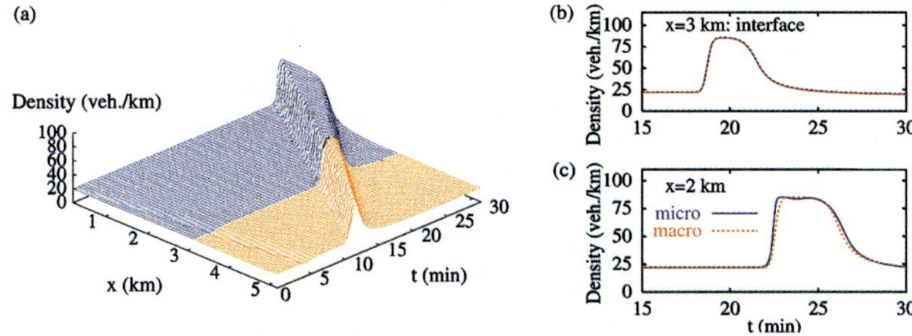

Fig. 3. Results of simulations with the microscopic IDM (dark grey) and its macroscopic counterpart (light grey), assuming identical boundary and interface conditions.

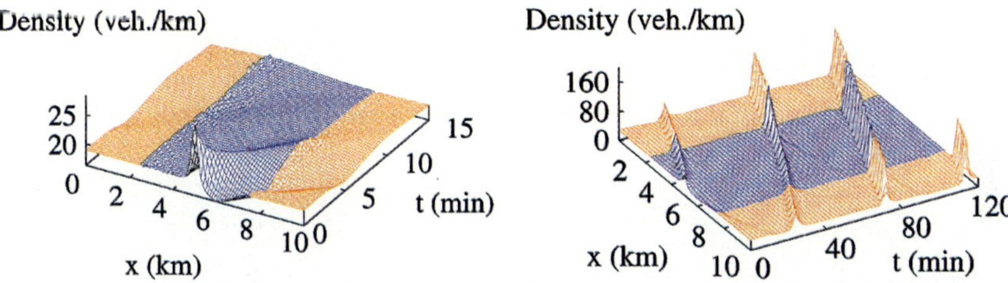

Fig. 4. Results of simulations of a circular road, half of which was simulated microscopically (dark grey), while the other half was simulated macroscopically (light grey). Whereas the left picture shows the forward propagation of a decaying initial perturbation at small vehicle density, the right one illustrates the development of a backwards moving traffic jam at medium density.

spective upstream and downstream boundary conditions at interfaces) *dynamically* during the simulation, i.e. *on-line.* Since information must flow through the interface in both directions, the formulation of dynamic interface conditions is a particularly tricky task which cannot be discussed here. Figures 4a and 4b show examples for the spatio-temporal evolution of the density on a circular road. One half is simulated with the microscopic IDM (dark grey) and the other half is simulated with the macroscopic counterpart of the same model (light grey). As can be seen, it is possible to connect both sections in a way that both, small perturbations propagating in *forward* direction, and developed density clusters propagating *backwards*, can pass the interfaces without any significant changes in the shape or propagation velocity.

While traffic simulations with microscopic single-lane models are more intuitive and detailed than macroscopic simulations, a microscopic implementation of on- and off-ramps would require rather complicated multi-lane models with lane-changing rules. Ramps can be treated much easier by macroscopic models, where they just enter by a simple source term in the continuity equation.

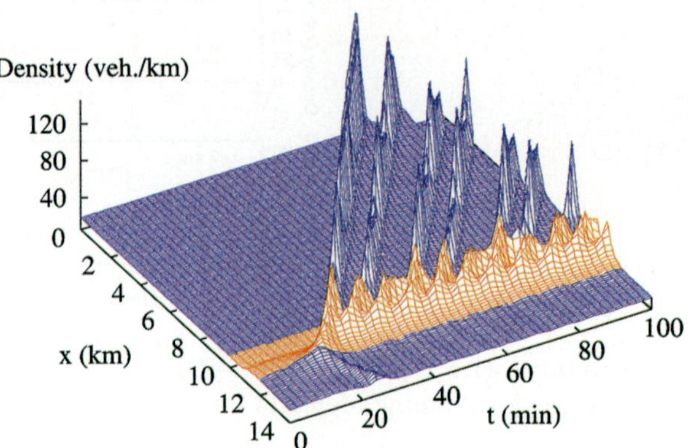

Fig. 5. Result of a simultaneous micro-macro-simulation of triggered stop-and-go traffic, which is caused by high inflow from the ramp at $x = 11$ km. While the ramp section (light grey) is implemented macroscopically, the other two sections (dark grey) were simulated microscopically.

Hence, the micro-macro link can be applied to combine both advantages. Figure 5 shows an open system that is simulated essentially with a microscopic single-lane model (dark grey). Only the ramp section (light grey) is simulated macroscopically. The on-ramp flow triggers a traffic breakdown to a triggered stop-and-go state, i.e. a cascade of traffic jams which propagate in the upstream microscopic section. The qualitative features of this complicated dynamics are in very good agreement with a purely macroscopic simulation of the same system.

Acknowledgement. The authors would like to thank for financial support by the BMBF (research project SANDY, grant No. 13N7092) and by the DFG (grant No. He 2789).

References

1. M. Treiber, A. Hennecke, and D. Helbing, *Derivation, properties, and simulation of a gas-kinetic-based, non-local traffic model,* Phys. Rev. E **59**, 239–253 (1999).
2. W.F. Ames, *Nonlinear Partial Differential Equations in Engineering,* Vols. I & II (Academic Press, New York, 1965).
3. W.H. Press, S.A. Teukolsky, W.T. Vetterling, and B.P. Flannery, *Numerical Recipes in C,* (Cambridge University Press, Cambridge, 1992).
4. D. Helbing, *From microscopic to macroscopic traffic models,* in: *A Perspective Look at Nonlinear Media,* J. Parisi, S.C. Müller, and W. Zimmermann, (Eds.), pp. 122–139 (Springer, Berlin, 1998).
5. M. Treiber et al., *Microsimulations of congested traffic,* in this volume.

Relating Car-Following and Continuum Models of Road Traffic

P. Berg and A. Woods

School of Mathematics, University of Bristol, University Walk, Bristol BS8 1TW, UK

Abstract. We derive a transformation that relates car-following models to their analogous continuum counterpart by using an integral representation of the density. A Taylor expansion of the latter yields an ordinary differential equation in terms of the density and the headway, hence a relation between the two characteristic variables in the corresponding models. This formal approach is supported by similar numerical solutions of the optimal-velocity (OV) model [1] and its continuum analogue. Moreover, the same stability criterion holds in both models. It can be seen that dispersive and anticipation terms of continuum models are a result of applying our transformation to car-following models.

1 Introduction

The mathematical modelling of road traffic is carried out by various traffic models. Continuum models give an overview of the global traffic flow, which is important for developing insights into macroscopic traffic quantities such as throughput, density distributions or the onset of jams, without detailed regard to the properties of each car. In addition, asymptotic solutions can be carried out most easily in the continuum formulation.

In contrast, car-following models give a microscopic picture of the traffic flow. This means each vehicle is described by an individual ordinary differential equation coupled to the surrounding flow. Hence, individual car characteristics can be attached to every single car and the complexity and variety of various vehicles can be simulated explicitly.

In much work to date car-following and continuum models have been considered as independent and as different approaches to simulate traffic flow. The purpose of this work is to develop a systematic method for relating these two models.

2 The Transformation

For uniform flow conditions, the density is simply given by the inverse of the headway

$$\rho = 1/b \ . \tag{1}$$

In the literature the density ρ is usually defined as in (1) for all traffic situations. But if we consider a set of cars positioned at $x = 1, 2, 4, 8, ...$, then the car at position x has headway $b = x$. Using (1), we obtain $\rho = 1/x$, which is extended into continuum domain by permitting x to take any positive, real value. According to this, the number of cars on the interval (1,y) is $\log_e y$. However, the actual answer is $\log_2 y$. We deduce that for non-homogeneous flow situations we cannot transform the car-following model simply using (1).

One approach is to require that

$$\int_{x_i}^{x_{i+1}} \rho(x,t)dx = 1, \tag{2}$$

for all i. We use the definition of headway $b = x_{i+1} - x_i$ to arrive at an equation involving the continuum variable ρ by extending (2) to all points along the road

$$\int_{x}^{x+b(x,t)} \rho(x',t)dx' = \int_{0}^{b(x,t)} \rho(x+y,t)dy \equiv 1. \tag{3}$$

Expanding the second integral in powers of y, we integrate to obtain an asymptotic series [6] that is truncated after the second order to yield

$$b\rho + \frac{1}{2!}b^2\rho_x + \frac{1}{3!}b^3\rho_{xx} = 1. \tag{4}$$

The first term corresponds to the usual definition of the density (1). We expand the series to this order for two reasons. We would like to obtain a continuum model, that is capable of describing some characteristic traffic parameters mentioned by Herrmann and Kerner [5]. Kerner showed, that a dispersive term has to be incorporated to do so. Second these higher order terms are needed to maintain the same stability criterion for the continuum model as for the car-following model [2].

It is assumed that each term is of smaller magnitude than the one preceding it. If we consider the cubic term to be much smaller than the linear and quadratic term, we can first solve the quadratic equation for b. Then by regarding the cubic term as a perturbation, we expand b in a perturbation series and approximate the solution as

$$b \approx \frac{1}{\rho} - \frac{\rho_x}{2\rho^3} - \frac{\rho_{xx}}{6\rho^4} + \frac{\rho_x^2}{2\rho^5}. \tag{5}$$

The first term represents the classic transformation for relating the headway and the density. The second term is similar to a *pressure term* in gas kinetics and acts to destabilise the traffic flow. The dispersive term ρ_{xx} smoothes variations in traffic density and has a stabilising effect on traffic flow, which counteracts the pressure term. We therefore retain terms up to this order.

To show that the velocity is transformed consistently we take a total time derivative of each side of (3) and obtain

$$\int_x^{x+b} \rho_t(y,t)dy + (x_t + b_t)\,\rho(x+b,t) - x_t\rho(x,t) \tag{6}$$

$$= \int_x^{x+b} \rho_t(y,t)dy + v(x+b,t)\rho(x+b,t) - v(x,t)\rho(x,t) \tag{7}$$

$$= \int_x^{x+b} [\rho_t(y,t) + (\rho(y,t)v(y,t))_y]dy = 0. \tag{8}$$

Hence, the conservation equation

$$\rho_t + (\rho v)_x = 0 \tag{9}$$

guarantees that the integral of density along the road from any vehicle to the vehicle it is following is 1. So in this sense, the definition of velocity v is consistent.

3 The OV-Model

Applying our transformation (5) to the second-order model of Bando *et al.* [1].

$$\dot{v}_n = a\left[V_B(b_n) - v_n\right], \tag{10}$$

we obtain the expression for the conservation of cars (9), coupled with a first order approximation of the car-following model

$$v_t + vv_x = a\left(\bar{V}(\rho) - v\right) + a\bar{V}'(\rho)\left[\frac{\rho_x}{2\rho} + \frac{\rho_{xx}}{6\rho^2} - \frac{\rho_x^2}{2\rho^3}\right]. \tag{11}$$

Here we have set

$$\bar{V}(\rho) = V_B(1/\rho). \tag{12}$$

Equation (11) shows some similarities to the Kerner-Konhäuser model [5]. However, an important difference between that model and the new model (11) lies in the coefficients of the higher order terms. In the Kerner-Konhäuser model the coefficients are assumed to be constant, while expression (11) reveals that they actually depend on ρ. By comparison with the discrete OV-model numerical simulations show that the dependence of these coefficients on the density ρ is necessary to match the length scale and qualitative behaviour of travelling wave solutions as in Fig. 1 and Fig. 2 [2,5]. The accuracy increases with further terms of the asymptotic series (5).

It is straightforward to show that the new model obeys the same stability criterion as the original discrete counterpart [2]. An initial homogeneous flow of density ρ_0 and velocity $v_0 = V(\rho_0)$ is stable if

$$-\frac{2\rho_0^2\bar{V}'(\rho_0)}{a} < 1. \tag{13}$$

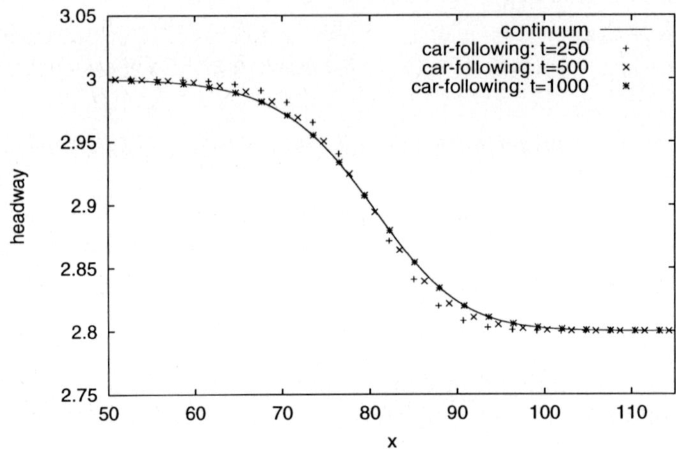

Fig. 1. The shock wave profile of the OV-model (10) approaching the travelling wave solution of the continuum model (11) for an initial jump in headway from $b_{-\infty} = 3.0$ to $b_{\infty} = 2.8$ [3].

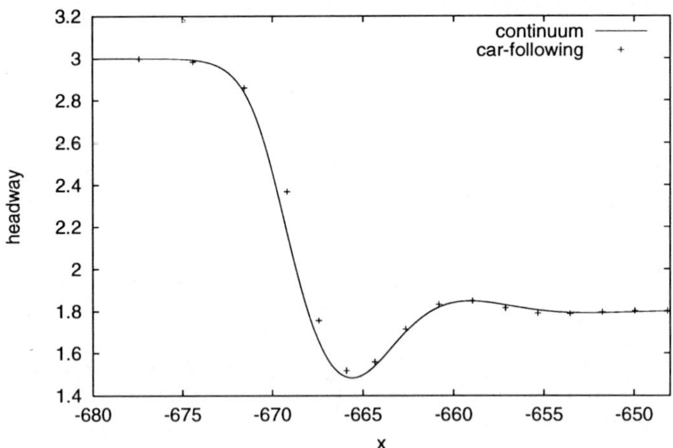

Fig. 2. The change of headway after $t = 1000$ for an initial jump from $b_{-\infty} = 3.0$ to $b_{\infty} = 1.8$.

4 The Inverse Transformation

It is also of interest to find the inverse transformation. This yields a relation between the density ρ of cars in terms of the headway b. We expect ρ to be a function of b and its spatial derivatives

$$\rho(x,t) = F\left(b(x,t), b_x(x,t), b_{xx}(x,t)\right). \tag{14}$$

The derivatives with respect to x are not easy to interpret. However, if we regard b as a continuous variable, the inverse transformation can be found formally by solving (4) for ρ. If we again suppose that each term is sufficiently smaller than the preceding order, it is possible to solve for ρ successively.

But we choose a more trivial way to find the solution and argue that only certain terms in b, b_x and b_{xx} have the right dimensions. To leading order once again ρ is the inverse of headway (1). To find the full solution up to second order and linear in b_x and b_{xx} the only possible ansatz is

$$\rho = \frac{1}{b} + a_1 \frac{b_x}{b} + a_2 b_{xx}, \tag{15}$$

with coefficients a_1 and a_2 to be determined. Inserting this into (4) leads to

$$a_1 = \frac{1}{2}\;, \quad a_2 = -\frac{1}{12} \tag{16}$$

yielding the transformation

$$\rho = \frac{1}{b} + \frac{b_x}{2b} - \frac{b_{xx}}{12}. \tag{17}$$

This is consistent with a straightforward, but more detailed derivation of the solution as mentioned above.

5 Analogous Models

Since both the transformation (5) relating car-following to continuum models and its inverse transformation (17) are known, it is possible to derive analogous models of various traffic flow models.

Car-following models are often described by delay differential equations

$$\dot{x}_n(t+T) = F(b_n, \dot{b}_n, v_n), \tag{18}$$

with parameters b_n, the headway of the n-th car to its preceding vehicle, its rate of change $\dot{b}_n$ and its velocity v_n. The delay time T causes trouble when applying (5) to obtain the continuum analogue. Since it can only be carried out by a Taylor expansion of $\dot{x}_n(t+T) = v_n(t+T)$ to a finite order, these models cannot be transformed exactly. But car-following models might deliver a future algorithm for autonomous cruise control systems and T will drop immensively. At least a first-order approach will then be sufficient to derive a macroscopic traffic flow model. However, for rather high values of T it shall be checked, whether the analogous continuum model corresponds with its original car-following model.

In continuum models inertia is often modelled by a relaxation term

$$\frac{1}{T}\left[V(\rho(x,t)) - v(x,t)\right], \tag{19}$$

with an optimal velocity function V. It is similar to various relaxation terms in physics. As these models are transformed into their car-following counterpart spatial derivatives of the headway occur. Discretizing them leads to terms depending on the headway of following cars similar to a recently published model proposed by Hayakawa *et al.* [4]. It accounts for the fact that dispersion was already incorporated implicitly or explicitly in the original continuum model.

6 Conclusion

We presented a transformation of the headway and the velocity into continuous variables. We supported our ideas by numerical simulations and stability analysis of the OV-model and its continuous counterpart. The inverse transformation enables one to derive car-following analogues of any macroscopic traffic flow model. It is characteristic that they lead to microscopic models which consist of spatial derivatives of the headway and include information of the traffic events further up- and downstream.

Acknowledgement. Peter Berg would like to thank the Alfried Krupp von Bohlen und Halbach-Stiftung and the EPSRC for their sponsorships of this project.

References

1. M. Bando, K. Hasebe, A. Nakayama, A. Shibata, and Y. Sugiyama, *Dynamical model of traffic congestion and numerical simulation*, Phys. Rev. E **51**, 1035–1042 (1995).
2. P. Berg and A.W. Woods, in press, Phys. Rev. E, (1999).
3. J. Bevan and A.W. Woods, submitted, (1999).
4. H. Hayakawa and K. Nakanishi, *Theory of traffic jam in a one-lane model,* Phys. Rev. E **57**, 3839–3845 (1998).
5. M. Herrmann and B.S. Kerner, *Local cluster effect in different traffic flow models*, Physica A **255**, 163–188 (1998).
6. A.D. Mason, *Mathematical problems of road traffic and related problems*, PhD Thesis, (University of Cambridge, 1998).

Microscopic Randomness in *Follow-the-Leader* Dynamics

H. Lehmann

GMD-FIRST
Rudower Chaussee 5, 12489 Berlin, Germany

Abstract. The "fundamental diagram" of macroscopic traffic flow modelling is related to the nearest-neighbour interaction of an exemplary *follow-the-leader* model, crucially taking into account microscopic fluctuations.

1 Introduction

Investigation of vehicular traffic flow by means of physical tools has produced a dichotomy well known to statistical physicists: the study of the explicit motion of individual vehicles [1–8] is contrasted by that of averaged quantities such as the mean flow [9–17]. The qualitative difference of these two schools of thought can be illustrated by the expressions for the interactions. In the microscopic models, the pair interaction of neighbouring vehicles is modelled by the two demands to proceed as fast as possible whilst avoiding crashes. In the macroscopic picture, the aggregate interaction is usually expressed as a relationship between mean density and mean flow. This is called the *fundamental diagram* (FD). Naturally, there has to be a profound relation between the microscopic *follow-the-leader* interaction and the FD. From similar relations between, say, Brownian particles and hydrodynamics it is well known that a pivotal role in this relation is played by the microscopic fluctuations. If V_0 describes an explicit microscopic interaction function and V_{FD} the FD, the purpose of this paper can be summarised as to study the relation

$$V_{\mathrm{FD}} = \langle V_0 \rangle_{\mathrm{fluctuations}} . \tag{1}$$

As a pedagogical note it can be underlined that the above relation has the same character as, say, the relation between Lennard-Jones potential and Carnahan-Starling pressure in conventional statistical mechanics.

The nature of this relation can be qualitatively studied by the respective time scales. In the traffic flow problem, the elementary time step τ_1 is given by the driver's reaction time to sudden perturbations. Obviously, no faster processes can be imagined. This time step is, however, the same for the deterministic and random microscopic motions. As a consequence, the conventional concept of very short, sudden acceleration (or deceleration) changes as in the philosophy of the Langevin equation does not apply; in the overall motion the effects of deterministic and stochastic acceleration cannot be clearly separated anymore. Granting the drivers this kind of indecisiveness then leads to the assertion that for the individual vehicle a steady state $\dot{x}_i = const.$ can never be reached. It is only in terms of the averaged quantities and, consequently, an appropriate time scale $\tau_2 \gg \tau_1$ that a steady state can be sensibly defined. Since such a steady state is the basis for the phenomenological relations of the macroscopic modelling school, τ_2 presents the frame for the so-called fundamental diagram.

2 Randomness in the Microscopic Motion: Model Build-Up

As a Newtonian microscopic model, the "Optimal Velocity Model" (OVM) [20,21] presents a set of coupled differential equations for the positions x_i of the ith vehicle involving as a crucial element the function V_0, the so-called "optimal velocity" function. It gives the optimal speed a driver should assume in dependence on the distance to the car in front:

$$V_0(\Delta x_i) = v_0 \left[\tanh\left(0.0860(\Delta x_i - 25)\right) + 0.97323\right] \quad (2)$$

(distances in meters). The action of V_0 becomes clear immediately: zero distance dictates stand-still whereas for infinite distances Δx_i the maximum speed v_0 is assumed. It is intuitively clear that the degree of determinism in the OVM is much too strong - it resembles a machine-like "cruise control". In reality, drivers will rather fluctuate around the "optimal" velocity with the important boundary condition, however, to avoid crashes.

In order to account for these fluctuations, the usual procedure would be to add a stochastic acceleration term. This term has to be designed, then, in such a way as to agree with the qualitative arguments given above: First, the minimum time scale on which the stochastic acceleration varies is given by the same τ_1 that governs the deterministic microscopic motion. Consequently, Langevin delta-type force but has to be replaced by an ordinary, continuous function of time. Second, the amplitude of the random acceleration has to scale in some form or other with the velocity itself in order to avoid crashes. Third, a time average on the macroscopic scale τ_2 equal to zero has to exist, thereby defining the steady state $\langle \dot{x}_i \rangle_{\tau_2} = const.$, the prerequisite for the existence of the fundamental diagram. These demands are met by the form:

$$\begin{aligned} \ddot{x}_i^{\text{random}} &= \frac{\alpha}{\tau_1} V_0\left(\Delta x_i\right) \xi_i(t) \\ \xi_i(t) &= C_\xi \sum_j^{N_\xi} \zeta_{ij} \sin\left[2\pi\omega_{ij} t + \phi_{ij}\right] \end{aligned} \quad (3)$$

with random frequencies $\omega_{ij} \in [\omega_-, \omega_+]$, random phases $\phi_{ij} \in [0, 2\pi]$ and random amplitudes $\zeta_{ij} \in [0,1]$ for a number N_ξ of modes . (Note, that the negative range of amplitudes $[-1,0]$ is subsumed into the random phases.) Since on the scale τ_2 the random forces must average to zero, the short-wave limit is given by $\omega_- = \tau_2^{-1}$, whereas the elementary time step presents the long-wave limit $\omega_+ = \tau_1^{-1}$. The numerical scaling factor α is to be determined from, desirably, experiments, or simulations.

The normalisation of ξ_i is best connected with the average of its mean square on the time scale τ_2 as is customary in the context of stochastic problems [18].

Now we are in the position to define the Random Optimal Velocity Model (ROVM):

$$\ddot{x}_i = \frac{1}{\tau_1}\left[V_0(\Delta x_i) - \dot{x}_i\right] + \frac{\alpha}{\tau_1} V_0(\Delta x_i)\, \xi_i(t) \quad . \tag{4}$$

In the following section, properties of the distribution spanned up by the random parts of the above dynamics are investigated.

3 Randomness in the Microscopic Motion: Ensemble Properties

In a steady state (as argued above, with respect to τ_2), P_d, the probability distribution of distances around some mean value $\bar{d}$, is stationary and allows to establish the desired connection between individual velocities and the fundamental diagram:

$$V_{\mathrm{FD}}(c) = \int dd'\, V_0(d')\, P_d(d'; \bar{d})$$
$$c = \frac{1}{\bar{d} + b} \tag{5}$$

where b is an effective car length. While it is impossible to extract P_d completely from the solution of the ROVM, essential information such as the width of the distribution can be obtained. From a linearised version the mean square average of the distance is found (Fig. 1). Figure 1 deserves several comments. First, it clearly displays the effects of coupling the random acceleration $\xi(t)$ to the characteristic velocity function $V_0(d)$: for small values of d the fluctuations have to vanish in order to leave the traffic crash-free in that limit. For large distances (i.e. the interaction-free limit) $P_d(d; \bar{d} \to \infty) = P^{(0)}(d)$ becomes independent of the mean distance. Consequently, the corresponding velocity distribution does not depend on the density, either. Secondly, it should be pointed out that a critical value for the microscopic relaxation time, $\tau_c \simeq 0.18\,\mathrm{sec}$, arises as a mathematical property of the solution underlying Fig. 1, which can be understood as a model-inherent vindication of the fitted value $\tau_1 = 0.5\,\mathrm{sec}$ [20]. However, the dependence on the actual value of the individual microscopic relaxation time τ_1 is rather weak. This becomes plausible by comparison of the microscopic relaxation processes and the random wavelengths in $\xi(t)$, ω_{ij}^{-1}: reactions to the fluctuations of the car in front are faster or of the same temporal order as the fluctuations themselves such that in the coarse-graining average the τ_1-effect remains small.

By further simplifying the ROVM to the assumption of instantaneous relaxation,

$$V_B(d_i) \simeq V_B(\bar{d}) \quad , \tag{6}$$

one arrives at the well known Ornstein-Uhlenbeck process (if ξ_i is taken as customary white noise). The corresponding Fokker-Planck equation yields a Gaussian for the local equilibrium fluctuations of the velocities which can, of course,

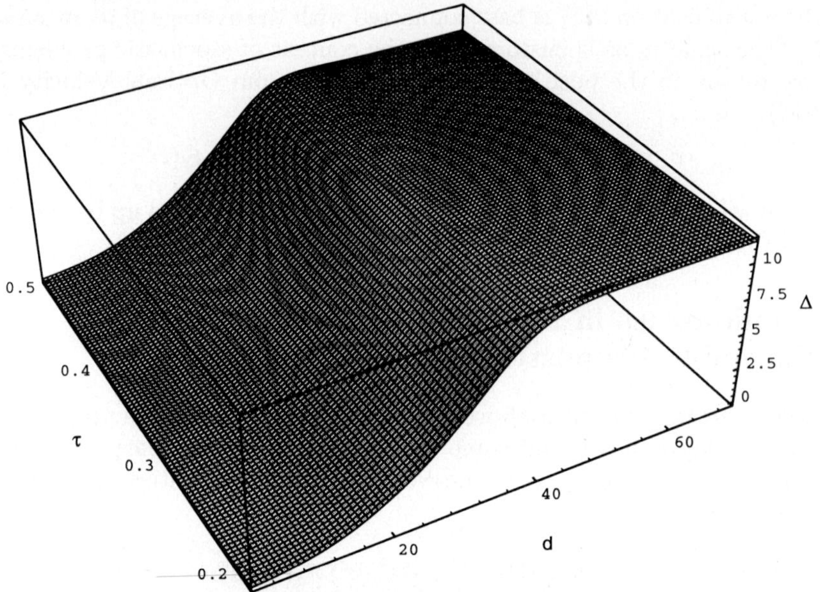

Fig. 1. distance variability $\Delta = \sqrt{\langle(\Delta d)^2\rangle}$ in m with numerical parameter $\alpha = 0.3$ and the frequency interval $\omega_- = 0.2\,\mathrm{sec}^{-1}$, $\omega_+ = 2\,\mathrm{sec}^{-1}$; the microscopic relaxation time τ ranges from the critical value τ_c (see text) to $\tau_1 = 0.5\,\mathrm{sec}$ as taken from [20]; headways d in m.

only be a rather crude approximation since for certain velocities the probability has to be strictly zero in order to remain crash-free. This failure becomes decisive in the region near the permissible maximum speed v_0. Nevertheless, from the Ornstein-Uhlenbeck argument and the result for $\sqrt{\langle(\Delta d)^2\rangle}$ it seems to be justified to guess the shape of the local equilibrium distance distribution P_d in first order as:

$$P_d(d;\,\bar{d}) \;=\; \sqrt{2\pi\langle(\Delta d)^2\rangle}\,\exp\left[-\frac{(d-\bar{d})^2}{2\,\langle(\Delta d)^2\rangle}\right] \tag{7}$$

whence it is possible, in the spirit of relation (1) to express the fundamental diagram in terms of microscopic features.

Figure 2 shows the thus obtained $V_{\mathrm{FD}}(\bar{d})$ in comparison with the function $V_0(\bar{d})$ (which corresponds to the idealised state of "total cruise control") and results of a direct simulation for intermediate headways (for very small and very large headways the velocity fluctuations vanish as argued elsewhere). It is clearly recognisable that, due to the fluctuations, the average velocity of the coarse-graining ensemble lies below that of the corresponding single-particle optimal velocity V_0. Furthermore, the estimate (7) agrees remarkably well with simulations of the full ROVM.

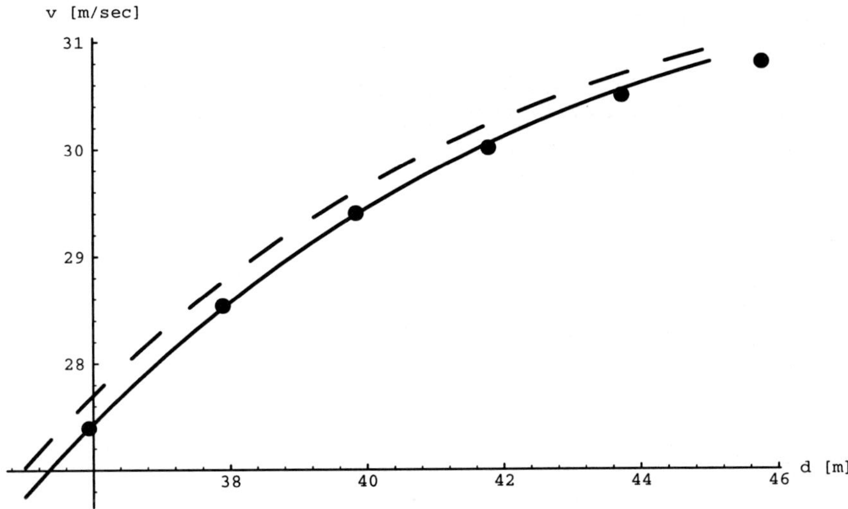

Fig. 2. average velocities in the ROVM for $\tau_1 = \tau_c$; solid line: fundamental diagram $V_{\mathrm{FD}}(\bar{d})$ according to (1) with estimate (7), dashed line: "total cruise-control" limit $V_0(\bar{d})$ as reference i.e. no microscopic randomness, points: simulations of the steady state average velocity

References

1. K. Nagel and M. Schreckenberg, J. Phys. I France **2**, 2221 (1992).
2. N. Rajewsky and M. Schreckenberg, Physica A **245**, 139 (1997).
3. O. Biham, A. Middleton, and D. Levine, Phys. Rev. A **46**, 6124 (1992).
4. T. Nagatani, Physica A **246**, 460 (1997).
5. T. Nagatani, H. Emmerich, and K. Naganishi, Physica A **255**, 158 (1998).
6. D.A. Kurtze and D. Hong, Phys. Rev. E **52**, 218 (1995).
7. D.C. Hong, S. Yue, J.K. Rudra, M.Y. Choi, and Y.W. Kim, Phys. Rev. E **50**, 4123 (1994).
8. M. Gerwinski and J. Krug, 188 (1999).
9. M.J. Lighthill and B.G. Whitham, Proc. Roy. Soc. London A **229**, 317 (1955).
10. R. Kühne, Physik in unserer Zeit, **15.3**, 84 (1984).
11. R. Kühne, (Proc. 10th Int. Symp. on Transportation and Traffic Theory, Delft, 1987).
12. B.S. Kerner and P. Konhäuser, Phys. Rev. E **50**, 54 (1994).
13. B.S. Kerner, S.L. Klenov, and P. Konhäuser, Phys. Rev. E **56**, 4200 (1997).
14. D. Helbing, *Verkehrsdynamik*, (Springer, Berlin, 1997).
15. I. Prigogine and R. Herman, *Kinetic Theory of Vehicular Traffic*, (American Elsevier, New York, 1971).
16. H. Lehmann, Phys. Rev. E **54**, 6058 (1996).
17. C. Wagner, C. Hoffmann, R. Sollacher, J. Wagenhuber, and B. Schürmann, Phys. Rev. E **54**, 5073 (1996).
18. C. W. Gardiner, *Handbook of Stochastic Methods*, (Springer, Berlin 1997).
19. B. Derrida, E. Domany, and D. Mukamel, J. Stat. Phys. **69**, 667 (1992).

20. M. Bando, K. Hasebe, A. Nakayama, A. Shibata, and Y. Sugiyama, Phys. Rev. E **51**, 1035 (1995).
21. M. Bando, K. Hasebe, A. Nakayama, A. Shibata, and Y. Sugiyama, Japan J. of Ind. and Appl. Math. **11**, 203 (1994).

"Car-SPH": A Lagrangian Particle Scheme for the Solution of the Macroscopic Traffic Flow Equations

S. Rosswog and P. Wagner

MS, German Aerospace Center (DLR), Köln-Porz, Germany

Abstract. A Lagrangian particle scheme for the solution of the macroscopic traffic flow equations is presented. The scheme transforms the partial Navier-Stokes-like differential equations into a set of ordinary differential equations that are given as sums over particle contributions. The continuity equation is fulfilled automatically by construction. The method is applied to the 'standard scenario' of a road loop where the traffic flow is governed by the macroscopic equations of Kerner and Konhäuser.

1 Introduction

Models for traffic flow are generally subdivided into three classes: microscopic, macroscopic and models that use concepts of both, which we will refer to as 'mesoscopic models'.

Microscopic models (e.g. [9,5]) intend to describe algorithmically the behavior of individual driver-vehicle units. Since they have to mimic individual decisions of the drivers as a response to the traffic situation created by the other drivers it is obvious that it is computationally very expensive to apply this approach to realistic road networks and thus efficiency is of highest priority. This underlines the need for models that are as simple as possible but are still able to reproduce the essential features observed in traffic flow.

The second class, the so-called *macroscopic models* (e.g. [6,10,4]) tries to model the evolution of aggregated quantities such as vehicle densities and flows. The corresponding equations of motion are largely similar to the equations governing the motion of compressible, viscous flows, i.e. they formally look like Navier-Stokes equations with a car-specific force term (an additional relaxation term) that accounts for the driver's wish to drive with a certain (density dependent) speed.

The third class, often referred to as *mesoscopic models* contains concepts of both previously mentioned classes. According to their motivations two subgroups can be distinguished. First there are the more practically motivated models that describe the dynamics of single vehicles by means of macroscopic equations (e.g. [3,13]). The second and to a certain extent opposite approach, is more theoretically motivated. In these *kinetic models* (e.g. [11,2,12]) one tries to define procedures that allow to aggregate quantities in a consistent way and then to evolve these aggregated quantities according to equations that have consistently been derived from a microscopic model. I.e., one evolves aggregated variables

according to microscopic laws rather than to evolve single vehicles according to a macroscopic dynamics.
The focus of our work here is to present a new, particle-based Lagrangian scheme for the solution of the macroscopic traffic flow equations. It is oriented at the well-known smoothed particle hydrodynamics method (SPH).

2 A Particle Scheme to Solve the Macroscopic Traffic Flow Equations

Macroscopic models are governed by equations of the form (in an Eulerian frame)

$$\frac{\partial \rho}{\partial t} + \nabla(\rho \boldsymbol{v}) = 0 \tag{1}$$

and

$$\rho \left(\frac{\partial \boldsymbol{v}}{\partial t} + \boldsymbol{v}(\nabla \boldsymbol{v}) \right) = -\nabla P + \boldsymbol{X} + \mu(\rho)\Delta \boldsymbol{v}, \tag{2}$$

where ρ is the vehicle density (vehicles per length), v is the macroscopic vehicle velocity, P denotes the "pressure", which is given as $P = \rho\Theta$, where Θ symbolizes the velocity variance. Θ also determines the traffic sound velocity c_s, which governs the propagation of perturbations. The relaxation force $\boldsymbol{X}$ models the drivers' tendency to drive with a "maximal and out of danger" velocity $V(\rho)$ and is usually assumed to be of the form

$$X = \rho \frac{V(\rho) - v}{\tau(\rho)}, \tag{3}$$

where τ is a relaxation constant. The last term on the right hand side is a diffusion term that contains the viscosity parameter μ. It has been mainly introduced for practical reasons since the macroscopic traffic flow equations allow for shock solutions whose steep density gradients may cause severe numerical problems. This is a set of partial differential equations that generally has to be solved by means of sophisticated numerical integration methods.
We want to suggest here a Lagrangian particle method to solve this set of equations. The proposed scheme is oriented at the well-known smoothed particle hydrodynamics method (SPH; see, e.g.,[1] and [8] for reviews). Its basic idea is to go from local quantities of a function $f(x)$ to local averages of this function $\langle f(x) \rangle$, where the averaging process is performed using a smoothing kernel W:

$$f(x) \rightarrow \langle f(x) \rangle = \int f(x')W(x - x', h)dx', \tag{4}$$

where h determines the length over which the function is to be smoothed. The smoothing kernel W has to fulfill the following properties:

$$\int W(x - x', h)dx' = 1, \tag{5}$$

i.e. it has to be normalized, and

$$\lim_{h \to 0} W(x - x', h) = \delta(x - x'), \tag{6}$$

i.e. in the limit that the smoothing is performed over a vanishing interval the local value should be retained. To estimate the accuracy of the smoothing process the function f can be expanded in a Taylor-series around x:

$$f(x') = f(x) + f'(x)\delta x + \frac{1}{2} f''(x)\delta x^2 + O(\delta x^3), \tag{7}$$

where $\delta x = x' - x$. Inserting this into equation (4), using (5) and using an even kernel function one finds:

$$\langle f(x) \rangle = f(x) + \frac{1}{2} f''(x) \int W(x' - x, h)\delta x^2 dx' + O(\delta x^3). \tag{8}$$

Since the smoothing is performed over $\sim h$ (depending on the choice of the kernel), $\delta x \sim h$ and thus

$$\langle f(x) \rangle = f(x) + O(h^2). \tag{9}$$

If the function f is known at a discrete set of points whose number density is given by $n(x) = \sum_j \delta(x - x_j)$ one can multiply the integrand in (4) with $n(x)/\langle n(x) \rangle (= 1 + O(h^2))$ to find

$$\langle f(x) \rangle = \sum_j \frac{f(x_j)}{\langle n(x_j) \rangle} W(x - x_j, h). \tag{10}$$

Thus, if a 'particle' at position x_j is associated with a weight m_j (e.g. it corresponds to m_j vehicles) one finds for the (vehicle) density ρ (not to be confused with the particle density n)

$$\langle \rho \rangle = \sum_j \frac{\rho(x_j)}{\langle n(x_j) \rangle} W(x - x_j, h) = \sum_j m_j W(x - x_j, h), \tag{11}$$

where $\rho(x_j) = m_j \langle n(x_j) \rangle$ has been used. The general prescription for a 'smoothed' quantity is given by:

$$\langle f(x) \rangle = \sum_j f_j \frac{m_j}{\rho_j} W(x - x_j, h), \tag{12}$$

where the index j means that the variable is to be taken at interpolation point j. This looks like evaluating the quantity f by summing over 'particles' at positions x_j. In order to perform the summation over neighboring particles only rather than over all particles of the system, it is recommendable to use a kernel with compact support. We use the spherical spline kernel of [7]. The essential point of this method (apart from being a Lagrangian one) is that derivatives do not have to be evaluated by means of finite differences on a grid. Since one is free to chose a differentiable kernel function, it is possible to perform an integration by parts and thus to transform the derivative to the kernel which then can be evaluated analytically. If the boundary term vanishes, the gradient of a function f can be expressed as (for reasons of convenience brackets are omitted in the sequel):

$$\nabla f(x) = \sum_j f_j \frac{m_j}{\rho_j} \nabla W(x - x_j, h). \tag{13}$$

The estimates for the smoothed values of a function and its gradient, equations (12) and (13), are the basic ingredients of the proposed scheme, which enable us to write down the macroscopic equations for a specified scheme within the framework of this particle scheme.

3 Application to the Kerner-Konhäuser Model

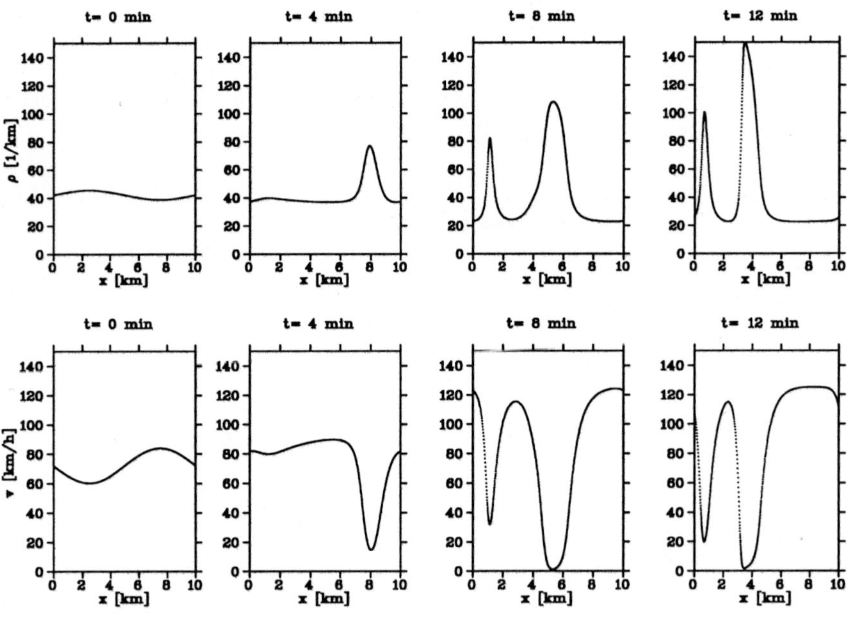

Fig. 1. Evolution of a slight perturbation of a homogeneous traffic state along a 10 km road loop ($\rho_{\max} = 168$ vehicles per km, $\rho_0 = 0.25 \cdot \rho_{\max}$, $\delta\rho = 0.02 \cdot \rho_{\max}$, $\mu = 121.11$ m s^{-1} and $\tau = 11$ s, 800 cars.)

Kerner and Konhäuser suggested a specific macroscopic model to study the formation of density clusters from initially homogeneous traffic flow ([4]). Their model is characterized by a constant velocity variance Θ which has to be regarded as a phenomenological relation since Θ should go to zero with vanishing velocity. They further assume a constant viscosity parameter μ and a specific velocity density relation $V(\rho)$ (see [4]). Using this notation the equation of motion reads in a Lagrangian frame:

$$\frac{dv}{dt} = -\frac{c_s^2}{\rho}\frac{\partial \rho}{\partial x} + \frac{V(\rho) - v}{\tau} + \frac{\mu}{\rho}\frac{\partial^2 v}{\partial x^2}. \tag{14}$$

The term with the density gradient is sometimes referred to as "anticipation term" since it accounts for the fact that drivers react on a *change* in the density with a reduction of their velocity even in the case where the local density would allow for a higher speed.
It is now straightforward to apply the above described scheme to the Kerner-Konhäuser model. The density ("cars per unit length") at position x_i reads:

$$\rho_i = \rho(x_i) = \sum_j m_j W_{ij}, \tag{15}$$

where $W_{ij} = W(x_i - x_j, h)$, the derivative with respect to the space coordinate x is

$$\left(\frac{\partial \rho}{\partial x}\right)_i = \sum_j m_j \frac{\partial}{\partial x_i} W_{ij}. \tag{16}$$

The only missing term now is the second derivative of the velocity with respect to the space coordinate x. To find a corresponding expression we use the following prescription (see [8]):

$$\eta_i \equiv \rho_i \left(\nabla v\right)_i = \sum_j (v_j - v_i) \nabla_i W_{ij}. \tag{17}$$

Applying equation (13) to equation (17) yields

$$\left(\frac{\partial^2 v}{\partial x^2}\right)_i = \left(\frac{\partial}{\partial x}\left(\frac{\eta}{\rho}\right)\right)_i = \sum_j \frac{m_j}{\rho_j^2} \eta_j \frac{\partial}{\partial x_i} W_{ij}. \tag{18}$$

Thus the full equations of motion read:

$$\frac{dv_i}{dt} = \frac{1}{\rho_i} \sum_j \left(\frac{\mu \eta_j}{\rho_j^2} - c_s^2\right) m_j \frac{\partial}{\partial x_i} W_{ij} + \frac{V(\rho_i) - v_i}{\tau} \tag{19}$$

Thus, we have to solve an ODE whose derivative terms can be evaluated analytically rather than a PDE by means of finite differencing.
To increase local resolution in high density regimes we solve an additional differential equation for the smoothing length, $dh/dt \sim div(v)$. Since the total number of vehicles is given by $N = \sum_j m_j$ and we do not evolve the weights m_j in time

there is no need for a continuity equation in this approach since it is fulfilled automatically. The equations are solved using a standard fourth order Runge-Kutta scheme with adaptive step-size control.
For reasons of illustration the scheme is applied to the 'standard scenario' of a road loop. The initially homogeneous traffic state (constant density and velocity everywhere) is slightly perturbed with a perturbation wavelength equal to the loop length. Since the initial state is in the linearly unstable regime the perturbation grows and finally leads to the formation of jams that move backwards against the traffic flow.

4 Summary

The presented Lagrangian particle scheme is a general method for the solution of the differential equations that are encountered in the context of macroscopic traffic flow models. It fulfills by construction the continuity equation and transforms the partial differential equations of the macroscopic traffic equations to ordinary differential equations. In addition it allows for the analytical calculation of derivatives from a differentiable kernel function rather than by finite differencing. Due to its Lagrangian nature it is especially well suited for the calculation of travel times and may provide a valuable link between macroscopic and microscopic models.

References

1. W. Benz, in: *The numerical Modelling of Nonlinear Stellar pulsations,* (Kluwer Academic Publishers, Dordrecht, 1990).
2. D. Helbing, Physica A **219**, 391 (1995).
3. R. Kates, (Traffic and Granular Flow, Duisburg, 1997).
4. B.S. Kerner and P. Konhäuser, Phys. Rev. E **48**, 2335 (1993).
5. S. Krauß, PhD Thesis, (German Aerospace Centre (DLR), Forschungsbericht 98-8, 1998).
6. M.J. Lighthill and G.B. Whitham, Proc. of the Royal Society A **229**, 317.
7. J.J. Monaghan and J.C. Lattanzio, Astron. Astrophys. **149**, 135 (1985).
8. J.J. Monaghan, Annu. Rev. Astron. Astrophys. **30**, 543 (1992).
9. K. Nagel and M. Schreckenberg, J. Phys. I France **2**, 2221 (1992).
10. H.J. Payne, in: *Mathematical Models of Public Systems*, Vol. 1, 51.
11. I. Prigogine and F.C. Andrews, Oper. Res. **8**, 789 (1960).
12. R. Wegener and A. Klar, Ber. Arbeitsgruppe Technomath., Report No. 138 (1995).
13. R. Wiedemann and T. Schwerdtfeger, Schriftenreihe: Forschung, Straßenbau und Straßenverkehrstechnik, Heft 500 (1987).

An Exactly Solvable Two-Way Traffic Model with Ordered Sequential Update

M.E. Fouladvand[1,3] and H.-W. Lee[2]

[1] Department of Physics, Sharif University of Technology,
P.O.Box 11365-9161, Tehran, Iran
[2] School of Physics, Korea Institute for Advanced Study,
207-43 Cheongryangri-dong, Dongdaemun-gu, Seoul 130-012, Korea
[3] Institute for Studies in Theoretical Physics and Mathematics,
P.O.Box 19395-5531, Tehran, Iran

Abstract. Within the formalism of the matrix product ansatz, we study a two-species asymmetric exclusion process with backward and forward site-ordered sequential updates. This model describes a two-way traffic flow with a *dynamic impurity* and shows a phase transition between the free flow and the traffic jam. We investigate characteristics of this jamming and examine similarities and differences between our results and those with the random sequential update [1].

1 Introduction

A variety of phenomena can be modeled by the one dimensional asymmetric simple exclusion process (ASEP) and its generalizations ([2,3] and references therein). The model has a natural interpretation as a description of traffic flow and constitutes a basis for more realistic ones [4,5]. In traffic flow theories, the formation of traffic jams due to "impurities" is one of the fundamental problems.

In the ASEP models, two kinds of impurities are discussed in the literature. The first one is "dynamic" impurities, i.e., defective particles which jump with a rate lower than others [1,6–8]. In the traffic terminology, such moving defects can be visualized as slow cars on a road. The other kind is "static" impurities such as imperfect links where the hopping rate is lower than in other links [9–12]. Both types of impurities can produce shocks in a system. Presently a limited amount of exact results is available for the shock formation and most of them are for models with the random sequential update [1,6,7].

The implementation of the update is an essential part of the definition of a model and it is of prime interest to determine whether distinct updating schemes can produce different behaviors. The aim of this work is to investigate consequences of changing the updating scheme of the model. Here, we study an exactly solvable traffic model with two types of ordered sequential updates, which is identical to the model studied in [1] except for the updating schemes.

2 Model Definitions and Matrix Product Ansatz

2.1 Two-Way Traffic Model

Consider two parallel one dimensional chains, each with N sites, and the periodic boundary conditions. There are M cars and K trucks in the first and the second chain, respectively, and cars move to the right while trucks move to the left. We introduce inter-chain interaction that forbids a car and a truck to occupy two parallel sites simultaneously. Then the state of the system can be described by a single set of occupation numbers $(\tau_1, \tau_2, \cdots, \tau_N)$ where $\tau_i = 0$ (empty site), 1 (occupied by a car), or 2 (occupied by a truck). There are three possible hopping processes:

$$\begin{array}{ll} \text{(i) Car hopping with rate } \eta & : (1,0) \to (0,1) \\ \text{(ii) Truck hopping with rate } \eta\gamma & : (0,2) \to (2,0) \\ \text{(iii) Car-truck exchange with rate } \frac{\eta}{\beta} & : (1,2) \to (2,1). \end{array} \tag{1}$$

The reduction factor β $(1 \leq \beta \leq \infty)$ is related to the width of roads: $\beta = 1$ corresponds to a very wide road or a highway with a lane divider, and $\beta = \infty$ corresponds to a one lane road. Thus the value $1 - 1/\beta$ can be used as a measure of the road narrowness.

Recently, the model is studied with the random sequential update (RSU). In an infinitesimal time interval dt, each link is updated with the probability that is the product of the relevant rate and dt (Since η does not affect the dynamics at all in the RSU scheme, we choose $\eta = 1$ for the RSU scheme).

Alternatively, the ordered sequential update (OSU) can be used. One first chooses a particular site, i.e. the site N, and updates the state of the links consecutively either in the backward direction $(N, N-1), (N-1, N-2), \cdots, (1, N)$ (backward sequential update BSU) or in the forward direction $(N, 1), (1, 2), \cdots, (N-1, N)$ (forward sequential update FSU). In contrast to the RSU update, the time is discrete in the BSU and FSU schemes, and in each time step, hopping occurs in links with the probabilities that are equal to the rates in (1).

2.2 Matrix Product State

The two-way traffic model is equivalent to a two-species ASEP and it can be solved exactly by the method of the matrix product state (MPS). In the RSU scheme, the steady state weight P_s of a given configuration $(\tau_1, \tau_2, \cdots, \tau_N)$ is proportional to the trace of the normal product of some matrices [1]:

$$P_s(\tau_1, \tau_2, \cdots, \tau_N) \sim \mathrm{Tr}(X_1 X_2 \cdots X_N) \tag{2}$$

where $X_i = D$ for $\tau_i = 1$; E for $\tau_i = 2$; and A for $\tau_i = 0$, and these matrices satisfy the quadratic algebra

$$DE = D + E, \quad \alpha AE = A, \quad \beta DA = A \quad (\alpha \equiv \beta\gamma) \, . \tag{3}$$

In a similar way, P_s in the BSU (FSU) scheme can be written as follows:

$$P_s(\tau_1, \tau_2, \cdots, \tau_N) \sim \mathrm{Tr}(X_1 X_2 \cdots \hat{X}_N) \tag{4}$$

where $X_i = D, E$, or A ($\hat{X}_N = \hat{D}, \hat{E}$, or $\hat{A}$) depending on τ_i (τ_N). Note that the matrices at the site N are different from those at other sites [13]. The presence of the hatted matrix in (4) breaks the translational invariance of the problem. In [14], it is shown that with a simple assumption,

$$\hat{A} = A + a, \quad \hat{D} = D + d, \quad \hat{E} = E + e, \tag{5}$$

where a, d, and e are carefully chosen real numbers, the relevant matrix algebra for A, D, E becomes identical to the one in (3) except for the renormalizations of α and β to $\tilde{\alpha}$ and $\tilde{\beta}$. In the BSU scheme, the choice $a = 0, d = -\eta/\beta, e = \eta/(\beta - \eta)$ gives $\tilde{\alpha} = \alpha(\beta - \eta)/(\beta - \alpha\eta)$, $\tilde{\beta} = \beta$, and in the FSU scheme, the choice $a = 0, d = \eta/(\beta - \eta), e = -\eta/\beta$ results in $\tilde{\alpha} = \alpha$, $\tilde{\beta} = (\beta - \eta)/(1 - \eta)$.

3 Average Velocities

In this section we consider the special case where there is only one truck and evaluate average velocities. The MPS method allows exact evaluations of velocities. We first present the car velocity in the thermodynamic limit. In the BSU scheme,

$$\langle v_{\mathrm{car}} \rangle = \begin{cases} \dfrac{\eta}{1 - \eta n}(1 - n) \text{ if } & n\tilde{\beta} \leq 1 \\[2ex] \dfrac{\eta}{\beta - \eta} \dfrac{1 - n}{n} \quad \text{if} & n\tilde{\beta} \geq 1 \, , \end{cases} \tag{6}$$

and in the FSU scheme,

$$\langle v_{\mathrm{car}} \rangle = \begin{cases} \dfrac{\eta(1 - n)}{1 - \eta(1 - n)} \text{ if } & n\tilde{\beta} \leq 1 \\[2ex] \dfrac{\eta}{\beta} \dfrac{(1 - n)}{n} \quad \text{if} & n\tilde{\beta} \geq 1 \, . \end{cases} \tag{7}$$

Figures 1a and 1b show the behaviors of $\langle v_{\mathrm{car}} \rangle$ in the two updating schemes as a function of the road narrowness $1 \ \frac{1}{\beta}$ while the values of $\eta, \gamma(= \frac{\alpha}{\beta})$ and n are fixed. In both schemes, a continuous phase transition is evident: While $\langle v_{\mathrm{car}} \rangle$ is constant below the critical narrowness, it begins to drop suddenly at the critical narrowness ($n\tilde{\beta} = 1$), generating a cusp. Thus above the critical narrowness, a single truck results in global effects. The functional dependence of the decrease differs in the two updating schemes.

The exact expressions for the truck velocity, on the other hand, are rather lengthy especially below the critical narrowness and we present the functional dependence on the road narrowness only through figures [Figs. 1c,d). For exact expressions, see [14]. The appearance of a continuous transition at the critical narrowness ($n\tilde{\beta} = 1$) is clear and the functional dependence of the truck velocity again differs in the two updating schemes.

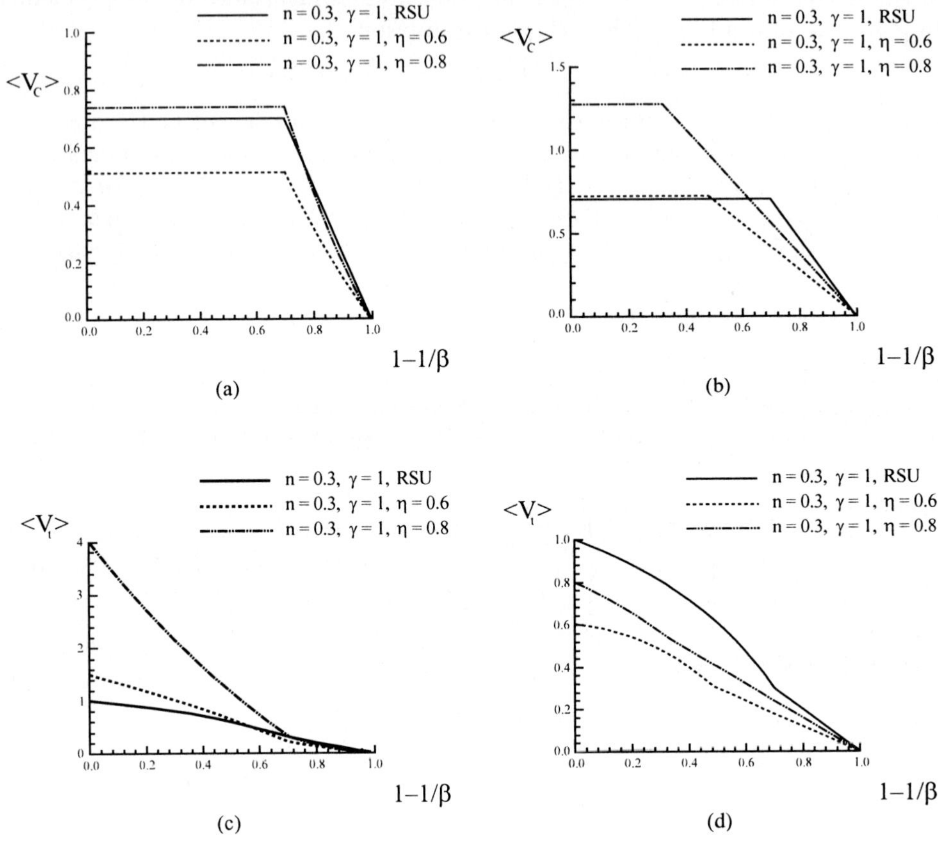

Fig. 1. (a) Average velocity of cars in the BSU and RSU schemes for $n = 0.3$ and different values of η. (b) Average velocity of cars in the FSU and RSU schemes for $n = 0.3$ and different values of η. (c) Average velocity of the truck in the BSU and RSU schemes for $n = 0.3$ and different values of η. (d) Average velocity of the truck in the FSU and RSU schemes for $n = 0.3$ and different values of η. After [14].

4 Density Profile

In the RSU scheme, the probability to find a car at a site depends only on its relative distance x to the truck due to the translational invariance. In [1], it is found that in the high density phase $n\beta \geq 1$, the probability or density profile $\langle n(x)\rangle$ becomes

$$\langle n(x)\rangle = \begin{cases} 1 \text{ for } \frac{x}{N} \leq \frac{n\beta-1}{\beta-1} \\ \frac{1}{\beta} \text{ otherwise .} \end{cases} \tag{8}$$

Note that the system consists of two regions, a traffic jam region in front of the truck and a free flow region behind it. In the low density phase $n\beta \leq 1$, on the other hand, the presence of the truck has only local effects. In the thermodynamic limit, the car density becomes

$$\langle n(x)\rangle = n\left\{1 + \frac{(\alpha+\beta-1)(1-n)}{1-n+\alpha n}(n\beta)^x\right\} \tag{9}$$

which shows that the disturbance by the truck decays exponentially with a characteristic length scale $|\ln(n\beta)|^{-1}$.

In the ordered sequential updates, complications occur in the definition of the probability itself. Since the choice of a particular site as a starting point of the update breaks the translational invariance, the probability to find a car at x sites in front of the truck depends not only on the relative distance x but also on the truck location. This unnecessary complication can be avoided by choosing the starting point in an even way. In [14], it is found that with this "homogeneous" choice, the expressions for the probability in the BSU and the FSU scheme become *identical* to (8,9) except for the replacement of the bare parameters α and β with the renormalized ones $\tilde{\alpha}$ and $\tilde{\beta}$.

5 Concluding Remarks

Characteristics of an exactly solvable two-way traffic model are investigated with the OSU scheme and both qualitative and quantitative differences are found compared to those with the RSU scheme [1]. Our approach is based on the so-called matrix product formalism which allows analytic solutions. In the OSU schemes, the choice of a particular site as a starting point of the update breaks the translational invariance of the steady state measure, which is also evident in the form of the MPS. Thus an averaging over the different choices of the update starting point is necessary to restore the translational invariance to the system. Performing the translationally invariant averaging, it is found that for characteristics such as the density profile of cars and the density-density correlation function, the differences in the updating schemes can be fully taken into account by the proper renormalization of the parameters α and β. However, this is not the case with average velocities. Changing the update scheme affects velocities in a more complicated manner and the renormalization of the parameters is not sufficient to account for different behaviors of $\langle v_{\rm car}\rangle$ and $\langle v_{\rm truck}\rangle$ in different updating schemes. Especially the dependence of $\langle v_{\rm car}\rangle$ and $\langle v_{\rm truck}\rangle$ on the road narrowness $1-\frac{1}{\beta}$ can vary qualitatively depending on the updating schemes.

Acknowledgement. M.E.F. would like to thank V. Karimipour for fruitful comments and D. Kim, R. Asgari for useful helps. H.-W.L. thanks D. Kim for bringing his attention to this problem. H.-W.L. was supported by the Korea Science and Engineering Foundation through the fellowship program and the SRC program at SNU-CTP.

References

1. H.-W. Lee, V. Popkov, and D. Kim, J. Phys. A. **30**, 8497 (1997).
2. V. Privman (Ed.), *Non equilibrium Statistical Mechanics in One Dimension,* (Cambridge University Press, Cambridge, 1997).
3. G.M. Schütz, to appear in: *Phase transitions and critical phenomena,* C. Domb and J. Lebowitz, (Eds.), (Academic Press, London, 1999).
4. M. Schreckenberg, A. Schadschneider, K. Nagel, and N. Ito, Phys. Rev. E **51**, 2939 (1995).
5. D.E. Wolf, M. Schreckenberg, and A. Bachem (Eds.), *Traffic and Granular Flow,* (World Scientific, Singapore, 1996).
6. B. Derrida, S.A. Janowsky, J.L. Lebowitz, and E.R. Speers, J. Stat. Phys. **73**, 813 (1993).
7. K. Mallick, J. Phys. A **29**, 5375 (1996).
8. M.R. Evans, Europhys. Lett. **36**, 1493 (1996); J. Phys. A. **30**, 5669 (1997).
9. S.A. Janowsky, J.L. Lebowitz, Phys. Rev. A **45**, 618 (1992); J. Stat. Phys. **77**, 35 (1994).
10. G. Schütz, J. Stat. Phys. **71**, 471 (1993).
11. H. Emmerich and E. Rank, Physica A **216**, 435 (1995).
12. S. Yukawa, M. Kikuchi, and S. Tadaki, J. Phys. Soc. Jpn. **63**, 3609 (1994).
13. N. Rajewsky, L. Santen, A. Schadschneider, and M. Schreckenberg, J. Stat. Phys. **92**, 151 (1998).
14. M.E. Fouladvand and H.-W. Lee, in press, Phys. Rev. **E**, (1999); [cond-mat/9906447].

Exact Traveling Cluster Solutions of Differential Equations with Delay for a Traffic Flow Model

K. Hasebe[1], A. Nakayama[2], and Y. Sugiyama[3]

[1] Faculty of Business Administration, Aichi University, Miyoshi, Aichi 470-0296, Japan
[2] Gifu Keizai University, Ohgaki, Gifu 503-8550, Japan
[3] Division of Mathematical Science, City College of Mie, Tsu, Mie 514-0112, Japan

Abstract. Exact solutions of the first order differential equation with delay are derived. The equation has been introduced as a model of traffic flow. The solution describes the traveling cluster of jam, which is characterized by Jacobi's elliptic function. The system is related to some soliton systems.

1 Model and Motivation

Temporal delay plays an important role in pattern formation of complex dynamical systems such as nonlinear physics, non-equilibrium chemical reactions, biological organization and social transport phenomena [1]. The influence of x_{n+1} on x_n is delayed in a reaction chain, where x_n represents physical, chemical or biological quantities.

In a simple case such a system can be formulated as the following first order differential-difference equation,

$$\frac{d}{dt}x_n(t+\tau) = f(x_{n+1}(t) - x_n(t)), \tag{1}$$

where τ is a real positive constant called "delay". The index n takes integer values. The set of equations of this type has been introduced for a car-following model of traffic flow [1–3]. In that case x_n is the position of the nth car. In several decades a lot of work have been done for the study in the traffic flow phenomena based on the equation of this type.

In 1994 we introduced a car-following model based on a second order differential-difference equation with no delay, called 'Optimal Velocity (OV) Model'.

$$\tau\frac{d^2}{dt^2}x_n(t) = f(x_{n+1}(t) - x_n(t)) - \frac{d}{dt}x_n(t). \tag{2}$$

We chose $f(x) = \tanh(x - c) + \text{const.}$ for the Optimal Velocity function $f(x)$, which determines the safe velocity as a function of the headway distance. The OV model has the stable traveling cluster solution as a traffic jam [4]. At a glance, the OV model may have some relation to (1) for relatively small τ [3]. As a model for traffic flow, $f(x) = \tanh(x)$ is the reasonable choice also for the type

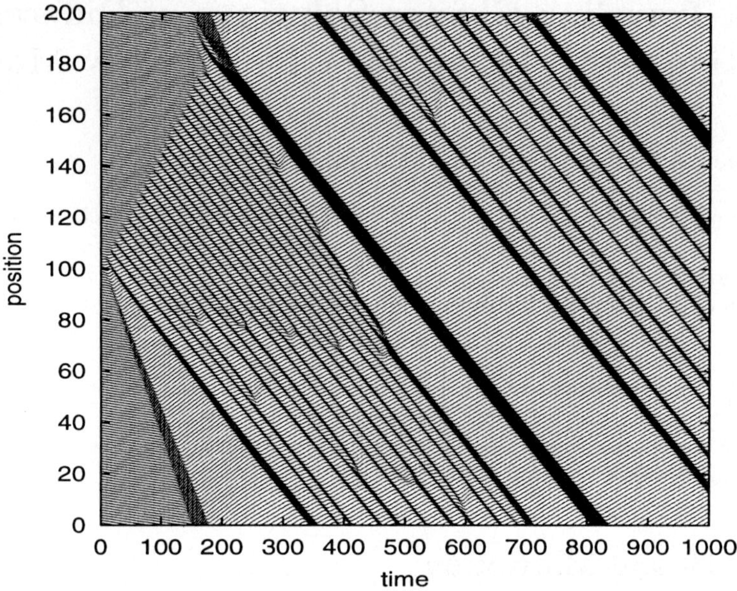

Fig. 1. The numerical simulation of the model (1) with $f(x) = \tanh(x - c) + \text{const}$. The initial condition is the homogeneous flow solution. The simulation is performed in the condition that the delay τ is appropriately large. All traveling clusters are moving with the velocity $v = 1/(2\tau)$.

of the formula (1). One interesting question is whether (1) with $f(x) = \tanh(x)$ has such traveling cluster solution or not.

Our simulation result suggest the existence of such a solution. Figure (1) shows that the generation of a jam cluster is very similar as is the case for the OV model, if we take τ for appropriately large value. Traveling clusters move with the same velocity $v = 1/(2\tau)$ opposed to the direction of the cars. The result indicates that the naive approximation of (1) for small τ is meaningless for the solution of the traveling cluster.

2 Exact Traveling Cluster Solution

In this paper, we derive a series of analytic solutions for a traveling cluster in (1), and confirm their stability by numerical simulations [5]. First, we introduce the new variable $h_n(t) = x_{n+1}(t) - x_n(t)$ and rewrite (1) as

$$\frac{d}{dt}h_n(t + \tau) = f(h_{n+1}(t)) - f(h_n(t)) \,. \tag{3}$$

2.1 Linear Analysis

We start at the linear theory assuming that the amplitude $h_n(t)$ is infinitesimal and $f(h_n) = f(0) + f'(0)h_n$. Equation (3) becomes

$$\frac{d}{dt}h_n(t+\tau) = h_{n+1}(t) - h_n(t). \tag{4}$$

Here we set $f'(0) = 1$ without loss of generality. We can obtain the solution of the form

$$h_n(t) = \exp(i\alpha(n + t/\tau)), \tag{5}$$

where a the real-valued variable α is given by

$$\frac{\sin\alpha/2}{\alpha/2} = \frac{1}{2\tau}. \tag{6}$$

Equation (6) has a solution if 2τ is larger than 1, which means $\tau = 1/2$ is critical. This represents a traveling wave solution with the velocity $1/(2\tau)$ in the space of index n, which is treated as a continuous variable. The wave moves backward against the numbering direction, which appears as the traveling wave in the real space moving backward in the flow of x_n. The above analysis is first given by Whitham [3].

2.2 Non Linear Analysis and Reduced Equations

Now let us investigate the exact traveling wave solution of (3). We treat the index n as a continuous variable, and change the notation $h_n(t)$ to $h(n,t)$. We introduce new variables in the moving frame of the traveling wave as $u = n + vt$, where v is the velocity of the traveling wave. We search the solution which does not change its form in this frame. We define the amplitude of the traveling wave

$$H(u) = H(n + vt) \equiv h(n,t). \tag{7}$$

Equation (3) for the amplitude $H(u)$ is expressed as

$$v\frac{d}{du}H(u + v\tau) = f(H(u+1)) - f(H(u)). \tag{8}$$

Replacing u by $u - 1/2$, we get the symmetric form

$$v\frac{d}{du}H(u) = f(H(u+\frac{1}{2})) - f(H(u-\frac{1}{2})), \tag{9}$$

where we fix

$$v = \frac{1}{2\tau}. \tag{10}$$

Here, we investigate the solution of the traveling wave, which propagates backward with just the same velocity as the linear theory, $v = 1/(2\tau)$. This assumption is supported by our simulation result.

Now we present the definite form of exact solutions giving concrete examples of $f(x)$. We take two cases: $(-)$type; $f(x) = \tanh(x)$,(a suitable choice for a model of traffic flow) and $\coth(x)$, $(+)$type; $f(x) = \tan(x)$ and $\cot(x)$. Changing $H(u)$ to $G(u)$ by $G = f(H)$, (9) is rewritten as

$$v \frac{dG(u)/du}{1 \mp G(u)^2} = G(u + \frac{1}{2}) - G(u - \frac{1}{2}), \tag{11}$$

respectively, for each type. We found that all Jacobi's elliptic functions satisfy (11).

2.3 Explicit Form of Exact Solutions

We can easily find a solution for the $(+)$ type of (11) in the form

$$G(u) = \pm k \frac{\alpha}{4\tau} \operatorname{sn}(\alpha u, k), \tag{12}$$

where sn is the Jacobi elliptic function with modulus k. The parameter α is determined by

$$\frac{\operatorname{sn}(\alpha/2, k)}{\alpha/2} = \frac{1}{2\tau}, \tag{13}$$

which has a real solution only if $1/(2\tau) < 1$. The modulus k is a free parameter of the solution, which indicates the existence of many solutions for the same traveling velocity $v = 1/(2\tau)$. The modulus k determines the period of elliptic functions, which is related to the number of traveling clusters and the boundary condition. In the case of $k = 0$, (13) reduces to the result of linear theory, (6). In the case of $k = 1$, (12) gives kink like solution corresponding to the boundary condition $G(-\infty) = -G(\infty)$. A solution for $(+)$ type of (11) is

$$G(u) = \pm k \frac{\alpha}{4\tau} \operatorname{cn}(\alpha u, k), \tag{14}$$

where α is given by

$$\frac{\operatorname{sd}(\alpha/2, k)}{\alpha/2} = \frac{1}{2\tau} . \tag{15}$$

We have found the family of solutions of (11). The solutions of this family are denoted in the similar form as (12) or (14), and all Jacobi elliptic functions are solutions [5]. Igarashi, Itoh and Nakanishi have found the exact solution of (1) in a form different from ours, which is characterized by the theta function [6].

3 Simulation and Discussion

We perform the simulation to check the stability of our analytic solution. Figure 2 shows an elliptic solution for $f(x) = \tanh(x)$ given by sn

$$H(u,k) = \operatorname{arc}\tanh(k\frac{\alpha}{4\tau}\,\mathrm{sn}(\alpha u,k)), \tag{16}$$

with $\tau = 0.501$, $k = 0.9965$ and $\alpha = 0.15522$, together with the result of a simulation for (1) performed with periodic boundary conditions and suitable initial conditions. We also show the form of this solution in the real space in Fig. 3.

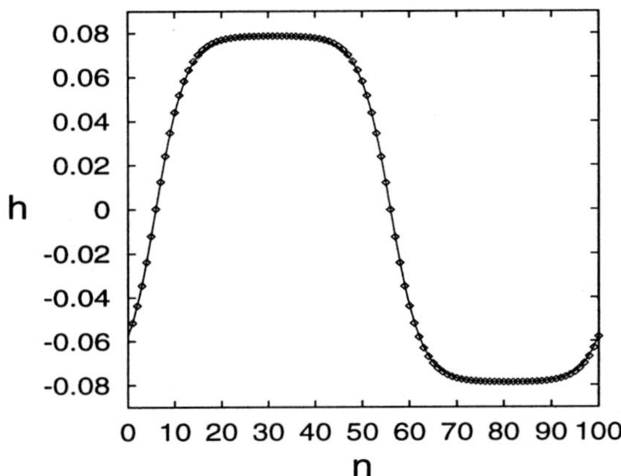

Fig. 2. The Simulation result of h_n for $f(x) = \tanh(x)$ is shown by diamonds together with the analytic solution given by sn. The curve of the analytic solution is shifted by $u \to u+$constant.

We remark that all these solutions have the common velocity $v = 1/(2\tau)$. Numerical simulations show that the traveling cluster solutions of (1) preserve their velocity $v = 1/(2\tau)$ in the deformation of $f(x)$ beyond $\tanh(x)$. This fact suggests that v is some invariant quantity of the structure in the set of the solutions.

Equations (11) are related to soliton systems. Our equations can be derived as the traveling wave equations from such soliton systems. (11) correspond to one of the evolution equations discussed by Ablowitz and Ladik [7]. The soliton systems related to the (+) type of (11) were widely discussed in the self-dual network equations of nonlinear inductors and capacitors by Wadati [8], Hirota and Satsuma [9]. Wadati showed the corresponding soliton system to the (+) type was derived from Lotka-Volterra system by a Bäcklund transformation [8].

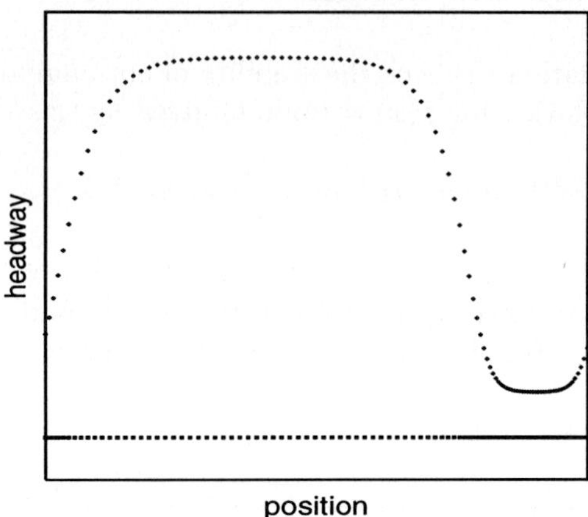

Fig. 3. The form in real space of the solution (16).

The traveling cluster solution of our model offers an information for the pattern formation of dissipative systems, which can be investigated in the relation with the above soliton systems.

Acknowledgement. This work was partly supported by a Grant-in-Aid for Scientific Research (C) (No. 10650066) of the Japanese Ministry of Education, Science, Sports and Culture.

References

1. D.E. Wolf, M. Schreckenberg, and A. Bachem, (Eds.), *Traffic and Granular Flow,* (World Scientific, Singapore, 1996); M. Schreckenberg and D.E. Wolf, (Eds.), *Traffic and Granular Flow '97,* (Springer Verlag, Singapore, 1998).
2. G.F. Newell, Oper. Res. **9**, 2209 (1961).
3. G.B. Whitham, Proc. R. Soc. Lond. A **428**, 49 (1990).
4. M. Bando, K. Hasebe, A. Nakayama, A. Shibata, and Y. Sugiyama, Phys. Rev. E **51**, 1035 (1995); Japan J. of Ind. and Appl. Math. **11**, 203 (1994), Y. Sugiyama, in [1].
5. K. Hasebe, A. Nakayama, and Y. Suigiyama, Phys. Lett. A **259**, 135 (1999).
6. Y. Igarashi, K. Itoh, and K. Nakanishi, J. Phys. Soc. Japan **68**, 791 (1999).
7. M.J. Ablowitz and J.F. Ladik, J. Math. Phys. **17**, 1011 (1976).
8. M. Wadati, Prog. Theor. Phys. Suppl. **59**, 36 (1976).
9. R. Hirota and J. Satsuma, Prog. Theor. Phys. Suppl. **59**, 64 (1976).

Stable and Metastable States in Congested Traffic

E. Tomer[1], L. Safonov[1,2], and S. Havlin[1]

[1] Minerva Center and Department of Physics, Bar–Ilan University, 52900 Ramat–Gan, Israel
[2] Department of Applied Mathematics and Mechanics, Voronezh State University, 394693 Voronezh, Russia

Abstract. A new single lane inertial car following model of traffic flow is presented. The model demonstrates the presence of three phases in traffic flow: free flow, non-homogeneous congested flow and homogeneous congested flow. In the non-homogeneous congested flow we find *many* periodic stable states with different values of flux and wavelength. We also find that states with relatively low and relatively high values of wavelength are metastable.

1 Introduction

Traffic dynamics has traditionally been modeled using either fluid dynamics approach or microscopic car-following rules [1–6]. In the present paper we investigate a new inertial car-following model recently introduced [7]. The results are in good agreement with experimental findings recently reported by Kerner and Rehborn [8–10].

We assume that the car acceleration is affected by four factors:

(a) aspiration to keep safety time gap T from the car ahead,
(b) pre-braking if the car ahead is much slower,
(c) aspiration not to exceed significantly the permitted velocity v_{per} (speed limit),
(d) random noise η.

The acceleration of the nth car a_n is therefore given by a sum of four terms depending on its coordinate x_n, velocity v_n, distance to the car ahead $\Delta x_n = x_{n+1} - x_n$ and the velocities difference $\Delta v_n = v_{n+1} - v_n$:

$$a_n = A\left(1 - \frac{\Delta x_n^0}{\Delta x_n}\right) - \frac{Z^2(-\Delta v_n)}{2(\Delta x_n - D)} - kZ(v_n - v_{\text{per}}) + \eta, \qquad (1)$$

where A is a sensitivity parameter, D is the minimal distance between consecutive cars, k is a constant, $\Delta x_n^0 = v_n T + D$ is the safety distance. The function Z is defined as $Z(x) = (x + |x|)/2$. In further analytical and numerical exploration of the model the noise term η is omitted unless otherwise stated.

In the following we discuss in more details the terms in the right side of (1):

The first term plays an important role when the velocity difference between consecutive cars is relatively small. In this case the nth car accelerates if $\Delta x_n > \Delta x_n^0$ and brakes if $\Delta x_n < \Delta x_n^0$.

The second term is essential when $v_n \gg v_{n+1}$. According to the first term a car getting close to a much slower car brakes only if $\Delta x_n < \Delta x_n^0$. This term enables it to start braking even at a bigger distance.

The dissipative third term represents a repulsive force acting when the velocity exceeds the permitted velocity.

In the deterministic case the motion of cars is described by the following system of ordinary differential equations

$$\begin{cases} \dot{x}_n = v_n, \\ \dot{v}_n = A\left(1 - \frac{v_n T + D}{x_{n+1} - x_n}\right) - \frac{Z^2(v_n - v_{n+1})}{2(x_{n+1} - x_n - D)} - kZ(v_n - v_{\text{per}}), \end{cases} \tag{2}$$

$n = 1, \ldots, N$ with periodic boundary conditions $x_{N+1} = x_1 + \frac{N}{\rho}$, $v_{N+1} = v_1$.

A solution of (2) which corresponds to the homogeneous flow is

$$v_n^0 = v^0 = \begin{cases} \frac{A(1-D\rho)+kv_{\text{per}}}{A\rho T + k}, & \rho \le \frac{1}{D+Tv_{\text{per}}}, \\ \frac{1-D\rho}{\rho T}, & \rho \ge \frac{1}{D+Tv_{\text{per}}}, \end{cases} \qquad x_n^0 = \frac{n-1}{\rho} + v^0 t. \tag{3}$$

In the following numerical analysis we use parallel updating rule and parameters values $A = 3$ m/s^2, $v_{\text{per}} = 25$ m/s, $T = 2$ s, $D = 5$ m, and $k = 2\,\text{s}^{-1}$. For the dimensionless model we have only three parameters: A, v_{per}, and k.

2 Three Phases in Traffic Flow

Results of the deterministic model simulations are presented in Figs. 1a,b. The data plotted on these figures correspond to the state of the system after a transient time, starting with random and (nearly) homogeneous initial conditions. The flux-density relation plotted in Fig. 1a (the fundamental diagram) consists of two curves: the first, increasing part starting at the origin, which represents free flow and the second part which corresponds to congested flow. Fig. 1b illustrates the dependence on density of mean square variation of velocities,

$$\sigma_v = \left[\frac{1}{N}\sum_{n=1}^{N}(v_n - \langle v \rangle)^2\right]^{1/2},$$

divided by the average velocity $\langle v \rangle$. From this figure it can be seen that congested traffic consists of two different phases, homogeneous and non-homogeneous.

Therefore, we can derive the existence of three phases in traffic flow which is in good agreement with experimental observations [9,10] and with results of other models. The three phases are:

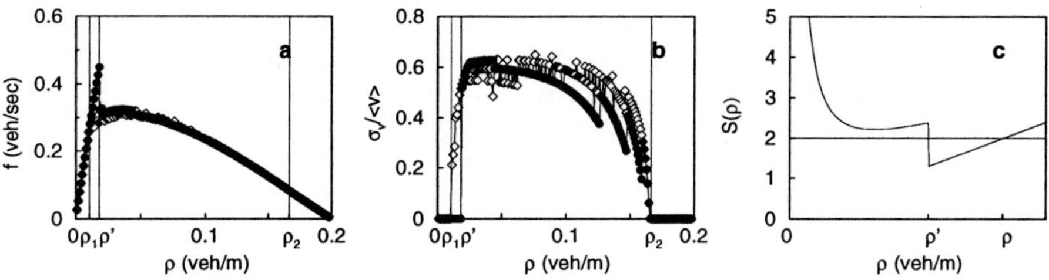

Fig. 1. Three phases of traffic flow, and bistability regime: (a) fundamental diagram and (b) mean square variation of velocities for homogeneous initial conditions (black circles) and non-homogeneous initial conditions (white diamonds). Parameters values are $A = 3$ m/s^2, $v_{\text{per}} = 25$ m/s, $k = 2$ s^{-1}, $T = 2$s , $D = 5$ m, and $N = 37$. (c) Qualitative plot of $S(\rho)$.

1. free flow (for $\rho < \rho'$),
2. non-homogeneous congested flow (for $\rho_1 < \rho < \rho_2$),
3. homogeneous congested flow (for $\rho > \rho_2$).

In the first and the third regimes $\sigma_v = 0$ which means that the flow is homogeneous. In the second regime the flow is non-homogeneous (even in absence of noise in drivers behavior). The characteristics of non-homogeneous congested flow are discussed in details in Sect. 4. In the range of densities $\rho_1 < \rho < \rho'$ both homogeneous and non-homogeneous states are stable, which is also in agreement with experimental results [8] and with results of other models (e.g., [3–5,11]).

3 Stability of Homogeneous Flow

In this section we give a stability analysis of the homogeneous flow solution (3) of (2). The linearization of (2) near the homogeneous flow solution (3) in variables $\xi_n = x_n - x_n^0$ has the form

$$\ddot{\xi}_n = -p\dot{\xi}_n + q(\xi_{n+1} - \xi_n), \quad n = 1, \dots, N, \tag{4}$$

where $\xi_{N+1} = \xi_1$, $p = AT\rho + k$, $q = \frac{AT+kTv_{\text{per}}+kD}{AT\rho+k} \cdot A\rho^2$ for $\rho \le \frac{1}{D+Tv_{\text{per}}}$ and $p = AT\rho$, $q = A\rho$ otherwise.

As in [3], a solution of (4) can be written as

$$\xi_n = \exp(i\alpha n + zt), \tag{5}$$

where $\alpha = \frac{2\pi}{N}\kappa$ ($\kappa = 0, \dots, N-1$) and z - a complex number. Substituting (5) into (4) we obtain the algebraic equation for z:

$$z^2 + pz - q(e^{i\alpha} - 1) = 0. \tag{6}$$

Each of the N equations (6) has two solutions. These $2N$ different complex numbers are the eigenvalues of system (4). One of them (which corresponds to

$\kappa = 0$) is equal to zero regardless of values of parameters. In this case all ξ_n in (5) are equal to a constant and belong to the one-dimensional subspace of equilibria of system (4) (defined by equations $\xi_1 = \ldots = \xi_N$, $\dot{\xi}_1 = \ldots = \dot{\xi}_N = 0$). This indicates that the disturbed state x_n for $z = 0$ is also homogeneous. For $z \neq 0$, ξ_n in (5) is a wave with increasing or decreasing amplitude. Therefore, if we find conditions under which other $2N - 1$ eigenvalues have negative real parts (the magnitude of wave (5) decreases with time) we can say that under these conditions the homogeneous flow solution (3) is stable.

Following the approach of [3] we can derive this condition as $\frac{p^2}{q} > 2$ or

$$S(\rho) > 2, \tag{7}$$

where

$$S(\rho) = \begin{cases} \frac{(AT\rho+k)^3}{\rho^2 A(AT+kv_{\mathrm{per}}T+kD)}, & \rho < \frac{1}{D+Tv_{\mathrm{per}}}, \\ A\rho T^2, & \rho > \frac{1}{D+Tv_{\mathrm{per}}}. \end{cases}$$

A qualitative plot of $S(\rho)$ is sketched in Fig. 1c. Condition (7) implies that the homogeneous flow state is stable for $\rho' < \rho < \rho''$, where $\rho' = \frac{1}{D+Tv_{\mathrm{per}}}$ and $\rho'' = \frac{2}{AT^2}$. Simulations show that $\rho'' \approx \rho_2$.

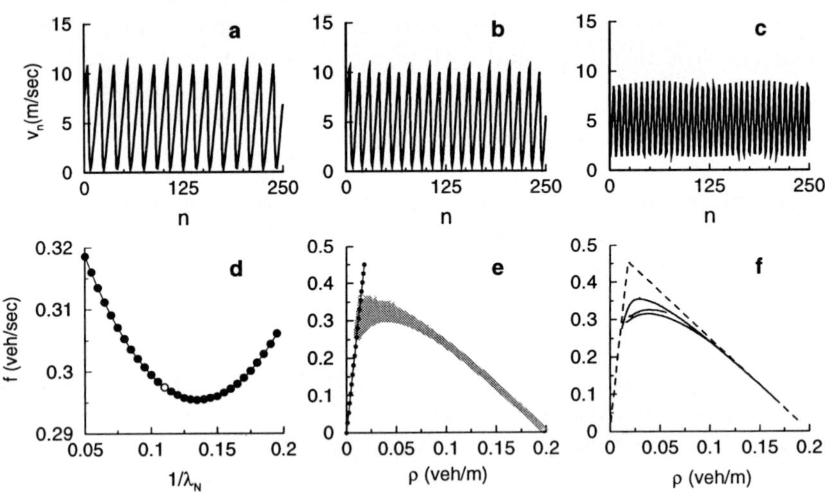

Fig. 2. Different stable states in the non-homogeneous congested flow regime, obtained from different initial conditions. (a-c) Cars velocities of three states with global density $\rho = 0.06$ veh/m. (d) Values of flux for many different states ($\rho = 0.06$ veh/m). Each circle corresponds to a different stable state. λ_N is the wavelength in units of number of cars. The empty circle denotes the most stable state. (e) Flux values observed for different harmonic initial conditions. (f) Fundamental diagrams for three different stable states with wavelengths 20, 5 and 6.67 cars (top to bottom). The dashed line corresponds to the homogeneous flow solution.

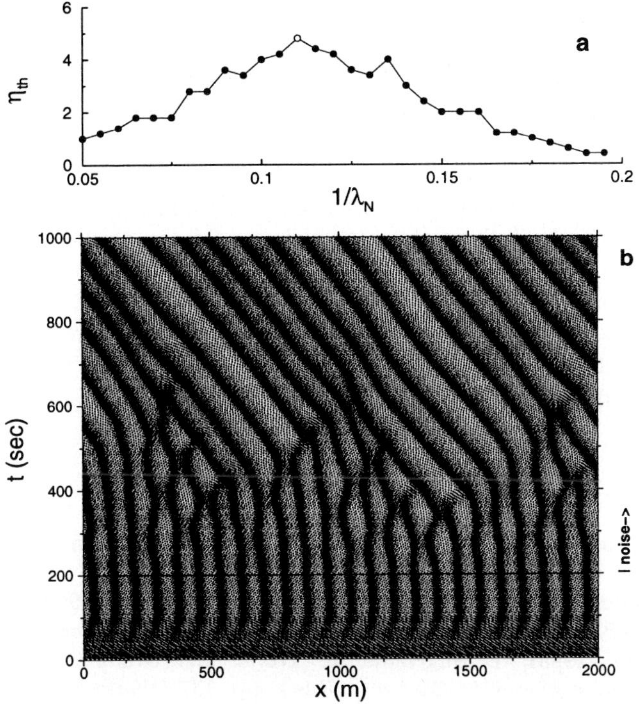

Fig. 3. (a) Noise sensitivity of stable states ($\rho = 0.06$ veh/m). The empty circle denotes the most stable state. (b) Transition from a metastable to a stable cycle in presence of noise. The global density is $\rho = 0.06$ veh/m, noise is added at $t = 200$ s.

4 Non-Homogeneous States and Their Stability

Our simulations show that for every given value of density in the non-homogeneous congested flow regime there exist many stable periodic states with different wavelengths. These states correspond to different limit cycles of system (2). Figs. 2a-c present three of these states for the same value of density. Shown are the cars velocities after the non-homogeneous flow regime has stabilized for three different initial conditions. The dependence of the flux on the wavelength λ_N is shown in Fig. 2d. Here λ_N is the wavelength in units of number of cars.

The last finding, namely the existence of a range of possible flux values for every given density, implies that non-homogeneous congested flow displays a two-dimensional region in the flux-density plane, as was found experimentally by B.S. Kerner [10]. This two-dimensional region is shown in Fig. 2e. This figure, which is qualitatively similar to corresponding figures in [10], was obtained by simulations with different harmonic initial conditions. But we also find that the two-dimensional region in the fundamental diagram consists of *many* curves and each of them corresponds to a different wavelength λ_N. Some of these curves are shown in Fig. 2f.

Simulations of the model with non-zero white noise term η in (1) show that different states have different sensitivity to noise. We define the noise threshold η_{th}, above which the average wavelength is not preserved. From Fig. 3a it can be seen that η_{th} is higher for the states with intermediate wavelengths than that for states with relatively high and relatively low values of wavelength. Applying noise with amplitude slightly above this threshold can shift the system from a periodic state to a more stable state which is less sensitive to noise. The space-time diagram on Fig. 3b illustrates this transition. After the noise term was applied at $t = 200s$ the system moved from a metastable state with $\lambda_N = 5$ to a more stable state with $\lambda_N = 8$. The latter state is closer to the most stable state, which is denoted with an empty circle in Figs. 3a and 2d.

References

1. R. Herman and R.W. Rothery, in: *Proc. of the 2nd Int. Symp. on the Theory of Traffic Flow*, London, 1963.
2. K. Nagel and M. Schreckenberg, J. Phys. I France **2**, 2221 (1992).
3. M. Bando, K. Hasebe, A. Nakayama, A. Shibata, and Y. Sugiyama, Phys. Rev. E **51**, 1035 (1995); Y. Sugiyama in: *Traffic and Granular Flow,* D.E. Wolf, M. Schreckenberg, and A. Bachem, (Eds.), pp. 137 (World Scientific, Singapore, 1996).
4. K. Nagel and M. Paczuski, Phys. Rev. E **51**, 2909 (1995).
5. S. Krauss, P. Wagner, and C. Gawron, Phys. Rev. E **55**, 5597 (1997).
6. D. Helbing and M. Schreckenberg, Phys. Rev. E **59**, R2505 (1999).
7. E. Tomer, L. Safonov, and S. Havlin, preprint, (1999).
8. B.S. Kerner and H. Rehborn, Phys. Rev. E **53**, R4275 (1996).
9. B.S. Kerner and H. Rehborn, Phys. Rev. Lett. **79**, 4030 (1997).
10. B.S. Kerner, Phys. Rev. Lett. **81**, 3797 (1998); Physics World **12**, 25 (1999).
11. H.Y. Lee, H.-W. Lee, and D. Kim, Phys. Rev. Lett. **81**, 1130 (1998).

Detailed Microscopic Rules to Simulate Multilane Freeway Traffic

A. Kittel, A. Eidmann, and M. Goldbach

Fachbereich Physik, Abteilung für Energie- und Halbleiterforschung, Universität Oldenburg, 26129 Oldenburg, Germany

Abstract. A simulation to model traffic on a multilane freeway is introduced starting from microscopic driving rules. The model takes each individual car into account with its individual features and actual situations so that a distribution of parameters as well as different behavior can easily be analyzed. Therefore, a detailed study of certain situations, driving tactics, vehicle properties, and their influence on the global traffic flow can be performed. The model and first results are discussed, namely, the influence of the driver behavior on the fundamental diagram.

1 Introduction

In recent years different attempts have been undertaken to develop fast and reliable algorithms, these attempts are ranging from minimal microscopic driving models [1–4], over car-following methods [5,6], to models based on gas kinetics [7–10]. In the present paper we introduce a model of traffic flow which is driven by the idea to be as close as possible to the microscopic behavior. During the simulation, the state variables of each car, i.e., position, speed, and acceleration, are updated depending on its neighboring cars. The model is, thus, car centered. Nevertheless, the computational efforts are reasonable and the speed of computation is considerably faster than real time even if a standard personal computer is used and 5000 cars are considered. There is no restriction of the initial situation, any choice is possible even a more interesting one, a situation measured from a real traffic situation.

In Sect. 2 we introduce the simulation concept and the microscopic model. Thereafter, in Sect. 3 the macroscopic features of the model are discussed and compared to measurements.

2 Microscopic Model

The idea of our model originates from the motivation to be as close as possible to the *real* world. The actual state and the individual features of each car, the specific traffic situation, and the behavior of the driver are taken into account during the simulation process. As a consequence, the individual car is represented by a list of variables and parameters characterizing the state and the features, respectively, of each car and driver (simply called car further on).

In the case of the realization presented in this paper we model a three lane freeway traffic. Each car indexed with i is represented by its actual speed $v^{(i)}$, its actual acceleration $a^{(i)}$, its actual position along the street $x^{(i)}$, and its actual lane $L^{(i)}$ as the state variables with the features, namely, desired traveling speed $v_d^{(i)}$, maximum acceleration $a_{\max}^{(i)}$, maximum deceleration $a_{\min}^{(i)}$, and a factor of no reaction $f_n^{(i)}$, explained in detail later. All cars are organized in a so called linked list ordered with respect to their position along the street even if they are driving on different lanes. The street, in the present case, is considered to be homogeneous without any local features, i.e., no local in- or outflow, constant number of lanes, no speed limit, sight is unlimited, no inclination, and so on. Actually, all these features can be easily incorporated within the model to simulate very special situations and, by means of that, to solve very specific problems. All changes of the cars during one time step ($\Delta t = 0.05$ s) are determined before changing the state of the entire street at once, this is necessary to avoid any artifacts. During all computation floating point operations are used. The model is, thus, continuous. The entire list is updated from the last car to the first one on the street.

At the beginning we initialize the desired traveling speed (DTS) using a Gaussian distribution with a mean value of 33 m/s, a standard deviation of 8 m/s, and a lower cut-off at 22 m/s (80 km/h). All cars have an increasing index i along the direction of driving and have the same length $s^{(i)} \equiv s = 5$ m. They are put on a three-lane street with a period of 10 m. Taking into account the space needed by a car of $s = 5$ m, this period corresponds to the spacing of 25 m for each lane. The initial speed $v^{(i)}$ is set to the minimum of the corresponding DTS $v_d^{(i)}$ and 40 m/s. Their individual speed $v^{(i)}$ is set to DTS in case it is lower than 40 m/s (144 km/h). The speed of cars with DTS higher than 40 m/s is set to this value in order to avoid accidents in this highly congested initial situation. In the present paper, the maximum acceleration, the maximum deceleration, and the no-reaction factor are chosen to be equal for all cars, $a_{\max}^{(i)} \equiv a_{\max} = 1$ m/s^2, $a_{\min}^{(i)} \equiv a_{\min} = -15$ m/s^2, and $f_n^{(i)} \equiv f_n$, respectively.

Each car is updated in the following way (see scheme Fig. 1). First, the distance $d^{(i)}$ to the car in front is compared to its safe distance $d_s^{(i)}$, i.e., the distance necessary to avoid an accident. The safe distance $d_s^{(i)}$ is defined as

$$d_s^{(i)} = f_s v^{(i)} + s; \qquad (1)$$

where f_s denotes a safety factor, which is 1.8 s.

If $d^{(i)}$ is smaller than the safe distance $d_s^{(i)}$, the car has to react in the following manner: If there is no possibility to change to the left lane, it has to slow down. We have chosen a detailed braking rule to determine the deceleration (negative $a^{(i)}$) to be sure to avoid accidents. The deceleration is dependent on the distance to the car in front $d^{(i)}$, the speed of the car in front $v^{(f)}$, the acceleration of the car in front $a^{(f)}$, and the speed of the car $v^{(i)}$. If the distance to the car in front becomes smaller than $d_s^{(i)}$, the car has to brake, i.e., the acceleration $a^{(i)}$ becomes negative. If, in addition, the car in front is braking ($a^{(f)} < 0$), the ith car is braking with $a^{(i)} = a_{\min}$. Such a behavior is plausible because the driver intensively brakes if $d^{(i)} < d_s^{(i)}$ and he can see the brake lights of the car in

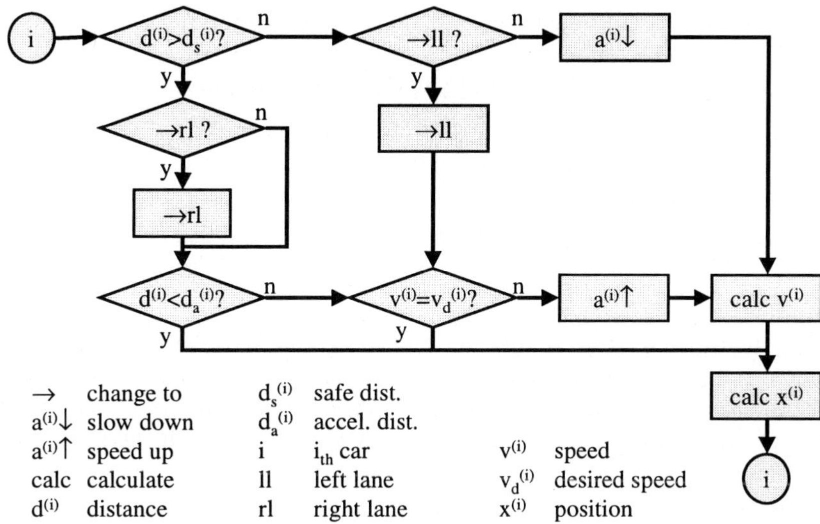

Fig. 1. Flow diagram which is processed for the car i^{th} in one time step. It has to be repeated for all cars during one time step (details see text).

front. If $d^{(i)} < d_s^{(i)}$ but $d^{(i)}$ is still larger than the stopping distance necessary to adjust $v^{(i)}$ to $v^{(f)}$ at $a_{\min}$, $a^{(i)}$ increases linearly with decreasing $d^{(i)}$. At a distance $d^{(i)}$ equal to the stopping distance the acceleration $a^{(i)}$ is set to $a_{\min}$. This rule guarantees that there are no accidents happening.

To be able to simulate a natural behavior of a driver we introduce a no-reaction factor $f_n \geq 1$. This denotes the fact that in reality a driver will never change from acceleration to deceleration at a precise separation to another car. It is more likely that there is an interval of distances where neither an acceleration nor deceleration takes place ($a^{(i)} = 0$). This interval is characterized by the factor of no-reaction. If the distance $d^{(i)}$ is less than $d_s^{(i)}$ the car decelerates, in the range $d_s^{(i)} < d^{(i)} \leq d_a^{(i)}$ it moves at a constant speed without acceleration or deceleration. If $d^{(i)}$ is larger than $d_a^{(i)}$ with

$$d_a^{(i)} = f_n d_s^{(i)} \tag{2}$$

and the car is driving at a speed $v^{(i)}$ smaller than its DTS it will accelerate ($a^{(i)} = a_{\max}$).

A change to the left lane is possible if the following conditions are fulfilled: first, the distance to the car following after a lane change has to be larger than $d_r^{(i)}$; second, the car ahead after a lane change has to be further apart than $d_f^{(i)} = d_s^{(i)}$; third, if the car is driving on the rightmost lane a car driving on the leftmost lane has to be further back than $d_r^{(i)}$, to avoid an accident after a lane change of both cars to the center lane. The rear distance $d_r^{(i)}$ is defined as

$$d_r^{(i)} = d_0 \exp(-v^{(i)}/v_0) + s. \tag{3}$$

The two constants d_0 and v_0 are set to 90 m and 27 m/s, respectively.

Furthermore, to obey the rule to drive on the rightmost possible lane as usual in most European countries, a lane change to the right is checked. To be able to change lanes to the right, it is necessary that the ith car and the car in front f^{th} and in rear r^{th} driving on the neighboring lane are further apart than $d_f^{(i)}$ and $d_r^{(i)}$, respectively.

After the possibility to change lanes has been checked, the actual acceleration, the actual position and the actual speed are calculated. This procedure is done for all the cars driving on the street. Thereafter, the calculated values are made the actual ones and the entire procedure is performed again as the next time step. In the present paper, we utilized the transient behavior starting with a high density evolving in time to scan through different densities.

3 Macroscopic Phenomena

Starting from the initial situation given in Section 2, the average local density for one lane n and the average local velocity $\bar{v}$ are calculated using the model described in Section 2. n and $\bar{v}$ are defined as $n \equiv N/(\lambda \Delta x)$ and $\bar{v} \equiv 1/N \sum_i v^{(i)}$ where N is the number of cars on the street located within each interval of $\Delta x = 0.25$ km, $\lambda = 3$ is the number of lanes, and $v^{(i)}$ is the velocity of cars indexed with i. The average local density develops in time. A typical situation can be observed which was also found for different initial conditions (e.g. for different values of initial average local densities or for different distributions of DTS). In the following, we discuss the dependence of characteristic macroscopic quantities on the no-reaction factor f_n. It turns out that the f_n significantly influences these quantities (such as the shape of the fundamental diagram, characteristic points of it, or the jam speed).

The corresponding fundamental diagram (Fig. 2) — the average local flow $q = n\bar{v}$ depending on density n — is shown as a density plot. For a certain n, we determined the conditional probability density $\mathcal{P}_n(q)$ of the flow q. Its intensity has to be interpreted only in the vertical line (along the fixed n). It is normalized to the corresponding maximum to achieve the best contrast in the plot. Due to the fact, we determine the local density on a certain interval and the integer number of the cars, the local density is discrete with a spacing of $\Delta n = 1.33$ veh/km resulting in a wavy structure of the plot. Figure 2 was calculated using $f_n = 1.2$. First, we discuss phenomena which are found for all values of f_n. Then we go into details which depend on f_n.

Two branches can be distinguished. The increasing branch shows a slight curvature which can be approximated by two lines with different slopes. We call the first part of the increasing branch *free traffic flow* and the second one *interacting traffic flow*. The decreasing branch is referred to as *congested traffic flow*.

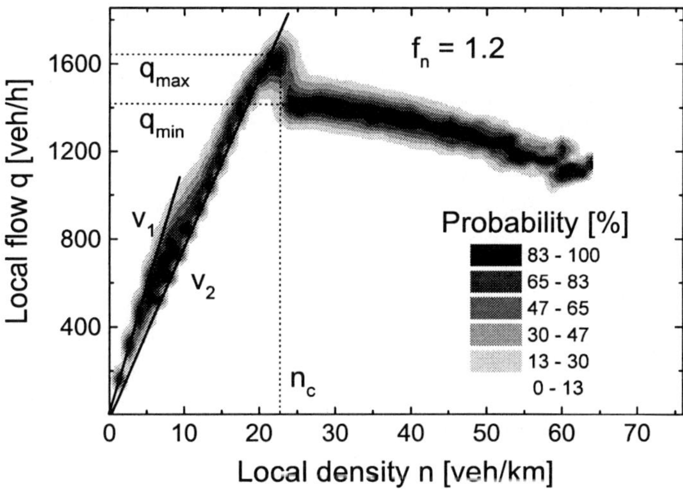

Fig. 2. Fundamental diagram for a three-lane freeway. The probability density $\mathcal{P}_n(q)$ for a given n is plotted versus the average local density n and is normalized to the maximum value to reach a maximal contrast. The local density is discrete due to its definition: $n = \text{veh}/\Delta x$ ($\Delta x = 0.25$km).

The transition region between both is characterized by *hysteretic traffic flow*. For small local densities $n < 8.33$ veh/km (i.e., an average distance $d > 115$ m between cars on a lane), the flow is determined by the DTS of the individual cars. The slope gives the mean velocity which is about $v_1 = 33$ m/s and corresponds to the mean value of the distribution $P(v_d)$ of the DTS. We refer to the region of low density as free flow because all cars are moving without any noticeable interaction. For higher average local densities (8.33 veh/km $< n <$ 16.67 veh/km) the slope is $v_2 = 22$ m/s and corresponds to the lowest DTS. It means for increasing local density, that the faster cars are more and more slowed down by the cars with a DTS at the lower end of the distribution (a similar result is reported in Ref. [13]). This region is called interacting traffic flow. Consequently, the distribution density $P(v_d)$ plays only a minor role. In the transition region up to the maximum flow $q_{\max}$ at the critical density n_c, the slope (i.e., the mean velocity) further decreases below the lowest value of the DTS [14]. The driving rules for the acceleration and deceleration process begin to determine the traffic flow. The mean spacing of cars is strongly influenced by the mean safe distance $\bar{d}_s$. If we compare the characteristic values to those reported in Ref. [11], we find a good agreement (see Tab. 1) for $f_n \approx 1.2$.

4 Conclusion

In the present paper we have introduced a model for multilane freeway traffic with rules which are very close to the real traffic situation. In the present pa-

Table 1. Comparison: simulation ($f_n = 1.2$) – measurement

	simulation	measurement
jam speed	10 km/h	15 km/h
in-/outflow	1400 veh/h	1100–1800 veh/h
average speed in the outflow	80 km/h	79–89 km/h
local density in the outflow	16.7 veh/km	17.7 veh/km

per we have restricted our investigations to a specific traffic situation, namely, the relaxation of a highly congested situation on a three lane freeway. We have discussed in detail the results gained for a distribution of desired speed, detailed rules for braking and lane changes, as well as a no-reaction factor. The no-reaction factor seems to play an important role. We were able to identify different types of traffic like free traffic flow, interacting traffic flow, the hysteretic transition region, and the congested, sometimes called synchronized, traffic flow. The results in case of a no-reaction factor of 1.2 coincide very well with the experimental data reported in literature.

References

1. M. Cremer and J. Ludwig, Math. Comput. Simulation **28**, 297 (1986).
2. K. Nagel and M. Schreckenberg, J. Physique I France **2**, 2221 (1992).
3. O. Biham, A. Middleton, and D. Levine, Phys. Rev. A **46**, 6124 (1992).
4. T. Nagatani, Phys. Rev. E **48**, 3290 (1995).
5. A.D. May, *Traffic Flow Fundamentals* (Prentice Hall, Englewood, NJ, 1990).
6. A.D. Mason and A.W. Woods, Phys. Rev. E **55**, 2203 (1997).
7. M. Lighthill and G. Whitham, Proc. Roy. Soc. of London A **229**, 317 (1955).
8. B. Kerner and P. Konhäuser, Phys. Rev. E **48**, 2335 (1993).
9. B. Kerner and P. Konhäuser, Phys. Rev. E **50**, 54 (1994).
10. D. Helbing, Phys. Rev. E **51**, 3164 (1995).
11. B.S. Kerner and H. Rehborn, Phys. Rev. E **53**, R1297 (1996).
12. B.S. Kerner, S.L. Klenov, and P. Konhäuser, Phys. Rev. E **56**, 4200 (1997).
13. E. Ben-Naim and P.L. Krapivsky, Phys. Rev. E **56**, 6680 (1997).
14. J. Krug and P. A. Ferrari, JPA **29**, L465 (1996).

CA Models for Traffic Flow: Comparison with Empirical Single-Vehicle Data

W. Knospe[1], L. Santen[2], A. Schadschneider[2], and M. Schreckenberg[1]

[1] Physik von Transport und Verkehr, Gerhard-Mercator-Universität, Duisburg, Germany
[2] Institut für Theoretische Physik, Universität zu Köln, Germany

Abstract. Although traffic simulations with cellular-automata models give meaningful results compared with empirical data, highway traffic requires a more detailed description of the elementary dynamics. Based on recent empirical studies we present a modified Nagel-Schreckenberg cellular automaton model which incorporates both a slow-to-start and an anticipation rule, which takes into account especially brake lights. The focus in this article lies on the comparison with empirical single-vehicle data.

1 Introduction

For a long time the modelling of traffic flow phenomena was dominated by two theoretical concepts (for a review, see e.g., [1]): Microscopic car-following models and macroscopic models based on the analogy between traffic flow and the dynamics of compressible viscous fluids. Both approaches are still used widely by traffic engineers but for practical purposes they are often not suitable, e.g., an efficient implementation for computer simulations of large networks is not possible. Macroscopic models use a large number of parameters which have partly no counterpart within empirical investigations. Moreover, the information which can be obtained using macroscopic models is incomplete in the sense that quantities concerning individual cars cannot be introduced or derived directly.

In order to fill this gap cellular automata (CA) models have been invented [2]. CA's are microscopic models which are by design well suited for large-scale computer simulations. A comparison of the simulations with empirical data shows that already very simple approaches give meaningful results. In particular they can be used to simulate dense networks like cities [3] which are controlled by the dynamics at the intersections. However, for highway traffic a more detailed description of the dynamics seems to be necessary.

Recent empirical studies show the existence of metastable states in traffic dynamics and the occurrence of synchronised flow [4–6], which can be identified by vanishing cross-correlations of the local density and the local flow [7]. Moreover, a detailed analysis of single-vehicle data [7] revealed important facts for the microscopic modelling of traffic. The time-headway distribution shows two characteristic peaks. Small time-headways ($\approx$ 0.8 sec) are a result of cars or clusters of cars moving with small headway but large velocity, a time-headway of 2 sec can be identified with the drivers efforts for safety: it is recommended to drive with a headway of 2 sec. Additionally, the distance-headway gives the most important information for the adjustment of the car's speed for the correct definition of the car-car interaction. This is introduced in several models by the so-called optimal velocity (OV) curve. It has been shown that one universal optimal velocity curve for all density regimes does not exist, but individual curves for different densities can be calculated [8]. In fact, some model extensions of the CA model proposed by Nagel and Schreckenberg (NaSch) [2] exist which are capable to reproduce metastable states [9] or small time-headways [10,11], but up to now it it not possible to generate synchronised traffic and the correct microscopic properties mentioned above.

Here we propose a new CA model generalising the NaSch model and some earlier extensions. We compare our simulations with the corresponding data used in [7]. The simulation data are evaluated by an artificial counting loop, i.e., we measure the speed and the time-headway of the vehicles at a given link of the lattice. This data set is analysed using the methods suggested in [7]. In particular, the density is calculated via the relation $\rho = J/v$ where J and v are the mean flow and the mean velocity of cars passing the detector in a time interval of 1 minute. This dynamic estimate of the density gives correct results only if the velocity of the cars between two measurements is constant, but for accelerating or braking cars, e.g., in stop-and-go traffic, the results do not coincide with the real occupation. In addition to the aggregated data also the single-vehicle data of each passing car are analysed. Although the empirical data have been obtained on a two-lane highway, the simulations are performed on a single-lane road with one type of cars, because the empirical results show no systematic lane dependence which is a consequence of the applied speed limit.

In Sect. 2 we give a brief description of the new model definition consisting of the NaSch-rules and some extensions. Section 3 compares the simulation results of the new model with the corresponding empirical data. Finally, Sect. 4 concludes with a short summary and discussion.

2 New Approach

Traffic networks can be classified as complex systems with a large number of individually interacting agents. In contrast to urban traffic where the flow is dominated by intersections, traffic lights etc., car characteristics like different maximum velocities, acceleration capabilities and car lengths become important on highways. In order to allow for a more realistic modelling of these characteristics we reduce the cell length of the standard NaSch model (see [2] for a detailed description of the model) to a length of $l = 1.5$ m. The acceleration and randomisation remains unaltered with one site per time step of 1 sec which leads to a velocity discretisation of 5.4 km/h which is slightly above "comfortable" acceleration of about 1 m/sec^2 [12].

The update rules of the new model combine the original NaSch model and some recent extensions, namely a slow-to-start rule [9] and an anticipation term [10,11]. The slow-to-start rule allows to tune the velocity of the upstream front of a traffic jam directly. It turns out that for a realistic choice of the parameters the outflow of a jam does not achieve the capacity of the road. This empirical fact is known to lead to the existence of metastable states.

The next step towards a more realistic description of especially highway traffic is to introduce anticipation effects, i.e., the adjustment of speed also takes into account the expected behaviour of the leading vehicle. Anticipation leads to a much more efficient lane-usage in multi-lane traffic. Although both modifications significantly increase the realism of the simulations a complete description of traffic highway traffic is not yet possible. The main problem with the existing discrete models is that they fail to reproduce platoons of slow moving vehicles. These patterns are not as stable as in real traffic, i.e., the models overestimate the probability to form large compact jams.

This deficiency motivated us to prolong the range of interactions if a braking maneuver of the leading vehicle occurs or, more figurative, we equipped the vehicles with brake lights. The event driven interaction leads to a timely adjustment of speeds and therefore to a more coherent movement of the vehicles in dense traffic. We implemented the reaction to a brake light simply by an increased randomisation parameter $p_b > p$.

3 Validation of the Model

Obviously, the fundamental diagram of the new model coincides very well with the empirical data (Fig. 1). In comparison we observe a more narrow distribution of densities. This further narrowing is simply an artifact of the discretisation of the velocities which determines the upper limit of detectable densities.

The slow-to-start rule has been introduced in order to reduce the outflow of a jam. This rule is responsible for the formation of large jams at high densities. To measure the outflow we used a megajam initialisation. Obviously, the outflow is reduced considerably (inset of Fig. 1). Using the auto-correlation function of the measured local density of a system initialised with a megajam it is possible to determine the jam velocity. A detailed analysis leads to a value of about 12.75 km/h.

As a next step for the validation of the model we compared single-vehicle data of the simulation with the corresponding empirical data. In order to give a correct comparison of our simulation data with the data of Neubert *et al.* we tried to identify the three traffic states described in [7]. We therefore analysed the local data by means of the average velocity. A contiguous time series of minute averages above 25 m/sec was classified as free flow, otherwise as congested flow. In Fig. 1 the cross-covariance $cc(J, \rho)$ of the flow and the local measured density for different traffic states is also shown. In the free-flow regime the flow is strongly coupled to the density indicating that the average velocity is nearly constant. Also for large densities, in the stop-and-go regime, the flow is mainly controlled by density fluctuations. In the mean density region there is a transition between these two regimes. At cross-covariances in the vicinity of zero the fundamental diagram shows a plateau which supports the interpretation of [7] that synchronised flow leads to $cc(J, \rho) \approx 0$. In the further comparison of our

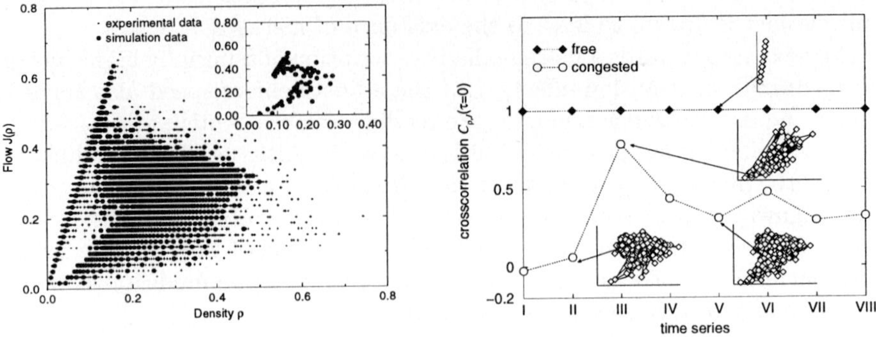

Fig. 1. Left: Comparison of the local fundamental diagram obtained by a simulation with the corresponding empirical fundamental diagram of Neubert *et al* [7]. The inset shows the outflow of a megajam. Right: Cross-covariance of the flow and the density for different global densities and homogeneous initialisation.

simulation with the corresponding empirical data we used these traffic states for synchronised flow data and congested states with $cc(J,\rho) > 0.7$ for stop-and-go data.

For the correct description of the car-car interaction the distance-headway (OV-curve) gives the most important information for the adjustment of the velocities. For densities in the free-flow regime it is obvious that the OV-curve (Fig. 2) deviates from the linear velocity-headway curve of the NaSch model. Due to anticipation effects smaller distances occur, so that driving with v_{max} is possible even within very small headways. This strong anticipation becomes weaker with increasing density and cars tend to have smaller velocities than the headway allows so that the OV-curve saturates for large distances. At headways of about 50 m the simulation data are in good agreement with the empirical ones, but for large distances the acceleration behaviour of the NaSch model cannot be suppressed so that the velocity increases with the headway. In Fig. 2 the time-headway distributions for different density regimes are shown. The time-headways are calculated via the relation $\Delta t = \Delta x/v$ with a resolution of 0.1 sec. Due to the discrete nature of the model large fluctuations occur. In the free-flow state the anticipation rule is responsible for time-headways smaller than 1 sec. The ability to anticipate the predecessors behaviour is getting weaker with increasing density so that small time-headways nearly vanish in the synchronised and the stop-and-go state. Two peaks arise in these states: The peak at 2 sec can be identified with the driver's efforts for safety: It is recommended to drive with a distance of about 2 sec. Nevertheless, with increasing density the characteristic peak of the NaSch-model at 1 sec (in the NaSch model the minimal time-headway is restricted to 1 sec) becomes dominant. The higher the density the stronger the peak structure is pronounced.

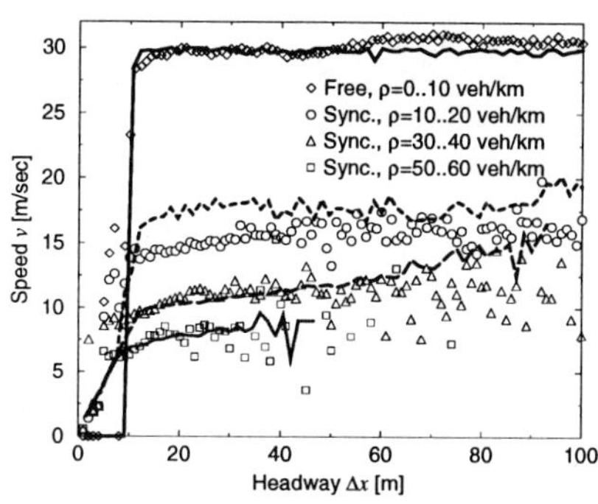

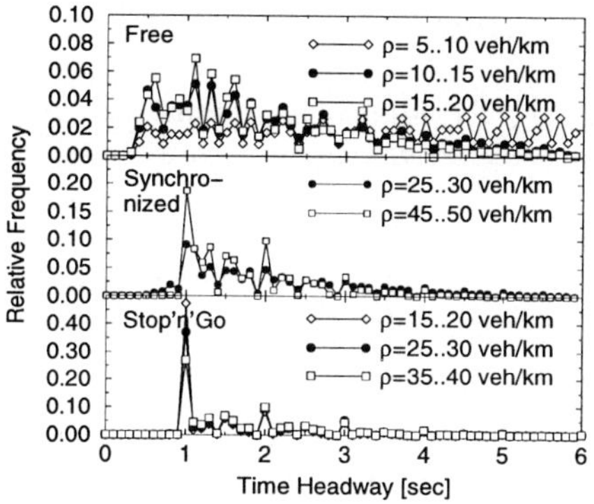

Fig. 2. Left: The mean speed chosen by the driver as a function of the gap to his predecessor. Comparison of simulations (lines) with empirical data (symbols) of Neubert *et al* [7]. Right: Time-headway distribution for different density regimes.

4 Summary and Discussion

Based on empirical data we tried to find a simple extension of the original NaSch model which is able to reproduce metastable states and synchronised flow as well as microscopic features like density-dependent OV-curves and characteristic time-headways.

First of all, the original NaSch cell length had to be reduced for a more realistic acceleration behaviour which is especially important on highways. For a more realistic car-car interaction anticipation terms seem to play a crucial role. On the one hand, anticipation of the predecessors movement in the next time step allows small time-headways and therefore high flows. On the other hand, braking anticipation by means of brake lights enables a driver to anticipate an imminent velocity reduction due to a jam. It is this braking anticipation which leads to synchronised states and to increased time-headways which results in plateaus in the local fundamental diagram.

Unfortunately, in this short contribution it is not possible to describe all the features of the new model [13]. For example, a finer discretisation of the NaSch model or the corresponding limit $v_{max} \to \infty$ leads to metastable states analogous to the VDR-model which can be characterised by an order parameter.

For a further validation of this approach it is necessary to extend the model to multi-lane traffic. The implementation of the new model in the online-simulation of the freeway network of North Rhine-Westphalia [14] should show its suitability for realistic traffic simulations.

References

1. D. Chowdhury, L. Santen, and A. Schadschneider, Curr. Sci. **77**, 411 (1999); in press, Phys. Rep.
2. K. Nagel and M. Schreckenberg, *A cellular automaton model for freeway traffic*, J. Physique I **2**, 2221 (1992).
3. J. Esser and M. Schreckenberg, *Microscopic simulation of urban traffic based on cellular automata*, Int. J. of Mod. Phys. C **8**, 1025 (1997).
4. B.S. Kerner, *Experimental features of self-organization in traffic flow*, Phys. Rev. Lett. **81**, 3797 (1998).
5. B.S. Kerner and H. Rehborn, *Experimental properties of phase transitions in traffic flow*, Phys. Rev. Lett. **79**, 4030 (1998).
6. B.S. Kerner and H. Rehborn, *Experimental features and characteristics of traffic jams*, Phys. Rev. E **53**, R1297 (1996).
7. L. Neubert, L. Santen, A. Schadschneider, and M. Schreckenberg, *Single-vehicle data of highway traffic: A statistical analysis,* Phys. Rev. E **60**, 6480 (1999).
8. L. Neubert, L. Santen, A. Schadschneider and M. Schreckenberg, *Statistical Analysis of Freeway Traffic*, in: *these proceedings.*
9. R. Barlovic, L. Santen, A. Schadschneider, and M. Schreckenberg, *Metastable states in cellular automata for traffic flow*, Eur. Phys. J **5**, 793 (1998).
10. W. Knospe, L. Santen, A. Schadschneider, and M. Schreckenberg, *Disorder effects in cellular automata for two-lane traffic*, Physica A **265**, 614 (1999).
11. C.L. Barrett and M. Wolinsky, *Emergent Local Control Properties in Particle Hopping Traffic Simulations*, in: *Traffic and Granular Flow*, D.E. Wolf, M. Schreckenberg, and A. Bachem, (Eds.), (World Scientific, 1996).
12. Institute of Transportation Engineers, (Traffic Engineering Handbook, Washington DC, 1992).
13. W. Knospe, L. Santen, A. Schadschneider, and M. Schreckenberg, in preparation.
14. O. Kaumann, K. Froese, R. Chrobok,J. Wahle, L. Neubert and M. Schreckenberg, *On-line Simulation of the Freeway Network of North Rhine-Westphalia*, in: *these proceedings.*

A New Cellular Automaton Model for City Traffic

A. Schadschneider[1], D. Chowdhury[1,2], E. Brockfeld[3], K. Klauck[1], L. Santen[1], and J. Zittartz[1]

[1] Institut für Theoretische Physik, Universität zu Köln, 50937 Köln, Germany
[2] Physics Department, I.I.T., Kanpur 208016, India
[3] Institut für Umweltsystemforschung, Universität Osnabrück, 49076 Osnabrück, Germany

Abstract. We present a new cellular automaton model of vehicular traffic in cities by combining ideas borrowed from the Biham-Middleton-Levine (BML) model of city traffic and the Nagel-Schreckenberg (NaSch) model of highway traffic. The model exhibits a dynamical phase transition to a completely jammed phase at a critical density which depends on the time periods of the synchronized signals.

1 Introduction

A one-dimensional cellular automaton (CA) model of highway traffic and a two-dimensional CA model of city traffic were developed independently by Nagel and Schreckenberg (NaSch) [1] and Biham, Middleton and Levine (BML) [2], respectively[1]. Highway traffic becomes gradually more and more congested in the NaSch model with the increase of density. Traffic jams appear because of the *intrinsic stochasticity* of the dynamics but no jam persists for ever. On the other hand, a first order phase transition takes place in the BML model at a finite non-vanishing density, where the average velocity of the vehicles vanishes discontinuously signaling complete jamming. In the BML model, the randomness arises only from the *random initial conditions*, as the dynamical rule for the movement of the vehicles is fully deterministic [2].

In the BML model a square lattice models the network of the streets. Each site of the lattice represents a crossing of an east-bound and a north-bound street; it can be empty or occupied by a vehicle moving either to the east or to the north. The dynamics is controlled by signals of period 1 such that at odd time steps only the east-bound vehicles are updated whereas at even time steps only the north-bound vehicles are updated, both in parallel. During the updating a vehicle moves one site forward if the next site ahead is not occupied by any other vehicle. In this simplest version of the model lane changes (e.g., by turning) are not possible and the number of vehicles on each street is conserved separately.

In the NaSch model a highway is represented by a one-dimensional lattice of cells that can accommodate not more than one vehicle at a time. Each vehicle

[1] For a review of the different approaches in modeling traffic flow we refer to [3] and references therein.

is characterized by a maximum velocity $v_{\max}$ and a randomization parameter p. The dynamics consists of four steps, each applied in parallel to all vehicles. In the first step all vehicles accelerate by 1 if they have not already reached the maximum velocity $v_{\max}$. Step 2 is the interaction step. Vehicles which have d empty cells in front and a velocity $v > d$ reduce their velocity to $v = d$ in order to avoid a crash. In step 3 the velocity is reduced by one unit with probability p. In step 4 the vehicles move forward v cells where v is the new velocity after the randomization step 3.

2 Definition of the Model

In the BML model the interplay between the vehicle dynamics and the time-scale set by the length of the signal period can not be studied. We therefore suggested a "unified" model [4] combining the BML model with the rules for the vehicle dynamics of the NaSch model. The lattice of our new model consists (in the simplest case) of N north-bound and N east-bound streets. The $N \times N$ crossings of these streets are arranged equidistantly. Between two consecutive crossings on a street there are $D - 1$ cells, i.e. each street has length $L = ND$ (see Fig. 1). The signals, installed at the crossings, are synchronized in such a way that all

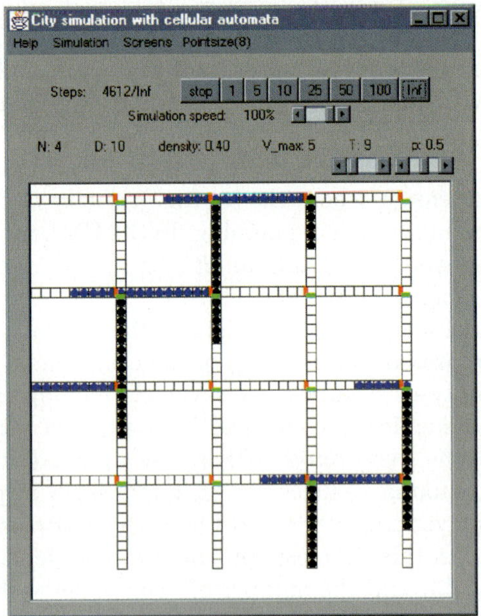

Fig. 1. Structure of the underlying lattice for $N = 4$ and $D = 10$. Shown is a typical jammed configuration of the vehicles. The east-bound and north-bound vehicles are represented by the blue and black symbols, respectively.

the signals remain green for the east-bound vehicles (and simultaneously, red for the north-bound vehicles) for a time interval T and then, simultaneously, all the signals turn red for the east-bound vehicles (and green for the north-bound vehicles) for the next T time steps before turning green again. This process is repeated so that there is a total time interval $2T$ between the beginning of two successive green (or red) phases of the signals.

As in the NaSch model the speed v of each vehicle can take one of the $v_{\max}+1$ integer values $v = 0, 1, ..., v_{\max}$. Suppose, v_n is the speed of the n-th vehicle at time t while moving either towards east or towards north. In the initial state of the system, N_x (N_y) vehicles are distributed among the east-bound (north-bound) streets. Here we only consider the case $N_x = N_y = N_v/2$ where N_v is the total number of vehicles. Since in the initial configuration the occupation of a crossing is strictly avoided, the global density is defined by $\rho = N_v/N^2(2D-1)$. Also, suppose d_n is the distance to the next vehicle in front while s_n denotes the distance to the nearest crossing in front of it. At each *discrete time* step $t \to t+1$, the arrangement of vehicles is updated *in parallel* according to the following "rules":

- Step 1 (Acceleration):
  $v_n \to \min(v_n + 1, v_{\max})$

- Step 2 (Deceleration due to other vehicles or signals):

  Case I: The signal is **red** for the n-th vehicle under consideration:
  $v_n \to \min(v_n, d_n - 1, s_n - 1)$
  Case II: The signal is **green** for the n-th vehicle under consideration:
  If the signal is going to turn to red in the next time-step then
  $v_n \to \min(v_n, d_n - 1, s_n - 1)$
  else $v_n \to \min(v_n, d_n - 1)$.

- Step 3 (Randomization):
  $v_n \to \max(v_n - 1, 0)$ with probability p

- Step 4 (Movement):
  $x_n \to x_n + v_n$.

Note that we have simplified Case II of Step 2 in comparison to [4]. This simplification does not change the overall behavior of the model [5].

These rules are not merely a combination of the BML and the NaSch rules but also involve some modifications. For example, unlike all the earlier BML-type models, a vehicle approaching a crossing can keep moving, even when the signal is red, until it reaches a site immediately in front of which there is either a halting vehicle or a crossing. Moreover, if $p = 0$ every east-bound (north-bound) vehicle can adjust speed in the deceleration stage so as not to block the north-bound (east-bound) traffic when the signal is red for the east-bound (north-bound) vehicles.

3 Results

The variations of $\langle v_x \rangle$ and $\langle v_y \rangle$ with time (see Fig. 2) as well as with c, D, T and p in the flowing phase are certainly more realistic than in the BML model [4]. In the case of $\rho < \rho_c$ and for $v_{\max} = 1$ the dynamics of the system can be described accurately by treating a single street with an improved version of the 2-cluster approximation [6], where the 2-cluster probabilities are equipped with a time and space dependence [5] (see Fig. 3).

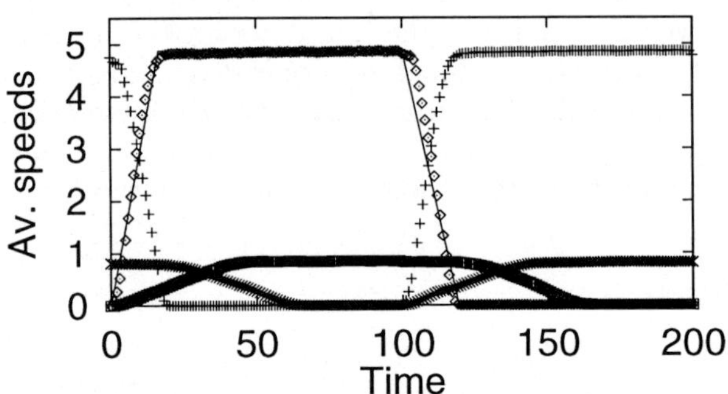

Fig. 2. Time-dependence of average speeds of vehicles. The symbols $+$, $\times$, $*$ and $\square$ correspond, respectively, to the average speeds $\langle v_x \rangle$, $\langle v_y \rangle$, and the fractions of vehicles with instantaneous speed $V = 0$, f_{x0} and f_{y0}, respectively. The common parameters are $v_{\max} = 5, p = 0.1, D = 100$, $T = 100$ and $c = 0.1$. The continuous line has been obtained from heuristic arguments given in [4].

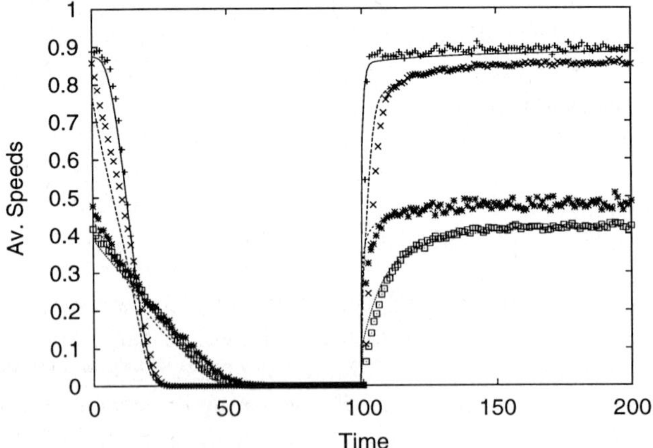

Fig. 3. Comparison between MC data and 2-cluster results. The common parameters are $v_{\max} = 1$, $D = 25$, $T = 100$ and $N = 4$. The solid lines correspond to 2-cluster results. The symbols $+, \times, *, \boxdot$ correspond to the ρ/p MC data sets $0.05/0.1, 0.25/0.1, 0.05/0.5, 0.25/0.5$ respectively.

The fundamental diagram (Fig. 4) also shows a rather complex behavior, at least for finite systems. E.g., the density corresponding to the maximum flux shifts to smaller densities with the decrease of T. Furthermore, the maximum throughput is a non-monotonic function of T in the "free-flowing" phase; this result may be of practical use in traffic engineering for maximizing the throughput.

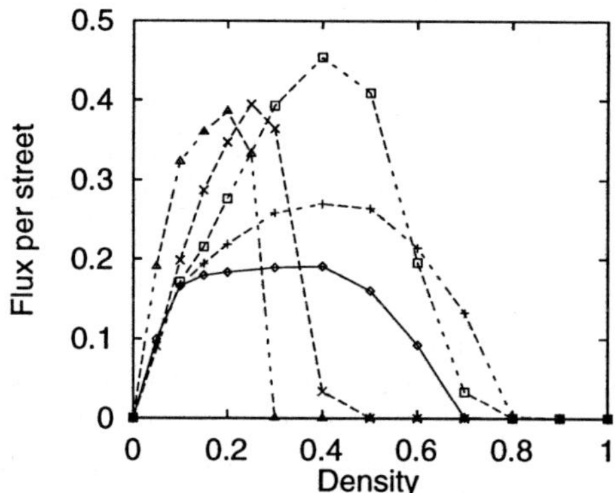

Fig. 4. Fundamental diagram for $v_{\max} = 5, p = 0.5$, $L = 100$, and $D = 20$. The symbols $\diamond, +, \Box, \times$ and Δ correspond, respectively, to $T = 100, 50, 20, 10, 4$.

A phase transition from the "free-flowing" dynamical phase to the completely "jammed" phase takes place in this model at a vehicle density ρ_c. The intrinsic stochasticity of the dynamics, which triggers the onset of jamming, is similar to that in the NaSch model, while the phenomenon of complete jamming through self-organization as well as the final jammed configurations (see Fig. 1) are similar to those in the BML model.

Due to the importance of finite-size and finite-time corrections it is not clear up to now how the critical density $\rho_c(D)$ depends on the dynamical parameters $v_{\max}$, p and T. It is possible that in the thermodynamic limit $N \to \infty$ the density ρ_c is completely determined by the structure of the underlying lattice, i.e. by D, as long as $p > 0$ which is necessary for the jamming transition to occur. In that case ρ_c would be independent of $v_{\max}$, p and T and the transition would be of 'geometrical' nature similar to the percolation transition. On the other hand, the transition could also be truly dynamical with ρ_c depending also on $v_{\max}$, p or T. The data obtained so far from the computer simulations (see Fig. 5) do not conclusively rule out either of these two possible scenarios.

The "unified" model has been formulated intentionally to keep it as simple as possible and at the same time capture some of the interesting features of the NaSch model as well the BML model. We believe that this model can be generalized (i) to allow traffic flow in both ways on each street which may consist of more than one lane, (ii) to make more realistic rules for the right-of-the-way

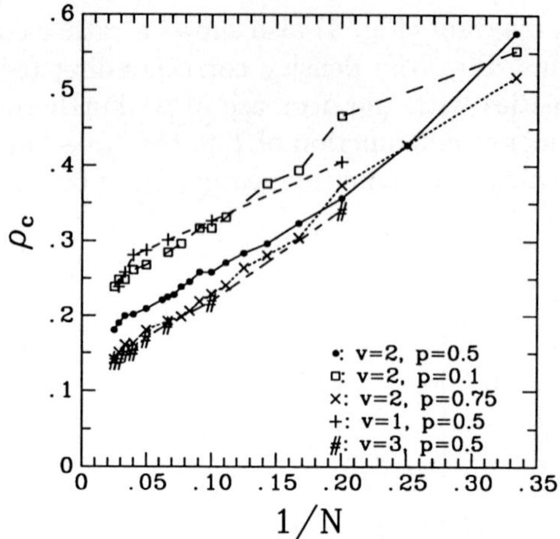

Fig. 5. Critical density ρ_c for different parameter combinations as function of the number of streets N. The common parameters are $D = 20$ and $T = 5$.

at the crossings and turning of the vehicles, (iii) to implement different types of synchronization or staggering of traffic lights, e.g. green-waves [7].

Acknowledgement. Part of this work has been supported by the SFB 341 (Köln-Aachen-Jülich).

References

1. K. Nagel and M. Schreckenberg, J. Phys. I France **2**, 2221 (1992).
2. O. Biham, A.A. Middleton, and D. Levine, Phys. Rev. A **46**, R6124 (1992).
3. D. Chowdhury, L. Santen, and A. Schadschneider, to be published, Phys. Rep.
4. D. Chowdhury and A. Schadschneider, Phys. Rev. E **59**, R1311 (1999).
5. D. Chowdhury, K. Klauck, L. Santen, A. Schadschneider, and J. Zittartz, to be published.
6. A. Schadschneider, in: *Traffic and Granular Flow '97,* M. Schreckenberg and D.E. Wolf, (Eds.), (Springer, Singapore, 1998).
7. P.M. Simon and K. Nagel, Phys. Rev. E **58**, 1286 (1998).

Stochastic Boundary Conditions in the Nagel-Schreckenberg Traffic Model

S. Cheybani[1,2], J. Kertész[2,3], and M. Schreckenberg[1]

[1] Physik von Transport und Verkehr, Gerhard-Mercator-Universität, 47048 Duisburg, Germany
[2] Department of Theoretical Physics, Technical University of Budapest, H-1111 Budapest, Hungary
[3] Laboratory of Computational Engineering, Helsinki University of Technology, FIN-02150 Espoo, Finland

Abstract. We consider the generalization of the asymmetric exclusion model (ASEP) with parallel update where cars can move with velocities $v \leq v_{max}$ and $v_{max} > 1$. For stochastic open boundary conditions we find a line of a first-order transition separating the free flow phase from the jammed phase. For maximum velocities $v_{max} \geq 3$ so-called "buffers" develop due to the hindrance an injected car feels from the front car at the beginning of the system. As a consequence, the phase diagram qualitatively differs from that for $v_{max} \leq 2$.

1 Introduction

Driven diffusive processes have been widely studied as prototypes of non-equilibrium systems [1–3]. A well-known modification of the basic one-dimensional diffusive system is the asymmetric exclusion process (ASEP) [4] which can be divided into four update classes (random-sequential, ordered-sequential, sub-lattice-parallel, and parallel) according to the order of the hopping of the cars as well as of injection and removal at the boundaries [5]. As it is common for traffic simulations [6–9] we will use parallel update in the following because this is the most effective among the four update types and shows the best congruence with real traffic data [10].

Comparing the ASEP with real traffic, however, it is obvious that phenomena like acceleration and slowing down are not included in the model. Here, cars either do not move at all or move one site per time step. It can therefore be said, that they move with maximum velocity $v_{max} = 1$. In order to get more realistic results, Nagel and Schreckenberg introduced a model for single lane traffic [11], where cars can move with different discrete integer velocities v, $0 \leq v \leq v_{max} > 1$.

According to the Nagel-Schreckenberg model the lane is divided into L cells of equal size and the time is also discrete. Each site is either empty or occupied by a car with velocity $v = 0, 1, \cdots, v_{max}$ (here: $v_{max} > 1$). All sites are simultaneously updated according to three successive steps (deterministic Nagel-Schreckenberg model):

1. Acceleration: increase v by 1 if $v < v_{max}$.
2. Slowing down: decrease v to v = d if necessary.
   (d: number of empty cells in front of the car).
3. Movement: move car v sites forward.

Open boundary conditions are defined in the following way: at site i = 0, that means out of the system a vehicle with probability α and with velocity $v = v_{max}$ is created. This car immediately moves according to the deterministic Nagel-Schreckenberg rules. If i = 1 is occupied by another car so that the velocity of the injected car on i = 0 is v = 0 then the injected car is deleted. At $i = L + 1$ a "blockage" occurs with probability 1 - β and causes a slowing down of the cars at the end of the system. Otherwise, with probability β, the cars simply move out of the system.

2 Phase Diagram

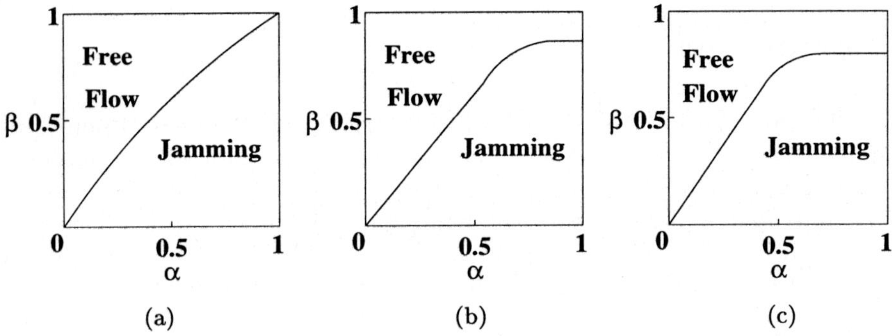

Fig. 1. Phase diagrams for the maximum velocity (a) $v_{max} = 2$, (b) $v_{max} = 3$, and (c) $v_{max} = 5$.

The phase diagrams for systems with maximum velocities $v_{max} = 2, 3, 5$ are shown in Figs. 1a-c. Figure 1a resembles the case $v_{max} = 1$ except for some deviations which are due to the fact that in systems with $v_{max} = 2$ we do not have a particle-hole symmetry as for $v_{max} = 1$. The course of the free flow - jamming border for the case $v_{max} = 3$, on the other hand, is very different (Fig. 1b). Here, the $\alpha = \beta$ - line does not separate the free flow and the jamming regime. Instead, the jamming regime is larger than the free flow regime, and for high extinction rates β cars freely move for *all* α. For the maximum velocity $v_{max} = 5$ these features are even stronger developed as it is obvious from Fig. 1c.

3 Current and Global Density

Considering the current q (Fig. 2a) and the global density $\overline{\rho}$ (Fig. 2b) for the injection rate $\alpha = 1$ and the system size $L = 1024$, we see that for $\mathrm{v}_{max} = 2$ these quantities behave similarly to the case $\mathrm{v}_{max} = 1$. For $\mathrm{v}_{max} \geq 3$ astonishing effects are observed which do not depend on the maximum velocity if $\mathrm{v}_{max} \geq 5$: Coming from low extinction rates β the current for $\mathrm{v}_{max} \geq 5$ increases proportionally to β and abruptly becomes constant at $\beta_c = 0.835$. For the global density, on the other hand, the transition seems to be continuous.

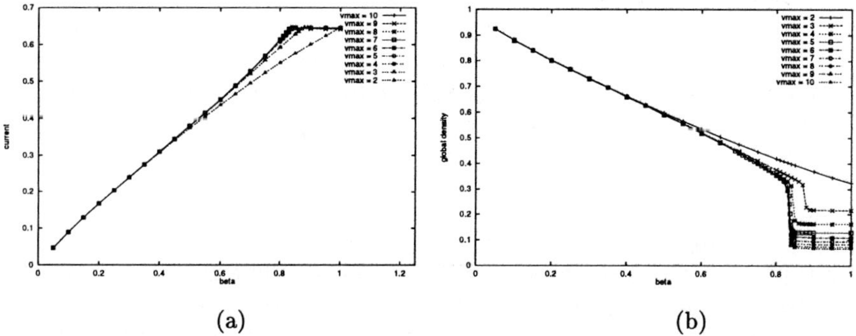

(a) (b)

Fig. 2. Current and global density for $\alpha = 1$, $L = 1024$, and $\mathrm{v}_{max} = 2, 3, \cdots, 10$.

Investigations of systems for large system sizes, however, show that the continuous change in the global density is just a finite size effect: although the curves are qualitatively the same as those in Fig. 2b the transition from free flow to jamming becomes more and more abrupt with increasing system size L what is a sign of a first-order phase transition. Furthermore, it turns out that the value of β_c is slightly smaller than for $L = 1024$. As a consequence from numerical investigations of systems with large L it is fair to assume that for $L \to \infty$ the current is described by:

$$\begin{aligned} \mathrm{q}(\alpha = 1, \beta < \frac{5}{6}, \mathrm{v}_{max} \geq 5) &= \frac{4}{5}\beta \qquad \text{jamming} \\ \mathrm{q}(\alpha = 1, \beta > \frac{5}{6}, \mathrm{v}_{max} \geq 5) &= \frac{2}{3} \qquad \text{free flow.} \end{aligned}$$

The corresponding global density is given by:

$$\begin{aligned} \overline{\rho}(\alpha = 1, \beta < \frac{5}{6}, \mathrm{v}_{max} \geq 5) &= 1 - \frac{4}{5}\beta \qquad \text{jamming} \\ \overline{\rho}(\alpha = 1, \beta > \frac{5}{6}, \mathrm{v}_{max} \geq 5) &= \frac{2}{3\mathrm{v}_{max}} \qquad \text{free flow.} \end{aligned}$$

For increasing system sizes current and global density converge against these values which can be calculated analytically.

4 Analytical Considerations

In order to get a better insight in the behavior of the current and the global density we consider the special case $\alpha = \beta = 1$. The car velocity is represented by numbers in brackets, (v) = (0), (1), ..., (v_{max}), and k connected unoccupied sites by the symbol x^k. The first number in brackets represents the car at i = 0 where cars are injected. Then we have for

$$
\begin{aligned}
t=0: &\quad (v_{max})\ x^L \\
t=1: &\quad (v_{max}-1)\ x^{v_{max}-1}\ (v_{max})x^{L-v_{max}} \\
t=2: &\quad (v_{max}-2)\ x^{v_{max}-2}\ (v_{max})\ x^{v_{max}}(v_{max})\ x^{L-2v_{max}} \\
&\quad \vdots
\end{aligned}
$$

After t = v_{max} time steps a self-repeating pattern establishes itself according to

$$
\begin{aligned}
&(2)\ x^2 \quad \cdots \quad (v_{max})\ x^{2(v_{max}-1)}\ (v_{max})\ x^{v_{max}}\ (v_{max})\ x^{2(v_{max}-1)}\ \ldots \\
&(1)\ x^1 \quad \cdots \quad (v_{max})\ x^{2(v_{max}-1)}\ (v_{max})\ x^{v_{max}}\ (v_{max})\ x^{2(v_{max}-1)}\ \ldots \\
&(0)\ (2)\ x^2 \cdots \quad (v_{max})\ x^{2(v_{max}-1)}\ (v_{max})\ x^{v_{max}}\ (v_{max})\ x^{2(v_{max}-1)}\ \ldots
\end{aligned}
$$

This is perhaps astonishing because we naively would expect v_{max} unoccupied sites between two neighboring cars for $\alpha = 1$. Actually, there are also spaces consisting of 2(v_{max}-1) sites what is a consequence of the hindrance the injected cars feel from the front car at the beginning of the system. In other words, v_{max}-2 additional sites - which we call "buffers" - occur playing an important role for systems with maximum velocity $v_{max} \geq 3$ as we will see below.

It follows from the self-repeating pattern that the distance between two neighboring cars driving with v_{max} is alternately $d_1 = v_{max}$ and $d_2 = 2(v_{max} - 1)$, i.e., buffers occur only for $v_{max} \geq 3$. $v_{max} = 2$ is a special case behaving similarly to $v_{max} = 1$. It is therefore no surprise that the corresponding phase diagram, the global density, and the current resembles the case $v_{max} = 1$. If finite size effects are left out of consideration the current is obviously given by

$$q(\alpha = \beta = 1, v_{max} > 1) = \frac{2}{3}$$

and the global density by

$$\overline{\rho}(\alpha = \beta = 1, v_{max} > 1) = \frac{2}{3v_{max}}$$

what coincides with numerical results.

In the following, the effect of the buffers for the injection rate $\alpha = 1$ is investigated which can be observed at the end of the system. For that purpose we start with the special case $\alpha = \beta = 1$. By simple analytical considerations it turns out that a self-repeating pattern

$$
\begin{aligned}
&\cdots\ x^{2(v_{max}-1)}\ (v_{max})\ x^{v_{max}}\ (v_{max})\ x^{2(v_{max}-1)}\ (v_{max}) \\
&\cdots\ (v_{max})\ x^{2(v_{max}-1)}\ (v_{max})\ x^{v_{max}}\ (v_{max})\ x^{v_{max}-1} \\
&\cdots\ (v_{max})\ x^{v_{max}}\ (v_{max})\ x^{2(v_{max}-1)}(v_{max})\ x^{v_{max}}
\end{aligned}
$$

establishes itself at the end of the system, too. It is important to mention that - due to $\beta = 1$ - no blockage occurs at all at the right boundary and that the buffers reach the right boundary with the rate $\alpha_{\text{buffer}} = \frac{1}{3}$.

(a)	$\cdots x^{2(v_{max}-1)}$	(v_{max})	$x^{v_{max}}$	(v_{max})	$x^{2(v_{max}-1)}$	(v_{max})	
	$\cdots (v_{max})$	$x^{2(v_{max}-1)}$	(v_{max})	$x^{v_{max}}$	(v_{max})	$x^{v_{max}-1}$	
	$\cdots (v_{max})$	$x^{v_{max}}$	(v_{max})	$x^{2(v_{max}-1)}$	(v_{max})	$x^{v_{max}}$	
	$\cdots x^{2(v_{max}-1)}$	(v_{max})	$x^{v_{max}}$	(v_{max})	$x^{2(v_{max}-1)}$	(0)	← blockage!
	$\cdots x^{2(v_{max}-1)}$	(v_{max})	$x^{v_{max}}$	$(v_{max}\text{-}2)$	$x^{v_{max}-2}$	(1)	
	$\cdots (v_{max})$	$x^{2(v_{max}-1)}$	$(v_{max}\text{-}2)$	$x^{v_{max}-2}$	$(v_{max}\text{-}1)$	x^1	
	$\cdots (v_{max})$	$x^{v_{max}}$	(v_{max})	$x^{2(v_{max}-2)}$	$(v_{max}\text{-}1)$	x^2	------------------
	$\cdots (v_{max})$	$x^{2(v_{max}-1)}$	(v_{max})	$x^{v_{max}}$	(v_{max})	$x^{v_{max}-1}$	from here on
	$\cdots (v_{max})$	$x^{v_{max}}$	(v_{max})	$x^{2(v_{max}-1)}$	(v_{max})	$x^{v_{max}}$	nothing reminds
	$\cdots x^{2(v_{max}-1)}$	(v_{max})	$x^{v_{max}}$	(v_{max})	$x^{2(v_{max}-1)}$	(v_{max})	of the disturbance

(b)	$\cdots (v_{max})$	$x^{2(v_{max}-1)}$	(v_{max})	$x^{v_{max}}$	(v_{max})	$x^{v_{max}-1}$	
	$\cdots (v_{max})$	$x^{v_{max}}$	(v_{max})	$x^{2(v_{max}-1)}$	(v_{max})	$x^{v_{max}}$	
	$\cdots x^{2(v_{max}-1)}$	(v_{max})	$x^{v_{max}}$	(v_{max})	$x^{2(v_{max}-1)}$	(v_{max})	
	$\cdots (v_{max})$	$x^{2(v_{max}-1)}$	(v_{max})	$x^{v_{max}}$	$(v_{max}\text{-}1)$	$x^{v_{max}-1}$	← blockage!
	$\cdots x^{v_{max}}$	(v_{max})	$x^{2(v_{max}-1)}$	$(v_{max}\text{-}1)$	$x^{v_{max}-1}$	(v_{max})	
	$\cdots (v_{max})$	$x^{v_{max}}$	(v_{max})	$x^{2v_{max}-3}$	(v_{max})	x^1	------------------
	$\cdots (v_{max})$	$x^{2(v_{max}-1)}$	(v_{max})	$x^{v_{max}}$	(v_{max})	$x^{v_{max}-1}$	from here on
	$\cdots (v_{max})$	$x^{v_{max}}$	(v_{max})	$x^{2(v_{max}-1)}$	(v_{max})	$x^{v_{max}}$	nothing reminds
	$\cdots x^{2(v_{max}-1)}$	(v_{max})	$x^{v_{max}}$	(v_{max})	$x^{2(v_{max}-1)}$	(v_{max})	of the disturbance

(c)	$\cdots (v_{max})$	$x^{v_{max}}$	(v_{max})	$x^{2(v_{max}-1)}$	(v_{max})	$x^{v_{max}}$	
	$\cdots x^{2(v_{max}-1)}$	(v_{max})	$x^{v_{max}}$	(v_{max})	$x^{2(v_{max}-1)}$	(v_{max})	
	$\cdots (v_{max})$	$x^{2(v_{max}-1)}$	(v_{max})	$x^{v_{max}}$	(v_{max})	$x^{v_{max}-1}$	
	$\cdots (v_{max})$	$x^{v_{max}}$	(v_{max})	$x^{2(v_{max}-1)}$	(v_{max})	$x^{v_{max}}$	← blockage!
	$\cdots x^{2(v_{max}-1)}$	(v_{max})	$x^{v_{max}}$	(v_{max})	$x^{2(v_{max}-1)}$	(v_{max})	
	$\cdots (v_{max})$	$x^{2(v_{max}-1)}$	(v_{max})	$x^{v_{max}}$	(v_{max})	$x^{v_{max}-1}$	
	$\cdots (v_{max})$	$x^{v_{max}}$	(v_{max})	$x^{2(v_{max}-1)}$	(v_{max})	$x^{v_{max}}$	

$\Rightarrow$ no effect of disturbance

Fig. 3. Disturbance in the space-time diagram for $\alpha = \beta = 1$ (at the end of the system). At one time step there is a blockage at i $= L + 1$.

We consider a slightly smaller extinction rate now by working a "disturbance" in the $\alpha = \beta = 1$ pattern, i.e., by placing a single blockage at the end of the system. As the self-repeating pattern consists of three time steps we have three possibilities to place the disturbance. In Fig. 3 the effect is illustrated for $v_{max} \geq 5$. Obviously, the movement of the cars does not change at all for possibility (c). For (a) and (b), however, the v_{max} - 2 additional sites (resulting from the hindrance the cars feel at the beginning of the system from the front car) play an important role at the end of the system as they have the effect of a "buffer" against the influence of the right boundary. It can be seen from Fig. 3 that two buffers are necessary to neutralize the blockage effect. Therefore, as long as (1-β) $< \frac{1}{2}\, \alpha_{\text{buffer}} = \frac{1}{6}$ a jamming wave cannot develop.

5 Conclusions

Systems with open boundaries where cars deterministically move with maximum velocity $v_{max} > 1$ show interesting features. As for the case $v_{max} = 1$ (ASEP) a free flow and a jammed phase can be distinguished from each other and the transition between the two phases is of first-order. There are, however, significant differences to the ASEP mainly resulting from the existence of so-called "buffers": As a consequence of the hindrance an injected car feels from the front car spaces $> v_{max}$ develop for high injection rates α. That means, in addition to the expected v_{max} sites further sites occur called "buffers" as they have a buffer effect at the end of the system. Due to the buffers the development of jamming waves is suppressed up to an extinction rate $\beta = \frac{5}{6}$ (for high α and $v_{max} \geq 5$) and this buffer effect is responsible for the characteristic course of the free flow – jamming border for phase diagrams with $v_{max} \geq 3$.

References

1. R.K.P. Zia, B. Shaw, B. Schmittmann, and R.J. Astalos, Phys. Rep. **301**, 45 (1998).
2. J. Krug, Phys. Rev. Lett. **67**, 1882 (1991).
3. J. Krug and P.A. Ferrari, J. Phys. A **29**, L465 (1996).
4. B. Derrida, E. Domany, and D. Mukamel, J. Stat. Phys. **69**, 667 (1992).
5. N. Rajewsky, L. Santen, A. Schadschneider, and M. Schreckenberg, J. Stat. Phys. **92**, 151 (1998).
6. L.G. Tilstra and M.H. Ernst, J. Phys. A **31**, 5033 (1998).
7. M.R. Evans, N. Rajewsky, and E.R. Speer, J. Stat. Phys. **95**, 45 (1999).
8. J. de Gier and B. Nienhuis, Phys. Rev. E **59**, 4899 (1999).
9. A. Benyoussef, H. Chakib, and H. Ez-Zahraouy, Eur. Phys. J. B **8**, 275 (1999).
10. M. Schreckenberg, A. Schadschneider, K. Nagel, and N. Ito, Phys. Rev. E **51**, 2939 (1995).
11. K. Nagel and M. Schreckenberg, J. Phys. I France **2**, 2221 (1992).

A Stochastic Multi–Cluster Model of Freeway Traffic

J. Kaupužs[1] and R. Mahnke[2]

[1] Institute of Mathematics and Computer Science, University of Latvia, 29 Rainja Boulevard, 1459 Riga, Latvia
[2] Universität Rostock, Fachbereich Physik, 18051 Rostock, Germany

Abstract. A stochastic approach based on the Master equation is proposed to describe the process of formation and growth of car clusters in traffic flow in analogy to usual aggregation phenomena such as the formation of liquid droplets in supersaturated vapour. We have extended the stochastic theory of freeway traffic allowing a coexistence of many clusters on a one–lane circular road. Analytical equations have been derived for calculation of the stationary cluster distribution and related physical quantities of an infinitely large system of interacting cars. If the probability per time to decelarate a car without an obvious reason tends to zero in an infinitely large system, our multi–cluster model behaves essentially in the same way as the one–cluster model presented at TRAFFIC AND GRANULAR FLOW '97. In particular, there are three different regimes of traffic flow (free jet of cars, coexisting phase of jams and isolated cars, highly viscous heavy traffic) and two phase transitions between them. In a general case, some qualitative differences in the behaviour of these two models have been observed.

1 Introduction

The formation and growth of clusters is a widely known phenomenon in physics. We mention condensation of liquid droplets in a supersaturated vapour [6]. The formation of car clusters (jams) at overcritical densities in traffic flow is an analogous phenomenon in the sense that cars can be considered as interacting particles [5], and the clustering process can be described by similar equations. In particular, the probability that the system has a given cluster distribution at a time t in both cases can be described by the stochastic Master equation. The transition probabilities depend on the specific physical model under consideration. It should be noted that the spontaneous emergence of car clusters has been studied by different authors (see Proceedings TGF [7,8]) based on different models and approaches. In spite of the complexity of real traffic [1], we believe that some general features, such as spontaneous formation of jams and some general scaling properties of traffic flow exist, which can be described and understood by relatively simple models.

Our purpose is to extend and improve the one–cluster stochastic model of a one–lane circular road, proposed and developed in Refs. [2–4], allowing the coexistence of many clusters.

2 The Model

Here we consider a model of traffic flow on a one–lane circular road of length L. For convenience, the length L is choosen

$$L = M(\ell + \Delta x_{\mathrm{clust}}), \tag{1}$$

where ℓ is the effective length of a car and $\ell + \Delta x_{\mathrm{clust}}$ is the distance between the front bumpers of two neighbouring cars in a jam (car cluster), as defined in our previous work [3], and M is a natural number. The distance between the front bumpers of two neighbouring cars, in general, is $\ell + \Delta x$. The total number of cars N is eliminated by $N \leq M$, where $N = M$ corresponds to the marginal case when the jam with car density $\varrho_{\mathrm{clust}} = 1/(\ell + \Delta x_{\mathrm{clust}})$ exists over the whole road. Similarly as in Ref. [3], we have used the optimal velocity approximation to describe the behaviour of individual drivers depending on the local density of cars or on headway Δx. The maximal velocity of each car is v_{max}. The desired (optimal) velocity v_{opt}, depending on the distance between two cars Δx, is given in dimensionless variables $w_{\mathrm{opt}} = v_{\mathrm{opt}}/v_{\mathrm{max}}$ and $\Delta y = \Delta x/\ell$ by the formula

$$w_{\mathrm{opt}}(\Delta y) = \frac{(\Delta y)^2}{d^2 + (\Delta y)^2}, \tag{2}$$

where the parameter $d = D/\ell$ is the interaction distance. D is the distance between two cars corresponding to the velocity value $v_{\mathrm{max}}/2$.

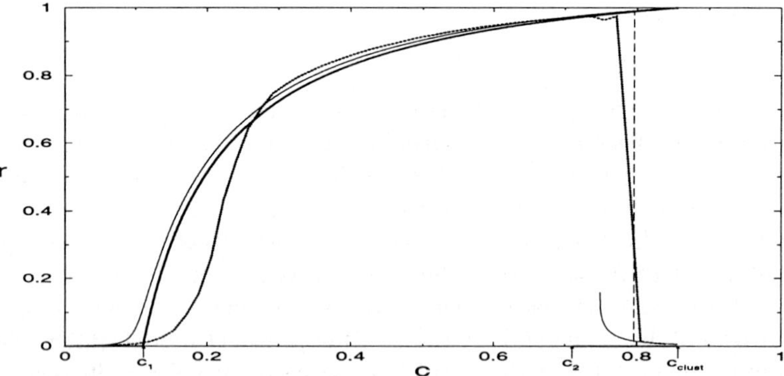

Fig. 1. The relative part of congested cars r vs. dimensionless density c. Thin solid line: $p = 0.001$, $M = \infty$; thick solid line: $p = +0$, $M = \infty$; dotted line: MC simulation for $p = 0.001$, $M = 50$; vertical dashed line indicates c_{jump}.

In distinction to our previous model [2–4], where one car cluster exists at any time, here we consider a more realistic multi–cluster case where the total number of clusters of congested cars (jams)

$$N_{\mathrm{cl}} = \sum_{k=1}^{N} N_k \tag{3}$$

may be varied. N_k is the number of clusters of size k, i.e., those consisting of k cars. In principle, we allow an absence of any congestion corresponding to $N_k = 0$ for all k. Some stochastic event or perturbation of the free traffic flow is necessary to initiate formation of a new cluster. Such stochastic events are simulated assuming that any car belonging to the free flow can reduce its velocity to $v_{\mathrm{opt}}(\Delta x_{\mathrm{clust}})$, i.e., become a single congested car or a cluster of size $k = 1$. The probability of such an event per time for a given free car is w_+^*. A cluster of size 1 appears also when a two–car cluster is reduced by one car. In this case cluster with $k = 1$ is a car which still have not accelerated after this event. In any case, the cluster of size 1 in our model is defined as a single car moving with velocity $v_{\mathrm{opt}}(\Delta x_{\mathrm{clust}})$. In such a way, the total number n of congested cars is

$$n = \sum_{k=1}^{N} k N_k \,, \tag{4}$$

and the number of free cars is $n_{\mathrm{free}} = N - n$. According to our definition, the length of the cluster of size k is $\ell k + (k-1)\Delta x_{\mathrm{clust}}$, which means that the total length of the congested part of the road is

$$L_{\mathrm{clust}} = \ell n + (n - N_{\mathrm{cl}})\Delta x_{\mathrm{clust}} \,. \tag{5}$$

Thus, with account for Eq. (1), the average distance $\Delta x_{\mathrm{free}} = \ell \Delta y_{\mathrm{free}}$ between two cars outside the jam (or free cars) distributed over the free part of the road with length $L_{\mathrm{free}} = L - L_{\mathrm{clust}}$ is given by

$$\Delta y_{\mathrm{free}}(n, N_{\mathrm{cl}}) = \frac{M - N + (M - n + N_{\mathrm{cl}})\Delta y_{\mathrm{clust}}}{N - n + N_{\mathrm{cl}}} \,, \tag{6}$$

where $\Delta y_{\mathrm{clust}} = \Delta x_{\mathrm{clust}}/\ell$. In the particular one–cluster case ($N_{\mathrm{cl}} = 1$) Eqs. (5) and (6) agree with corresponding formulae in Ref. [3].

The traffic flow is described as a stochastic process where adding a vehicle to a given car cluster (any of N_{cl} clusters) is characterized by a transition frequency (attachment probability per time unit) $w_+(n, N_{\mathrm{cl}})$ and the opposite process by a frequency $w_-(n, N_{\mathrm{cl}})$. The stochastic variables are N_k with $k = 1, 2, \ldots, N$, whereas the transition frequencies depend on n and N_{cl}, as discussed below. We have assumed that the free cars are distributed uniformly over the spacings between clusters, i.e., all these parts of the free road are characterised by the same mean headway $\Delta y_{\mathrm{free}}(n, N_{\mathrm{cl}})$ defined by Eq. (6), which allows us to use the ansatz for transition frequencies proposed in Ref. [3]. However, at $n_{\mathrm{free}} < N_{\mathrm{cl}}$ the ansatz for w_+ is corrected, taking into account that some of N_{cl} parts of the free road contain no cars. We have introduced the probability, represented by theta (step) function,

$$R(n, N_{\mathrm{cl}}) = 1 + (n_{\mathrm{free}}/N_{\mathrm{cl}} - 1)\,\theta(N_{\mathrm{cl}} - n_{\mathrm{free}}), \tag{7}$$

that at a given time moment there exists at least one car in the considered part of the free road, assuming that the distribution of free cars is maximally uniform. In such a way, our ansatz for the transition frequencies reads

$$w_+(n, N_{\rm cl}) = \frac{b}{\tau} \frac{w_{\rm opt}(\Delta y_{\rm free}(n, N_{\rm cl})) - w_{\rm opt}(\Delta y_{\rm clust})}{\Delta y_{\rm free}(n, N_{\rm cl}) - \Delta y_{\rm clust}} R(n, N_{\rm cl}) \tag{8}$$

$$w_-(n, N_{\rm cl}) = 1/\tau = \text{const} , \tag{9}$$

where $b = v_{\max}\tau/\ell$ denotes a dimensionless parameter, and τ is a time constant which can be interpreted as the waiting time for the escape (detachment) of the first car out of the jam into free flow [3]. Eqs. (7) and (8) ensure that $w_+(N, N_{\rm cl}) = 0$, therefore n cannot become larger than N. We have excluded any merging and splitting of clusters which is usually also done to describe aggregation in supersaturated systems like droplets [6]. Because in our model $\Delta y_{\rm clust}$ is strictly constant, such processe are impossible due to purely geometrical aspects.

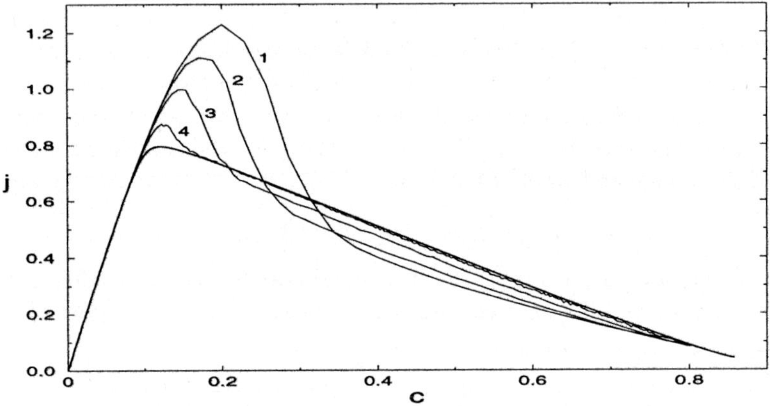

Fig. 2. The finite size effect on the flux–density diagram. Results of Monte–Carlo simulations at $p = 0.001$ and different sizes of the system: $M = 30$ (curve 1), $M = 50$ (curve 2), $M = 100$ (curve 3), and $M = 300$ (curve 4). The thicker smooth line, calculated from analytical equations, corresponds to $p = 0.001$, $M = \infty$.

The stochastic variables N_k may be considered as components of an N–dimensional vector $\mathbf{q} = \sum_k N_k \mathbf{q}_k$ where $\mathbf{q}_k$ is a unit vector the i–th component of which is $\delta_{i,k}$. In such a notation, the stochastic Master equation describing the evolution of the probability distribution function $P(\mathbf{q}, T)$ with the dimensionless time $T = t/\tau$ reads

$$\begin{aligned}\frac{1}{\tau}\frac{dP(\mathbf{q},T)}{dT} &= (N - n + 1)\, w_+^*\, P(\mathbf{q} - \mathbf{q}_1, T) \\ &+ \sum_{k=2}^{N} (N_{k-1} + 1) w_+(n-1, N_{\rm cl}) P(\mathbf{q} + \mathbf{q}_{k-1} - \mathbf{q}_k, T)\end{aligned}$$

$$+ (N_1 + 1)\, w_-(n+1, N_{\mathrm{cl}}+1)\, P(\mathbf{q}+\mathbf{q}_1, T) \tag{10}$$
$$+ \sum_{k=1}^{N-1} (N_{k+1}+1) w_-(n+1, N_{\mathrm{cl}}) P(\mathbf{q}+\mathbf{q}_{k+1}-\mathbf{q}_k, T)$$
$$- \left[(N-n)\, w_+^* + N_{\mathrm{cl}}\, (w_+(n, N_{\mathrm{cl}}) + w_-(n, N_{\mathrm{cl}}))\right] P(\mathbf{q}, T)\,.$$

3 Results and Discussion

Here we discuss our analytical results for an infinite system and the results of Monte–Carlo (MC) simulation of stochastic trajectories for finite systems. Based on Eq. (10), the stationary cluster distribution function at $M \to \infty$

$$C(k) = \langle N_k/M \rangle = C(0)\, p\, Q^{k-1} \tag{11}$$

has been obtained from the balance conditions between concentrations of clusters of different sizes, where $C(0) = n_{\mathrm{free}}/M$, $p = \tau w_+^*$ is the stochastic perturbation parameter, and Q is given by the equation $Q(\Delta y_{\mathrm{free}} - \Delta y_{\mathrm{clust}}) = b[w_{\mathrm{opt}}(\Delta y_{\mathrm{free}}) - w_{\mathrm{opt}}(\Delta y_{\mathrm{clust}})]$ as a function of the dimensionless density $c = N\ell/L$. The relative part of the congested cars $r = n/N = p/(p + (1-Q)^2)$ and the average cluster size $s = 1/(1-Q)$ correspond to the distribution (11). The same set of dimensionless control parameters $b = 8.5$, $d = 13/6$, and $\Delta y_{\mathrm{clust}} = 1/6$ has been used in all calculations. These values have been found in Ref. [3] by matching the theory with experimental data from German highways [1]. The stationary value of r vs. c is shown in Fig. 1. In this Figure the analytical solutions in the limit $\lim\limits_{p\to 0} \lim\limits_{M\to\infty}$ (first the limit $M \to \infty$ is found at a given $p > 0$), are shown by thick solid lines. The obtained analytical results in this limit clearly shows the existence of three different regimes of traffic flow, like in the one–cluster model [2,3], i.e., free flow of cars at small densities ($c < c_1$), congested traffic or coexisting (cluster) phase at intermediate densities ($c_1 < c < c_{\mathrm{jump}}$, where c_{jump} is shown by a vertical dashed line), and highly viscous overcrowded homogeneous state at high densities ($c_{\mathrm{jump}} < c < c_{\mathrm{clust}}$, where $c_{\mathrm{clust}} = 1/(1 + \Delta y_{\mathrm{clust}})$). Note that c_1 and c_2 in Fig. 1 are given by $c_{1,2} = 1/(1 + \Delta y_{1,2})$ where $\Delta y_{1,2}$ are roots of the equation $Q = 1$ solved with respect to Δy_{free}.

It is evident that there is a breakpoint at the first critical density $c = c_1$, and spontaneous formation of infinitely large clusters takes place at $c > c_1$. This phenomenon can be understood in analogy with the formation of liquid droplets in supersaturated vapour [6]. The singularity appears only if $p \to 0$, whereas at finite values of p (i.e., at $p = 0.001$ shown by thin solid lines) there is no sharp phase transition in the vicinity of $c = c_1$ and, in distinction to the one–cluster model, the average cluster size is always finite. It is evident that both at $p \to 0$ and at a finite p there are two solutions at large densities ($c_2 < c < c_{\mathrm{clust}}$): that with larger values of r corresponds to the cluster phase, whereas another one reflects the homogeneous state.

Following an analogy with the one–cluster model [2,3], a jump–like phase transition between these two states takes place at some density c_{jump}. We believe that the value $c_{\text{jump}} \simeq 0.796$ extracted from the one–cluster model and shown in Fig. 1 by a vertical dashed line represents a reasonable estimate for our multi–cluster model. The existence of jump at $c \simeq 0.796$ is confirmed by the results of MC simulation shown in the Figure by the dotted line. The simulation results refer to a finite–size system with $M = 50$ and $p = 0.001$ and reflects the values of r obtained by an averaging over a time interval $T = 200000$ to 500000 with the initial condition $n = N$ and $N_{\text{cl}} = 1$.

The stationary flux–density (fundamental) diagram of traffic flow has been calculated. In the limit $p \to 0$ at $M = \infty$ this diagram is identical to that for the one–cluster model presented in Refs. [2,3]. We have revealed an interesting finite–size effect, which has not been observed in the one–cluster model. In Fig. 2 the dimensioless flux $j = J\tau$ vs. c (J is the flux) is shown at $p = 0.001$. The solution for an infinite system is depicted by a smooth relatively thicker solid line. By thin solid lines, results of MC simulation are shown for finite roads of several lengths ($M = 30$, 50, 100, and 300) obtained by averaging over a time interval $T = 200000$ to 500000. As we see, in smaller systems the average flux is remarkably increased at densities slightly above c_1. We think that this is due to the stochastic fluctuations which are more important in finite systems as compared to the case $M \to \infty$ where the stationary cluster distribution (11) is determined and does not fluctuate.

References

1. B.S. Kerner and H. Rehborn, *Experimental properties of complexity in traffic flow*, Phys. Rev. E **53**, R4275 (1996).
2. R. Mahnke and J. Kaupužs, *One more fundamental diagram of traffic flow*, Presentation at second workshop on: *Traffic and Granular Flow*, Duisburg, October 1997, see [7], p. 439, 447.
3. R. Mahnke and J. Kaupužs, *Stochastic theory of freeway traffic*, Phys. Rev. E **59**, 117 (1999).
4. R. Mahnke and N. Pieret, *Stochastic master–equation approach to aggregation in freeway traffic*, Phys. Rev. E **56**, 2666 (1997).
5. I. Prigogine and R. Herman, *Kinematic Theory of Vehicular Traffic,* (Elsevier, New York, 1971).
6. J. Schmelzer, G. Röpke, and R. Mahnke, *Aggregation Phenomena in Complex Systems,* (WILEY–VCH, Weinheim, 1999).
7. M. Schreckenberg and D.E. Wolf, (Eds.), *Traffic and Granular Flow '97,* (Springer, Singapore, 1998).
8. D.E. Wolf, M. Schreckenberg, and A. Bachem, (Eds.), *Traffic and Granular Flow,* (World Scientific Publ., Singapore, 1996).

Granular Dynamics

Jamming Patterns and Blockade Statistics in Model Granular Flows

E. Clément, G. Reydellet, F. Rioual, B. Parise, V. Fanguet, J. Lanuza, and E. Kolb

Laboratoire des Milieux Désordonnés et Hétérogènes, UMR7603 - Université Pierre et Marie Curie - Boîte 86, 4 Place Jussieu, 75005 Paris, France

Abstract. We report experimental results on two model granular media submitted to gravity flows, i.e., a bidimensional hopper and a vertical column. For the hopper, the statistics of the flow blockade is monitored. For the vertical column, we study the flowing dynamics and we show, for narrow openings, regimes of density waves and full blockade depending on the dissipative character of the grains. For the least dissipative grains the density, velocity and temperature fields are measured and compared with the prediction of a granular hydrodynamics theory.

1 Introduction

Discharge of grains in silos, hoppers and tubes is an important practical issue. The correct assessment of the flow fields needed to achieve an optimal design and to avoid nuisance such as blockage or partial discharge, is currently limited by a lack of fundamental knowledge in the granular rheological properties [1]. For dry granular materials, a very versatile behavior is observed: the flow status depends on the dissipative nature of the granular interactions, the modes of solicitation and the boundary conditions. In the limit of quasi-static solicitations and dense packing, flows are often described using an adaptation of the Coulomb plasticity theory, which is the basis for most practical engineering approaches [2]. For rapid and moderately dense flows, the pioneering work of Bagnold [3] describes transfer of momentum essentially mediated by collisions. This regime last was rigorously derived using an adaptation of the gas kinetic theory to the case of locally dissipative materials such as colliding grains [4,6]. This approach leads to a well defined "granular hydrodynamics" [5] with a continuous description in terms of density, velocity and temperature fields.

For flows in containers, experimental and numerical works show many surprising features. For example density waves were evidenced inside hoppers [17] and in closed tubes [18,19]. In the last case, the presence of air was identified as a crucial parameter to trigger density waves flows. On the other end, numerical simulations of colliding dissipative particles (without a surrounding fluid) have displayed complicated wave patterns [20] and subsequent analysis was proposed, based on an effective traffic-jam theory [7] or on a cellular automaton simulations [8]. Experiments on 2D granular layers rolling down an incline also exhibit complex density wave patterns [9,21], but in this case, the frustrations of the rolling motion due to the multiple particle interactions is at the origin of the

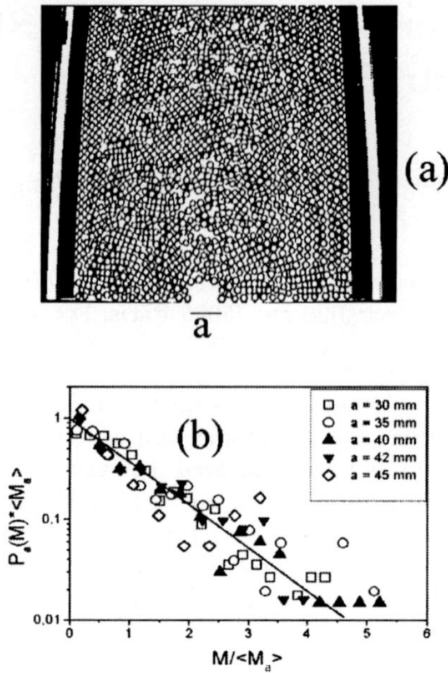

Fig. 1. Experiment in a model hopper: (a) picture of a blocking arch. (b) rescaled mass distributions below the blocking transition; $P_a(M)$ is the probability density at a constant opening size a of a discharge with a mass M smaller than the full mass M_∞. The average mass is $< M_a >$. The solid line is $y = \exp(-x)$.

clogging dynamics. Recently, starting from granular hydrodynamic equations, Valance et al. [10] have demonstrated the possibility of complex wave patterns stemming from the non-linear structure of these equations. So far, there was no evidence that a real experimental system could exhibit such patterns in the absence of air and solely due to the presence of dissipative collisions but in a recent contribution [12], we have presented a clear evidence of such a phenomenon in the context of bidimensional granular chutes. In this paper, we summarize some recent results we obtained in our group for model granular chutes. More precisely, we address the issue of the transition to the flow blockade regime in a hopper and the formation of density patterns in a vertical chute.

2 Stable Vault Formation from the Statistical Point of View

The issue of flow blockade and permanent arch formation is of course of real practical interest to design industrial hoppers [13]. For practical purpose, this type of incident is solved using different techniques such as air blow or repeated shocks to break the vault. The presence of a typical opening for which the flow

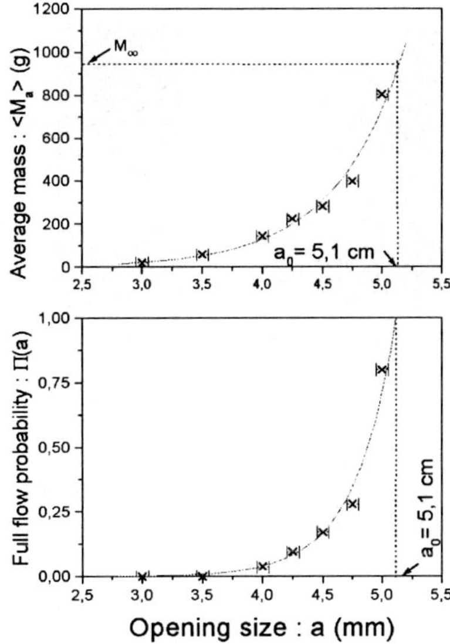

Fig. 2. Approach of the blocking transition threshold: (a) average discharged mass $< M_a >$ (different from the full discharge M_∞) as a function of the hopper opening: a. (b) probability density $\Pi(a)$ for having a full hopper discharge. The solid lines are the best exponential fits.

is bound to stop was studied extensively from the flowing regime. The relation between granular flow and the hopper opening (Beverloo's law) derives from a simple scaling argument stating that the flow J coming out of a hopper scales like the free fall velocity on a distance comparable to the opening size a, multiplied by the outlet surface. This argument yields in spatial dimension Δ:

$$J \approx \sqrt{g\,a}\,a^{\Delta-1} \approx a^{\frac{2\Delta-1}{2}} \tag{1}$$

Empirically, for various opening sizes a and grain sizes d one obtains relations of the type:

$$J^{\frac{2}{2\Delta-1}} \approx a - \alpha d, \tag{2}$$

where is α is a constant with values around $4-6$. The distance $a_0 = \alpha d$ is called the blocking size and is obtained by a linear interpolation at $J = 0$ using relation (2). Note that flow measurements are usually performed for openings much larger than this distance. In spite of the importance of the issue, there is in fact no fundamental understanding of the real nature of this transition. Here we expose briefly some results of a preliminary study in dimension $\Delta = 2$. We aim to approach the transition from the flow-blocking side. We designed

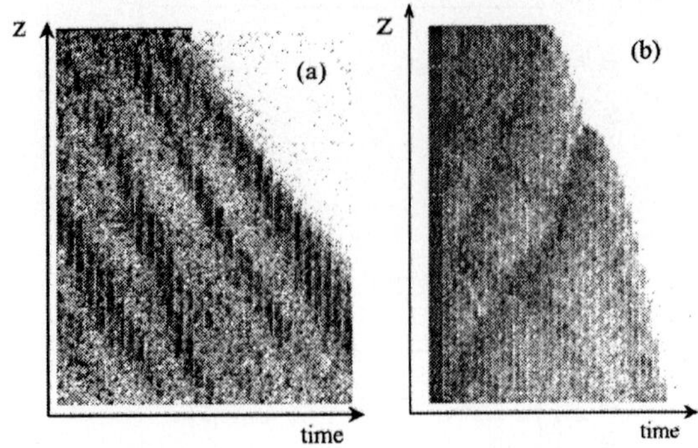

Fig. 3. Spatio-temporal representation of the density variations along the chute for the outlet opening $a = 10mm$; a) steel spheres and b) aluminum spheres.

a model hopper using a pile of hexagonal bolts lying on a glass plane covered with Teflon. The falling angle is around 30°. The pile is made of a mixture of bolts with different sizes to create a geometrical disorder (around $d = 8$ mm). For a given opening, we monitor the mass M which has flown between successive destabilizations of the blocking arches (Fig. 1a). After each discharge, the hopper is refilled with the fallen grains. Hence, we obtain a stationary state and we show that the fallen mass probability is well characterized by an exponential decrease and thus a well defined characteristic mass $< M_a >$ (Fig. 1b).

On the other hand, we monitor the fraction of events corresponding to a full discharge of the hopper (with a remaining part bearing a value M_∞ in the average). Of course, to be really meaningful the points obtained in the vicinity of the blockade transition should be calculated with a larger number of realizations (here we only took systematically a maximum of 200 realizations for each opening, leaving only several tenths of realizations in the close vicinity of the threshold). Nevertheless, the fitted law seem to indicate the presence of a transition when the average mass is *exactly* equal to M_∞ and such that this typical size is exactly the one for which the full discharge probability equals one (Fig. 2). What is remarkable here is the fact that the fits are made with opening values rather far away from the transition size and both measurements (average mass + full flow probability) give an identical blocking size: a_0. We argue that this could reveal the presence of a rather universal statistical framework describing the blockade for all hopper openings. In fact, we propose to check whether the blockade transition size a_0 could be described by a relation such as:

$$a_0/d \approx \ln N, \tag{3}$$

where N represents the number of grains contained in the hopper and that are likely to flow through the exit; this number may depend on the hopper aspect

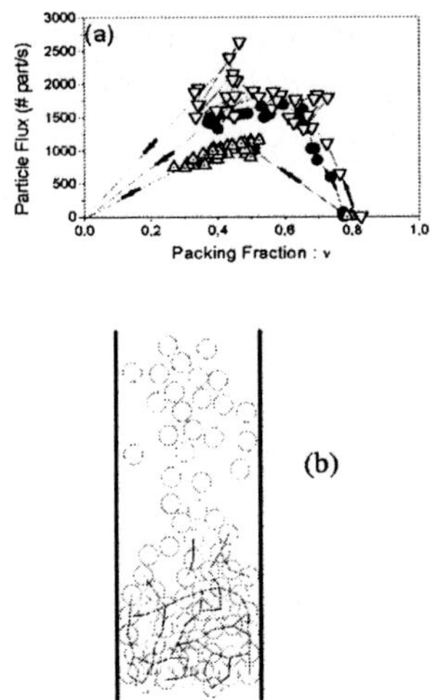

Fig. 4. Kinematics of the density waves regime: (a) flux/density diagrams for $a = 10$ mm , steel spheres ($\triangle$), brass spheres ($\bullet$) and aluminum spheres ($\triangledown$). (b) zoom inside a clogged region of aluminum spheres. The solid lines indicate long lasting contacts.

ratio and on its geometrical dimensions. In our case, for a hopper of size L and an aspect ratio 1 we would have: $N \approx (L/d)^2$. A relation such as (3) would mean that for any opening, a full blockade is always possible provided that the reservoir of grains is large enough to give a chance for this realization to happen. We are at the moment in the process of checking the validity of this relation for different grain sizes and also in three dimensions.

3 Traffic Jam Formation in a Model Granular Flow Experiment

Here, we present the results of an experimental work on a model granular system consisting of a 2D vertical granular chute exhibiting complex flow patterns [12]. We show the presence of a density wave regime depending on the dissipative character of the grains. We identify a regime of steady flow and compared it with the result of granular hydrodynamic equations. The cell consists of two glass plates confining two vertical posts of height $H = 50$ cm separated by an horizontal distance a. The walls are made out of a metal-saw which teeth are 1 mm deep and distant. The presence of a wall roughness with a size comparable to the bead size is here crucial to obtain a steady state. The space between the glass plates is $e = 1.55$ mm and is large enough to let $d = 1.5$ mm metallic beads flow with minimal friction. We use metallic beads, not only because the

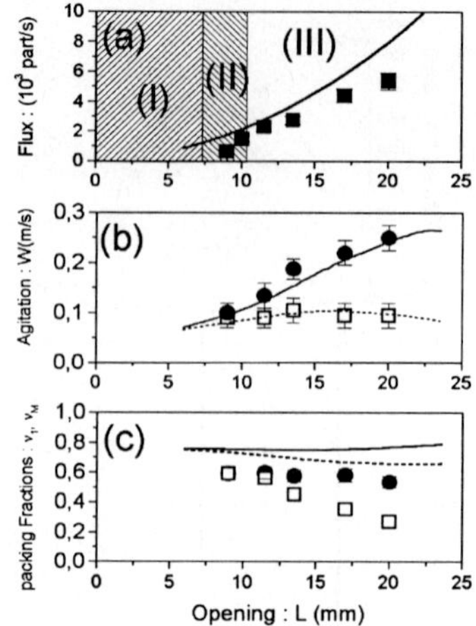

Fig. 5. Experimental and theoretical field values for steel beads as a function of openings a. (a) particle flux; region I: permanent blockade; region II: density waves; region III: steady regime. (b) agitation values at the edges: $W_1(\bullet)$ and in the center: $W_0(\boxdot)$. (c) packing fraction at the edges: $\nu_1(\boxdot)$ and average packing fraction: $\overline{\nu}(\bullet)$. The solid $(\overline{\nu})$ and dashed (ν_1) lines are the theoretical derivations with $\varepsilon = 0.925$ and boundary conditions of Fig 7 (see text).

diameters are defined with a good precision, but also because their weight is large enough to make hydrodynamic forces irrelevant.

Experiments are performed with three types of beads: steel beads with a restitution coefficient around $\varepsilon \simeq 0.9$, brass beads with $\varepsilon \simeq 0.8$ and aluminum beads with a restitution coefficient of $\varepsilon \simeq 0.6$ (values measured for a typical impact velocity of 1m/s). The chute is started using a horizontal gate triggered by an electro-magnetic switch. A CCD camera, hooked to a frame grabber inside a computer records 25 images per second. The stroboscopes lighting up the set-up, are synchronized by the video signal coming out of the camera. The triggering pace is 50 Hz and therefore, each video image contains two images of the chute separated by 0.02 s.

Afterwards, a computer program separates the intertwined frames. Here, we present results made with two different set-ups. In the first display, the camera focuses on a small region situated at the bottom third of the bin which encompasses a window of typically 10x30 beads size. The stroboscopes, flashing at 50 Hz, are situated in front of the cell on both sides of the camera. A small disyncronization is adjusted electronically to probe the displacement of the bead during a time gap of $\delta t = 0.5$ ms (typically). The left stroboscope delivers a big flash and the right one delivers a small flash in order to facilitate the subsequent computer analysis of each image. Hence, we monitor the position and the velocity fields during the chute. In the second set-up, the cell is uniformly lighten

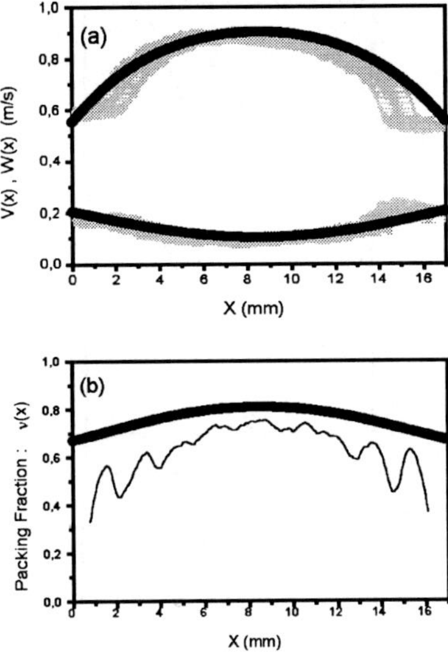

Fig. 6. Experimental horizontal fields for steel beads at $a = 17$ cm. (a) upper curve is the velocity: $V(x)$ and lower curve is the agitation $w(x) = \sqrt{T(x)}$; (b) packing fraction: $\nu(x)$ (upper curve) and packing fraction at the walls. The solid lines are the theoretical profiles (see text).

from behind by two stroboscopes on the top of each others and flashing simultaneously. Another computer program is designed to display next to each others the recorded images of the chute. Thus, a spatio-temporal diagram of the density variations along the chute is easily obtained.

In a way, which is very analogous to the hopper case, we evidence the presence of permanent vaults for openings $a \simeq 5 - 6$ grain sizes. But interestingly, for opening sizes slightly larger, we see the occurrence of density waves as displayed in Fig. 3. For the least dissipative grains (steel beads), we observe downwards moving pulses (Fig. 3a) but for very dissipative grains we have upwards moving waves (Fig. 3b). In Fig. 4a we displayed the kinematic flux/density relations for steel, brass and aluminum spheres at a constant width: $a = 10$ mm. These relations are indeed reminiscent of traffic jam kinematic waves and we observe (i) that the more dissipative the particles are, the larger is the flux and (ii) the wave velocities seem to decrease with the restitution coefficient. For steel beads, the wave velocity is positive (downwards), for brass beads the dissipation is such that the velocity is close to zero and for aluminum beads, it is negative (backwards, Fig. 3b). For aluminum, a close inspection at a clog shows an interesting behavior (Fig. 4b).

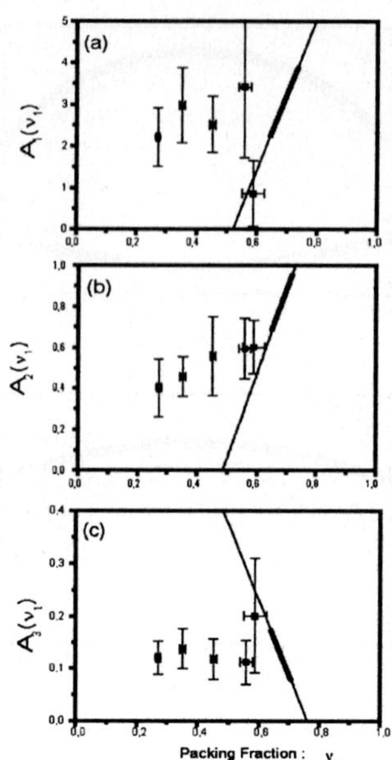

Fig. 7. Empirical determination of the boundary conditions: (a) $A_1(\nu_1)$; (b) $A_2(\nu_1)$; (c) $A_3(\nu_1)$. The dashed lincs correspond to the boundary conditions used in the theory, the solid lines indicate the domain of use of the theoretical boundary conditions.

We identify the following mechanism. Due to the rain of particles falling on its top, the clog has the tendency to grow upwards. The impact momentum is transferred to the edges via long lasting contacts between the grains. These contacts are represented by solid lines between the centers of the spheres. This can be called dynamical arches which are slowing down the particles falling velocity. Meanwhile, at the bottom of the clog, the dynamical arches desegregate into a rain which feeds the next clog downwards. This complicated dynamics is at the origin of kinematic waves going backwards.

In the following, we investigate the case of the least dissipative beads, i.e., the steel beads and we show a mechanism of pattern formation which is seemingly radically different from the one obtained in the aluminum case. We show that this can be understood in the context of granular hydrodynamics. For larger openings and after a small transient of the order of 0.1 s, the wave pattern disappears and a steady state is reached. The average fluxes are measured and the limit of the different phases is displayed in Fig. 5a as a function of the chute width a. We also measure the experimental profiles for the volume fraction $\nu(x)$, the vertical velocity $V(x)$ and the kinetic temperature $T(x)$ for various openings. The fields average values (denoted $\langle . \rangle$ in the following) are obtained from both a spatial average using a bin of 1 bead diameter width and 10 diameters large and an average over several images at steady state (between 10 and 30). Note that we use the kinetic definition of temperature, i.e., $T(x) = 1/2\langle(\vec{v} - V(x)^2)\rangle$, where $\vec{v} = (v_x, v_z)$ is a particle velocity and $\vec{u}(x) = \langle\vec{v}\rangle$ is the average. We note $V(x) = u_z(x)$ and the agitation velocity: $w(x) = \sqrt{T(x)}$. In Figs. 5b and 5c, particular values of those fields are displayed for different openings. In Fig. 6, the field profiles are displayed for $a = 17$ cm; the features observed are similar for all widths. We observe that the velocity $V(x)$ increases from the edges to the center and shows a slipping velocity at the wall. The role of this slipping velocity is central here since it creates a particle agitation at the boundaries which is evidenced from the granular temperature profile. The agitation decreases from the edges to the center probing the existence of a "heat flux" originating from the boundaries which is responsible for the whole system fluidization. The density profiles shows a decompaction at the edges due to shearing. To describe this situation more precisely, we use an hydrodynamic theory stemming from a kinetic theory derived for dissipative hard-spheres [4,6]. The general transport equations are:

$$\begin{aligned} \frac{\partial \rho}{\partial t} + \boldsymbol{\nabla}(\rho \boldsymbol{u}) &= 0 \\ \rho \frac{D\boldsymbol{u}}{Dt} &= \rho g + \boldsymbol{\nabla}\overline{\overline{P}} \\ \frac{3}{2}\rho \frac{DT}{Dt} &= -\boldsymbol{\nabla}\boldsymbol{Q} + \mathrm{Tr}\left(\overline{\overline{P}} \cdot \overline{\overline{E}}\right) - \gamma, \end{aligned} \tag{4}$$

where $\overline{\overline{E}}$ is the deformation rate tensor, $\overline{\overline{P}}$ is the stress tensor and γ is the local energy loss rate which makes *the* fundamental difference with classical Euler equations for thermalized fluids. We pursue further by developing a theoretical approach derived in the collisional limit. Though here, the neglect of kinetic terms in the constitutive relations is somehow borderline (as we will see later), we deliberately choose this simple approach since at this point, we are not seeking for an ultimate confrontation between theory and experiments. We are just interested in learning if a kinetic theory is able to produce semi-quantitatively the features we observe experimentally. Moreover, seeking for more quantitative comparison could be at the end useless since we might have to introduce the facts that the restitution coefficient varies (slowly) with the impact velocity and to include the problem of momentum transfer with frontal boundaries as well as the coupling with particle rotations. The constitutive equations derived from the kinetic theory[11], are for the stress tensor $\overline{\overline{P}} = P_h\overline{\overline{I}} + 2\mu\overline{\overline{E}}$ and the heat flux $\vec{Q} = \kappa\vec{\nabla T}$ are standard for hard sphere fluids where P_h is the hydrodynamic pressure, μ the granular viscosity and κ is the heat diffusivity. In the limit of dominating collisions, the constitutive coefficients in 2D are:

$$\begin{aligned} p_h &= \alpha_0\, \rho\, G(\nu)\, T \\ \mu &= \alpha_1\, \rho\, d\, G(\nu)\sqrt{T} \\ \kappa &= \alpha_2\, \rho\, d\, G(\nu)\sqrt{T} \end{aligned}$$

and the local energy loss is $\gamma = (1-\varepsilon)\frac{\alpha_3}{d}\rho G(\nu)T^{3/2}$. The constants α_i depend on the restitution coefficient $\varepsilon = 2r - 1$ and were derived by Jenkins et al. [11]. We have in the collisional limit the values : $\alpha_0 = 2r$, $\alpha_1 = r(3\pi r^2 - 2(\pi+6) + 20)/(5-3r)/4\sqrt{\pi}$, $\alpha_2 = r\sqrt{\pi}(9r^2 - (30/\pi + 27/4)r + 34/\pi)/(17-15r)$ and $\alpha_3 = 8r/\sqrt{\pi}$. The function $G(\nu)$ is describing the contact probability between two spheres. We use the Verlet's [16] form of this equation in 2D, i.e., $G(\nu) = \nu\frac{1-7\nu/16}{(1-\nu)^2}$. Hence, in the limit of these approximations, a set of equations for the fields at steady-state is obtained:

$$\begin{aligned} p_h &= 2\rho G(\nu)T = const. \\ \rho G(\nu)T^{1/2}d\partial_x V &= -g\int_0^x \rho(\zeta)d\zeta \\ \partial_{xx}T^{1/2} - KT^{1/2} &= 0, \\ \text{with } K &= \frac{1}{d^2}\left(\frac{(1-\epsilon)\alpha_3}{2\alpha_2}\frac{2\alpha_1}{\alpha_2}\left(\frac{g\int_0^x \rho(\zeta)d\zeta}{p_h}\right)^2\right). \end{aligned} \tag{5}$$

To solve these equations a set of boundary conditions is needed. Deriving from the saw tooth boundary conditions values of momentum transfer at the wall is *a priori* quite complicated. For a fixed array of spheres glued at the wall, Jenkins [14] has proposed a formula introducing a geometrical bumpiness parameter but this is not really practicable here. Moreover, the exact status of these boundary conditions here is a rather tricky issue as it was noticed recently by Goldhirsch [15]. We rather propose an approach based on a dimensional argument. Balancing the horizontal/vertical momentum fluxes and heat flux on both sides of a surface close to the boundaries we obtain the dimensional relations: $V_1 = A_1(\nu_1)w_1$, $\left(\frac{\partial V}{\partial x}\right)_1 d = -A_2(\nu_1)w_1$ and $\left(\frac{\partial w}{\partial x}\right)_1 d = A_3(\nu_1)w_1$, where $A_1(\nu_1)$,$A_2(\nu_1)$ and $A_3(\nu_1)$ are dimensionless functions and subscript 1 means values taken at positions $x = \pm L/2$. We propose to integrate numerically equations (5) using these three empirical boundary conditions and a dissipation value : $\varepsilon = 0.925$. Optimizing the agreement for the velocity and agitation fields we use the linear estimates: $A_1(\nu_1) = 18\nu_1 - 9.5$, $A_2(\nu_1) = 4\nu_1 - 2$ and $A_3(\nu_1) = -1.4\nu_1 + 1$. These functions are displayed in Fig. 7 and are compared with the experimental measurements of A_1, A_2, A_3. Note that the magnitudes are correct but a strong discrepancy exists for the volume fraction dependence. In Fig. 6b, we display the agitation field and the velocity field as a function of the horizontal coordinate x for $a = 17$ cm. The quality of agreement is similar for all widths as witnessed by the results displayed on Fig. 4b. On the other hand, for the density fields (see Fig. 6b, Fig. 5a and Fig. 5c for fluxes and packing fractions) , the experimental densities are lower than the densities we derive from the kinetic theory in the collisional limit. A clear reason for this discrepancy is that the actual transfer of momentum and heat from the edges is not properly described by the collisional part of the constitutive relations. A full implication of all the kinetic terms should provide a better theory. On the other hand, our impression is that the collisional limit we present here provides a simple theory which is able to reproduce rather well all the features we observed experimentally and as a consequence, it could be a fair base for preliminary theoretical derivations. In a recent theoretical work, investigating the stability of the transport equations in the collisional regime, Valance et al. [10] have shown that this flow is unstable for narrow gaps and dissipative particles. A non-linear analysis shows that beyond this threshold, non-linear terms may generically control the density dynamics and in particular, the leading equation is a mixture of a Kuramoto-Shivasinsky equation and a Kortweg de Vries equation. When the former term is dominant, pulse sequences are predicted which indeed resembles to our experimental results for the narrow openings at the limit of the full blockade domain.

4 Conclusion

In conclusion, we presented experimental results on two different granular flows demonstrating the complexity of the flow patterns and also the striking analogy with traffic jam issues In a first model system representing a hopper, we investigated the transition to full flow blockade as a function of the outlet opening. We study the real nature of the transition and we propose that the notion of a well defined clogging size could be revisited under the scope of a statistical theory of extreme events. In the second example, we expose some recent results on the dynamics of a vertical chute. We investigate the steady regime for rather weakly dissipative particles falling in a bidimensional column with very rugous wall. Then, we present results showing that the flow can be qualitatively well described using a set of hydrodynamic equations stemming from a kinetic theory which is derived for dissipative hard spheres and taken in the collisional regime. We obtain in the limit of narrow openings, just before the clogging transition, density waves in the form of regular pulses going downwards which are similar to non-linear waves predicted by Valance et al. [10]. Moreover, we find that the wave direction may change with the dissipative character of the grains. Ultimately for the more dissipative particles, a wave regime is obtained but its dynamics experiences a quantitative change: we evidence kinematic waves going backwards in the form of compact clogs with a flow structure controlled by dynamic arches. Understanding the dynamics of this regime would in principle require a theoretical treatment different from a standard kinetic theory since the energy and momentum transfer picture inside a clog is obviously non-local. This phenomenon is a challenge for theoretical treatment and falls in the general issues concerned with the rheology of dense granular materials which is so far an open problem.

Acknowledgement. We acknowledge fruitful discussions with A. Valance, Prof. I. Goldhirsch and Prof. J. Jenkins.

References

1. H. J. Herrmann, J.-P. Hovi, and S. Luding, (Eds.), *Physics of Dry Granular Media,* (Kluwer, Dordrecht, 1998); R.P. Behringer and J.T. Jenkins, (Eds.), *Powders and Grains '97,* (A.A. Balkema, Rotterdam, 1997); H.M. Jaeger, S.R. Nagel, and R.P. Behringer, Rev. Mod. Phys. **68**, 1259 (1996).
2. R.M. Nedderman, *Statics and kinematics of granular materials,* (Cambr. Univ. Press, Cambridge, 1992).
3. R.Bagnold, Proc. Roy. Soc. Cond. A **295**, 219 (1954).
4. S.B. Savage, J. Fluid Mech. **92**, 53 (1979).
5. P.K. Haff, J. Fluid Mech. **134**, 401-430 (1983).
6. J.T. Jenkins and S.B. Savage, J. Fluid Mech. **130**, 187-202 (1983).
7. D.A.Kurtze and D.C.Hong, Phys. Rev. E **52**, 218 (1995).
8. G. Peng and H. Herrmann, Phys. Rev. E **49**, 1796 (1994).
9. T. LePennec, M. Ammi, and A. Valance, Eur. Phys. J. B **7**, 657 (1999).
10. A. Valance and T. LePennec, Eur. Phys. J. B **5**, 223 (1998).
11. J.T. Jenkins and M.W. Richman, Phys. of Fluids **28**, 3485-3494 (1985).
12. G. Reydellet, F. Rioual, and E. Clément, *Granular hydrodynamics and density wave regimes in a vertical chute experiment*, preprint, (1999).
13. R.L. Brown and Richard, *Principle of Powder Mechanics*, (Pergamon, New York, 1966).
14. J. Jenkins, Appl. Mech. Rev. **47**, S240 (1994).
15. Private com.: see paper by I. Goldhirsch in these proceedings.
16. L. Verlet and D. Levesque, Molec. Phys. **46**, 969 (1982).
17. G.W. Baxter, R. Leone, and R.P. Behringer, Europhys. Lett. **21**, 569 (1993).
18. K. Chick and A. Verveen, Nature **251**, 599 (1974).
19. T. Raafat and J.P. Hulin, and H.J. Herrmann, Phys. Rev. E **53**, 4345 (1996); J.-L. Aider, N. Sommier, T. Raafat, and J.P. Hulin, Phys. Rev. E **59**, 778 (1999).
20. K. Nagel, Phys. Rev. E **53**, 4655 (1996); T. Pöschel, J. Phys. I **4,** (1994); J. Lee and M. Liebig, J. Phys. I **4** , 507 (1994).
21. C.T. Veje and P. Dimon, Phys. Rev. E **54,** 4329 (1996); Phys. Rev. E **56,** 4376 (1997); S. Horluck and P. Dimon, Phys. Rev. E **60,** 671 (1999).
22. R.K.P. Zia, B. Shaw, B. Schmittmann, and R.J. Astalos, Phys. Rep. **301**, 45 (1998).
23. J. Krug, Phys. Rev. Lett. **67**, 1882 (1991).
24. J. Krug and P.A. Ferrari, J. Phys. A **29**, L465 (1996).
25. B. Derrida, E. Domany, and D. Mukamel, J. Stat. Phys. **69**, 667 (1992).
26. N. Rajewsky, L. Santen, A. Schadschneider, and M. Schreckenberg, J. Stat. Phys. **92**, 151 (1998).
27. L.G. Tilstra and M.H. Ernst, J. Phys. A **31**, 5033 (1998).
28. M.R. Evans, N. Rajewsky, and E.R. Speer, J. Stat. Phys. **95**, 45 (1999).
29. J. de Gier and B. Nienhuis, Phys. Rev. E **59**, 4899 (1999).
30. A. Benyoussef, H. Chakib, and H. Ez-Zahraouy, Eur. Phys. J. B **8**, 275 (1999).
31. M. Schreckenberg, A. Schadschneider, K. Nagel, and N. Ito, Phys. Rev. E **51**, 2939 (1995).
32. K. Nagel and M. Schreckenberg, J. Phys. I France **2**, 2221 (1992).

Kinetics of Granular Gases

I. Goldhirsch

Tel-Aviv University, Department of Fluid Mechanics and Heat Transfer,
Faculty of Engineering, Ramat-Aviv, Tel-Aviv 69978, Israel

Abstract. It is shown how the Boltzmann equation corresponding to a model granular gas can be systematically analyzed in order to obtain constitutive relations as well as boundary conditions. It appears that the formulation leading to the establishment of boundary conditions is the first systematic approach to the problem even in the realm of molecular gases. Limitations on the results, in particular the restriction of their validity to near elastic systems are explained and shown to follow from the mesoscopic nature of granular gases.

1 Introduction

Granular gases are defined as states of granular matter in which the grains interact by practically instantaneous collisions. As this picture is similar to that of the classical model of a molecular gas it has spurred the application of kinetic theoretical methods to the study of this problem. Pioneering kinetic studies of the dynamics of granular gases [1,2] employed assumed forms of the single particle distribution function and closures based on the requirement that moments calculated using the left hand side of the Boltzmann equation equal the corresponding moments calculated from its right hand side (the collisional term). These (Enskog) relations were rightfully referred to as 'balances' (and usually mass, momentum and energy balances were studies). This approach resulted in constitutive relations which have proven useful for the study of some real granular flows. One of the properties of granular flows which the first studies missed was the 'normal stress difference', i.e. the fact that granular materials exhibit anisotropic pressures. It was later realized [2] that by conjecturing anisotropic Maxwellian distributions (following a similar conjecture in astro physics) one could account for the anisotropy of the stresses but the true source of this phenomenon remained unclear.

The derivation of boundary conditions for molecular gases dates back to Maxwell [3]. His basic assumption was that molecules colliding with a solid wall could be characterized by a distribution function which is identical to that in the bulk. Many years later, Kramers [4], using Maxwell's assumption, improved upon his derivation and obtained a formula for the slip length at the wall. The establishment of boundary conditions requires information on the properties of the wall. As realistic geometric and dynamics properties of walls are hard to measure and even if they were known they would be hard to implement, it is common to model walls by accommodation functions [5]. These functions provide statistical information on the distribution function of particles emerging from a collision with the wall, given their incoming velocity. Maxwell, for instance, assumed that part of the particles colliding with a wall are reflected specularly and the rest are thermalized. Many improvements of Maxwell's approach have been proposed over the years. However, it seems that unlike in the derivation of constitutive relations, where the Chapman-Enskog expansion [6] (abbreviated below as CE) provides a powerful and systematic tool, no systematic approach has been proposed for the derivation of boundary conditions. The reason for this state of affairs is, as shown below, that the greatest justifiable simplification of the problem (which corresponds to the regime of very low Knudsen numbers) reduces it to a complicated integrodifferential equation which is not characterized by any small parameter. Boundary conditions have also been derived in the realm of granular gases [7] by employing Maxwell's assumption and balancing low order moments of this distribution against the appropriate fluxes in the bulk.

It is shown below that one can devise a systematic method for the study of granular gas dynamics [8] and that this method produces constitutive relations, which are different (in some cases in a minor way and in others in a significant manner) from those obtained by phenomenological means. The proposed method is based on a generalization of the CE expansion to granular systems which involves a double expansion, the small parameters of which are the Knudsen number (the small parameter of the classical CE expansion) and the degree of inelasticity, which is defined below. The advantage of employing this double expansion is that the double limit in which both small parameters vanish (no gradients, i.e. zero Knudsen number and no inelasticity, i.e. an elastic gas) is the equilibrium limit of an elastic gas and thus the zeroth order distribution function is Maxwellian. Once a zeroth order distribution function is established it is a (rather tedious but) straightforward matter to apply the machinery of perturbation theory to obtain a systematic theory. The resulting perturbation theory suggests its own limitations, the major restriction being to relatively low degrees of inelasticity or, in other words, to near elastic gases. This limitation is not only a consequence of the chosen approach to the problem, it is first and foremost a result of the physics of granular systems. It can be shown [9] that granular gases lack scale separation in an inherent way (i.e. the absence of scale separation is not necessarily a 'practical' result of the fact that typical granular systems include far less than an Avogadro number of grains), i.e. any granular gas, no matter how 'large' lacks scale separation.

As the derivation of boundary conditions involves the study of an integrodifferential equation which has no small parameter, it seems to be fundamentally harder than the problem of deriving constitutive relations. The solution proposed below is based on the observation that molecular systems approach equilibrium in a matter of a few (or $\mathcal{O}(1)$) collisions per particle. This is suggested both by the nature of the eigenvalues of the linearized Boltzmann equation [10] (which, strictly speaking, describe only the decay to equilibrium of a near-equilibrium state) and Molecular Dynamics (MD) simulations. It thus seems that if one could devise a perturbative method in which each 'next order' involves one more collision per particle, the method should converge very rapidly. This is the philosophical basis for the method presented below and indeed it converges rather rapidly. The above statements concerning the derivation of boundary conditions refer to molecular systems. They should be relevant to granular gases as well provided the deviation from elasticity is not too severe (since then an $\mathcal{O}(1)$ number of collisions does not change the energy significantly and thus the system may be thought of as slowly progressing from one near-equilibrium state to another). Thus the main restriction on the validity of the boundary conditions turns out to be the same as on the constitutive relations.

The structure of this paper is as follows. Section 2 presents a brief explanation of the lack of scale separation [9] in granular gases. Section 3 is devoted to an outline of the method for obtaining constitutive relations for granular gases. Section 4 presents an outline of a systematic approach to the derivation of boundary conditions for granular gases. Section 5 presents a brief summary as well as some thoughts on required future work.

2 Lack of Scale Separation in Granular Gases

The present section is devoted to the demonstration of the mesoscopic nature of rapid granular flows [9]. To this end consider a stationary mono disperse granular system, whose collisions are characterized by a fixed coefficient of normal restitution, e. Assume that (locally) the macroscopic velocity field, $\mathbf{V}$, is given by $\mathbf{V} = \gamma\, y\, \hat{\mathbf{x}}$, where γ is the shear rate, y is the span-wise coordinate and $\hat{\mathbf{x}}$ is a unit vector in the stream wise direction. It can be shown on the basis of phenomenological considerations or directly from (any of) the continuum equations (proposed for) describing the dynamics of granular gases that the (local) granular temperature, T, is given by:

$$T = C\frac{\gamma^2\, l_0^2}{\epsilon}, \tag{1}$$

where $\epsilon \equiv 1 - e^2$ is the degree of inelasticity, l_0 is the mean free path and C is a volume fraction dependent prefactor, whose whose value at low volume fractions can be shown to equal approximately 0.6 in two dimensions and 3 in three dimensions [8]. It is important to stress that the granular temperature is defined as the average of the square of the *fluctuating* parts of the particle velocities.

The change of the macroscopic velocity over a distance of a mean free path, in the y direction, is given by: $\gamma\, l_0$. A shear rate can be considered small if $\gamma\, l_0$ is small with respect to the thermal speed, $\sqrt{T}$. Here, following (1): $\gamma\, l_0/\sqrt{T} = \sqrt{\epsilon}/\sqrt{C}$, i.e. the shear rate is not 'small' unless the system is nearly elastic (notice that for e.g. $e = 0.9$: $\sqrt{\epsilon} = 0.44$). Thus, except for very low values of ϵ, the shear rate is always 'large'. In other words, the spatial gradients in the system are 'large' as the velocity changes significantly over the scale of a mean free path. This result implies that the Chapman-Enskog expansion of kinetic theory (which is basically a gradient expansion) must be carried out beyond the Navier-Stokes order, the lowest next order being the Burnett and super-Burnett orders.

While the resulting hydrodynamic equations are suitable for the description of steady states they are generally ill-posed [11]! Thus, unless a resummation or regularization scheme that tames this ill posedness is applied, one cannot obtain useful results from the 'higher orders' (except in steady states). The method developed by Rosenau [11] and recently further developed by Slemrod [11] shows some promise in this direction.

Consider next the mean free time, τ, i.e. the ratio of the mean free path and the thermal speed: $\tau \equiv l_0/\sqrt{T}$. Clearly, τ is the microscopic time scale characterizing the system at hand and γ^{-1} is the macroscopic time scale characterizing this system. The ratio $\tau/\gamma^{-1} = \tau\gamma$ is a measure of the temporal scale separation in the system. Since $\tau\gamma = \sqrt{\epsilon}/\sqrt{C}$, it is an $\mathcal{O}(1)$ quantity. It follows that (unless $\epsilon \ll 1$) there is no temporal scale separation in this system, *irrespective of its size or the size of the grains.* Consequently, one cannot a-priori employ the assumption of "fast local equilibration" and/or use local equilibrium as a zeroth order distribution function (both for solving the Boltzmann equation and for the study of generalized hydrodynamics [12] of these systems [13]) unless the system is nearly elastic (in which case, scale separation is restored) and (in unsteady states) the rate of change of the external parameters (e.g.the shear rate) is sufficiently slow. The latter condition severely limits the applicability of the hydrodynamic description. For instance, consider the application of these equations to a stability study. As expected (and is well known) some of the eigenvalues of the granular stability problem (including those corresponding to instabilities) must be of the order of the only "input" inverse time scale (in the absence or irrelevance of gravity), i.e. $1/\gamma$. Since, as explained above, $\tau \propto 1/\gamma$, one obtains instabilities whose characteristic times are comparable with the mean free time. If one adopts the conservative view that hydrodynamics should be valid only on time scales which significantly exceed the mean free time, one encounters the paradoxic situation in which the hydrodynamic equations predict instabilities on time scales which they are not supposed to resolve. It can also be shown that the mean free paths in granular gases can be of macroscopic dimensions [9] but this takes us beyond the scope of the present article.

3 On the Derivation of Constitutive Relations

One of the main problems one encounters when developing a perturbative approach to the kinetics of rapid granular flows is the absence of a finite temperature equilibrium state in free (unforced) systems. Indeed, when a granular gas is left to its own fate, its energy decays (asymptotically) to zero due to the inelasticity, i.e. the only "equilibrium state" is that of vanishing temperature. It is obviously inconvenient to employ such a state as a zeroth order in a perturbation theory for a system at a finite granular temperature. One solution of this problem is to devise a perturbation theory around a decaying granular flow [14]. A different way is based on the aforementioned observation that in the limit of vanishing gradients (formally, the Knudsen number, K) *and* inelasticity, a granular gas becomes elastic and it possesses an equilibrium state [8]. Therefore, one can employ K and ϵ as small parameters in a perturbation expansion applied to the pertinent Boltzmann equation. The limit $\epsilon \to 0$, $K \to 0$ is not singular since local equilibration occurs, as mentioned, on time scales which are of the order of the mean free time. In other words, when ϵ is sufficiently small the effects of dissipation during the $\mathcal{O}(1)$ collisions (per-particle) required for local equilibration can be small enough so that an elastic local distribution function can serve as an excellent approximate description, corrections being rightfully described as perturbations.

The expansion of the solution of the Boltzmann equation in ϵ and K begets constitutive relations which differ, both qualitatively and quantitatively, from those obtained in previous studies [1,2,15] (see however [14]). In particular, the Navier-Stokes (order) terms have a different dependence on the degree of inelasticity and the number density than in previously derived constitutive relations; for instance the expression for the heat flux contains a term which is proportional to $\epsilon\nabla \log n$, where ϵ is a measure of the degree of inelasticity and n denotes the number density. A similar term, i.e. one that is proportional to $\epsilon\nabla n$, has been obtained by using the Enskog correction [15], but this term is $\mathcal{O}(n)$ and it vanishes in the Boltzmann limit. In addition, some minor quantitative differences between our results and previous ones stem from the fact that in our work an isotropic correction to the leading Maxwellian distribution, which has not been considered before, is taken into account and also because the full dependence of the corrections on the (fluctuating) speed is computed and employed in the calculations.

It is rather remarkable that, given the differences between granular and elastic systems, some of which are presented in this paper, methods borrowed from the kinetic theory of gases are so useful in the realm of granular gases. The agreement of constitutive relations derived by employing these methods with numerical and physical experiments for which ϵ cannot be considered to be small is truly remarkable; it is possible that this ϵ expansion is an asymptotic expansion (like many perturbative expansions), in which case one expects a larger range of validity than naively expected.

Below we consider a mono disperse collection of smooth, inelastically colliding, spheres of diameter d, whose collisions are characterized by a *constant* coefficient of normal restitution, e, which satisfies $0 < e \leq 1$. A Boltzmann equation corresponding to this system is easy to derive [8]. As mentioned, the small parameters in the CE-like expansion employed to solve the Boltzmann equation are the Knudsen number, K, and the degree of inelasticity, $\epsilon \equiv 1-e^2$. The hydrodynamic fields considered below are [3,5,6,8] the number density field, $n(\mathbf{r},t)$, the macroscopic velocity field, $\mathbf{V}(\mathbf{r},t)$, and the granular temperature field, $T(\mathbf{r},t)$. These are given by the averages of $1, \mathbf{v}$ (the velocity of a particle) and $\mathbf{v}^2 - \mathbf{V}^2$ with respect to the single particle distribution function, f. Multiplication of the Boltzmann equation by 1, $\mathbf{v}$ and $\mathbf{v}^2$, and subsequent integration over $\mathbf{v}$, yields the standard continuum equations for the hydrodynamic fields [3,5,6,8], where the equation of motion for the temperature includes a sink term, reprenting inelasticity, which vanishes in the elastic limit. Below, the mass, m, of a particle, is normalized to unity.

The Knudsen number is given by $K \equiv l/L$, where l is the mean free path(here: $l = 1/\pi n d^2$) and L is the macroscopic length scale, i.e. the length scale which is resolved by hydrodynamics, not necessarily the system size. It is convenient to non-dimensionalize all quantities in the problem studied here, as follows: spatial gradients are rescaled as $\nabla \equiv L^{-1}\tilde{\nabla}$, the rescaled fluctuating velocity is $\tilde{\mathbf{u}} \equiv \sqrt{\frac{3}{2T}}(\mathbf{v} - \mathbf{V})$ and $f \equiv n\left(\frac{3}{2T}\right)^{\frac{3}{2}} \tilde{f}(\tilde{\mathbf{u}})$. In terms of the rescaled quantities, the pertinent Boltzmann equation assumes the form

$$\tilde{\mathbf{D}}\tilde{f} + \tilde{f}\tilde{\mathbf{D}}\left(\log n - \frac{3}{2}\log T\right) =$$

$$\frac{1}{\pi}\int_{\hat{\mathbf{k}}\cdot\tilde{\mathbf{u}}_{12}>0} d\tilde{\mathbf{u}}_2 d\hat{\mathbf{k}}(\hat{\mathbf{k}}\cdot\tilde{\mathbf{u}}_{12})\left(\frac{1}{e^2}\tilde{f}(\tilde{\mathbf{u}}_1')\tilde{f}(\tilde{\mathbf{u}}_2') - \tilde{f}(\tilde{\mathbf{u}}_1)\tilde{f}(\tilde{\mathbf{u}}_2)\right) \equiv \tilde{\mathbf{B}}(\tilde{f},\tilde{f},e), \quad (2)$$

where

$$\tilde{\mathbf{D}} \equiv K\sqrt{\frac{3}{2T}}\left(L\frac{\partial}{\partial t} + \mathbf{v}\cdot\tilde{\nabla}\right). \quad (3)$$

Notice that $\tilde{\mathbf{D}}$ is not a material derivative since the velocity $\mathbf{v}$ is not the hydrodynamic velocity but rather the particle's velocity. The rescaled distribution function, $\tilde{f}$ can be expressed as follows: $\tilde{f}(\tilde{\mathbf{u}}) = \tilde{f}_0(\tilde{u})(1+\Phi)$, where $\tilde{f}_0(\tilde{u}) = 1/\pi^{\frac{3}{2}}e^{-\tilde{u}^2}$ and Φ is considered to be a 'small' perturbation for $K \ll 1$ and $\epsilon \ll 1$.

Employing the above definition of $\tilde{u}$ one obtains from (2):

$$(1+\Phi)\left(\tilde{\mathbf{D}}\log n + 2\sqrt{\frac{3}{2T}}\tilde{u}_i\tilde{\mathbf{D}}V_i + \left(\tilde{u}^2 - \frac{3}{2}\right)\tilde{\mathbf{D}}\log T\right) + \tilde{\mathbf{D}}\Phi = \frac{1}{\tilde{f}_0}\tilde{\mathbf{B}}(\tilde{f},\tilde{f},e). \quad (4)$$

The action of $\tilde{\mathbf{D}}$ on $\log n$, $\mathbf{V}$ and T can be computed from the definition of $\tilde{\mathbf{D}}$ and the general equations of motion satisfied by the hydrodynamic fields. Details can be found in [8]. The next step is to define an expansion in powers of K and ϵ. The idea is to expand Φ in these parameters, as follows: $\Phi = \Phi_K + \Phi_\epsilon + \Phi_{KK} + \Phi_{K\epsilon} + \ldots$ where here and below, such subscripts indicate the order of the corresponding terms in the small parameters, e.g. $\Phi_K = \mathcal{O}(\mathcal{K})$. It is reiterated that the $\mathcal{O}(\mathcal{K})$ corrections to the single particle distribution function are the Navier-Stokes or CE terms, the $\mathcal{O}(\mathcal{K}^{\in})$ corrections are known as the Burnett terms [6] (irrespective of the order in ϵ) and the $\mathcal{O}(K^3)$ contributions are the super-Burnett terms. The operation of $\tilde{\mathbf{D}}$ on any function of the fields, ψ, can be formally expanded as follows: $\tilde{\mathbf{D}}\psi = \tilde{\mathbf{D}}_K\psi + \tilde{\mathbf{D}}_\epsilon\psi + \tilde{\mathbf{D}}_{KK}\psi + \tilde{\mathbf{D}}_{K\epsilon}\psi + \tilde{\mathbf{D}}_{\epsilon\epsilon}\psi + \ldots$, where e.g. $\tilde{\mathbf{D}}_{K\epsilon}\psi$ is the $\mathcal{O}(\mathcal{K}\epsilon)$ term in the expansion of $\tilde{\mathbf{D}}\psi$ in powers of K and ϵ. Since this expansion is well defined, the symbols $\tilde{\mathbf{D}}_K$, $\tilde{\mathbf{D}}_\epsilon$ etc. are referred to as operators in their own right. This expansion leads to a series of inhomogeneous integral equations, each corresponding to a given order in K and ϵ. A typical equation encountered at order (n, m) is of the form $\tilde{\mathbf{L}}\, \Phi_{K^n \epsilon^m}$ = a function of the derivatives of the hydrodynamic fields and of $\tilde{\mathbf{u}}$. The solution involves the inversion of $\tilde{\mathbf{L}}$ (a singular operator, as it has zero eigenvalues) and the addition of a particular inhomogeneous solution (up to a free parameter, which is determined as explained below). As is always the case with such equations the right hand side must be orthogonal to the (left) eigenfunctions of the operator $\tilde{\mathbf{L}}$, this being the solubility condition. This condition fixes, among other things, the free parameter mentioned above. The physical meaning of the solubility condition is explained next.

The local equilibrium distribution function, f_0, is defined in such a way that the hydrodynamic fields are *given* by its appropriate moments. The CE expansion is designed to find the distribution of fluctuating velocities when the macroscopic, or continuum fields are given; in other words the local distribution function corresponds, by construction, to the true values of the macroscopic fields: the corrections to f_0 should not change the values of the macroscopic fields. It follows that the contribution of the correction, Φ, to the above mentioned moments, should vanish, i.e. *Φ should be orthogonal (with respect to the weight function f_0) to the invariants of the (linearized) Boltzmann operator* (the eigenfunctions which correspond to zero eigenvalues): 1, $\tilde{\mathbf{u}}$ and $\tilde{u}^2$, whose respective averages are the density, the velocity and the temperature field. This orthogonality property should hold to all orders in perturbation theory [8]; it is also the reason the (generalized) CE expansion can be systematically carried out to all orders [8]. For full details of the theory, the precise constitutive relations to Burnett (i.e. K^2) order and the problem of normal stress differences the reader is referred to [8].

4 Boundary Conditions

A systematic method for deriving boundary conditions for granular gases [7], which is also relevant to molecular gases, has been developed [16]. Only the essential ingredients of the method are presented in this section. The method is based on an expansion in the number of particle collisions. It should be clear from the formulation presented below that it is not restricted to the simple case (flat boundary) treated below for sake of demonstration.

Consider a solid boundary situated at $z = 0$ and a (granular) gas occupying $z > 0$. The solution of the pertinent Boltzmann equation can be written as $f = f_{ce} + f_0 \Phi_w = f_0(1 + \Phi_{ce} + \Phi_w)$, where Φ_w represents the effect of the wall and $f_0(1 + \Phi_{ce})$ is the CE (or, in principle, an exact solution of the Boltzmann equation far enough from the boundary). The function Φ_w should vanish far away from the boundary (in practice, a few mean free paths away from it). Notice that the value of Φ_{ce} at the boundary is determined by extrapolating the CE solution to the boundary, since the CE solution itself is not correct near the boundary. It is convenient to choose a frame of reference in which the solid boundary is stationary (when the boundary moves at constant velocity this involves only a Galilean transformation; below we implicitly assume this to be the case). Inside the domain influenced by the boundary, i.e. the *Knudsen layer*, the macroscopic flow field $\mathbf{V}$ (rescaled by the square root of the temperature field, T) is of the order of the Knudsen number (see also below), hence it is justified to expand the (extrapolated) CE solution f_{ce} in the Knudsen layer in powers of the rescaled velocity field.

The boundary conditions apply to the hydrodynamic fields extrapolated to the boundary, not the true values of the fields there. The role of these conditions is to ensure that the solutions of the hydrodynamic equations outside the Knudsen layer, whose width is a few mean free paths, are compatible with the kinetics near the boundary. In other words, the hydrodynamic equations provide 'outer solutions' that should match the 'inner' kinetic solution next to the boundary. Since Φ_w vanishes when the hydrodynamic fields are space independent it follows that Φ_w is $\mathcal{O}(K)$.

The method we have developed for obtaining boundary conditions is outlined below at $\mathcal{O}(\epsilon^0 K)$; results are presented to $\mathcal{O}(\epsilon K)$. At order $\epsilon^0 K$ it is sufficient to consider the following linearized Boltzmann equation:

$$v_z \frac{\partial \Phi_w}{\partial z} = \mathbf{L}\, \Phi_w, \tag{5}$$

whose solubility conditions are: $\int d\mathbf{v}\, v_z \psi_i(\mathbf{v}) e^{-v^2}\, \Phi_w(z=0) = 0$, where ψ_i are the invariants of $\mathbf{L}$. These relations require that Φ_w does not contribute to the mass, momentum and energy flux in the z direction. The latter fluxes are completely determined by the CE solution. The physical role of these requirements is to ensure the continuity of the fluxes, i.e. the values of the fluxes in the bulk, given by the CE expansion, should match the rate of the transfer of the corresponding moments to the boundary. Equation (5) is not easy to solve since it is a non-trivial integrodifferential equation.

The basis of the method described below is the observation that initial distribution functions converge rapidly (for $\epsilon \ll 1$) to local equilibrium or equilibrium-like distributions (in a matter of a few collisions [10]). Indeed, as the results presented below indicate, this approach is justified since the contributions of multiple particle collisions to the transport coefficients are increasingly smaller. It is convenient to use the Fredholm [17] form of the linearized Boltzmann operator, $\mathbf{L}$: $\mathbf{L} = \mathbf{A} - q$, where

$$\mathbf{A}\Phi = \pi^{-\frac{3}{2}} \int d\mathbf{v}'\, e^{-v'^2} \left(\frac{2}{R} e^{w^2} - R \right) \Phi(\mathbf{v}') \equiv \int d\mathbf{v}'\, K(\mathbf{v},\mathbf{v}')\, \Phi(\mathbf{v}'), \tag{6}$$

$R = |\mathbf{v} - \mathbf{v}'|$, $w = \frac{\mathbf{v}\times\mathbf{v}'}{R}$ and $q(v) = \frac{1}{\sqrt{\pi}} \left(e^{-v^2} + \frac{\sqrt{\pi}}{2} \left(2v + \frac{1}{v}\right) \mathrm{erf}(v) \right)$ is a positive definite function which depends on the speed v alone. The operator $\mathbf{A}$ includes the full 'gain term' of $\mathbf{L}$ and part of the 'loss term'. The second part of the 'loss term' is q. The function q basically represents the rate of 'loss' of particles, having velocity $\mathbf{v}$, due to collisions and it is trivially related to the (velocity dependent) mean free path. The operator $\mathbf{A}$ represents the rate of 'creation' of particles with velocity $\mathbf{v}$ at a given point in space. Using this representation one can transform (5) as follows:

$$\frac{\partial}{\partial z}\left(e^{\frac{q}{v_z}z}\Phi_w\right) = \frac{1}{v_z}e^{\frac{q}{v_z}z}\mathbf{A}\Phi_w \tag{7}$$

The solution of (7) can be formally written as follows. When $v_z > 0$:

$$\Phi_w(z) = e^{-\frac{q}{v_z}z}\Phi_w(0) + \frac{1}{v_z}\int_0^z dz' e^{\frac{q}{v_z}(z'-z)}\mathbf{A}\Phi_w(z') \equiv \mathbf{G}\Phi_w(0) + \mathbf{Q}\Phi_w. \tag{8}$$

where only the dependence of Φ_w on z is explicitly spelled out. When $v_z < 0$:

$$\Phi_w(z) = -\frac{1}{v_z}\int_z^\infty dz' e^{\frac{q}{v_z}(z'-z)}\mathbf{A}\Phi_w(z') \equiv \mathbf{S}\Phi_w. \tag{9}$$

The definitions of the operators $\mathbf{G}$, $\mathbf{Q}$ and $\mathbf{S}$ can be read off (8) and (9). Let $\mathbf{P}$ and $\mathbf{N}$ be projection operators on the $v_z \geq 0$ and $v_z < 0$ velocity subspaces, respectively (with $\mathbf{P}+\mathbf{N} = \mathbf{I}$). It follows that $\mathbf{P}\Phi_w = \mathbf{PGP}\Phi_w(0)+\mathbf{PQP}\Phi_w+\mathbf{PQN}\Phi_w$, and $\mathbf{N}\Phi_w = \mathbf{NSP}\Phi_w+\mathbf{NSN}\Phi_w$. The operator $\mathbf{NS}$ represents the events in which a particle whose velocity is $\mathbf{v}$, with $v_z < 0$, emerges from a collision at a point $z' > z$ and moves to the point z without further collision. Similarly, the operator $\mathbf{PQ}$ represents the events in which a particle whose velocity is $\mathbf{v}$, with $v_z > 0$, collides at $z' < z$ and proceeds, without further collision, to the point z. The operator $\mathbf{G}$ is the propagator corresponding to the motion of a particle from the boundary, $z = 0$, to z, without collision. Straightforward algebra yields:

$$\Phi_w = (\mathbf{I}-\mathbf{C})^{-1}\mathbf{PGP}\Phi_w(0) = (\mathbf{I}+\mathbf{C}+\mathbf{C}^2+...)\mathbf{PGP}\Phi_w(0), \tag{10}$$

where $\mathbf{C} \equiv \mathbf{NS}+\mathbf{PQ}$. The interpretation of (10) is rather simple: the function Φ_w is determined from its value at the boundary, $\Phi_w(0)$, via successive processes of collisions and free motions. For example, the nth order term $\mathbf{C}^n\mathbf{PGP}\Phi_w(0)$ is the contribution of the particles that come from the boundary (with positive z-component velocity), collide n times, following which their velocity is $\mathbf{v}$ and they move without collision to the point z. The next step is to invoke the 'microscopic' boundary conditions:

$$(\mathbf{v}\cdot\hat{\mathbf{n}}(\mathbf{x}))f(\mathbf{v},\mathbf{x}) = -\int_{\mathbf{v}'\cdot\hat{\mathbf{n}}<0} d\mathbf{v}'(\mathbf{v}'\cdot\hat{\mathbf{n}}(\mathbf{x}))f(\mathbf{v}',\mathbf{x})W(\mathbf{v}'\to\mathbf{v};\mathbf{x}), \tag{11}$$

where W is the accommodation which determines the distribution of outgoing velocities of a particle that collides with the boundary with a given velocity and $\hat{\mathbf{n}}$ is the vector normal to the boundary at a point $\mathbf{x}$ on it (in the case considered here it is the unit vector in the z direction). The accommodation function accounts for unresolved corrugation (of scales less than l). Upon employing the representation $f = f_0(1+\Phi_{ce}+\Phi_w)$ one can write the boundary condition in operatorial form as follows: $\mathbf{P}\Phi_w(0) = \mathbf{PRN}\Phi_w(0)+\mathbf{PB}f_{ce}(0)$, where $\mathbf{R}$ and $\mathbf{B}$ are operators that can be read off the result of the substitution of the decomposition of f in (11). Next, defining the operator $\mathbf{Z}$ as the projection on the value of a function at $z = 0$, e.g. $\mathbf{Z}\phi(z) = \phi(0)$, one obtains: $\mathbf{N}\Phi_w(0) = \mathbf{ZNS}(\mathbf{I}-\mathbf{C})^{-1}\mathbf{PGP}\Phi_w(0)$, where use has been made of the identity $\mathbf{N}(\mathbf{I}-\mathbf{C})^{-1}\mathbf{P} = \mathbf{NS}(\mathbf{I}-\mathbf{C})^{-1}\mathbf{P}$, which follows from $\mathbf{NP} = 0$ and $\mathbf{NC} = \mathbf{NS}$. Combining these expressions in a straightforward way one obtains: $\mathbf{N}\Phi_w(0) = \mathbf{ZNS}(\mathbf{I}-\mathbf{C})^{-1}\mathbf{PGP}(\mathbf{PRN}\Phi_w(0)+\mathbf{PB}f_{ce}(0))$, or, solving for $\mathbf{N}\Phi_w(0)$: $\mathbf{N}\Phi_w(0) = (\mathbf{I}-\mathbf{ZNS}(\mathbf{I}-\mathbf{C})^{-1}\mathbf{PGPR})^{-1}\mathbf{ZNS}(\mathbf{I}-\mathbf{C})^{-1}\mathbf{PGPB}f_{ce}(0)$. This is an explicit expression for the wall contribution to the distribution function. Next, the solubility conditions for (5) are invoked:

$$\int \left(\Psi_i f_0\right)\left(v_z \frac{\partial}{\partial z}\Phi_w\right) d\mathbf{v} = \int \left(\Psi_i f_0\right)\left(\mathbf{L}\Phi_w\right) = \int \left(\mathbf{L}\Psi_i\right) f_0 \Phi_w d\mathbf{v} = 0, \quad (12)$$

i.e. $\frac{\partial}{\partial z}\int \Psi_i f_0 v_z \Phi_w d\mathbf{v} = 0$, hence (since $\Phi_w \overset{z\to\infty}{\longrightarrow} 0$): $\int \Psi_i v_z \Phi_w f_0 d\mathbf{v} = 0$. The physical significance of the latter formula is that the value of each flux in the bulk equals its value at the boundary, hence Φ_w cannot contribute to any of the fluxes. The resulting conditions are truly conditions on f_{ce} since Φ_w has been expressed as a functional of f_{ce}. In practice, the orthogonality (or solubility) conditions create relations between the fields and their derivatives. A similar formulation holds for the case of an inelastic gas.

Upon employing the above formulation one obtains for the case of an inelastic gas and a diffusely (but elastically) reflecting boundary that the (slip) velocity parallel to the boundary is given by: $V_x = \alpha l \frac{\partial}{\partial z} V_x$, where $\alpha \approx 0.728 + 0.130\epsilon$, to second order in the collisions. When the boundary is characterized by a degree of inelasticity (for the normal part of the velocity), ϵ_w, and a thermal gradient in the z direction is present (as must be the case for an energy absorbing boundary), one obtains the following result: $V_z = -\zeta l \sqrt{T}\left(\frac{\partial}{\partial z}\log n + \frac{1}{2}\frac{\partial}{\partial z}\log T\right)$ where $\zeta \approx 0.044\epsilon$, to lowest order in the collisions. This boundary condition for V_z is quite surprising. It implies that V_z does not necessarily vanish in the general case, unless a specific relation between the gradients of the number density and granular temperature is satisfied. The above expression for V_z does not vanish in the case of steady shear, hence mass conservation seems to be violated. This result pertains only to inelastically colliding systems as V_z is predicted to vanish for $\epsilon = 0$. The resolution of this "paradox" can be found by recalling that in a steady sheared state the orders $K\epsilon$ (which is the order of the above expression for V_z) and K^3 are the same (since, following (1), $\gamma^2 \propto \epsilon$ at given T or $K^2 \propto \epsilon$), hence a correction that is of super-Burnett order should be added to the above expression for V_z. When this is done the value of V_z vanishes in steady states, as it should. In the same situation, the boundary condition for the temperature, T, is $\epsilon_w T = \beta l \frac{\partial}{\partial z} T + \delta l \frac{T}{n}\frac{\partial}{\partial z} n$, where $\beta \approx 2.671 + 1.945\epsilon$ and $\delta \approx 3.810\epsilon$, to second order in the collisions.

5 Conclusion

It has been shown how a systematic perturbative approach to the derivation of equations of motion and boundary conditions for granular gases can be designed and carried out. The formulation for the boundary conditions is relevant to molecular gases as well and it is the first systematic expansion devised for this purpose. These expansions are formally limited to near elastic cases though in practice they seem to hold for moderate degrees of inelasticity, a fact that needs to be better understood. The ill posedness of the Burnett and super-Burnett equations of motion creates a serious problem: on one hand these contributions are sizeable in granular systems and they cannot be ignored but their incorporation leads to a fundamental problem. A possible solution to this problem is formal resummation or regularization. This technique needs to be further developed. The root cause of all of these 'technical' problems arising in the kinetic theory of granular gases is the lack of scale separation in these systems. This is also a property of dense systems. The latter have so far not been studied beyond the level of phenomenological (or mean field) modeling, though some attempts in this direction have been noted. In summary, there are numerous open questions and fundamental problems one encounters while attempting to produce a first-principles theory for granular matter; these exciting and important problems will no doubt be part of many future investigations.

Acknowledgement. This work has been partially supported by the National Science Foundation, the Department of Energy, the United-States - Israel Binational Science Foundation (BSF) and the Israel Science Foundation (ISF).

References

1. J.T. Jenkins, S.B. Savage, *A Theory for Rapid Granular Flow of Identical, Smooth, Nearly Elastic, Spherical Particles*, J. Fluid Mech. **130**, 187–202 (1983); C.K.K. Lun, S.B. Savage, D.J. Jeffrey, and N. Chepurniy, *Kinetic Theories of Granular Flow: Inelastic Particles in a Couette Flow and Slightly Inelastic Particles in a General Flow Field*, J. Fluid Mech. **140,** 223–25 (1984).
2. J.T. Jenkins and M.W. Richman, *Plane Simple Shear of Smooth Inelastic Circular Disks: the Anisotropy of the Second Moment in the Dilute and Dense Limits*, J. Fluid Mech. **192**, 313–328 (1988).
3. M.K. Kogan, *Rarefied Gas Dynamics*, (Plenum Press, New York, 1969).
4. H.A. Kramers, *On the Behavior of a Gas Near a Wall,* Suppl. Vol. VI, Nuovo Cimento **9**, 297–304 (1949).
5. C. Cercignani, *Theory and Applications of the Boltzmann Equation*, (Scottish Academic Press, Edinburgh and London, 1975).
6. S. Chapman and T.G. Cowling, *The Mathematical Theory of Nonuniform Gases*, (Cambridge University Press, Cambridge, 1970, and refs. therein).
7. J.T. Jenkins and E. Askari, *Boundary Conditions for Rapid Granular Flows,* J. Fluid Mech. **223**, 497–508 (1991), and refs. therein .
8. I. Goldhirsch, N. Sela, and S.H. Noskowicz, *Kinetic Theoretical Study of a Simply Sheared Granular Gas to Burnett Order*, Phys. Fluids **8**, 2337–2353 (1996); I. Goldhirsch and N. Sela, *Origin of Normal Stress Differences in Rapid Granular Flows*, Phys. Rev. E **5**, 4458–4461 (1996); N. Sela and I. Goldhirsch, *Hydrodynamic Equations for Rapid Flows of Smooth Inelastic Spheres, to Burnett Order*, J. Fluid Mech. **361**, 41–74 (1998), and refs. therein.

9. M.-L. Tan and I. Goldhirsch, *Rapid Granular Flows as Mesoscopic Systems,* Phys. Rev. Lett. **81**, 3022–3025 (1998).
10. F. Schurrer and G.Kugerl, *The Relaxation of Single Hard Sphere Gases*, Phys. Fluids A **2**, 609–618 (1990).
11. A.V. Bobylev, *Exact Solutions of the Nonlinear Boltzmann Equation and the Theory of Maxwell Gas Relaxation*, Theor. Math. Phys. **60**, 280–310 (1984); P. Rosenau, *Extending Hydrodynamics via the Regularization of the Chapman-Enskog Expansion,* Phys. Rev. A **40**, 7193–7196 (1989); M. Slemrod, *Renormalization of the Chapman-Enskog Expansion: Isothermal Fluid Flow and Rosenau Saturation*, J. Stat. Phys. **91**, 285–305 (1998).
12. I. Oppenheim, *Nonlinear Response Theory,* in: *Correlation Functions and Quasiparticle Interactions in Condensed Matter,* J. Woods Halley, (Ed.), pp. 235–258 (Plenum Press, 1978).
13. I. Goldhirsch and T.P.C. van Noije, *Green-Kubo Relations for Granular Fluids*, Phys. Rev. E, submitted, (1999).
14. J.J. Brey, J.W. Dufty, C.S. Kim, and A. Santos, *Hydrodynamics for Granular Flow at Low Density*, Phys. Rev. E **58**, 4638–4653 (1998), and refs. therein.
15. E.J. Boyle and M. Massoudi, *A Theory for Granular Materials Exhibiting Normal Stress Effects Based on Enskog's Dense Gas Theory*, Int. J. Eng. Sci. **28**, 1261–1275 (1990).
16. N. Sela and I. Goldhirsch, *Boundary Conditions for Granular and Molecular Gases*, unpublished, (1999).
17. C.L Pekeris, *Solution of the Boltzmann-Hilbert Integral Equation,* Proc. Nat. Acad. Sci. **41,** 661–664 (1955).

Particle Segregation in the Context of the Species Momentum Balances

B.Ö. Arnarson and J.T. Jenkins

Department of Theoretical and Applied Mechanics, Cornell University, Ithaca, NY 14853, USA

Abstract. We outline the development of a kinetic theory for particle segregation in collisional flows of a binary mixture of nearly elastic spheres. We take care to place the segregation mechanism in the context of an appropriately weighted difference of the momentum balances for the two species.

1 Introduction

Theories for the prediction of particle segregation in collisional shearing flows make use of the analogy between the molecules of a dense gas and the agitated macroscopic grains that exists, provided that the collisions between grains do not dissipate too much energy [1]. The most refined theory for segregation in dense molecular gases is a kinetic theory for mixtures of elastic spheres based on the assumption of molecular chaos and the correct extension to mixtures of Enskog's characterization of the influence of the finite volume of the particles on their frequency of collision [2,3]. The derivation of the theory employs the Chapman-Enskog expansion to obtain the velocity distributions to first order in spatial gradients. The Enskog approximation is then employed in order to get explicit expressions for the velocity distribution functions. The results of the theory are consistent with irreversible thermodynamics, have been presented graphically, and are available numerically.

In our application of this theory to mixtures of inelastic grains, we have characterized the amount of permitted dissipation and we have obtained analytical results valid to second order in the Enskog approximation for binary mixtures [1,4]. Analytical expressions permit the rational approximation of the coefficient governing transport, segregation, and dissipation over a range of mixture and species volume fractions and particle size and mass ratios.

The presence of dissipation in collisions permits collisional shearing flows of inelastic grains to achieve a steady balance in which the rate at which the kinetic energy of the velocity fluctuations increases, due to collisions driven by gradients in the mixture mean velocity, is equal to the rate at which it decreases, due to the energy lost in each collision, and the rate at which it is transported, due to inhomogeneities in the flow. The experimental study of special, simple types of binary mixtures of spheres in inhomogeneous steady shearing flows permits various aspects of the phenomena of segregation to be studied separately and helps in developing intuition about the behavior of more complicated mixtures.

In order to carry out the analysis of such experiments, we have extended existing theory for shearing flows of identical spheres in several ways. We have accounted for friction in collisions between particles in the flow by incorporating the additional energy loss into an effective coefficient of restitution [5]. We have extended existing boundary conditions for a single species interacting with either a bumpy or a frictional boundaries to apply to a binary mixture interacting with a boundary that is both bumpy and frictional [6]. We have generalized existing theory for a dense shearing flow of a single species to apply to the prediction of the profiles of mixture mean velocity and mixture mean kinetic energy in dense shearing flows of a binary mixture between bumpy, frictional boundaries.

In order to incorporate friction into the kinetic theory, a simple but realistic model of a frictional collision must be employed. The simplest such model distinguishes between collisions in which the relative velocity of the points of contact is momentarily zero during a collision and those in which it is not. The former are called sticking collisions, the latter are called sliding collisions. The model employs three parameters: a coefficient of normal restitution for both sticking and sliding collisions, a coefficient of friction for sliding collisions, and a coefficient of tangential restitution for sticking collisions [7]. We have employed the model to interpret the results of experiments on binary collisions between identical spheres and collisions between a single sphere and a flat or a bumpy boundary. The parameters determined in this way provide an excellent fit to the data [8,9].

2 Theoretical Framework

Here we sketch the structure of the theory for segregation in a binary mixture of spheres and focus on how segregation can be understood in the context of the balances of forces for each species in the context of the Revised Enskog Theory (RET) developed by van Beijeren and Ernst [10]. This theory overcomes the difficulties associated with the choice of the point along the line of centers of two colliding particles at which the radial distribution functions in the Enskog factor is evaluated in the Standard Enskog Theory (SET). In the SET, different choices result in different theories for segregation, none of which is consistent with irreversible thermodynamics [2]. The detailed forms of the terms not given explicitly are provided by Jenkins and Mancini [1] and Arnarson and Willits [4].

The two species are labeled A and B. The spheres have radii r_i, masses m_i, and number densities n_i, where $i = A, B$.

2.1 Momentum Balance

The balance of momentum for the mixture as a whole may be written as

$$\rho \frac{\partial \mathbf{u}}{\partial t} + \rho (\mathbf{u} \cdot \nabla) \mathbf{u} = \nabla \cdot \mathbf{t} + \rho \mathbf{F}$$

respectively, where $\rho \equiv m_A n_A + m_B n_B$ is the mixture mass density, $\mathbf{u}$ is the mass-center mixture velocity, $\mathbf{t}$ is the symmetric mixture stress tensor, and $\mathbf{F}$ is the force per unit mass on the mixture.

The mixture stress tensor is given by

$$\mathbf{t} = -P\mathbf{1} + 2\eta\left[\nabla\mathbf{u} + (\nabla\mathbf{u})^T\right],$$

where the superscript T denotes the transpose.

The mixture pressure P is proportional to the mixture fluctuation energy T:

$$P = \left(n + \sum_{i=A,B}\sum_{j=A,B} K_{ij}\right)T,$$

where $n \equiv n_A + n_B$ is the mixture number density, the K_{ij} are functions of the sums $r_{ij} \equiv r_i + r_j$, the number densities n_i, and the radial distribution functions g_{ij} of contacting pairs:

$$K_{ij} \equiv \frac{2}{3}\pi n_i n_j r_{ij}^3 g_{ij},$$

with [11]

$$g_{ij} \equiv \frac{1}{1-\xi_3} + 3\frac{r_i r_j}{r_{ij}}\frac{\xi_2}{(1-\xi_3)^2} + 2\left(\frac{r_i r_j}{r_{ij}}\right)^2\frac{\xi_2^2}{(1-\xi_3)^3},$$

and

$$\xi_\alpha \equiv \frac{4\pi}{3}(n_A r_A^\alpha + n_B r_B^\alpha).$$

The mixture fluctuation energy T is two-thirds the sum of the average kinetic energy associated with the velocity of each species relative to the mixture velocity, weighted by the number fractions n_A/n or n_B/n of that species.

The shear viscosity η is

$$\eta \equiv \frac{1}{2}\sum_{i=A,B} b_{i0}\left(n_i + \frac{4}{5}\sum_{j=A,B} K_{ij}\frac{m_j}{m_{ij}}\right)T + \frac{1}{5}\sum_{i=A,B}\sum_{j=A,B}\left(\frac{2}{\pi}\frac{m_i m_j}{m_{ij}}T\right)^{1/2} K_{ij} r_{ij},$$

where $m_{ij} = m_i + m_j$ are the sums of the masses and the b_{i0}, associated with the perturbation to the Maxwellian velocity distribution function, are known analytic functions of the number densities, diameters, and masses of the two species that are given by the Chapman-Enskog procedure. The fluxes of mixture momentum and mixture fluctuation energy are the same in the RET and SET [10].

2.2 Energy Balance

Similarly, the balance of fluctuation energy for the mixture is

$$\frac{3}{2}n\dot{T} - \frac{3}{2}T\nabla \cdot \mathbf{j} = -\nabla \cdot \mathbf{Q} + tr(\mathbf{t} \cdot \nabla \mathbf{u}) + (\mathbf{F}_A + \mathbf{F}_B) \cdot \mathbf{j} - \gamma,$$

where $\mathbf{Q}$ is the flux of mixture fluctuation energy, $\mathbf{j} \equiv n_A \mathbf{v}_A + n_B \mathbf{v}_B$ is the particle flux associated with the diffusion velocities, and γ is the rate of dissipation of fluctuation energy in the mixture due to the inelasticity of the collisions. The species diffusion velocities $\mathbf{v}_i$ are defined as the difference between the average velocity of a species $\mathbf{u}_i$ and the mass-center velocity $\mathbf{u}$ of the mixture.

When the influence of diffusion on the flux of fluctuation energy is neglected, it is proportional to ∇T:

$$\mathbf{Q} = -\kappa \nabla T,$$

where the thermal conductivity κ is

$$\kappa \equiv -\frac{5}{4} \sum_{i=A,B} a_{i1} \left(n_i + \frac{12}{5} \sum_{j=A,B} K_{ij} \frac{m_i m_j}{m_{ij}^2} \right) \left(\frac{2T}{m_i} \right)^{1/2}$$
$$+2 \sum_{i=A,B} \sum_{j=A,B} \left(\frac{2}{\pi} \frac{m_i m_j}{m_{ij}^3} T \right)^{1/2} K_{ij} r_{ij},$$

and the a_{i1}, associated with the perturbation to the Maxwellian velocity distribution function, are known analytic functions of the number densities, radii, and masses of the two species. The more elaborate form of the energy flux that includes the effects of diffusion is provided by Jenkins and Mancini [1].

The rate of collisional dissipation γ is proportional to $T^{3/2}$:

$$\gamma \equiv \sum_{i=A,B} \sum_{j=A,B} K_{ij} \frac{m_j}{m_{ij}} (1 - e_{ij}) \left(\frac{2}{\pi} \frac{m_{ij}}{m_i m_j} \right)^{1/2} T^{3/2},$$

where e_{ij} are the effective normal coefficients of restitution for a collision between spheres of species i and j. The presence of the collisional dissipation in the energy balance distinguishes the macroscopic system from its molecular counterpart.

2.3 Segregation

The species diffusion velocities may be calculated using the distribution function determined in the Chapman-Enskog procedure in both the SET and RET. In either case, their difference is given by

$$\mathbf{v}_A - \mathbf{v}_B = -\frac{n^2}{n_A n_B} D_{AB} (\mathbf{d}_A + K_T \nabla T), \tag{1}$$

where D_{AB} and K_T are called, respectively, the coefficients of ordinary and thermal diffusion, and $\mathbf{d}_A$ is called the diffusion force. These have the forms:

$$D_{AB} \equiv \frac{3}{2}\frac{1}{n g_{AB}}\left(\frac{2 m_{AB} T}{m_A m_B}\right)^{1/2}\frac{1}{8 r_{AB}^2},$$

$$K_T \equiv \frac{4}{3}\pi^{1/2}\frac{n_A n_B}{n} g_{AB} r_{AB}^2\left[\left(\frac{m_B}{m_{AB}}\right)^{3/2} a_{A1} - \left(\frac{m_A}{m_{AB}}\right)^{3/2} a_{B1}\right],$$

and

$$\begin{aligned}
\mathbf{d}_A \equiv & -\frac{\rho_A}{\rho}\frac{1}{nT}\left[\nabla P + \rho_B\left(\frac{\mathbf{F}_A}{m_A} - \frac{\mathbf{F}_B}{m_B}\right)\right] \\
& + \frac{1}{nT}\left(n_A + K_{AA} + 2\frac{m_A}{m_{AB}} K_{AB}\right)\nabla T \\
& + \frac{1}{n}\left(E_{AA}\nabla n_A + E_{AB}\nabla n_B\right). \qquad (2)
\end{aligned}$$

In the SET,

$$E_{ij} \equiv \delta_{ij} + \frac{4\pi}{3} n_i r_{ij}^3 g_{ii} + \sum_{k=A,B}\frac{4\pi}{3} y_{ij} n_i n_k r_{ik}^3 \frac{\partial g_{ik}}{\partial n_j},$$

where the quantities y_{ij}, with $y_{ij} + y_{ji} = 1$ and $0 \leq y_{ij} \leq 1$, determine the distance along the line of centers of two colliding spheres at which the radial distribution functions are evaluated; in the RET,

$$E_{ij} \equiv \frac{n_i}{T}\frac{\partial \mu_i}{\partial n_j},$$

where μ_i is the chemical potential of species i:

$$\begin{aligned}
\frac{\mu_i}{T} = & \ln n_i - \ln(1 - \xi_3) + \frac{4\pi r_i^3 P}{3T} + \frac{3\xi_2 r_i}{1 - \xi_3} + \frac{3\xi_1 r_A^2}{1 - \xi_3} + \frac{9\xi_2^2 r_i^2}{2(1 - \xi_3)^2} \\
& + 3\left(\frac{\xi_2 r_i}{\xi_3}\right)^2\left[\ln(1 - \xi_3) + \frac{\xi_3}{1 - \xi_3} - \frac{\xi_3^2}{2(1 - \xi_3)^2}\right] \\
& - \left(\frac{\xi_2 r_i}{\xi_3}\right)^3\left[2\ln(1 - \xi_3) + \frac{\xi_3(2 - \xi_3)}{1 - \xi_3}\right].
\end{aligned}$$

Here we wish to consider segregation in the context of the species momentum balances:

$$\rho_i\frac{\partial \mathbf{u}_i}{\partial t} + \rho_i(\mathbf{u}_i \cdot \nabla)\mathbf{u}_i = \nabla \cdot \mathbf{t}_i + n_i \mathbf{F}_i + \phi_{ij},$$

where ϕ_{ij} is the interaction force exerted on species i by species j. The interaction forces can, in principle, be calculated in the RET by evaluating the collision integral for the momentum exchange using the distribution function provided by the Chapman-Enskog procedure and the appropriate Enskog factor. Here, we determine the form of the interaction force in a different but equivalent way.

Following Jenkins and Mancini [12], we consider the difference between the momentum balance for species A, weighted by $1/\rho_A$, and the momentum balance for species B, weighted by $1/\rho_B$. To be faithful to the Chapman-Enskog procedure, we disregard all terms involving derivatives higher than the first. We obtain

$$\frac{1}{\rho_A}\phi_{AB} - \frac{1}{\rho_B}\phi_{BA} = \frac{1}{\rho_A}\nabla p_A - \frac{1}{\rho_B}\nabla p_B - \frac{\mathbf{F}_A}{m_A} + \frac{\mathbf{F}_B}{m_B}, \tag{3}$$

where p_i are the partial pressures:

$$p_i = (n_i + K_{ii} + K_{ij})T,$$

for $i \neq j$. Because $\phi_{AB} = -\phi_{BA}$, we have

$$\phi_{AB} = \frac{\rho_B}{\rho}\nabla p_A - \frac{\rho_A}{\rho}\nabla p_B - \frac{\rho_A\rho_B}{\rho}\left(\frac{\mathbf{F}_A}{m_A} - \frac{\mathbf{F}_B}{m_B}\right). \tag{4}$$

The gradients in partial pressure are

$$\nabla p_i = (n_i + K_{ii} + K_{ij})\nabla T + T\left(E_{ii} + \frac{1}{n_i}K_{ij}\right)\nabla n_i \\ + T\left(E_{ij} - \frac{1}{n_j}K_{ij}\right)\nabla n_j, \tag{5}$$

for $i \neq j$, with the E_{ij} are to be taken in SET or RET, as appropriate. We carry out the calculation using the SET and interpret the result in terms of the RET at its conclusion.

Because $\mathbf{d}_A + \mathbf{d}_B = \mathbf{0}$, we can add

$$nT\left(\mathbf{d}_A - \frac{\rho_B}{\rho}\mathbf{d}_A + \frac{\rho_A}{\rho}\mathbf{d}_B\right) \tag{6}$$

to the right-hand side of (4) without influencing the equation. Upon employing (2) and expanding the last two in terms in (6), we find that

$$\phi_{AB} = nT\mathbf{d}_A - \left(\frac{m_A}{m_{AB}} - \frac{m_B}{m_{AB}}\right)K_{AB}\nabla T + K_{AB}T\left(\frac{1}{n_A}\nabla n_A - \frac{1}{n_B}\nabla n_B\right).$$

With the exception of $\mathbf{d}_A$, all terms on the right-hand side of this equation are the same in the SET and RET and we know precisely how $\mathbf{d}_A$ differs from the SET and RET; so we conclude that the expression is also valid in the RET. Then, with $\mathbf{d}_A$ given by (1),

$$\phi_{AB} = -\frac{n_A n_B}{n}TD^{-1}_{AB}(\mathbf{v}_A - \mathbf{v}_A) - nK_T\nabla T \\ - \left(\frac{m_A}{m_{AB}} - \frac{m_B}{m_{AB}}\right)K_{AB}\nabla T + K_{AB}T\left(\frac{1}{n_A}\nabla n_A - \frac{1}{n_B}\nabla n_B\right). \tag{7}$$

Equation (7) is the desired expression for the inter-particle force in the RET. With it and the RET version of the gradients of partial pressure (5), the weighted difference of the approximate species momentum balances (3) can be considered to describe segregation as an equivalent alternate to (1), and particle segregation in collisional flows can be regarded as having been placed in the context of the balance of momentum.

Acknowledgement. This research has been supported by NASA's Microgravity Science and Applications Division under contract NCC3-468.

References

1. J.T. Jenkins and F. Mancini, *Kinetic theory for smooth, nearly elastic spheres,* Physics of Fluids **A1**, 2050-2057 (1989).
2. M. López de Haro, E.G.D. Cohen, and J.M. Kincaid, *The Enskog theory for multicomponent mixtures. I. Linear transport theory*, J. Chem. Physics **78**, 746-2759 (1983).
3. J.M. Kincaid, E.G.D. Cohen, and M. López de Haro, *The Enskog theory for multicomponent mixtures. IV. Thermal diffusion,* J. Chem. Physics **86**, 963-975 (1987).
4. B.Ö. Arnarson and J.T. Willits, *Thermal diffusion in binary mixtures of smooth, nearly elastic spheres with and without gravity,* Physics of Fluids **10**, 1324-1328 (1998).
5. J.T. Jenkins and C. Zhang, *Kinetic theory for identical, slightly frictional, nearly elastic spheres,* under review, Physics of Fluids.
6. J.T. Jenkins, *Boundary conditions for collisional grain flow at a bumpy, frictional wall,* in: *Granular Gases*, H.J. Herrmann, S. Luding, and T. Pöschel, (Eds.), in press, (Springer, Berlin).
7. O.R. Walton, *Numerical simulations of inelastic, frictional particle interactions,* in: *Particulate Two-Phase Flow*, M.C. Roco, (Ed.), (Butterworth-Heinemann, Boston, 1992).
8. S.F. Foerester, M.Y. Louge, H. Chang, and K. Allia, *Measurements of the collision properties of small spheres*, Physics of Fluids **6**, 1108-1115 (1994).
9. A. Lorenz, C. Tuozzolo, and M.Y. Louge, *Measurement of the impact properties of small, nearly spherical particles*, Exp. Mechanics **37**, 292-298 (1997).
10. H. Van Beijeren and M.H. Ernst, *The modified Enskog equation,* Physica **68**, 437-456 (1973).
11. G.A. Mansoori, N.F. Carnarhan, K.E. Starling, and T.W. Leland, Jr, *Equilibrium thermodynamic properties of a mixture of hard spheres,* J. Chem. Physics **54**, 1523-1525 (1971).
12. J.T. Jenkins and F. Mancini, *Balance laws and constitutive relations for plane flows of a dense, binary mixture of smooth, nearly elastic disks,* J. Appl. Mechanics **109**, 27-34 (1987).

Avalanche Parameters: Dependence on the Size of the Granular Packing

M.A. Aguirre[1], N. Nerone[1], A. Calvo[1], I. Ippolito[2], and D. Bideau[2]

[1] Grupo de Medios Porosos, Facultad de Ingeniera, Universidad de Buenos Aires, Paseo Coln 850, 1063 Buenos Aires, Argentina
[2] Groupe Matiere Condensée et Matériaux, U.M.R. 6626, Université de Rennes 1, Campus de Beaulieu, 35042 Rennes Cedex, France

Abstract. The influence of the granular packing length on the avalanche parameters, such as its mass and the critical angles at which it begins and stops, is studied for packings of mono size glass beads under a controlled humidity environment.
In order to understand the dynamics of this kind of systems, experiments are performed in boxes of two different dimensions to see the influence on the parameters. For both boxes, the critical angles show the same qualitative behavior. While the number of layers involved in the avalanche are determined by the box dimensions.

1 Introduction

Surface flows on a sand heap have been the subject of a tremendous number of scientific works [3], particularly in the last ten years. The equilibrium of the heap is defined by (at least) two critical angles (Fig. 1): The maximum angle of stability, θ_M, which is here defined as the angle at which, on average, the avalanche starts and the angle of repose, θ_r, which is the angle at which, on average, it stops. The difference between these two angles $\delta = \theta_M - \theta_r$ can be more easily measured than θ_r so measuring θ_M and δ will be enough to know the behavior of these critical angles.

2 Experimental Set Up and Procedure

To study the influence of the packing size on the stability of the granular system, experiments are done in two different boxes: 32 cm long, 26 cm wide and 64 cm long, 13 cm wide. In order to insure disorder, the bottom bed is made by gluing, over a flat piece of glass, the same glass beads (2.2 ± 0.2 mm diameter) that fill both systems. The quantity of mass necessary to cover the bottom of the box with a 2D packing fraction of 0.7 (widely used in previous works [1,2]) is calculated. This quantity (230 g) is taken as the mass of one layer which insures a regular increase of the height of the system. We have also placed a blocking bar at the outlet of the box whose upper side is at the level of the center of the balls of the upper layer of the system. Our experiments are done under a controlled humidity of 50% provided by a cold air flux. Humidities out of the interval 45%–60% can have a great influence in the experiments due to cohesion effects provoked by capillary and electrostatic forces [3].

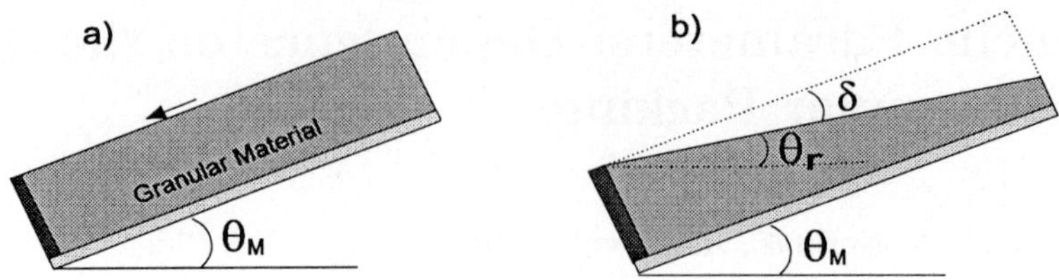

Fig. 1. a) Maximum angle of stability θ_M, the avalanche begins. b) Final configuration for a number of layers large enough. The white zone corresponds to the total displaced mass during the avalanche.

For a system with a given number of layers, N, 10 identical experiments are performed. The number of layers is varied between 1 and 34. A given box is secured on a heavy inclinable plane and the angle with respect to the horizontal is increased until the system destabilizes at θ_M [4]. We measure θ_M and the mass of the avalanche, M. δ can be measured directly for $N \geq 10$. For $N < 10$ an effective δ value, $\delta_{\text{effective}}$, can be obtained considering the avalanche mass and the final geometry reached by the system, as will be explained in Sect. 3.3. Finally mean values $\langle\theta_M\rangle$, $\langle\delta\rangle$ and $\langle M\rangle$ are computed.

3 Experimental Results

3.1 Experimental Observations

Independently of the packing length the following observations can be remarked:

- Before a large avalanche is produced at θ_M, small surface rearrangements, involving few grains, take place.
- This avalanche starts at any place of the free surface, but very quickly, its size increases by a domino like effect, and becomes of the order of the size of the system.
- Depending on the thickness of the packing two kinds of flow can be observed:
  - Thin avalanches ($N \leq 8$): the avalanche starts anywhere on the free surface. A bouncing flow takes place and grains interact directly with the fixed rough bed.
  - Thick avalanches: ($N \geq 13$) avalanches also start anywhere on the free surface. In this case grains creep down and direct observation shows that the grains rolling in upper layers move faster than those of lower layers.
  - Finally, a transition regime from thick to thin avalanches can be observed ($8 < N < 13$). In this case both bouncing and creeping flows are observed.
- A critical value for the number of layers, N_c, is found. For a packing with $N > N_c$ only the N_c superficial layers are affected by the avalanche and the final free surface of the system is flat and extends over the full length and width of the box leaving an empty volume that is almost a perfect

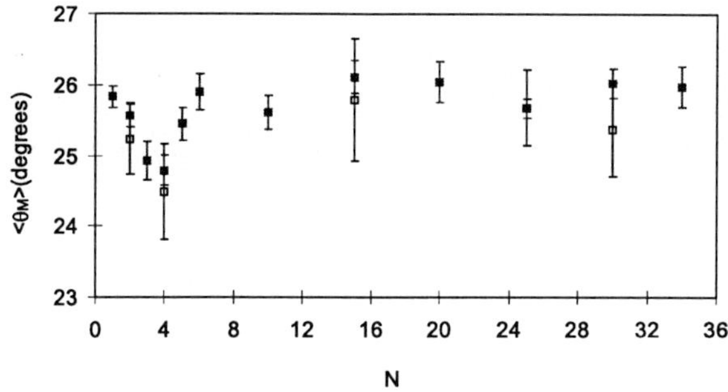

Fig. 2. $\langle\theta_M\rangle$ as a function of the number of layers for both boxes. Filled squares correspond to the short box and open squares correspond to the longer one.

wedge. On the contrary, for granular systems with $N < N_c$, all N layers are involved in the avalanche process. We found $N_c^{\text{short}} \cong 13$ for the short box and $N_c^{\text{long}} \cong 26$ for the box twice longer.

3.2 Maximum Angle of Stability

In the following sections the average over the 10 measurements for a given box and a given number of layers is presented. The maximum angle of stability presents the same behavior for both boxes (Fig. 2). Three regimes are observed for the mean maximum angle of stability depending on the number of layers of the granular packing:

a) $N \leq 4$ (thin avalanche regime): the system becomes more unstable as the number of layers increases ($\langle\theta_M\rangle$ decreases). This fact is predictable because the system is not able to dilate: for few layers the grains must interact with the fixed rough bed and as the number of layers is larger this constraint gets weaker and dilatancy can take place, allowing an easier displacement of grains. According to this, we can consider that dilatancy only requires 4 layers to develop. Nevertheless, as the system gets larger and denser, this dilatancy effect begins to be shielded and gives place to the following regime.
b) $4 < N < 10$ (transition to thick avalanche regime): the system becomes more stable. This transition may be due to an increase of the packing fraction of the granular system which means a decrease in dilatancy [5,6]. Thicker systems are more packed and therefore more stable.
c) $N \geq 10$ (thick avalanche regime): the stability of the system is independent of the number of layers and $\langle\theta_M\rangle = 25.9° \pm 0.2°$. In this regime the packing fraction reaches an asymptotic value and the stability of the system is independent of its thickness.

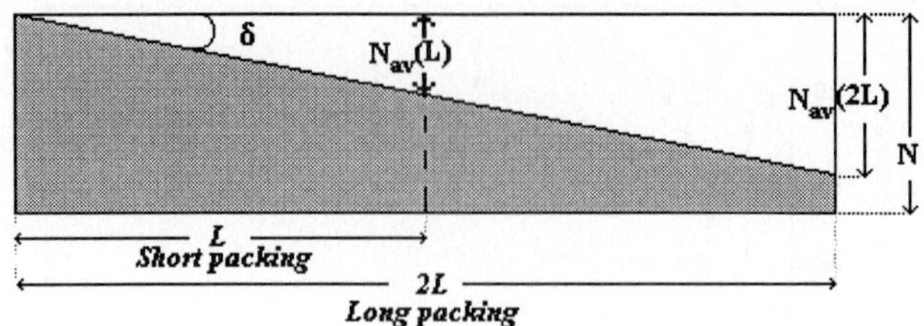

Fig. 3. Visualization of the number of layers (N_{av}) affected by the avalanche. For a fixed N layer packing, δ is almost the same for both granular packings but not the number superficial layers involved in M.

3.3 Angle $\langle\delta\rangle$

Angle $\langle\delta\rangle$ presents two behaviors as a function of N. For $N < 10$, the total mass and the bouncing flow during the avalanche do not allow the packing final free surface to take a clear profile. Nevertheless, a $\langle\delta_{\mathrm{effective}}\rangle$ can be obtained considering that the retained mass is placed near the blocking bar forming a wedge of angle $\delta_{\mathrm{effective}}$. This $\langle\delta_{\mathrm{effective}}\rangle$ increases with N. For $N > 10$, the packing final free surface is flat and forms a wedge of angle δ that is directly measured with a goniometer. In this case, $\langle\delta\rangle$ reaches the same constant value for both boxes: $\langle\delta\rangle = 4.7° \pm 0.8°$. This means that after the packing destabilizes, it evolves toward the same $\langle\delta\rangle$.

The latter explains the existence of different N_c values depending on the system length. As δ takes the same value independently of the box length (Fig. 3), it means the avalanche has affected a larger number of layers in the longer box. When the wedge is reached ($N > 10$): $N_c = \frac{L\tan(\delta)}{d}$ and as $\langle\delta\rangle$ is constant, $N_c \propto L$, which explains why $N_c^{\mathrm{long}} = 2N_c^{\mathrm{short}} \cong 26$. In this way, we verify that the avalanche dynamics is not only determined by the number of layers of the system but strongly depends on its length. Also, notice that the angle of repose is $\theta_r = \theta_M - \delta$, then it will also reach a constant value for $N > 10$.

3.4 Avalanche Mass

The mean avalanche mass presents two different behaviors depending on the number of layers involved in the avalanche. Remember that we have experimentally found that $N_c^{\mathrm{short}} \cong 13$ and $N_c^{\mathrm{long}} \cong 26$.

$N < N_c$: The mean mass of the avalanche $\langle M\rangle$, is proportional to the number of layers of the system and all N layers are involved in the avalanche but only a portion of them flows out of the box.

$N > N_c$: The mass of the avalanche is approximately constant. In these experiments the final free surface of the system is almost flat (the wedge is reached) and only N_c superficial layers are involved in the avalanche. Then, the mass of the avalanche must be equal to half the parallelepiped containing

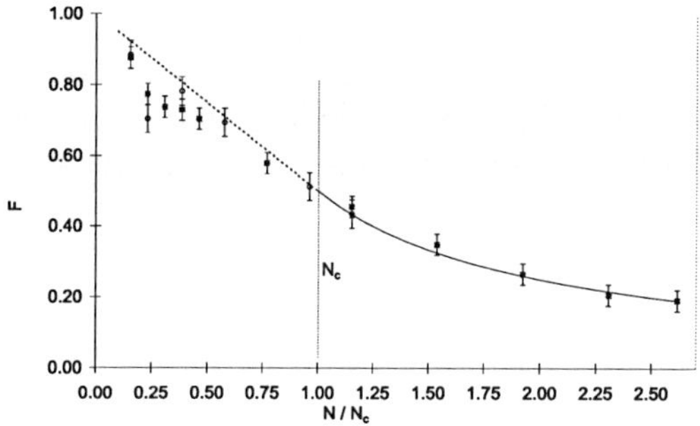

Fig. 4. Fraction of avalanche mass as a function of the number of layers scaled with N_c for both boxes. Filled squares correspond to the short box and open circles correspond to the longer one.

these N_c layers: $\langle M\rangle = mN_c/2$. Considering the asymptotic values of $\langle M\rangle$ for both boxes ($\langle M^{\mathrm{long}}\rangle = 3000 \pm 30$ g and $\langle M^{\mathrm{short}}\rangle = 1524 \pm 70$ g) one finds: $N_c^{\mathrm{short}} = 13 \pm 1$ and $N_c^{\mathrm{long}} = 26 \pm 1$ which agrees quite well with experimental values.

For an easier comparison of the results obtained for both boxes we have plotted (Fig. 4) the ratio, F, between the avalanche mass M and the total packing mass, M_T, as a function of N/N_c. We observe that both sets of data collapse in the same curve, displaying a decreasing behavior. Considering that $M_T = m\,N$ and $\langle M\rangle = m\,N_c/2$ we obtain: $F = \langle M\rangle/M_T = \frac{1}{2}N_c/N$ for $N/N_c > 1$. This variation is drawn, in Fig. 4, with a solid line showing an excellent agreement with experimental data. For $N/N_c < 1$ and assuming a constant value of $\langle\delta\rangle$ for all N and not only for $N\rangle 10$ as actually occurs, the fraction F will vary as $F = \langle M\rangle/M_T = 1 - \frac{1}{2}N/N_c$. This variation is shown in Fig. 4 with a dashed line.

Experimental values are well fitted for $N > 10$ ($N/N_c^{\mathrm{short}} > 0.77$, $N/N_c^{\mathrm{long}} > 0.38$). As was expected, for $N < 10$ the experimental points are not fitted by the equation. In fact, the values are below the curve which indicates that the rough bottom affects the dynamics of the system producing a larger retention of mass.

4 Conclusions

For θ_M, the same qualitative results, due to dilatancy effects, were observed for all the studied packings independently of the system length.

For $N > N_c$: the avalanches always involve the same number N_c of superficial layers so $\langle M\rangle$ is constant.

For $N < N_c$: the avalanche regime depends critically on the thickness of the packing. All layers are involved in the avalanche and its mass is proportional to the thickness of the packing.

For $N > 10$: the critical angles $\langle\theta_M\rangle$, $\langle\delta\rangle$ and $\langle\theta_r\rangle$ are constant.

Angle δ is found to be constant for all the studied systems, it is independent of the system size (L and N). This leads us to think that, as in the case of one ball rolling down an inclined plane [7], the avalanche dynamics is also influenced by the roughness ϕ (ratio between the upper layer grain diameter and the lower one seen by the moving grains, in this case $\phi = 1$) and by the rough bed nature. That is, the bottom friction coefficient μ (in this case corresponding to glass spheres), so δ is constant for all the studied packs. Its seems logical to think that increasing μ will only increase cohesion between grains so that the system stability and the mass retention will be higher (larger θ_M and δ), but results would qualitatively be the same: δ will remain constant. Experiments are in progress to verify this hypothesis.

Acknowledgement. We thank very valuable collaboration of J.P. Hulin. This work was supported by Ecos-Sud A97-E03, PICS CNRS- CONICET 561 and TI-07 SECyT UBA. One of us (D. B.) was also financially supported by FIUBA.

References

1. F.X. Riguidel, PhD Thesis, (Université de Rennes 1, France, 1994).
2. L. Samson, PhD Thesis, (Université de Rennes 1, France, 1997).
3. J. Duran, *Sables, poudres et grains,* (Eyrolles Sciences, France, 1997).
4. M.A. Aguirre, N. Nerone, A. Calvo, I. Ippolito, and D. Bideau, *Influence of the number of layers in the equilibrium of a granular packing*, submitted to Phys. Rev. E, (1999).
5. P. Evesque, D. Fargeix, P. Habib, M.P. Luong, and P. Porion, Phys. Rev. E **47,** 2326–2332 (1993).
6. S.R. Nagel, Rev. Mod. Phys. **64,** 321 (1992).
7. M.A. Aguirre, I. Ippolito, A. Calvo, C. Henrique, and D. Bideau, Powder Tech. **92,** 75–80 (1996).

Ripple Formation in a Saltation-Avalanche Model

S. Galam[1], N. Vandewalle[2], and H. Caps[2]

[1] Laboratoire des Milieux Désordonnés et Hétérogènes, Tour 13, Case 86, 4 place Jussieu, 75252 Paris Cedex 05, France
[2] GRASP, Institut de Physique B5, Université de Liège, B-4000 Liège, Belgium

Abstract. A stochastic model is proposed to describe the ripple formation. Saltation and avalanches are the unique ingredients of the model. The dynamics of ripple formation is studied using a cellular automata. The "ripple state" turns out to be metastable. An extension of the model to the case of binary mixtures is also discussed.

1 Introduction

The ripple formation due e.g., the wind blowing across a sand bed [1] has recently received much attention in the statistical physics community [2–7]. Indeed, the physical mechanisms involved like saltation, creeping and avalanching are complex phenomena of granular transport. The latter implies collective effects.

Experimental works as well as natural observations [1] have underlined the primary role played by saltation in the emergence of ripples and dunes. Along this line, various models for ripple formation have been proposed in the past. Theoretical models which consider the hopping and rolling of grains have generally led to travelling ripple structures [2]. Simulations [3,4] have also considered various additional effects like the screening of ridges, the grain reptation and the existence of a grain ejection threshold.

We have recently proposed a stochastic model for ripple formation driven by both saltation and avalanches [5]. The associated dynamics of our model is rather complex. First, ripples appear which then coalesce into "giant dunes" after logarithmic waiting times [5]. We present herein new simulations of the model as well as some extension to binary granular mixtures.

2 Saltation-Avalanche Model

The model is defined as follows. A one-dimensional granular landscape is expressed through the height $h(i)$ of a column of identical grains on site i. This variable takes integer values only. Similarly to classical sandpile models [6], a stable surface should satisfy the condition

$$|h(i) - h(i+1)| < z_c, \tag{1}$$

for all sites i where z_c is a constant. It means that the surface local slope is always less than the granular medium critical angle $\tan^{-1}(z_c)$. The value of z_c

is an intrinsic parameter of the system. When condition (1) is not verified, an avalanche is initiated on site i, relaxing the surface till a stable configuration is reached.

At the beginning of the simulation, the surface is assumed to be flat, i.e., $h = 0$ everywhere. At each time step, the surface is first slightly modified (perturbed) by e.g., displacing a stable grain from site i to a site in the neighbourhood $j = i + 2$ (creeping). Thereafter, two relaxation mechanisms can occur: either saltation or avalanching. Indeed, an unstable grain exposed to the wind coming from the left, i.e., a grain on site j such that $h(j) - h(j - 1) \geq z_c$, will jump to the column $j + \ell$ where ℓ is the saltation length. After the jump, the grain relaxes via avalanches. If the perturbed grain on site j is not exposed to the wind, i.e., when $h(j) - h(j + 1) \geq z_c$, only avalanches occur. A dilemma occurs when the grain is located on a crest, i.e., when $h(j) - h(j-1) \geq z_c$ and $h(j) - h(j+1) \geq z_c$. In such a case, the saltation mechanism is prefered.

Besides z_c, the other parameters of the model are, respectively, the saltation length ℓ and the system size L on which periodic boundary conditions are used. In the real world, saltating grains may eject new grains from the site where the former grain lands. This effect may be introduced easily in the model. It should be also noted that the time variable t is a discrete Monte-Carlo time. The relationship between t and real time is not obvious.

3 Numerical Results

Figure 1 presents a typical evolution of the granular landscape $h(i)$ from a flat surface to a ripple state. The wind blows from left to right. One observes small ripples growing, travelling, and merging.

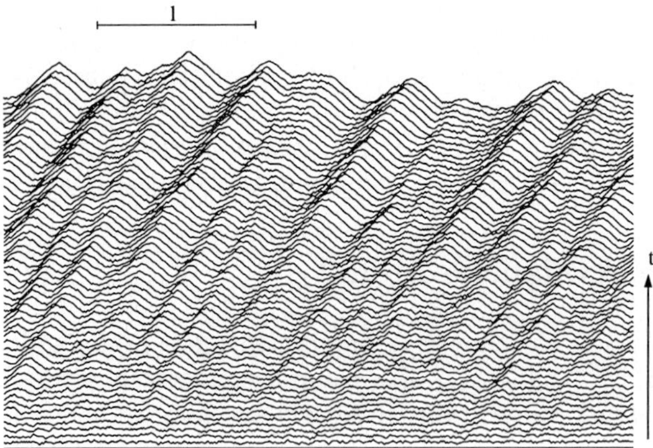

Fig. 1. Snapshot of the granular landscape $h(i)$ for various successive time steps. The length of the system is $L = 512$. The critical local slope is $z_c = 2$. The saltation length is $\ell = 128$.

In fact, the small ripples are travelling faster than the larger ones leading to the merging of these structures [4]. At intermediate stages, ripples can coexist with large dunes or else, can be observed on the dunes themselves, as it is in the real world [1]. When ℓ is large, the transient regime is characterised by ripples with various lengths like ℓ, $\ell/2$, $\ell/3$, etc... as observed in Fig. 1 for $z_c = 2$ and $L = 512$. Therefore, some specific modes are selected by the process similarly to what is observed in nature [1]. Indeed, a second mode $\ell/2$ is sometimes observed in addition to the primary ripples of size ℓ [1].

Our results imply that dunes emerge from an unstable ripple state. However, ripples and dunes are always distinguished in geomorphology due to the difference of their respective length scales even though they originate from the same cause: the wind. We will see below that various length scales separated by large gaps can naturally appear. It is worth noticing that previous numerical works [3,4] reported a slow growth of ripples. These studies corroborate our findings.

The present work shows that these patterns are two aspects of the same phenomenon since the ripple state is metastable. The metastability of ripples corroborates a very recent experimental result [7].

Fig. 2. Two different cases for a system size $L = 256$. (top) $\ell_1 = 32$, $\ell_2 = 5$, $z_c^{11} = 2$, $z_c^{12} = 2$, $z_c^{21} = 2$, $z_c^{22} = 2$. (bottom) $\ell_1 = 24$, $\ell_2 = 24$, $z_c^{11} = 3$, $z_c^{12} = 3$, $z_c^{21} = 2$, $z_c^{22} = 2$. Wind is blowing from left to right.

4 Binary Mixtures

The above model is readily extended to the case of binary mixtures [8]. Each granular species $\alpha \in \{1, 2\}$ can be characterised by different saltation lengths

ℓ_α. Moreover, a local height differences $z_c^{\alpha\alpha}$ or equivalently by a repose angle $\tan^{-1}(z_c^{\alpha\alpha})$ can be associated to each species α. The angle of repose may be different when the species are mixed such that the parameters $z_c^{\alpha\beta}$ should also be considered. The parameter $z_c^{\alpha\beta}$ corresponds to the maximum slope (more precisely, the local height difference) on which a particle of type α can remain on the top of a particle β without starting to roll down. Thus, six parameters should be considered in that generalisation of the former model. Of course, a wide variety of cases can be investigated. Let us present two typical cases which are illustrated in Fig. 2.

When the saltation lengths are quite different (see top of Fig. 2), a phase segregation takes place. Granular species are found in different parts of the landscape. The species having the smallest ℓ is always located at the top of ripples while the other species is located inside the ripples. This type of phase segregation is usually reported for aeolian sand ripples found at various places like the Namibian desert [9].

When saltation length are the same but the angles of repose are different (see bottom of Fig. 2), an other kind of phase segregation takes place: ripples of each pure species are first formed. Then, they are travelling with different speeds and they merge forming thereafter horizontal layers in the landscape.

5 Conclusion

In summary, we have presented a model for ripple and dune formation. The ripple state has been found to be metastable. The model emphasises the fact that ripples and dunes are closely related and interact with each other.

Moreover, we have investigated binary mixtures. Phase segregations are formed and are quite similar to binary ripples observed in nature.

Acknowledgement. N.V. thanks the FNRS (Brussels, Belgium). N.V. thanks the LMDH for their hospitality during the progress of this work. We acknowledge J. Rajchenbach for fruitfull discussions.

References

1. R.A. Bagnold, *The physics of blown sand and desert dunes,* (Chapman and Hall, London, 1941).
2. R.B. Hoyle and A.W. Woods, Phys. Rev. E **56**, 6861 (1997).
3. W. Landry and B.T. Werner, Physica D **77**, 238 (1994).
4. H. Nishimori and N. Ouchi, Phys. Rev. Lett. **71**, 197 (1993).
5. N. Vandewalle and S. Galam, in press, Int. J. Mod. Phys. C **10**, (1999).
6. P. Bak, C. Tang, and K. Wiesenfeld, Phys. Rev. Lett. **59**, 381 (1987).
7. A. Betat, V. Frette, and I. Rehberg, Phys. Rev. Lett. **83**, 88 (1999).
8. N. Vandewalle and S. Galam, in preparation, (1999).
9. R.E. Hunter, Sedimentology **24**, 361 (1977).

Particle Diffusion and Segregation in Rotating Cylinders

G.H. Ristow

Fachbereich Physik, Universität des Saarlandes
Postfach 15 11 50, 66041 Saarbrücken, Germany

Abstract. The interface dynamics in rotating cylinders of an initially well-segregated binary particle configuration is studied numerically. The process is characterized by calculating diffusion coefficients for different friction and density ratios. A transient *segregation wave* is observed in the mixing regime.

1 Introduction

Even though granular materials are an integral part of our everyday life, many surprising and even puzzling pattern formation processes are observed during their handling and processing [1].

A rather striking example can be observed when attempting to mix particles of different material properties like size, density or shape in a roughly half-filled rotating cylinder. After only a few rotations, the denser or smaller particles will *segregate* instead of mix and they will form a radially segregated cluster, called *core*, close to the rotation axis [2]. This core formation takes place everywhere along the rotation axis in a long cylinder and it can become unstable after many rotations, leading to the formation of visible bands along the rotation axis [3]. In the long run, these bands can even become pure, i.e. each band consists only of one particle component if a binary mixture is used. However, it is still an open question for binary mixtures, what kind of materials will form axial bands and if these bands will eventually become pure.

To address this question, we will use numerical simulations on a discrete particle basis in order to study the front propagation in an initially fully-segregated configuration. This allows us to vary the material parameters over a wide range and to trace each particle individually.

2 Numerical Model

Each particle is approximated by a sphere and interacts with other particles and with the cylinder boundaries only via contact forces. This is motivated by the fact that the particles in mind are uncharged and have a diameter in the mm- or cm-range. The forces in the normal direction are modeled via a spring and a dash-pot and in the shear direction via a viscous-friction force for particle–particle and via a static-friction force for particle–wall interactions [1,7–9].

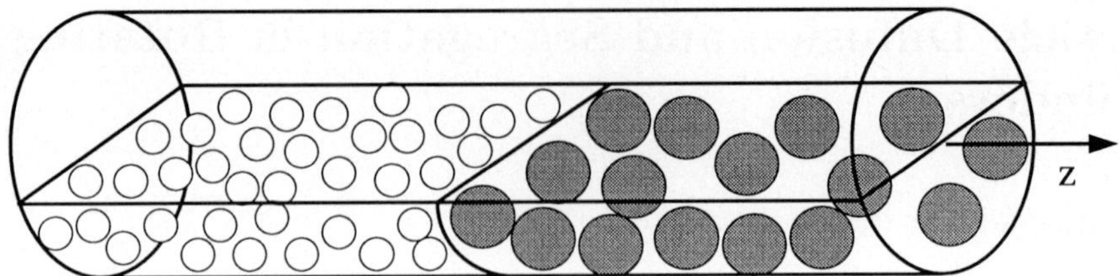

Fig. 1. Sketch of the initial configuration to determine the diffusion coefficient: Large particles are all in the right half of the cylinder and shown in grey

3 Particle Diffusion across an Interface

Since pure bands seem to be a stable configuration in some experiments, it is very instructive to study their stability for different material properties. In order to do so, we start with the system depicted in Fig. 1, where the left half is filled with small and the right half with large particles. The large particles are taken as a reference and have a diameter of 3 mm and a density of $\rho_l = 1.3\,\mathrm{g/cm^3}$. The material properties of the large particles were chosen to correspond to the measured values of mustard seeds [4], giving a friction coefficient of $\mu_l = 0.2$ and a inter-particle restitution coefficient $e_n = 0.58$.

The small particles have a diameter of 2 mm and a variable density and a variable friction coefficient.

3.1 Approximation through 1D Diffusion Process

Assuming random particle motion along the mixer axis (z axis in Fig. 1), *one*-component systems could be well described by a one-dimensional diffusion process [5]. It can be speculated that the interface of a *two*-component system can also be studied in this fashion and the diffusion equation reads

$$\frac{\partial C(z,t)}{\partial t} = \frac{\partial}{\partial z}\left(D\frac{\partial C(z,t)}{\partial z}\right) , \qquad (1)$$

where $C(z,t)$ denotes the relative concentration by volume of the smaller particles. Here D stands for the corresponding diffusion coefficient which does not have to be constant and the influence of assuming a concentration dependence is discussed in [6]. The initial conditions for a drum with length L are

$$C(z,0) = \begin{cases} 1, & -L/2 \le z < 0 \\ 0, & 0 < z \le L/2 \,, \end{cases} \qquad (2)$$

whereas the boundary conditions read

$$\left.\frac{\partial C}{\partial z}\right|_{z=-L/2} = \left.\frac{\partial C}{\partial z}\right|_{z=L/2} = 0 \,. \qquad (3)$$

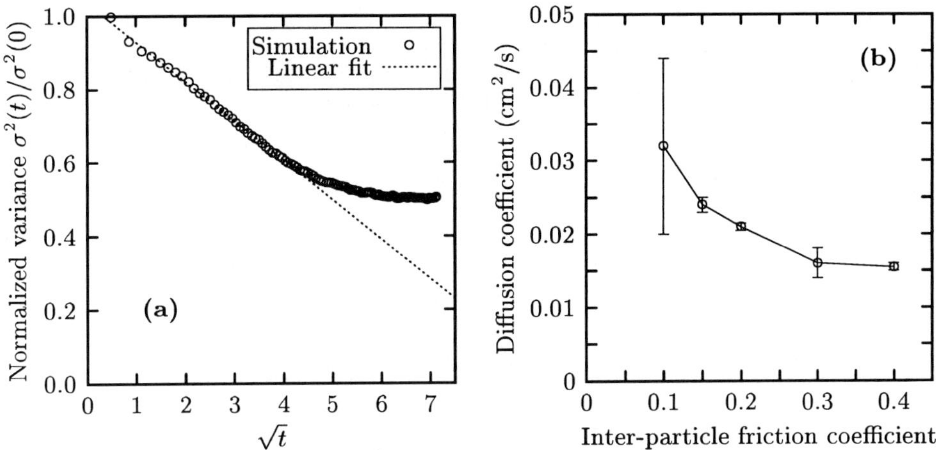

Fig. 2. **(a)** Normalized variance $\sigma^2(t)/\sigma^2(0)$ vs. $\sqrt{t}$. The linear fit is used to determine D and **(b)** diffusion coefficient as function of inter-particle friction μ.

3.2 Calculation of Diffusion Coefficients

In order to study the short-time behavior, one can solve (1) for the diffusion in an infinite long cylinder. This approximation is valid as long as the concentrations at the real cylinder boundaries have their initial values.

It is desirable to have a single parameter which characterizes the mixedness of the system at any given time. This can be done via the *true variance* $\sigma^2(t)$ of particles within the cylinder of length L defined as [5]

$$\sigma^2(t) \equiv \int_{-L/2}^{0} \left[C(z,t) - C(z,\infty)\right]^2 dz \ , \tag{4}$$

where $C(z,\infty) = 1/2$ denotes the steady-state concentration.

For small times t one obtains as an approximation for (4) [5,7]

$$\sigma^2(t) = \sigma^2(0)\left(1 - \frac{4}{L}\sqrt{\frac{2Dt}{\pi}}\right) \ . \tag{5}$$

The highest value of $\sigma^2(t)$ is given for $t = 0$ and a decrease linear in $\sqrt{t}$ is expected for short times. This is shown in Fig. 2a, taken from [7], where we plot $\sigma^2(t)$, normalized by the initial value $\sigma^2(0)$, as a function of $\sqrt{t}$. From the slope of the linear fit shown as a dotted line in Fig. 2a, we can calculate a constant diffusion coefficient based on our approximations, which gives $D = 0.022 \pm 0.002\,\mathrm{cm^2/s}$ and agrees quite well with values extracted from experiments [6]. When small particles are close to the opposite wall, our approximation of an infinite long cylinder does not hold any more, which leads to a systematic deviation from the $\sqrt{t}$ behavior, visible for times larger than 20 s in Fig. 2a.

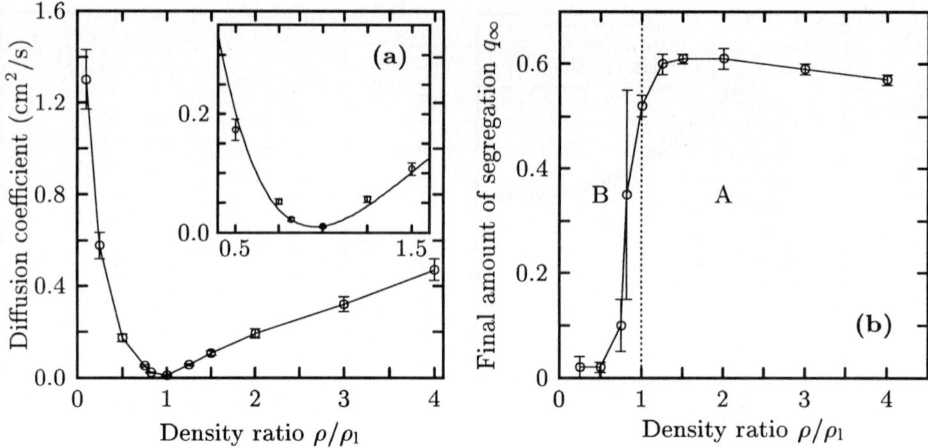

Fig. 3. Dependence on the density ratio of **(a)** the diffusion coefficient D and **(b)** of q_∞, where different regions are separated by the dotted line.

3.3 Dependence on Friction

The dependence of the constant diffusion coefficient, calculated from plots similar to Fig. 2a, on the friction coefficient of the small particles is quite small and shown in Fig. 2b. The diffusion coefficient decreases with increasing inter-particle friction, which persists up to quite large friction coefficients where the small particles have a much higher angle of repose than the large particles (for $\mu = 0.2$ and $\Omega = 15\,\mathrm{rpm}$, the angle of repose is the same for large and small particles). This dependence is rather weak and can be explained by the so-called *roller coaster* effect [7].

3.4 Dependence on Density

The particle motion also depends on the density ratio ρ/ρ_l, which is illustrated in Fig. 3a for a constant value of $\mu = 0.2$. The diffusion constant is plotted as a function of this density ratio, showing a minimum value for $\rho/\rho_l = 1$ and a large increase for lower and higher values. In general, smaller and denser particles will segregate radially, so increasing the density ratio will enhance radial segregation, but when decreasing the density ratio, the larger particles become denser and eventually the large particles will segregate into the radial core. Also shown in the same graph as an inset is a magnification of the region close to $\rho/\rho_l = 1$ with a non-linear fit as a solid line. This inset shows that our numerical model always gives a diffusion coefficient larger than zero, thus indicating that the front is *not* stable, regardless of the density ratio of the two particle components.

In order to quantify the segregation process, we plot in Fig. 3b the final amount of segregation q_∞ [7,8] as a function of the density ratio. Two regions can be distinguished clearly: (A) a very high radial segregation parameter for

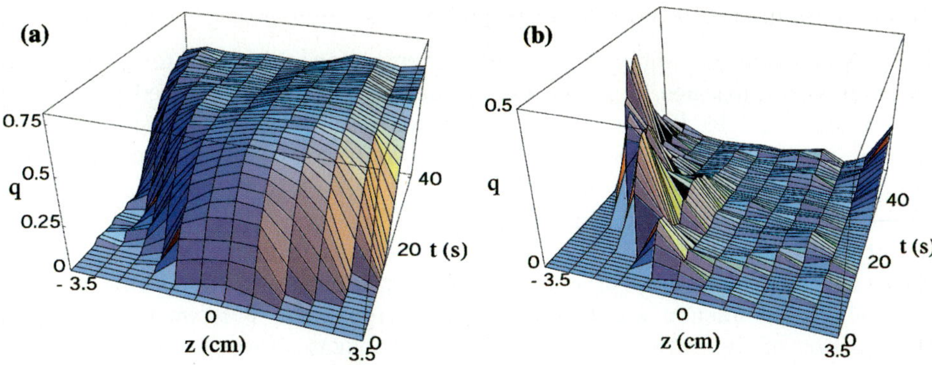

Fig. 4. Time evolution of the radial segregation for a density ratio of **(a)** $\rho/\rho_l = 2$ and **(b)** $\rho/\rho_l = 0.5$. The latter picture shows a *segregation wave* moving to the left.

values of $\rho/\rho_l \geq 1$ to the right and (B) a very low radial segregation parameter for values of $\rho/\rho_l < 0.8$ to the left. In region (A), the final amount of segregation increases with density ratio until $\rho/\rho_l \approx 2$ and then it seems to decrease, for more details see [7]. The latter region (B) corresponds to the regime where the size segregation can be partially counter-balanced by density segregation and we get a nearly perfect mixing of small and large particles indicated by a small value of q_∞ in Fig. 3b. However, the segregation dynamics are quite different in the two regions, which we will illustrate by discussing the time evolution of the segregation order parameter $q(t, z)$.

3.5 Propagation with Segregation

In region A, we obtain a very fast radial segregation which initiates at the initial interface at $z = 0$. In Fig. 4a, we show the time evolution of the segregation parameter q as a function of position along the rotation axis for a density ratio of $\rho/\rho_l = 2$. For $t = 0$, we get $q(0, z) = 0$ throughout the system, since in each of the slices where we computed q, either only large or only small particles are present and $q \equiv 0$ by definition, i.e. there is *no* radial segregation since only one particle type is present. For later times, we obtain a very fast radial segregation when the two particle components start to mix in the axial direction. This can be seen in Fig. 4a since the slope of $q(t, z)$ starts very steeply in the t direction everywhere along the rotation axis and saturates to a value of $q_\infty \approx 0.6$ throughout the system.

After approximately 30 s, we found a radial core of smaller particles everywhere in the cylinder despite the fact that the concentration profile had not yet reached its steady state [7]. This was verified by visualizing the central core on the computer screen.

3.6 Propagation without Segregation

For particles that only differ in size, radial segregation is observed for an arbitrarily small size difference [9] which can be partially counter-balanced by making the smaller particles lighter [1]. In our case of a size ratio of 2:3, we found that the density ratio for the least radial segregation is 1:2. This gives a final value of q_∞ close to zero, see Fig. 3b.

In Fig. 4b, we show the time evolution of the segregation parameter q for a density ratio of $\rho/\rho_l = 0.5$. The picture is strikingly different from Fig. 4a and shows a clear wave in the segregation parameter moving to the left. This new and surprising phenomenon was termed a *segregation wave*. This wave starts from the position of the initial front, $z = 0$, and propagates into the region initially occupied by small particles. During this process, the amount of segregation in the region behind the wave starts to decrease. The wave reaches the cylinder end cap at a time around $t \approx 40\,\mathrm{s}$, after which it dissolves completely leading to a final value of $q_\infty \approx 0$ everywhere. The origin of the wave can be understood in the following way: Since the larger particles are the denser ones, they push into the region of the smaller particles (to the left) below the free-particle surface, for more details see [9].

On the other hand, no pronounced wave is visible in the right half of the cylinder, indicating that the system is always well mixed in the region originally occupied by large particles. In this case the small particles flow in the fluidized surface layer and get directly mixed into the large ones.

In order to better visualize the spatial dynamics of this *segregation wave*, three cross-sectional views along the rotation axis in the region originally occupied by small particles for different times were given in [9], which confirmed the above-described particle dynamics.

4 Conclusions

Diffusion constants were calculated numerically in binary particle mixtures that differ in size, density and in their frictional properties.

References

1. G.H. Ristow, *Pattern Formation in Granular Materials,* (Springer, Heidelberg, 1999).
2. C.M. Dury and G.H. Ristow, J. Phys. I France **7**, 737 (1997).
3. K.M. Hill, A. Caprihan, and J. Kakalios, Phys. Rev. Lett. **78**, 50 (1997).
4. C.M. Dury, G.H. Ristow, J.L. Moss, and M. Nakagawa, Phys. Rev. E **57**, 4491 (1998).
5. R. Hogg, D.S. Cahn, T.W. Healy, and D.W. Fuerstenau, Chem. Eng. Sci. **21**, 1025 (1966).
6. G.H. Ristow and M. Nakagawa, Phys. Rev. E **59**, 2044 (1999).
7. C.M. Dury and G.H. Ristow, Granular Matter **1**, 151 (1999).
8. C.M. Dury and G.H. Ristow, Phys. Fluids **11**, 1387 (1999).
9. C.M. Dury and G.H. Ristow, Europhys. Lett. **48**, 60 (1999).

Molecular Dynamics Simulation of Cohesive Granular Materials

A. Schinner[1] and H.-G. Matuttis[2]

[1] FNW/ITP, Otto-von-Guericke University Magdeburg, Universitätsplatz 2, 39016 Magdeburg, Germany
[2] ICA I, University of Stuttgart, Pfaffenwaldring 27 70569, Stuttgart, Germany

1 Introduction

The experimental motivation for this study are recent publications on cohesive granular materials [2–4,10]. Our central question is, in which regime and by which mechanism the the movement of grains changes from movement of independent particles to a movement of small clusters with increasing cohesion. Cohesion introduces an additional length scale, so that the effects become size-dependent. The cohesive force acting on a volume element of size $l \times l \times l$ is proportional to its surface, or $\propto l^2$. The repulsive force generated by the mass of the volume element is $\propto l^3$. The strength of the cohesion and the density of the particles determine the size for which repulsion and cohesion are in equilibrium for a certain characteristic length d.

2 Simulation Method

The idea was to model the cohesive force on the particle level, without any macroscopic modeling. The particles are represented by polygons to allow arbitrary shape and size dispersions. They move according to phenomenological interactions in a molecular dynamics simulation.

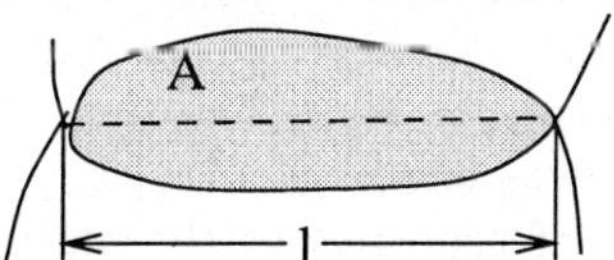

Fig. 1. Undeformed (full line) and deformed (dashed line) particles in a contact. The force resulting from the deformation is assumed to be proportional to the area overlap of the colliding particles. The penetration depth is exaggerated in comparison to simulations with realistic parameters.

The repulsive contact force in normal direction is proportional to the particle overlap and to Young's modulus. The overlap represents the deformation of the overlapping polygons in the "real world". Additional damping in normal direction as well as a model for static friction in tangential direction [7] are also present. The implementation of static friction is indispensable for the heap formation, without static friction the grains behave like a fluid. For the simulations presented in this article, a friction coefficient of $\mu = 0.6$ and Young's modulus of Y= $4 \cdot 10^7$ N/m was used.

The cohesion was modeled proportional to the contact length (see Fig. 1) and chosen proportional to a cohesion parameters k_{coh}. In two dimensions, k_{coh} has the units [N/m], so that the attractive force F_{coh} is proportional to the contact length l:

$$F_{\text{coh}} = k_{\text{coh}} \cdot |l|. \tag{1}$$

Further details on the model can be found in [11].

3 Setup of the Simulation

The simulation is performed with the so-called "draining-crater method" [5]. An upper vessel filled with cohesive granulate is emptied via an outlet, see Figs 2 and 3. The outcome for angle of repose, correlation time etc. is studied in dependence of the cohesion.

The simulation was performed using particles which had about the same diameter as the experiment [10,2,3]. The box size in the experiments was about 80-250 particle diameters, in our simulation it was about 160-200 particle diameters. The size of the outlet was about 12-25 particles in the experiment and 12-40 particles in the simulation. One series of measurements was taken with mono-disperse regular polygons with 15 faces, one series with a poly-disperse mixture, and a linear distribution of the radius within the interval $[0.75 \cdot r, 1.25 \cdot r]$. Another series was taken with the same size dispersion, but with regular polygons with 63 faces to monitor the effect of size dispersion and particle shape. All series give consistent data for medium to strong cohesion. For weak cohesion, the mono-disperse grains have a strong tendency to order on a triangular grid which dominates the entire physics of the system. In the simulation, the static and dynamic friction coefficients were chosen as $\mu_{\text{stat.}} = \mu_{\text{dyn.}} = 0.6$, Young's modulus was $Y = 10^7$ N/m, the particle diameter was 1 mm for the mono-disperse, and 0.6-1 mm for the poly-disperse particles. The time step for the simulations was dt=$0.2 \cdot 10^{-5}$ s, the density was 5000 kg/m^2.

We computed the angle of repose ϕ of the material in the upper vessel by calculating the two-dimensional analogue from [10] so that

$$\tan\phi = 2 \cdot \frac{\text{area}}{\text{base length}^2}. \tag{2}$$

The advantage of calculating ϕ from the area below the slope is that it yields an integral criterion. This smoothes out any effects from jagged surfaces of the slopes, which are typical for strongly cohesive materials such as in Fig. 3. Different layers of the grains during initialization of the particles are denoted by different shadings. For non-cohesive materials like in Fig. 2, the angle of repose could also be computed using the tangent of the slope. Differential criteria for ϕ, e.g., via the local inclination of the slope become ambiguous for increasing cohesion, and additional averaging or smoothing is necessary.

3.1 Weak Cohesion

For no or weak cohesion < 0.005 N/m single particles flow through the outlet. The slopes of the heap above and below the flow through the outlet are straight and smooth, the irregularities are of the size of up to two particle diameters. The good mixing is indicated by the colors of the particles, see Fig. 2.

Heaps built from a point source with non- or weakly-cohesive poly-disperse material show a pressure dip in the middle of the heap. For mono-disperse non-cohesive particles, there is no pressure minimum [11], but the pressure distribution is similar to that of a regular packing of particles like the one in [9].

3.2 Strong Cohesion

For strong cohesion > 0.005 N/m not single particles but whole clusters leave the outlet. The slopes of the resulting heaps above and below the outlet are ragged and rounded. The suppression of mixing is indicated by layers of particles of the same color, see Fig. 3.

For strongly cohesive materials even heaps built from a point source with mono-disperse particles exhibits a pressure minimum, see Fig. 5.

4 Implications for the Modeling of Cohesion and Friction

Fig. 6 (left) shows an increased correlation time between neighboring particles. The time correlation between particles is measured as the percentage of particles which were nearest neighbors at the beginning of the simulations and within one average particle diameter at the end of the simulation. The data indicates as well the cluster movement as well as the suppressed mixing observed in Fig. 3 for strong cohesion.

The graphic view of this correlation means that the particles are "chained together" (see Fig. 7) by the cohesive forces. Actually, this is reminiscent of simulations of non-spherical particles made from connected round particles [8]. In the same way, an attempt was used in modeling "rough" particles via connected spheres to mimic static friction [12].

In powder technology, the strength of the cohesion is classified by a single parameter together with the the surface roughness of the grains. This parameter is independent of the angle of repose, as proposed by Carr [6] see, e.g., [1]. The fact that the static friction for bulk solids can be independent of the particle roughness is already mentioned in the textbook by Rabinowicz [13]. This means that roughness/cohesion on the one hand is distinct from friction on the other hand. For systems of connected spherical particles, namely dimers, we also observed a linear increase of the angle of repose on a rough surface.

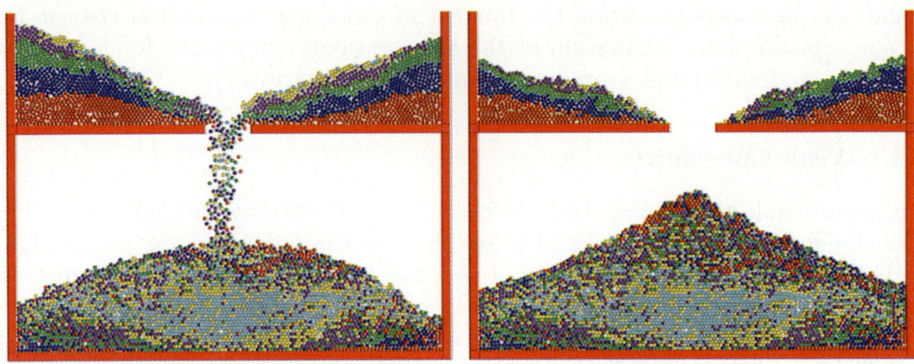

Fig. 2. Snapshots of the outflow with weak cohesion (left) and final configuration ($k_{\mathrm{coh}} = 4 * 10^5$ N/m)

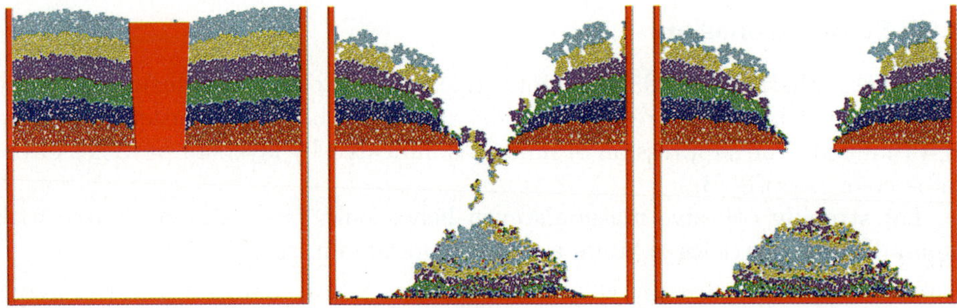

Fig. 3. Snapshot of the starting configuration before the stopper is removed (left), outflow of particle clusters and the final configuration with ragged surfaces (right) for strong cohesion. ($k_{\mathrm{coh}} = 2 * 10^3$ N/m).

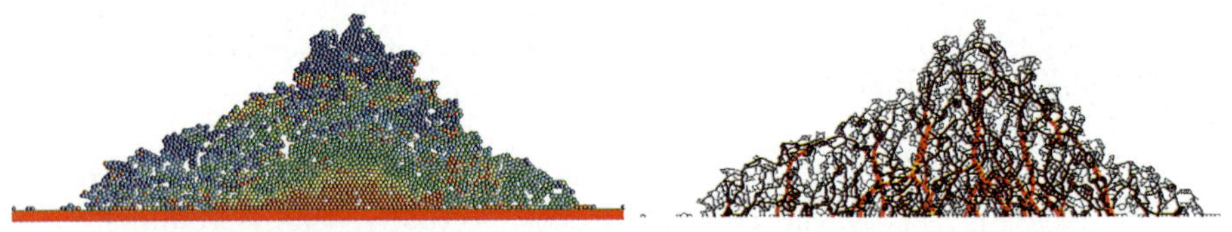

Fig. 4. Heap made from cohesive particles (left) and the corresponding force network (right). For increasing strength of the force, the color changes from black to red.

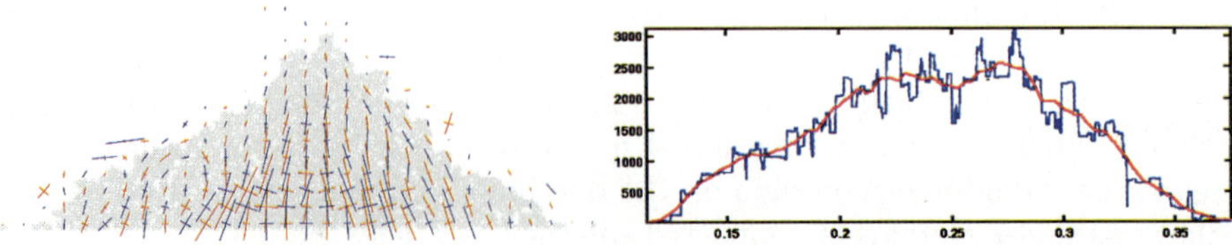

Fig. 5. The stress inside the heap is calculated and the main axis of the stress tensor is plotted (left). The pressure on the ground (lower right) shows a pronounced dip.

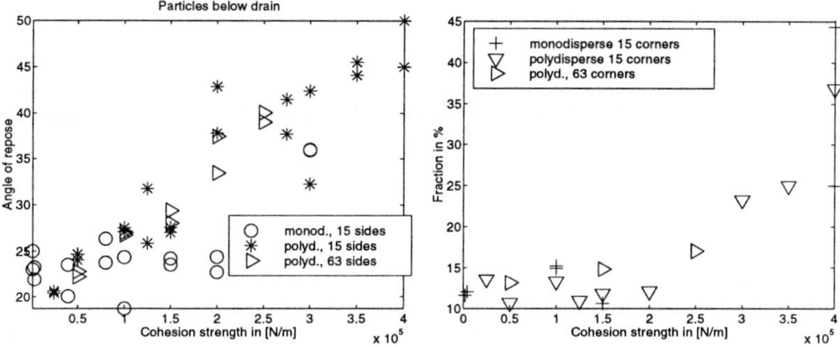

Fig. 6. Left: Angle of repose in the lower box of Fig. 3. Right: Percentage of particles which were nearest neighbors at the beginning of the simulation and less than 1 particle diameter apart at the end in the lower box (see Fig. 3). Statistics for next nearest neighbors give similar, but more noisy data.

Fig. 7. Modeling of "rough" granular materials as polymers of "smooth" round monomers. The coupling can either be introduced by constraints like in [8] or with "soft" springs [12].

From our cohesive modeling and the approach of [6], one can reinterpret the results of [8,12] as the modeling of rough and/or cohesive granular materials without static friction. The smallest angle of repose was well below 20 degrees (14 degrees for dimers in [8] on a rough bottom, 14 degrees for particles made from 5 balls in [12] in a drum), which is definitely below the angles which were observed by us and experimentally in [10] of about 22 degrees. Therefore, "rough" or "connected smooth" particles without Coulomb friction can be expected to behave very much like cohesive particles without friction. As cohesion and particle roughness cannot be distinguished by macroscopic parameters, we propose our modeling of cohesion also to mimic an effective "roughness" of grains in computer simulations without implementing additional geometric information for the particles.

References

1. T. Akiyama and Y. Tanijiri, *Criterion for re-entrainment of particles,* Powder Technology **57**, 21–26 (1989).

2. R. Albert, I. Albert, D. Hornbaker, P. Schiffer, and A.-L. Barabási, *Maximum angle of stability in wet and dry spherical granular media,* Phys. Rev. E **56**, R6271–R6274 (1997).
3. R. Albert, M. A. Pfeifer, and A.-L. Barabási, *Drag force in a granular medium,* (1998).
4. L. Bocquet, E. Charlaix, S. Ciliberto, and J. Crassous, *Moisture induced ageing in granular media,* (1998).
5. R. L. Brown and J. C. Richards, *Principles of Powder mechanics*, (Pergamon, Oxford, 1970).
6. R. L. Carr, J. Chem. Eng. **18**,163 (1965).
7. P. A. Cundall and O. D. L. Strack, *A discrete numerical model for granular assemblies,* Géotechnique 29, 47–65 (1979).
8. J.A.C. Gallas and S. Sokołowski. *Grain non-sphericity effects on the angle of repose of granular material,* Int. J. Mod. Phys. B **7**, 2037–2046 (1993).
9. D.C. Hong, *Stress distribution of a hexagonally packed granular pile*, Phys. Rev. E **47**, 760–762 (1993).
10. D.J. Hornbaker, R. Albert, I. Albert, A.-L. Barabasi, and P. Schiffer, *What keeps sandcastles standing?* Nature **387**, 765 (1997).
11. H.-G. Matuttis, *Simulations of the pressure distribution under a two dimensional heap of polygonal particles,* Granular Matter **1**, 83–91 (1998).
12. T. Pöschel and V. Buchholtz, *Static friction phenomena in granular materials: Coulomb law vs. particle geometry,* Phys. Rev. Lett. **71**, 3963 (1993).
13. E. Rabinowicz, *Friction and Wear of Materials,* (John Wiley, New York, London, Sidney, 1965).

A Model for Slowly Moving Granular Matter

K.P. Hadeler and C. Kuttler

Biomathematik, Universität Tübingen, Auf der Morgenstelle 10, 72076 Tübingen, Germany

Abstract. A system of two coupled partial differential equations is proposed as a model for the standing and rolling layers of slowly moving granular matter with external sources. The model describes the accumulation of matter on tables and in silos with prescribed areas and cross sections. For these boundary conditions stationary and similarity solutions are characterized and explicitly presented.

1 Introduction

The motion of granular matter can be described on various levels, and these descriptions are all useful for particular applications. One can follow collisions of individual particles [8], derive Boltzmann type equations for particle densities based on detailed assumptions, formulate dynamical models as in fluid dynamics [11], and finally consider phenomenological model systems for slow motion of large amounts of granular matter. In granular matter theory these latter models play the same role as the heat equation plays for heat transfer or the diffusion equation for the motion of particles dispersed in a fluid. In fact, as we shall show, there are some analogies between diffusion of a drop of ink in a beaker of water and the accumulation of dry sand in a silo, but there are also striking differences.

In the diffusion model we have a density and a flow and two laws: conservation of mass and the first Fickian law. From these the diffusion equation follows, i.e., the second Fickian law. The only parameter is the diffusion coefficient D. In models for slowly moving granular matter there are two dependent variables: the height of the standing layer and the thickness of the rolling layer. There are three parameters: the angle of repose φ, usually given as $\alpha = \tan\varphi$, the rate γ of exchange between the rolling layer and the standing layer, and β which relates the speed of the rolling layer to the gradient of the standing layer.

The model equation proposed in [6] reads

$$\begin{aligned} v_t &= \beta \,\mathrm{div}\,(v \,\mathrm{grad}\, u) - \gamma(\alpha - |\mathrm{grad}\, u|)v + f \\ u_t &= \gamma(\alpha - |\mathrm{grad}\, u|)v. \end{aligned} \tag{1}$$

Here f describes granular matter entering the rolling layer from external sources without impact. The system (1) is an extension of the BCRE model [1-4]. The BCRE system is formulated in 1D and the velocity of the rolling layer is assumed to be constant. In spite of the term appearing on the right hand side of (2), the system (1) is not a diffusion system but a hyperbolic system. Indeed, if $|\mathrm{grad}\, u| \ll \alpha$ then the equation says approximately

$$u_{tt} = \gamma\alpha v_t = \gamma\alpha\beta \mathrm{div}\,(v \,\mathrm{grad}\, u) - \gamma\alpha v.$$

Hence the equation has the form of a wave equation with velocity $(\gamma\alpha\beta v)^{1/2}$. If we add the two equations, we get $(u+v)_t = \beta\text{div}\,(v\text{grad}\,u)+f$ which is a conservation law for $u+v$ (in case $f=0$). If we would neglect the time derivative of the rolling layer v_t then we would get a diffusion equation $u_t = \beta\text{div}\,(v\text{grad}\,u)+f$. In [9,10] an equation of this form has been proposed as a model for granular matter.

2 Boundary Conditions

For the system (1) there are several geometrical problems, in any space dimension. Dimension one corresponds to experiments in which dry sand is accumulating between two vertical glass plates. Dimension two describes experiments with sand on planes, tables, and in silos.

The first experimental setup is an infinite plane (or line) onto which sand is falling from sources high up and forming a sand landscape. Sand is entering the rolling layer from the sources. When the sources are shut down, the rolling layer disappears and the standing layer forms a landscape. As proposed in [10], one can also study sand falling onto a rugged solid landscape. Of course, this solid landscape can have any gradient.

The second experiment is a table of bounded area onto which sand is poured from sources high up. The standing layer will form a landscape on the table. The rolling layer will leave the table over the edge.

The third experiment is that of a silo with bounded cross section (1D or 2D) into which granular matter is shot from sources high up. One expects that matter is accumulating in the form of a similarity solution where the standing layer rises at a constant rate and the rolling is constant in time.

In [6] for these problems the appropriate boundary conditions have been found. For the infinite plane (or line) with sources in some bounded domain (or interval) we require a boundary condition at infinity: both the standing layer u and the rolling layer v vanish at infinity. We assume that the table is given by a bounded domain Ω (or interval). Then the boundary condition says that the standing layer vanishes at the boundary: $u=0$ on $\partial\Omega$.

For the silo with vertical walls the cross section is a bounded domain Ω (or interval). Then the boundary condition says that the normal derivative of the standing layer vanishes at the boundary: $\partial u/\partial\nu = 0$ on $\partial\Omega$.

Of course one can consider geometric settings composed from the above, e.g. a table where part of the boundary is formed by vertical walls and the remainder of the boundary is free [9,10]. A typical problem in space dimension one is the half-line $\Omega = [0,\infty]$ with a vertical wall at $x=0$ and a point source at $x=0$ (the standard setup for stratification experiments).

3 Maximal Solution for a Given Source

Consider the diffusion equation $u_t = D\Delta u + f(x)$ on a bounded domain Ω with absorbing boundary condition (zero Dirichlet condition) $u=0$ on $\partial\Omega$. For any initial data the time-dependent solution approaches the same stationary solution which satisfies

$$-\Delta u = f(x)/D, \quad u = 0 \quad \text{on } \partial\Omega. \tag{2}$$

Using the Green's function of the domain Ω, the solution u can be represented as $u(x) = \frac{1}{D}\int_\Omega G(x,y)f(y)dy$, i.e., as a superposition of solutions for point sources. The boundary value problem (2) for the Poisson equation also describes the membrane with load f (then D is the tension), and u is the deviation from the horizontal position. The Green's function is non-negative and vanishes at the boundary. In dimension one G is the well-known triangle function, in dimension two it has a logarithmic singularity.

In [6] an analogy between the stationary diffusion equation (or the loaded membrane) and the granular matter problem has been found. We start from the following observation. If we pour granular matter from a point source onto a table, then the standing layer forms a circular cone which grows in a symmetric fashion by exchange with the rolling layer. But once the foot of the heap touches the edge of the table then the rolling layer forms avalanches running down from the source to the contact point and the cone does not grow any more. After this stage has been reached, we can shut down the source, the rolling layer disappears, and the standing layer stays the same. This experiment can be repeated with more general sources. It yields a certain class of maximal stationary solutions for given sources.

We show, at the example of an interval with three sources, how a maximal solution develops (Fig.1a). Of course, in an actual experiment, some stray grains fall into the gap between sources 1 and 2, and eventually fill it up. But the latter process occurs on a much longer time scale and it does not lead to a well-defined stationary solution (Fig.1a). The maximal solution can be constructed as follows: For $x \in \Omega$ let $\delta(x)$ be the distance to the boundary. Consider a point source at $y \in \Omega$. It produces a circular cone with radius $\delta(y)$ described by the function

$$\Gamma(x,y) = \begin{cases} \alpha(\delta(y) - |x-y|)\,, & |x-y| < \delta(y) \\ 0\,, & \text{otherwise.} \end{cases} \tag{3}$$

For a source f we define the support $\text{supp}(f) = \overline{\{x \in \Omega : f(x) \neq 0\}}$ and the characteristic function as $\chi_f(x) = 1$ for $x \in \text{supp}(f)$ and $\chi_f(x) = 0$ otherwise. Then the standing layer of the maximal solution for the source f is

$$u(x) = \max_{y \in \text{supp}(f)} \Gamma(x,y) = \max_{y \in \Omega} \Gamma(x,y)\chi_f(y). \tag{4}$$

This formula looks quite similar to the formula for the membrane, the main difference being that superposition is accomplished by taking the maximum rather than the integral, due to the nonlinear structure of the problem.

4 The Maximal Volume Solution

Now we pose the inverse question. Given some function $u : \Omega \to \mathbb{R}_+$ with $u(x) = 0$ for $x \in \partial\Omega$. Can this function be produced by some distribution of

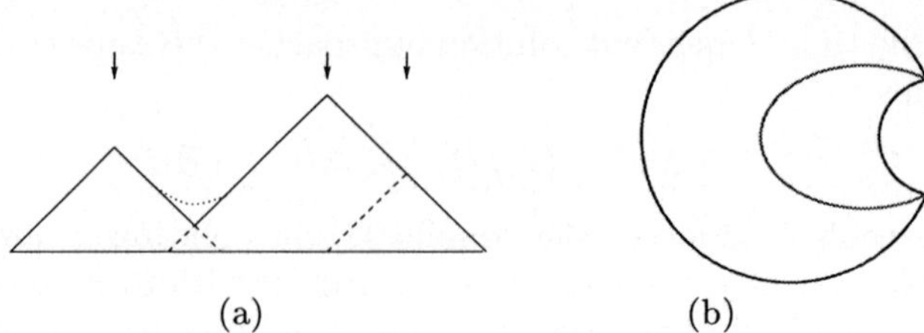

Fig. 1. a) 1D table with three sources. b) The singular set for a circular disc with circular indentation.

sources? Notice that the corresponding question for the loaded membrane has a simple answer: The function u must be super-harmonic, i.e., $-\Delta u \geq 0$.

In the 1D case maximal solutions have one or two peaks. One can ask whether by appropriately positioned sources one can place a maximal amount of sand on a given table. This problem has a unique solution $\bar{u}$. The function $\bar{u}$ is a solution of the boundary value problem of the eikonal equation

$$|\operatorname{grad} u| = \alpha \quad \text{in } \Omega, \quad u = 0 \quad \text{on } \partial\Omega. \tag{5}$$

It has necessarily singularities. If we allow piecewise differentiable solutions, then the solution of the problem (5) is not unique. On the other hand, the maximal volume solution is unique, it can be represented "explicitly" as

$$\bar{u}(x) = \alpha \operatorname{dist}(x, \partial\Omega). \tag{6}$$

Let $S \subset \Omega$ be the set of all points in Ω where the function $\bar{u}$ is not differentiable. If the boundary $\partial\Omega$ is piecewise smooth then the singular set is a rather small set, it corresponds to the set of ridges and hilltops formed by $\bar{u}$, see Fig.1b). The singular set characterizes the maximal volume solution $\bar{u}$. If $x \in \Omega$ is not in S and not on the boundary $\partial\Omega$, then there is a unique point y on $\partial\Omega$ that has minimal distance from x. Then draw the line through y and x and continue until it meets the singular set at same point z. The ray from z to y is called a transport ray. Then the standing layer is $\bar{u}(x) = \alpha|x - y|$. The rolling layer flows along these transport rays. If matter leaves a source at the point x then it runs down along the transport ray. To produce the maximal volume solution it suffices to place sources at all points of the singular set.

In the case of a square Ω the maximal volume solution is a regular pyramid. The singular set consists of the center and the four rays running to the corners. For a point source at the center of the square the maximal solution is a circular cone touching the four edges.

For any distribution of sources and the corresponding stationary solution the rolling layer at x, i.e. $v(x)$ can be computed by integrating the rolling layer along a transport ray from x upward to $z \in S$, see [6].

5 The Silo Problem

The problem of granular matter in a silo has been considered by several authors, e.g., in [4] (based on the constant speed assumption) and in [5] (experiments and approximate profiles). For the silo problem with a nonzero source there cannot be a stationary solution but only a similarity solution describing a rising level of matter. For space dimension one the standing layer and the rolling layer can be explicitly computed (as integrals over the sources). In particular, for a point source placed at the center of the silo with $\Omega = [-R, R]$, the profile of the standing layer is, up to an additive constant, (here we give the correct scaling of formula (47) in [6]))

$$u(r) = -\alpha\left(r + \frac{\beta}{\gamma}\log\left(\frac{\beta}{\gamma} + R - r\right)\right). \tag{7}$$

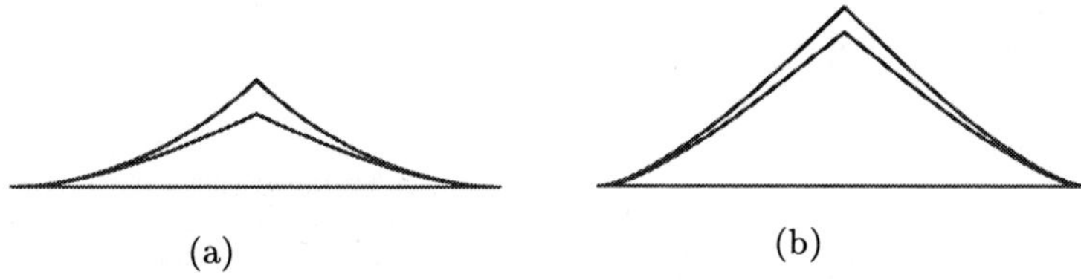

Fig. 2. Profiles of the standing layer in the 1D and 2D silo, normalized to $u(R) = 0$. $R = 1$, $\alpha = 1$. The lower curve is the 1D case. a) $\beta/\gamma = 1$, strongly concave profile. b) $\beta/\gamma = 0.2$, almost straight profile, concave boundary layer.

Of course the surface rises in proportion to time with a factor that is the average of the source over the cross section.

Also in space dimension two there is a unique similarity solution. It can be characterized by a boundary value problem [6] and, of course, it can be computed numerically. Here we consider the standard situation of a circular silo of radius R and a point source at the center. We consider a rotationally symmetric solution $(u(r), v(r))$ with $u_r(r) < 0$. Then the equations

$$\begin{aligned} 0 &= \beta \mathrm{div}\,(v \mathrm{grad}\, u) - \gamma(\alpha - |\mathrm{grad}\, u|)v + \delta_0(x) \\ c &= \gamma(\alpha - |\mathrm{grad}\, u|)v \end{aligned} \tag{8}$$

assume the form

$$\frac{c}{\beta} = u_r v_r + v\frac{1}{r}(r u_r)_r, \quad \frac{c}{\gamma} = (\alpha + u_r)v. \tag{9}$$

We solve for v in the second equation and replace it in the first, put $w = u_r$, $z = w/(\alpha - w)$, to get $(rz)_r = -(\gamma/\beta)r$ which can be integrated twice to give (see [7] for details), up to an additive constant,

$$u(r) = -\alpha \left[r + \frac{\beta}{\gamma} \log\Big(\frac{2\beta r}{\gamma} + R^2 - r^2\Big) - \frac{\beta^2}{\gamma^2} \frac{1}{\sqrt{R^2 + \beta^2/\gamma^2}} \log \frac{\sqrt{R^2 + \beta^2/\gamma^2} - \beta/\gamma + r}{\sqrt{R^2 + \beta^2/\gamma^2} + \beta/\gamma - r} \right]. \tag{10}$$

Taylor expansion in formulae (7) and (10) gives

$$u(r) = -\alpha r \frac{\gamma R}{\gamma R + \beta} + o(r) \quad (1D); \quad u(r) = -\alpha r + o(r) \quad (2D). \tag{11}$$

Hence in the 2D case the slope near the top of the heap is given by the angle of repose and, up to order r^2, does not depend on the other parameters. If $\beta << \gamma$ then the radial section is approximately a straight line with a concave tail. It agrees with experimental results, e.g. [5] (up to flattening by impact near the funnel). If $\beta \approx \gamma$ then it looks concave everywhere. In the 1D case the slope near the top depends on the parameters β, γ and on the radius R. Only for large R this boundary effect disappears.

References

1. J.-P. Bouchaud, M.E. Cates, J.R. Prakash, and S.F. Edwards, *A model for the dynamics of sandpile surfaces,* J. Phys. I France **4,** 1383-1410 (1994).
2. J.-P. Bouchaud, M.E. Cates, J.R. Prakash, and S.F. Edwards, *Hysteresis and metastability in a continuum sandpile model,* Phys. Rev. Lett. **74**, 1982-1985 (1995).
3. T. Boutreux and P.-G. de Gennes, *Surface flows of granular mixtures: I. General priciples and minimal model,* J. Phys. I France **6,** 1295-1304 (1996).
4. P.-G. de Gennes, *Dynamique superficielle d'un matériau granulaire,* C. R. Acad. Sci. **321**, 501-506 (1995).
5. Y. Grasselli and H.J. Herrmann, *Shapes of heaps and in silos,* Eur. Phys. J. B **10,** 673-67 (1999).
6. K.P. Hadeler and C. Kuttler, *Dynamical models for granular matter,* Granular Matter **2,** 9-18 (1999).
7. K.P. Hadeler and C. Kuttler, *Granular matter in a silo,* (Draft, University of Tübingen, October 1999).
8. S. Luding, *Die Physik kohäsionsloser granularer Medien,* (Logos Verlag, Berlin, 1998).
9. L. Prigozhin, *Sand piles and river networks: Extended systems with nonlocal interactions,* Phys. Rev. E **49,** 1161-1167 (1994).
10. L. Prigozhin, *Variational model of sandpile growth,* Eur. J. of Applied Math. **7,** 225-235 (1996).
11. S.B. Savage and K. Hutter, *The motion of a finite mass of granular material down a rough incline,* J. Fluid Mech. **199**, 177-215 (1989).

Pressure Distribution of a Two-Dimensional Sandpile

S. Inagaki

Dept. of Mathematical Sciences, Ibaraki University, Mito, 310-8512, Japan

Abstract. Using the discrete element method (DEM), I have constructed a numerical code to investigate the static properties of granular particles. In this model, only the gravitational and the contact forces are considered. I focus on piles made up of grains in the two-dimensional geometry. After depositing grains, a static configuration of the grains is obtained, with a reasonable angle of repose. In addition, the network structure of the pile is visualized by connecting centers of grains in contact, showing an interesting dispersion of densities. Furthermore, the horizontal and vertical pressure components along the bottom of the pile are measured. The existence of a dip, still a disputed issue, is inconclusive in this study.

1 What's a Dip?

When the vertical pressure distribution is measured along the bottom of a sandpile, it may show a local minimum right under the apex of the pile where one usually expects the global maximum. Such a minimum is called a *dip*.

Some experimental results about a dip have been reported, e.g., in [1], [2], and [3]. In particular, the experiment performed by Vanel et al. [3] shows the clearest evidence of the existence of a dip. In addition, it also shows the dependency of the final pressure distribution on the history of a pile.

Meantime, many theoretical researchers have been trying to explain this interesting phenomenon with various models, e.g., in [4–7]. Many of these models average the discrete system in order to derive simpler equations, for instance, a continuum approximation. Here in this article, I consider the direct contact among granular particles using the discrete element method (DEM) to study the mechanism by which the force is transmitted through the granular media, possibly creating a dip.

2 Discrete Element Method

2.1 What is the DEM?

The DEM is a method commonly used, for instance, in manufacturing engineering. Since the paper by Cundall and Struck [8], it has also been used from the scientific point of view in simulating granular dynamics. The main idea of the method is to solve many-body problems by considering two-body interactions at the moment of contact.

The following is the equation of motion for the j-th grain.

$$m_j \frac{dv}{dt} = \boldsymbol{F}_j + \sum_{i \neq j} \boldsymbol{F}^{i \to j} \tag{1}$$

In this case, the first term on the right hand side expresses the force of gravity, and the second term expresses all of the contact forces acting on the j-th grain.

2.2 Contact Forces

In this model, I take three kinds of contact forces into account. They are as follows.

1. ELASTIC FORCE that is proportional to the relative displacement between two grains in contact.
2. VISCOUS FORCE that is proportional to their relative velocity, not only of the translational motion but also due to their rotations.
3. FRICTIONAL FORCE for which only dynamical friction is considered.

3 The Calculation with DEM

3.1 How to Calculate Forces

I follow the algorithm of Atsuko Shimosaka [9]. The normal and tangential forces are calculated, respectively, as follows.

$$\begin{cases} \boldsymbol{F}_n^{i \to j} = k_n \boldsymbol{nn} \cdot (\boldsymbol{x}_i - \boldsymbol{x}_j) + \eta_n \boldsymbol{nn} \cdot (\dot{\boldsymbol{x}}_i - \dot{\boldsymbol{x}}_j) \\ \boldsymbol{F}_t^{i \to j} = k_t \boldsymbol{tt} \cdot (\boldsymbol{x}_i - \boldsymbol{x}_j) + \eta_t \boldsymbol{t}[\boldsymbol{t} \cdot (\dot{\boldsymbol{x}}_i - \dot{\boldsymbol{x}}_j) + (r_i \omega_i + r_j \omega_j)] = \boldsymbol{F} \end{cases} \tag{2}$$

We then supplement the equations by

$$\boldsymbol{F}_t^{i \to j} = \begin{cases} \boldsymbol{F} & [\boldsymbol{F} < \mu \boldsymbol{F}_n^{i \to j}] \\ \boldsymbol{t}|\mu \boldsymbol{F}_n^{i \to j}| sign(\boldsymbol{F}) & [\boldsymbol{F} \geq \mu \boldsymbol{F}_n^{i \to j}] \end{cases} . \tag{3}$$

Equation (3) means, if the dynamical friction acting on the j-th grain is large enough, then the grain slips against the i-th grain. Here the symbols are as follows.

k_n, k_t	: spring constants
η_n, η_t	: coefficients of viscosity
μ	: coefficient of dynamical friction
$\boldsymbol{x}_i, \boldsymbol{x}_j$	: positions of the i-th and j-th grains
$\boldsymbol{F}_n^{i \to j}, \boldsymbol{F}_t^{i \to j}$	: forces acted on the j-th grain by the i-th grain
r_i, r_j	: radii of the i-th and j-th grains
ω_i, ω_j	: angular velocities of the i-th and j-th grains
$\boldsymbol{n}, \boldsymbol{t}$	: unit vectors of normal and tangential directions

Actual values of the constants used in this calculation are taken from the engineering literatures [9].

3.2 How to Update

The equations for time integration of the positions and the velocities are of second order accuracy, as follows.

$$\boldsymbol{v}_j(t+\delta t) = \boldsymbol{v}_j(t) + \frac{\delta t}{2m_j}(3\boldsymbol{F}_j(t) - \boldsymbol{F}_j(t-\delta t)) \tag{4}$$

$$\boldsymbol{r}_j(t+\delta t) = \boldsymbol{r}_j(t) + \frac{\delta t}{2}(\boldsymbol{v}_j(t+\delta t) + \boldsymbol{v}_j(t)). \tag{5}$$

The equations for updating the angles and the angular velocities are obtained similarly. The gravitational acceleration is normalized to 1, and the maximum radius is also scaled to unity. When calculating all of the contact forces acting on the j-th grain, only the grains in its vicinity are tested whether they are in contact with it or not.

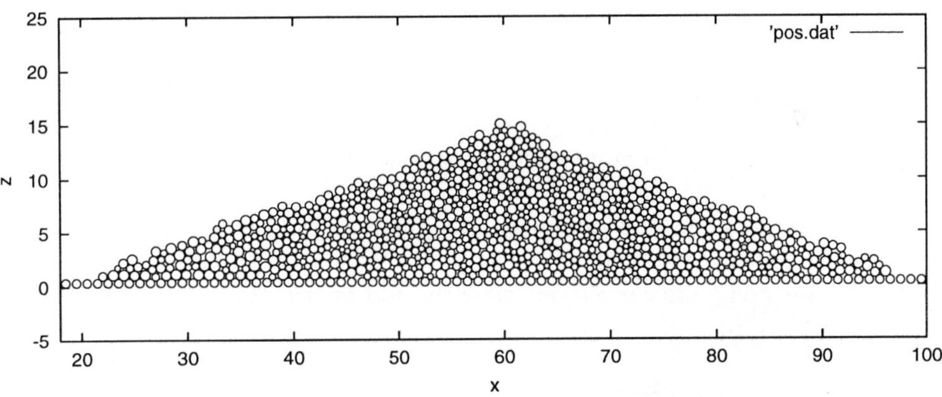

Fig. 1. The configuration of the accumulated grains after settling. The angle of repose is about 20 degrees, a reasonable value. In addition, one finds that the packing is not uniform.

3.3 Making a Pile

In this article, I focus on making two-dimensional piles in the x-z coordinate plane. I deposit grains calmly to avoid breaking the network of a pile. The details of the procedure are as follows.

1. Place horizontally 120 immobile grains that are perfectly circular and of the uniform size to represent a bottom. The bottom is wide enough for the following pile.
2. Let 1000 free grains, also perfectly circular but of distributed sizes, fall along the central axis from the position one unit higher than the current height of

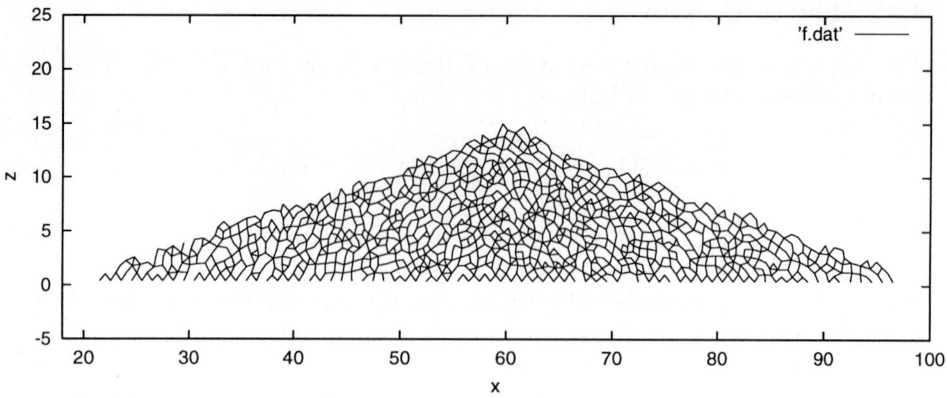

Fig. 2. How grains touch each other is visualized by connecting the centers of the grains in contact. There are some polygons with many corners, where arches are formed and the densities are low. On the other hand, there are regions with many triangles in which grains are densely packed.

the pile. The reason that I put them that way is to let them fall as gently as possible. Here, each grain is deposited one by one with a long time interval (3 time units) in between.

3. After the final grain is deposited, the pile is allowed sufficient time (15 time units, typically) to relax.
4. Measurements are carried out after the stationary state is obtained.

4 Results

I show some graphs here. Figure 1 shows the configuration of the accumulated grains after settling. The angle of repose is about 20 degrees, a reasonable value. In addition, one finds in the figure that the packing is not uniform.

Figure 2 shows this non-uniformity more clearly. How grains touch each other is visualized by connecting the centers of the grains in contact. There are some polygons with many corners, where arches are formed and the densities are low. On the other hand, there are regions with many triangles in which grains are densely packed.

Figure 3 shows the pressure distribution along the bottom of the pile. The pressure is defined to be the amplitude of the force acting on each grain fixed on the bottom. To smoothen the fluctuations, the amplitudes are averaged over the neighboring 9 grains. The solid curve shows the horizontal component of the pressure, and the dotted line shows the magnitude of the downward pressure component. Unfortunately, it is not conclusive whether there is a dip in this profile.

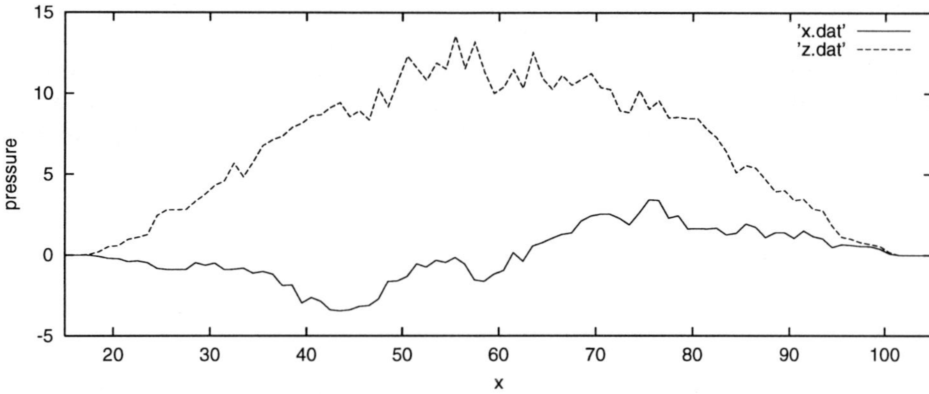

Fig. 3. The pressure distribution along the bottom of the pile. The pressure is defined to be the amplitude of the force acting on each grain fixed on the bottom. The solid curve shows the horizontal component of the pressure, and the dotted line shows the magnitude of the downward pressure component.

5 Conclusion

In this paper, I have developed the numerical code using DEM to investigate the properties of two-dimensional static sandpiles. With the idea of a DEM, I was able to obtain a reasonable static configuration that shows an interesting dispersion of local density inside the network. However, this preliminary study has not shown the existence of a dip clearly. A further work is in progress.

Acknowledgement. I thank Hisao Hayakawa for his unpublished note about DEM and his invaluable advice, and Shinya Watanabe for his careful reading and comments.

References

1. J. Smid and J. Novosad, *Pressure distribution under heaped bulk solids,* I. Chem. E. Symposium Series 63, D3/V/1.
2. R. Brockbank, J.M. Huntley, and R.C. Ball, *Contact force distribution beneath a three-dimensional granular pile,* J. Phys. 2 France **7**, 1521-1532 (1997).
3. L. Vanel, D. Howell, D. Clark, R.P. Behringer, and E. Clément, *Memories in sand: Experimental tests of construction history on stress distributions under sandpiles,* Phys. Rev. E **60**, (1999).
4. S.F. Edwards and C.C. Mounfield, *A theoretical model for the stress distribution in granular matter,* Physica A **226**, 1-33 (1996).
5. J.P. Wittemer, M.E. Cates, and P. Claudin, *Stress Propagation and Arching in Static Sandpiles,* J. Phys. 1 France **7**, 39-80 (1997).
6. D.H. Trollope and B.C. Burman, *Physical and numerical experiments with granular wedges,* Géotechnique **30**, 137-157 (1980).

7. H.G. Matuttis, *Simulation of the pressure distribution under a two dimensional heap of polygonal particles,* Granular Matter **1**, 83-91 (1998).
8. P.A. Cundall and O.D.I. Strack, *A discrete numerical model for granular assemblies,* Géotechnique **29**, 47-65 (1979).
9. Edited by the society of the powder engineering of Japan, *Introduction to granular simulation,* (Funtai simulation Nyumon, in Japanese, 1998).